Sustainability Sciences in Asia and Africa

Sustainable Agriculture and Food Security

Series Editor

Rajeev K. Varshney, Centre for Crop and Food Innovation, Food Futures Institute, Murdoch University, Perth, WA, Australia

This book series support the global efforts towards sustainability by providing timely coverage of the progress, opportunities, and challenges of sustainable food production and consumption in Asia and Africa. The series narrates the success stories and research endeavors from the regions of Africa and Asia on issues relating to SDG 2: Zero hunger. It fosters the research in transdisciplinary academic fields spanning across sustainable agriculture systems and practices, post- harvest and food supply chains. It will also focus on breeding programs for resilient crops, efficiency in crop cycle, various factors of food security, as well as improving nutrition and curbing hunger and malnutrition. The focus of the series is to provide a comprehensive publication platform and act as a knowledge engine in the growth of sustainability sciences with a special focus on developing nations. The series will publish mainly edited volumes but some authored volumes. These volumes will have chapters from eminent personalities in their area of research from different parts of the world.

Series Preface

The global food production systems require expanding their capacity to produce almost twice the current levels to safeguard the food security of the burgeoning population worldwide. More than 800 million people suffering from undernourishment worldwide pose a great risk to the attainment of sustainable development goal (SDG) 2 of the UN, which targets "End hunger, achieve food security and improved nutrition and promote sustainable agriculture" within the next 7 years. The challenge is further exacerbated by the rising weather extremities, unpredictability in rainfall patterns, and pest–pathogen dynamics associated with global climate change, which have a profound negative impact on agricultural productivity and farm incomes worldwide. Also, the future targets of food production should be secured in resource-constrained agricultural settings and with the least environment footprint, thus calling for sustainable innovations in agri-farming systems and enhanced participation of women in agriculture. The challenge to reduce hunger is alarming in developing nations, particularly Asia and Africa, which house the largest proportion of people suffering from malnutrition and other nutrition-related issues. Furthermore, the agri-food systems in Asia and Africa are severely constrained by the subsistence nature of farming, declining land and other agricultural resources, increasing environmental pollution, soil and biodiversity degradation, and climate change. Therefore, this book series, *Sustainable Agriculture and Food Security*, has been planned to support the global efforts toward sustainability by providing timely coverage of the progress, opportunities, and challenges of sustainable food production and consumption in Asia and Africa. The series narrates the success stories and research endeavors from the regions of Africa and Asia on issues relating to SDG 2: Zero Hunger. It fosters research in transdisciplinary academic fields spanning sustainable agriculture systems and practices, post-harvest, and food supply chains. The focus of the series is to provide a comprehensive publication platform and act as a knowledge engine in the growth of sustainability sciences with a special focus on developing nations.

As the global population grows, so does the demand for food. To meet the food demands of an estimated 9.7 billion people by 2050, global food production must increase by 70%. Thus far, this demand has largely been met through scientific and technological advances, often with negative environmental consequences. With the threat of climate change and resource scarcity becoming more pressing, it is crucial to adapt food production systems to increase output while maintaining environmental sustainability. Diversified farming systems offer a sustainable, environmentally friendly approach by enhancing agrobiodiversity. Practices such as crop rotations, cover cropping, intercropping, and agroforestry increase the variety of crop cultivars or species in farming systems, benefitting biodiversity, soil health, and yields. Diversified agriculture, therefore, can play a key role in achieving global food security and environmental stability. However, designing these systems poses challenges, as each farming system is unique and will require tailored planning and development to achieve optimal results.

While conventional tillage-based agriculture has helped build toward food security, it has also degraded essential natural resources such as soil, water, biodiversity, and the associated ecosystems. Conservation agriculture offers an alternative approach, aiming to save on energy and mineral nitrogen use in farming, reduce greenhouse gas emissions, enhance biological activity in soils, and increase overall system-level biomass production. This approach is based on three key principles: (1) minimal soil disturbance, (2) maintaining permanent biomass soil mulch cover, and (3) diversifying crop species. Another approach to sustainable agriculture is diversified farming, which shares similarities with conservation agriculture by focusing on creating, maintaining, and regenerating biodiversity in farm systems through practices like intercropping and agroforestry. Soil health is critical to sustainable agriculture, requiring investment in techniques that maintain and enhance soil quality. Farmers can improve soil health by supplementing the soil with conditioners like biochar or adopting "carbon farming," which helps retain nutrients and sequester carbon into the soil, contributing to climate change mitigation. Agroforestry, an approach that integrates trees into farming systems, further supports soil health by stabilizing soil, encouraging aggregate formation, storing carbon, enhancing nutrient availability, and promoting healthy soil biota. Additionally, agroforestry plays a significant role in achieving carbon neutrality.

In view of this, this book, *Agricultural Diversification for Sustainable Food Production*, edited by Subhash Babu, Raghavendra Singh, Sanjay Singh Rathore, Anup Das, and Vinod Kumar Singh, provides a platform to discuss the present status and problems for agricultural diversification at the global scale. This book presents 18 chapters written by various experts and practitioners, each containing up-to-date and important insights into the topic of agricultural diversification. I would like to congratulate the editors, Subhash Babu, Raghavendra Singh, Sanjay Singh Rathore, Anup Das, and Vinod Kumar Singh and all authors for their valuable contributions. I am sure this book is a great resource for students, researchers, farmers, educators, extension agents, policymakers, and other stakeholders engaged in diversifying agriculture for sustainable food production around the world.

I wish to extend my sincerest thanks and gratitude to the Springer staff, particularly Sanchi Bhimrajka, Associate Editor (Books, Medicine and Life Sciences), and Naren Aggarwal, Editorial Director (Books, Asia, Medicine and Life Sciences) for their constant support in the accomplishment of this compendium. The cooperation

Subhash Babu • Raghavendra Singh •
Sanjay Singh Rathore • Anup Das •
Vinod Kumar Singh
Editors

Agricultural Diversification for Sustainable Food Production

Springer

Editors
Subhash Babu
Division of Agronomy
ICAR-Indian Agricultural Research Institute
New Delhi, Delhi, India

Sanjay Singh Rathore
Division of Agronomy
ICAR-Indian Agricultural Research Institute
New Delhi, Delhi, India

Vinod Kumar Singh
ICAR-Central Research Institute for
Dryland Agriculture
Hyderabad, Telangana, India

Raghavendra Singh
Cropping Systems and Resource
Management
ICAR-Indian Institute of Farming Systems
Research
Modipuram, Uttar Pradesh, India

Anup Das
ICAR-Research Complex for
Eastern Region
Patna, Bihar, India

ISSN 2730-6771 ISSN 2730-678X (electronic)
Sustainability Sciences in Asia and Africa
ISSN 2730-6798 ISSN 2730-6801 (electronic)
Sustainable Agriculture and Food Security
ISBN 978-981-97-7516-3 ISBN 978-981-97-7517-0 (eBook)
https://doi.org/10.1007/978-981-97-7517-0

This Springer imprint is published by the registered company Springer Nature Singapore Pte Ltd.
The registered company address is: 152 Beach Road, #21-01/04 Gateway East, Singapore 189721, Singapore

received from my supervisors and research leaders such as Prof. Andrew Deeks (Vice Chancellor), Prof. Peter Eastwood (Deputy Vice Chancellor—Research), Prof. Peter Davies (Pro Vice Chancellor), and my laboratory colleagues, Rhys Copeland, Anu Chitikineni, and Vanika Garg from Murdoch University, Australia, is also gratefully acknowledged. I would like to thank my family members, Monika Varshney, Prakhar Varshney, and Preksha Varshney, for their love and support in discharging my duties as Series Editor.

Perth, WA, Australia Rajeev K. Varshney

Preface

Ensuring nutritional security while sustaining the environmental resource base are critical challenges of the twenty-first century. To meet global food security while harmonizing the scare resources and fluctuating climate, agricultural production systems must become more productive and adaptable. Thus, ecosystem management along with ecological harmony is the future perspective of food security and environmental sustainability. Conventional resource-intensive agricultural systems are not able to feed the global population sustainably. Energy-intensive, conventional agricultural production systems not only accelerate environmental footprints, resource mining, and biodiversity losses but also contribute to multiple human health issues besides reducing soil functionality. Diversified farming offers a promising solution by enhancing yield stability, food security, and climate resilience. Therefore, while assuring long-term sustainability in agricultural production systems, diversification is the core field of action. Diversified agriculture helps to achieve the twin objectives of advancing food security and environmental sustainability through efficient resource use and profit maximization. Despite the enormous positive outlooks, there are several challenges in the adoption of diversified farming. Diversified farming is a multiproduct-oriented production system that needs multiple specialties and marketing avenues. Moreover, the development of a diversified model is highly individualistic and location-specific and needs proper resource optimization and characterization. Hence, developing site-specific diversified farming models to maintain food quality with productivity improvement is a major challenge to the researchers, which warrants adequate management of agroecosystem and food production resources. Very few resources and reference books are available in the global and regional market that address specific issues on diversified agriculture and its impact on farm production and environmental sustainability. This book encompasses holistic concepts of agricultural diversification and integrated management of natural resources for nutritional and environmental sustainability which will be useful for students, research scholars, scientists, academicians, policy planners, and environmentalists as a ready reckoner for rationalized and updated information. Thus, the book would help to address multipronged challenges of food and nutritional security, preserving dwindling natural resources in a comprehensive manner. It aims to consolidate concepts related to agricultural diversification and its impact on food production and environmental sustainability. The objectives include: Assessing the current status and challenges in adopting agricultural diversification, identifying key areas linking agricultural diversification with nutritional security and environmental sustainability, and enhancing understanding and providing new insights to stakeholders, aiding in policy planning for societal betterment. This comprehensive book includes contributions to authoritative and cutting-edge scientific information from different researchers and academicians. The publication will serve as a handy resource book for preparing a roadmap for policy planning and advancing knowledge in sustainable and resilient farming, helping to address future nutritional security and environmental sustainability challenges aiming to achieve sustainable development goals.

New Delhi, India — Subhash Babu
Modipuram, India — Raghavendra Singh
New Delhi, India — Sanjay Singh Rathore
Patna, India — Anup Das
Hyderabad, India — Vinod Kumar Singh

Contents

Series Editor and Editorial Board Members

About the Series Editor

Rajeev Kumar Varshney is a renowned agricultural scientist with over two decades of experience in genomics and molecular breeding. Currently, he holds multiple prestigious positions at Murdoch University in Australia. Prof. Varshney has made a significant impact on global food security by creating genomic resources for major "orphan" tropical crops, using these resources to identify genetic loci and candidate genes for drought and pest tolerance in staple crops in Sub-Saharan Africa and India, and spearheading international programs that deliver superior crop varieties to some of the world's most impoverished farmers. Prof. Varshney is an accomplished author, with more than 600 publications to his name and has been recognized as a Clarivate Highly Cited Researcher for ten consecutive years (2014–2023).

Food Futures Institute; Director Centre for Crop and Food Innovation; Director State Agricultural Biotechnology Centre; International Chair in Agriculture and Food Security, Murdoch University, Australia

Editorial Board Members

About the Editors

Subhash Babu is currently working as a Senior Scientist at the Division of Agronomy, ICAR-Indian Agricultural Research Institute, New Delhi, India. He is known for his work on carbon management in agroecosystems, soil health, integrated farming systems, and the integration of innovative climate resilient technology in agriculture. His research and initiatives aim to improve food security and environmental sustainability. Dr. Babu collaborates with various organizations to promote knowledge sharing making a substantial impact on the agricultural sector. Dr. Babu has 315+ publications and 6 books.

Raghavendra Singh is currently working as a Head, Division of Cropping Systems and Resource Management, ICAR-Indian Institute of Farming Systems Research, Modipuram, Meerut, UP (India). He has made significant contributions to soil health, nutrient management, and the development of innovative farming systems. Dr. Singh's research focuses on enhancing crop productivity, resource use efficiency, and environmental sustainability. Dr. Singh is associated with the development of high yielding varieties of rice, maize, and rajmash for organic conditions of Sikkim. He has published 120+ research articles and 3 books.

Sanjay Singh Rathore is currently working as a Head, Division of Agronomy, ICAR-IARI, New Delhi, and National Coordinator of World Bank-funded National Agricultural Higher Education Project (NAHEP). He has developed IFS models under shifting cultivation, efficient irrigation water management, and resource conservation technologies for mustard-based cropping systems, suitable agri horti, and potential crop diversification options for maize-based systems under limited water conditions. Dr. Rathore's research focuses on enhancing crop yields, improving resource use efficiency, and promoting environmentally sustainable farming techniques. He has published 160+ international and national research papers and 6 books.

Anup Das is currently serving as Director, ICAR Research Complex for Eastern Region, Patna, India, and Vice President, ISA. His extensive research works led to the development of the conservation agriculture-based package of practices for production of pulses and oilseeds in rice and maize fallow areas; organic production packages of important crops in system approach and developed climate resilient integrated farming system and land use models. He has conceptualized and executed an Integrated Organic Farming System in a cluster approach for safe and healthy food production in northeast India. His research focuses primarily on developing innovative agronomic approaches for managing soil health and environmental security and providing practical solutions to challenges faced by small farmers. He has published 250+ research articles and 45 books and bulletins.

Vinod Kumar Singh is currently serving as a Director, ICAR-Central Research Institute for Dryland Agriculture, Hyderabad, India. Dr. Singh has made significant contributions to soil health and nutrient management, particularly in the rice-wheat system (RWS) and other intensive cropping systems. His innovations include site-specific nutrient management (SSNM) and induced defoliation technology. As the leader of the National Innovation for Climate Resilient Agriculture (NICRA) program, he has developed technologies to mitigate climatic stress in high-risk districts. Dr. Singh has over 200 research publications.

Contributors

Akram Ahmed ICAR-Research Complex for Eastern Region, Patna, Bihar, India

S. K. Asraful Ali Division of Agronomy, ICAR-Indian Agricultural Research Institute, New Delhi, India

Anjana J. Atapattu Agronomy Division, Coconut Research Institute, Lunuwila, Sri Lanka

Subhash Babu Division of Agronomy, ICAR-Indian Agricultural Research Institute, New Delhi, India

Koushik Bag Agronomy, DSRE, ICAR Research Complex for NEH Region, Umiam, Meghalaya, India

Shreyas Bagrecha Agronomy Section, ICAR-National Dairy Research Institute, Karnal, India

Anamika Barman Division of Agronomy, ICAR-Indian Agricultural Research Institute, New Delhi, India

Biswaranjan Behera ICAR-Indian Institute of Water Management, Bhubaneswar, Odisha, India

B. Bhargavi ICAR-Central Research Institute for Dryland Agriculture, Hyderabad, India

ICAR-Central Institute for Cotton Research, Nagpur, Maharashtra, India

Jayant S. Bhat ICAR-Indian Agricultural Research Institute, Dharwad, India

T. L. Bhutia ICAR-Research Complex for NEH Region, Sikkim Centre, Gangtok, Sikkim, India

ICAR-Indian Institute of Soil and Water Conservation Research, Dehradun, Uttarakhand, India

Niraj Biswakarma Agronomy, DSRE, ICAR Research Complex for NEH Region, Umiam, Meghalaya, India

G. Ravindra Chary ICAR-Central Research Institute for Dryland Agriculture, Hyderabad, India

Anup Das ICAR-Research Complex for Eastern Region, Patna, Bihar, India

Debarati Datta ICAR-Central Research Institute for Jute and Allied Fibres, Barrackpore, West Bengal, India

Man Mohan Deo ICAR-Indian Institute of Pulses Research, Kanpur, Uttar Pradesh, India

Ramgopal Devadas ICAR-Research Complex for NEH Region, Sikkim Centre, Gangtok, Sikkim, India

ICAR-Indian Institute of Soil and Water Conservation Research, Dehradun, Uttarakhand, India

E. L. Devi ICAR-Research Complex for NEH Region, Sikkim Centre, Gangtok, Sikkim, India

ICAR-Indian Institute of Soil and Water Conservation Research, Dehradun, Uttarakhand, India

Khushboo Devi Division of Agronomy, ICAR-Indian Agricultural Research Institute, New Delhi, India

Savita Dewangan Department of Agronomy, Institute of Agricultural Sciences, Banaras Hindu University, Varanasi, Uttar Pradesh, India

M. K. Dhakar ICAR Research Complex for Eastern Region, Farming System Research Centre for Hill and Plateau Region, Ranchi, Jharkhand, India

Nuwandhya S. Dissanayaka Agronomy Division, Coconut Research Institute, Lunuwila, Sri Lanka

A. P. Dwivedi ICAR-Indian Institute of Sugarcane Research, Lucknow, Uttar Pradesh, India

Ananya Gairola Division of Agronomy, ICAR-Indian Agricultural Research Institute, New Delhi, India

Sonaka Ghosh ICAR Research Complex for Eastern Region, Patna, Bihar, India

Sourav Ghosh ICAR-Central Research Institute for Jute and Allied Fibres, Barrackpore, West Bengal, India

K. A. Gopinath ICAR-Central Research Institute for Dryland Agriculture, Hyderabad, India

Prabhu Govindasamy Division of Agronomy, ICAR-Indian Agricultural Research Institute, New Delhi, India

K. V. Gowtham ICAR-Indian Agricultural Research Institute, New Delhi, India

B. A. Gudade Spices Park Chhindwara, Spices Board of India, Chhindwara, Madhya Pradesh, India

Nitin N. Gudadhe Navsari Agricultural University, Navsari, Gujarat, India

N. C. Gulleibi Agronomy, DSRE, ICAR Research Complex for NEH Region, Umiam, Meghalaya, India

Mohammad Hashim ICAR-Indian Institute of Pulses Research, Kanpur, Uttar Pradesh, India

K. K. Hazra ICAR-Indian Institute of Pulses Research, Kanpur, Uttar Pradesh, India

Hejbina M. Hussain ICAR – Agricultural Technology Application Research Institute, Umiam, Meghalaya, India

Twinkle Jena Division of Agronomy, ICAR-Indian Agricultural Research Institute, New Delhi, India

Mrunalini Kancheti ICAR-Indian Institute of Pulses Research, Kanpur, Uttar Pradesh, India

Namrata Kashyap Krishi Vigyan Kendra, Lakhimpur, Assam Agricultural University, Jorhat, India

Amit Kumar ICAR-Research Complex for NEH Region, Sikkim Centre, Gangtok, Sikkim, India

ICAR-Indian Institute of Soil and Water Conservation, Dehradun, Uttarakhand, India

Narendra Kumar ICAR-Indian Institute of Pulses Research, Kanpur, Uttar Pradesh, India

Sanjeev Kumar Department of Soil Science & Agricultural Chemistry, College of Agriculture, Sardar Vallabhbhai Patel University of Agriculture and Technology, Meerut, Uttar Pradesh, India

Shamal Shasang Kumar Crop Research Division, Ministry of Agriculture & Waterways (MOA & W), Suva, Fiji

Vipin Kumar Division of Agronomy, ICAR – Indian Agricultural Research Institute, New Delhi, India

Megha Kumari Division of Agronomy, ICAR-Indian Agricultural Research Institute, New Delhi, India

Kamlesh K. Lakhena Division of Agronomy, ICAR-Indian Agricultural Research Institute, New Delhi, India

Jayanta Layek Agronomy, DSRE, ICAR Research Complex for NEH Region, Umiam, Meghalaya, India

Badapmain Makdoh Agronomy, DSRE, ICAR Research Complex for NEH Region, Umiam, Meghalaya, India

S. S. Mali ICAR Research Complex for Eastern Region, Farming System Research Centre for Hill and Plateau Region, Ranchi, Jharkhand, India

Santosh S. Mali Farming Systems Research Center for Hill and Plateau Region, ICAR-Research Complex for Eastern Region, Ranchi, Jharkhand, India

Ab Raouf Malik Division of Fruit Science, Faculty of Horticulture, Sher-e-Kashmir University of Agricultural Sciences and Technology of Kashmir, Srinagar, Jammu and Kashmir, India

Rohit Kumar Mishra School of Crop Health Biology, ICAR – National Institute of Biotic Stress Management, Raipur, Chhattisgarh, India

A. K. Mohanty ICAR – Agricultural Technology Application Research Institute, Umiam, Meghalaya, India

S. Mondal ICAR-Central Citrus Research Institute, Nagpur, India

Ganapati Mukri ICAR-Indian Agricultural Research Institute, New Delhi, India

S. K. Naik ICAR Research Complex for Eastern Region, Farming System Research Centre for Hill and Plateau Region, Ranchi, Jharkhand, India

C. P. Nath ICAR-Indian Institute of Pulses Research, Kanpur, Uttar Pradesh, India

Dileswar Nayak Navsari Agricultural University, Navsari, Gujarat, India

Somanath Nayak AICRP on MULLaRP, Central Agricultural University, Imphal, Manipur, India

Kadagonda Nithinkumar Division of Agronomy, ICAR-Indian Agricultural Research Institute, New Delhi, India

Tharindu D. Nuwarapaksha Agronomy Division, Coconut Research Institute, Lunuwila, Sri Lanka

Alapati Nymisha Division of Soil Science and Agricultural Chemistry, ICAR – Indian Agricultural Research Institute, New Delhi, India

Venkatesh Padimi Department of Agricultural Biology, Colorado State University, Fort Collins, CO, USA

Aastika Pandey Division of Agronomy, ICAR-Indian Agricultural Research Institute, New Delhi, India

Alok Kumar Pandey Department of Soil Science and Agricultural Chemistry, College of Agriculture, Acharya Narendra Deva University of Agriculture and Technology, Ayodhya, Uttar Pradesh, India

Anjali Patel Division of Agronomy, ICAR – Indian Agricultural Research Institute, New Delhi, India

Roopa Patel Department of Soil Science & Agricultural Chemistry, Institute of Agricultural Sciences, Banaras Hindu University, Varanasi, Uttar Pradesh, India

Department of Soil Science & Agricultural Chemistry, Banaras Hindu University, Varanasi, Uttar Pradesh, India

P. K. Pathak Krishi Vigyan Kendra, Nalbari, Assam Agricultural University, Jorhat, India

Vijay Pooniya Division of Agronomy, ICAR-Indian Agricultural Research Institute, New Delhi, India

Ved Prakash ICAR-Research Complex for Eastern Region, Patna, Bihar, India

Sandesh Rai Department of Agronomy, Uttar Banga Krishi Vishwavidyalaya, Pundibari, West Bengal, India

Rishi Raj Division of Agronomy, ICAR-Indian Agricultural Research Institute, New Delhi, India

Namrata Rajbonshi Agronomy, DSRE, ICAR Research Complex for NEH Region, Umiam, Meghalaya, India

V. Ramamurthy ICAR-National Bureau of Soil Survey & Land Use Planning, Regional Centre, Bangalore, India

G. I. Ramkrushna ICAR-Central Institute for Cotton Research, Nagpur, Maharashtra, India

Sanjay Singh Rathore Division of Agronomy, ICAR-Indian Agricultural Research Institute, New Delhi, India

Swarna Ronanki ICAR-Indian Institute of Millets Research, Hyderabad, Telangana, India

Saurav Saha ICAR-Research Complex for NEH Region, Sikkim Centre, Gangtok, Sikkim, India

ICAR-Indian Institute of Soil and Water Conservation Research, Dehradun, Uttarakhand, India

Nilutpal Saikia Division of Agronomy, ICAR-Indian Agricultural Research Institute, New Delhi, India

Rahul Saikia Agronomy, DSRE, ICAR Research Complex for NEH Region, Umiam, Meghalaya, India

Kapila Shekhawat Division of Agronomy, ICAR – Indian Agricultural Research Institute, New Delhi, India

Shilpa ICAR-Indian Agricultural Research Institute, New Delhi, India

Reshma Shinde ICAR Research Complex for Eastern Region, Farming System Research Centre for Hill and Plateau Region, Ranchi, Jharkhand, India

S. K. Shukla ICAR-Indian Institute of Sugarcane Research, Lucknow, Uttar Pradesh, India

Saurbh Shukla Department of Silviculture and Agroforestry, Dr. Yashwant Singh Parmar University of Horticulture and Forestry, Solan, Himachal Pradesh, India

A. K. Singh Department of Agricultural Chemistry and Soil Science, SAS, Nagaland University, Medziphema, Nagaland, India

Alok Kumar Singh Rajiv Gandhi South Campus, Banaras Hindu University, Mirzapur, Uttar Pradesh, India

Bipin Kumar Singh Department of Agroforestry, Acharya Narendra Deva University of Agriculture and Technology, Ayodhya, Uttar Pradesh, India

Chandu Singh ICAR-Indian Agricultural Research Institute, New Delhi, India

Manoj Kumar Singh Department of Agronomy, Institute of Agricultural Sciences, Banaras Hindu University, Varanasi, India

Raghavendra Singh ICAR-Indian Institute of Farming Systems Research, Meerut, Uttar Pradesh, India

Vinod Kumar Singh ICAR-Central Research Institute for Dryland Agriculture, Hyderabad, India

Shashi S. Udumann Agronomy Division, Coconut Research Institute, Lunuwila, Sri Lanka

Owais Ali Wani Division of Soil Science and Agricultural Chemistry, Faculty of Agriculture, Sher-e-Kashmir University of Agricultural Sciences and Technology of Kashmir, Srinagar, Jammu and Kashmir, India

Abha Yadav Department of Soil Science and Agricultural Chemistry, College of Agriculture, Acharya Narendra Deva University of Agriculture and Technology, Ayodhya, Uttar Pradesh, India

Devideen Yadav Division of Agronomy, ICAR – Indian Agricultural Research Institute, New Delhi, India

ICAR-Indian Institute of Soil and Water Conservation, Dehradun, Uttarakhand, India

Pooja Yadav ICAR-Indian Institute of Sugarcane Research, Lucknow, Uttar Pradesh, India

S. K. Yadav ICAR-Indian Institute of Sugarcane Research, Lucknow, Uttar Pradesh, India

V. P. Yadav ICAR-Indian Institute of Sugarcane Research, Lucknow, Uttar Pradesh, India

Md Yeasin ICAR-Indian Agricultural Statistics Research Institute, New Delhi, India

R. R. Zhiipao Department of Agronomy, Rajiv Gandhi University, Papum Pare, Arunachal Pradesh, India

Chapter 1
Crop Diversification in India: Implications to Food Security and Soil Health

Anjali Patel, Subhash Babu, Sanjay Singh Rathore, Ananya Gairola, Vipin Kumar, Kapila Shekhawat, Rishi Raj, and Rohit Kumar Mishra

Abstract Sustainable cultivation is recognized as a pivotal strategy to address the various crises confronting rural households in developing nations. In India, farming continues to be the basis of livelihood, providing nutrition, food, income, and employment to the majority of the rural areas' population. In agriculture, one of the critical challenges is increasing yield without compromising its foundations such as a healthy environment, particularly soil health. Hence, there is an immediate necessity to pinpoint management practices that simultaneously support soil health and production, striving for environmentally sustainable agricultural systems. Crop diversification, an approach involving the cultivation of diverse crops within a specific area or region, has gained significant attention for its profound implications on food security and soil health enhancement. Adoption of crop diversification will reduce the incidence of crop failure and pest and disease infestation in monoculture systems while also improving soil quality and crop yields. Utilizing a diverse array of scholarly sources, empirical studies, and agricultural reports, the chapter seeks to clarify the pivotal role of crop diversification in sustainable agriculture and global food systems. Its primary objective is to comprehend the benefits, as well as the environmental and socioeconomic challenges, associated with the practice of crop diversification.

Keywords Climate change · Crop diversification · Food security · Soil health

A. Patel · S. Babu · S. S. Rathore · A. Gairola · V. Kumar · K. Shekhawat (✉) · R. Raj
Division of Agronomy, ICAR – Indian Agricultural Research Institute, New Delhi, India

R. K. Mishra
School of Crop Health Biology, ICAR – National Institute of Biotic Stress Management, Raipur, Chhattisgarh, India

S. Babu et al. (eds.), *Agricultural Diversification for Sustainable Food Production*, Sustainability Sciences in Asia and Africa,
https://doi.org/10.1007/978-981-97-7517-0_1

1.1 Introduction

Agriculture has traditionally been the paramount livelihood source in developing nations. The key emphasis of agricultural development over the past 70 years has been to enhance agricultural production and ensure nutrition security. Fortunately, with the collaborative efforts of farmers, researchers, and policymakers, milestones like the "Green Revolution" and "Rainbow Revolution" were achieved. With a net cultivated area of approximately 140 million hectares (Mha), food grain production increased from 82 million tons (Mt) before the Green Revolution era in 1960–61 to 330 Mt in 2022–23. However, challenges like increasing population, rising living standards, industrial development, and urbanization have significantly impacted the limited natural resources. The population growth is reducing the availability of per capita land. The land-to-people proportion has fallen from 0.39 hectares in 1950 to 0.10 hectares in 2021 (Das et al. 2022b). Despite the successes, strategies have not adequately addressed the imperative to increase farmers' income or directly promote farmers' welfare. As agriculture modernizes and industrializes, it has also become not so diverse (Baldwin-Kordick et al. 2022). Small and marginal farmers find themselves economically unviable, unable to generate sufficient income that maintains a reasonable standard of living in the current technological and socioeconomic system. Traditional agriculture is dependent on farmers (Roesch-McNally et al. 2018), which perpetuates the simplified principle that traditional methods are more efficient and effective. To resolve these issues, there has been a call to increase food production without endangering the environment by adopting the integrated approach of farming and resource management (Rathore et al. 2022; Babu et al. 2020c). However, successful integration requires a proper understanding of the links between soil health and food security (Singh et al. 2021; Das et al. 2022a; Yadav et al. 2023).

Crop diversification involves enhancing the variety of crops grown through techniques like crop rotation, multiple cropping, or intercropping. This approach aims to boost productivity, promote sustainability, and maintain ecological balance in agriculture, in contrast to specialized farming (Garbach et al. 2017; Babu et al. 2023). It is a demand-driven, dynamic, and iterative concept based on the specific context, including value addition, spatial, temporal, and resource reuse, in addition to shifting from less remunerative and traditional crops (Fig. 1.1). The idea involves substituting crops that either yield low output or demand high inputs and harm the environment with more suitable alternatives. For instance, replacing rice with millet and pulses would be a viable option. Farm productivity in diversified cropping is generally greater than in single or double cropping; this phenomenon is known as "the rotation effect" (Bowles et al. 2020; Stanger and Lauer 2008). In this context, diversification in agriculture as a way to advance the quality of life should be changed from low-value crops to more lucrative alternatives. It is noted that crop diversity affects food security through its supply chain by generating food self-sufficiency for the population, comprehensive food patterns that include food grains and other protein foods, and its demand chain by increasing people's buying power

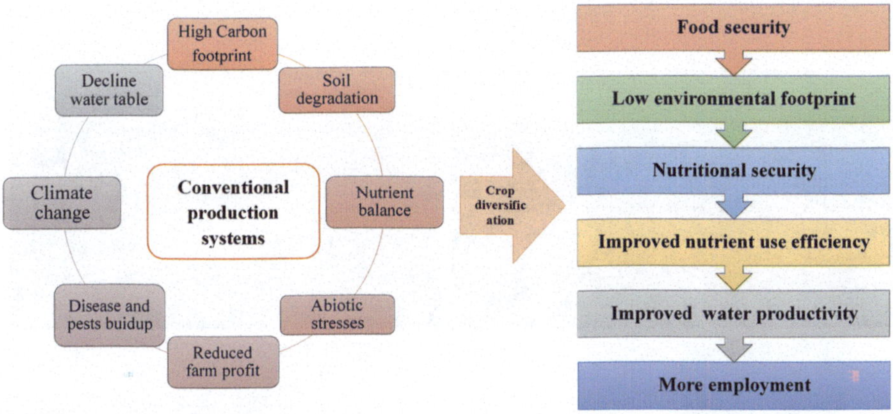

Fig. 1.1 Conventional cropping vis- à-vis diversified cropping

(Paul et al. 2016). Diversification of crops not only contributes to food security but also allows farmers to produce more crops to sell in the market, thus increasing their income (Singh et al. 2023; Yadav et al. 2023).

Agricultural divergence involves three situations: (a) transitioning from farming to non-farming activities, (b) shifting from less profitable enterprises to more profitable ones, and (c) using resources in diverse but complementary activities. Some examples of intensified cropping systems are depicted in Fig. 1.2.

The cultivation of trees alongside crops offers a promising strategy to diversify and enhance overall land productivity. Agroforestry, the practice of growing perennial trees as well as crops and livestock for spatial arrangements or temporal sequences, currently covers 25.32 million hectares. It is viewed as an approach to adapt to climate alteration while solving issues such as soil fertility depletion due to intensive agriculture. Agroforestry systems contribute significantly to the supply of fuelwood, small timber, plywood material, paper and pulp, and green fodder while also fulfilling household subsistence needs (Yadav et al. 2021; Babu et al. 2020b). Diversification of crops in an area is thought to be directly related to both food availability (or supply) and food affordability (or demand), which are central concepts for food security. In the 1970s, the focus was initially on the physical availability of food as an important factor in achieving food security (Frankenberger and Maxwell 1992). Far ahead, food availability gave way to rights, as analysis showed that hunger was not caused by a lack of food but by a lack of respect for the "right" to food (Sen 1981).

In the 1980s, the concept of food insecurity was prominent. By the twenty-first century, food security encompasses not only food availability but also its biological utilization. Although crop diversification has long been practiced in rural communities in developing countries and developing economies, research in this context is scarce. Only few studies have explored the indirect effects of crop diversification on nutriment security (Paul et al. 2016). Particularly in the framework of agriculture in India, systematic research explaining crop diversity as a determining factor of food

Fig. 1.2 (**a**) Intercrops of chickpea (*Cicer arietinum*) + Indian mustard (*Brassica juncea*), (**b**) wheat (*Triticum aestivum*) + Indian mustard, (**c**) bund utilization by planting jackfruits (*Artocarpus heterophyllus* Lam.), (**d**) intercropping of drumstick (*Moringa oleifera*) + Indian mustard, (**e**) brinjal (*Solanum melongena*) + karonda (*Carissa carandas*), and (**f**) intercropping of pomegranate (*Punica granatum*) + spinach (*Spinacia oleracea* L.)

security is lacking. However, numerous scientific papers in agronomy focus on agricultural diversification such as intercropping or crop rotation (Babu et al. 2016). In view of this research gap in the prevailing literature, this chapter aims to address the questions like: How can a relation between food security and crop diversification be established? What is the general decision that affects farmer's economics? Does using different types of diversified cropping affect the soil health of a region?

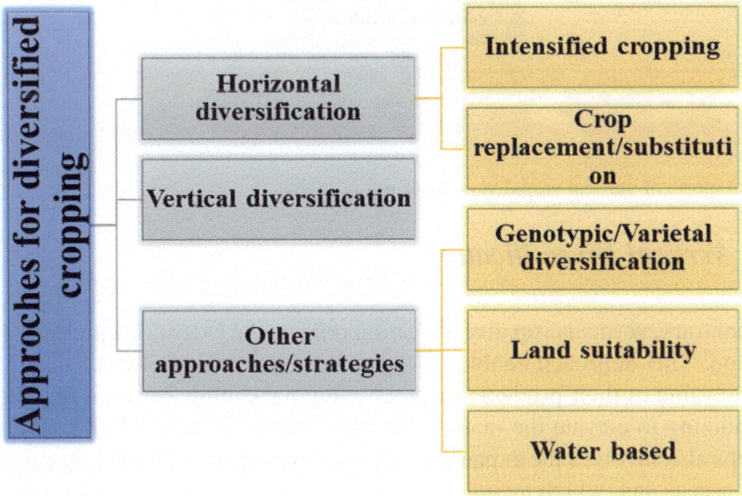

Fig. 1.3 Crop diversification approaches

1.2 Crop Diversification Approaches

The following subsections examine various crop diversification methods based on water availability, land suitability, and economic needs like seasonal, temporal, and regional (Nishan 2014). Various crop diversification methods are shown in Fig. 1.3.

1.2.1 Horizontal Diversification

Horizontal diversification can be achieved through two primary strategies: crop substitution, which entails replacing continuously grown crops, and crop intensification. For example, during the Green Revolution, the focus was primarily on cultivating single crops. Today, farmers, especially in developing countries, are transitioning from monoculture of cereal-based staples to high-value crops such as vegetables and spices while also dealing with pests and disease infestation. In contrast, crop utilization entails integrating new useful crops into prevailing crops or cropping systems to boost the overall yield of agriculture. To reap the benefits of diversified agriculture, it is necessary to move from simple cropping systems to more complex and intensive ones, such as intercropping, relay cropping, multiple cropping, and other practices. Crop intensification also contributes to increased profitability, job opportunities, and enhanced energy use efficiency (Babu et al. 2016).

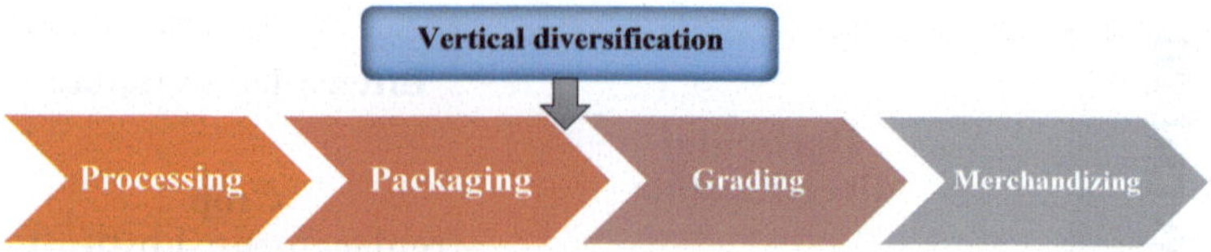

Fig. 1.4 Aspects of vertical crop diversification

1.2.2 Vertical Diversification

On the contrary, vertical crop diversification reflects the extent of agricultural industrialization. This approach enables farmers and other stakeholders to increase the economic value of their products by processing, packaging, regional branding, and merchandizing to elevate the marketable value of crops. Vertical diversification of food crops also involves the expansion of postharvest actions, including transformation and processing, which includes sorting, grading, processing into both industrial and food products, packing, storage, and transportation to export or domestic markets (Hedley 1987). The development of processing and transformation industries is a crucial factor in rural areas, fostering revenue generation and job creation. A combination of diversification (i.e., horizontal diversification or multi-cropping and vertical diversification or agribusiness) is considered to be needed to increase crop production and generate income in the region and country. Figure 1.4 illustrates the concept of vertical diversification.

1.2.3 Other Approaches

- Approach based on soil
- Approach based on aquatic system
- Diversification toward nutritional security
- Diversification toward nutrient management
- Diversification toward pest and disease management
- Diversification to alleviate and adapt to climate change

1.3 Measures of Crop Diversification

1.4 Importance of Crop Diversification

1.4.1 Optimizing Smallholders Food Security Through Crop Diversification

One of the challenges confronting the agricultural sector in India is prevalence of the small as well as marginal-scale farmers. In India, about 86% of farmers fall under the classification of smallholder farmers. These predominantly tribal farmers are particularly vulnerable to climatic and economic shocks due to their cultural beliefs and practices. Traditional monocropping practices, prevalent among most smallholder farmers, have adverse effects on yield, soil health, and food security. Seasonal yields are essential for food and economic returns, and low yields can negatively impact family income. Investing in crop diversification can help mitigate food insecurity, enhance yields, and provide a buffer against crop failures. Diversification of agriculture can meet the objectives of sustainability and lead to higher farm incomes through adjustments that look to benefit the small farm holder (Table 1.1). According to Michlera and Josephson (2017), cultivating various species of crop can assist small farmers in managing production risks and prices. Agronomic aspect of crop diversity is presented in Table 1.2.

1.4.2 Mitigating Risks Associated with Monocropping

Monocropping, which includes the year-after-year cultivation of a single crop on the same piece of land, has its disadvantages. It leaves the soil depleted of essential nutrients, making it less productive over time. Additionally, it renders the soil vulnerable to pests and diseases, further degrading its quality and impacting crop yield. Diversification, an increasingly supported concept in the agricultural sector, provides a solution. Growing a variety of plants on the same plot of land can significantly reduce susceptibility to pests, improve soil health, and ultimately boost production. This practice not only addresses concerns about food security but also enhances farmers' income by providing more options for selling their produce. Additionally, it enhances the resilience of the agricultural ecosystem to climate change. To mitigate risks of crop production or manage scarced resources during seeding to harvesting, farmers often employ diversification by choosing a variety of crops, preceded by crop rotation. Ongoing financial uncertainties, rapid demand fluctuations, and advancements in technology provide a compelling reason to diversify crops, which helps minimize risk in the agricultural system.

Table 1.1 Different measures of crop diversification and their description

Measures of crop diversification	Description
1. Temporal diversification in crops	
Catch crop	Utilizing the window period by the cultivation of short-duration crops between the two main crops
Relay cropping	Cultivation of the second crop in standing crop before the first crop is harvested
Crop rotation	The cultivation of two or more crops in succession on a given piece of land in a year
Multiple or double cropping	Cultivation of two and/or more crops in a single season
2. Spatial diversification in crops	
Trap crops	A crop planted to appeal to insect pests from another crop
Strip cropping	Cultivation of crops in narrow strips either following the natural contours of the terrain or at right angles to the direction of the prevailing wind to prevent wind and water erosion
Intercropping	Concurrent cultivation of two or more crops on the same land in a definite row arrangement
Alley cropping	Food crops grown in alleys formed by trees
Mixed cropping	Two or more crops grown concurrently on the same piece of land by mixing their seeds
Varietal/cultivar mixture	Two or more varieties/cultivar of the same species being grown together

Table 1.2 Agronomic benefits of crop diversification

Influence	Key drivers	Instances
Main effects	Effect of break crop	Cereal cultivation can potentially decrease due to induction of oil seeds and pulses; however, it is location-specific
	Better resource utilization	Diverse crop growth habits can potentially use the solar radiation, nutrients, and moisture efficiently
	N fixation by legumes	Owing to nitrogen fixation ability, pulses improve overall soil health, which improves the productivity of succeeding crops in system mode
Secondary effects	Weed pressure	Diversified cropping facilitates herbicide rotation, reduction in weed seed bank, and eventually plant competition
	Insect infestation	Species diversification mitigates pest infestations by disrupting host dynamics and interrupting pest life cycles
	Disease incidence	Cereal legume integration can potentially suppress the disease incidence
Other impacts	Nutrient use efficiency	Different rooting behaviors facilitate the varied nutrient extraction rates by crops from the soil
	Water use efficiency	Replacing water grueling ones with less water-demanding crops can optimizes water use efficiency and resource allocation as well as diverse produce demands in long-run
	Allelopathy	Crop residue retention/incorporation under diversified cropping can have a potential negative impact on other plants

1.4.3 Economic Benefits for Farmers

Agricultural diversification can play an important role in generating higher incomes, increasing employment, stabilizing incomes, and protecting natural resources. Crop diversification can help reduce the risks of low income, food insecurity, and nutrition insecurity (Mango et al. 2018). It might be an efficient idea to keep farmers financially healthy and better integrate them into local economy. Several researchers have emphasized that diversification can enhance the sustainable development of rural regions by establishing links between agriculture and other economic sectors (Iocola et al. 2020). Crop diversification may improve the economics of small farms (Manjunatha et al. 2013). Increasing commercialization of small farms with different crops provides relative income stabilization (Berg et al. 2007). Crop diversity can surge the small farms' income, provide an alternate source of earning, and make them more resilient to price fluctuations.

According to Makate et al. (2016), a diverse crop rotation system increases crop yields, resulting in higher annual income for farmers. Likewise, Bravo-Ureta et al. (2006) showed that farms experienced a 21% increase in revenue, while a positive relationship was identified by Perz (2004) between diversification and income levels in the Brazilian Amazon. In addition, cultivating multiple crops within the same region can boost farmers' income and provide greater financial stability (Hansen et al. 2019; Jat et al. 2012).

1.4.4 Employment Generation/Creation

The agriculture sector has a noteworthy role in employment generation. However, The use of conventional cropping systems such as rice–wheat results in unemployment during offseasons. Some studies have shown that severe seasonal unemployment exists in many parts of the country, leading to the seasonal movement of workers and agricultural laborers to surrounding cities/towns in pursuit of employment opportunities (Behera and Jha 1999). Crop diversification helps rural families by providing additional full-time employment opportunities.

1.4.5 Environmental Benefits

Soil health refers to "the ability of soil to function as a vital living system, within ecosystem and land-use boundaries, to sustain plant and animal productivity, uphold or improve water and air quality, and promote plant and animal health" (Doran 1996). An improved soil health helps increase fiber, fuel, and food production; it will also increase ecosystem services of soil (SESs) while improving sequestration of carbon (C), purification of water, habitat and resource provision, nutrient cycling,

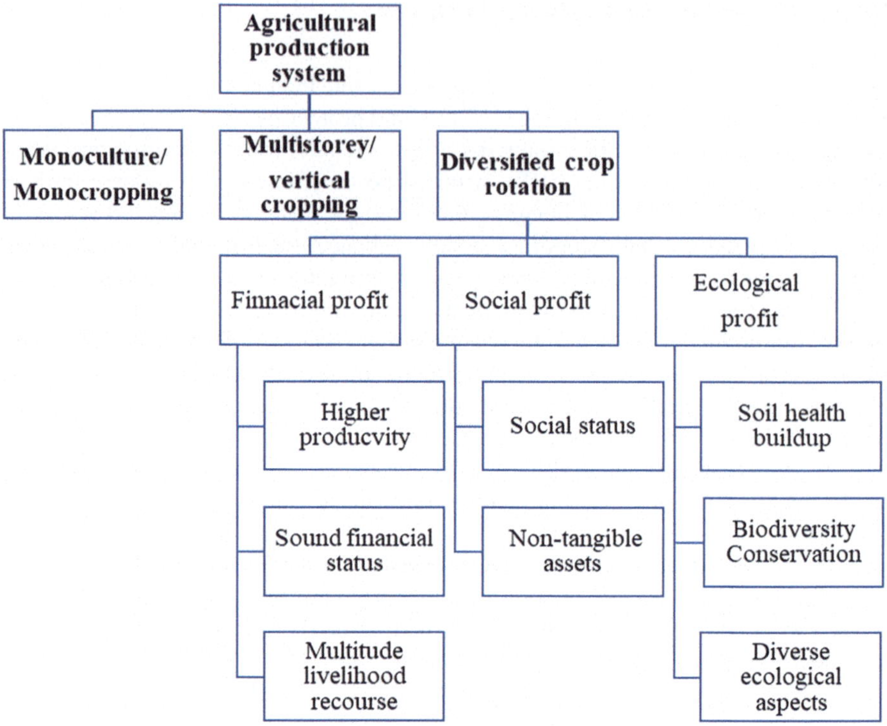

Fig. 1.5 Food security and soil health associated with crop diversification

and flood control (FAO 2015). Diversified cropping potentially reduces soil erosion and improves nutrient recycling under diverse agroecosystems (Yadav et al. 2023, 2024b; Das et al. 2022a; Rathore et al. 2022). Implementing practices that will improve SESs is important for the future of environmental quality, agriculture, and well-being of human (Fig. 1.5).

1.5 Implications for Food Security

Food security is realized by ensuring that everyone consistently has social, economic, and physical access to healthy, safe, and adequate food that fulfills their dietary requirements and preferences for an active and healthy lifestyle (FAO 1996). Figure 1.6 and Table 1.3 illustrate the connection between crop diversification and food security.

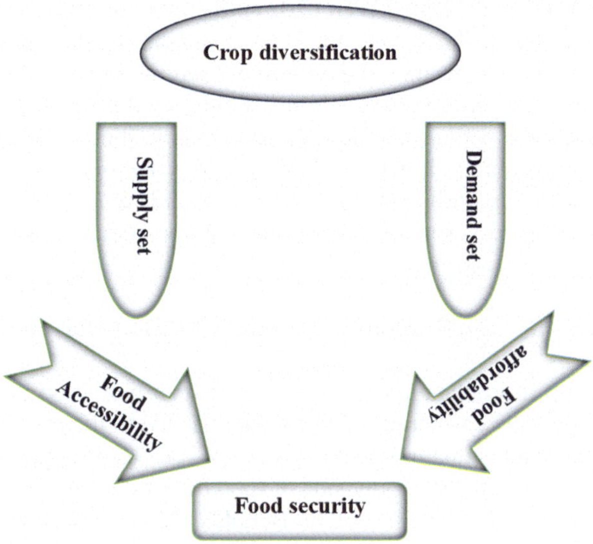

Fig. 1.6 Crop diversification connecting with food security

Table 1.3 Crop diversification in relation to food security

Crop diversification	Location	Food security benefit	References
Rice–yellow sarson–groundnut	Uttarakhand, India	Increased rice yield by 10.09% over rice–wheat system	Chaudhary et al. (2020)
A 4-year maize-soybean-oat+alfalfa (*Medicago sativa* L.)-alfalfa sequence	Minnesota, USA	Increased maize and soybean yield by 5% and 21% over maize and soybean rotation, respectively	Baldwin-Kordick et al. (2022)
Maize–toria Maize–rajmash Maize (green cobs)–urdbean–buckwheat maize–buckwheat	Sikkim, India	Higher production and relative production efficiencies and land-use efficiency	Babu et al. (2016)
Maize + sweet potato–wheat	Uttarakhand, India	Maize + sweet potato system had 0.77 Mg/ha higher maize productivity over maize–wheat	Yadav et al. (2023)

1.5.1 Addressing Nutritional Deficiencies

Food insecurity and undernutrition are significant barriers to the progress of developing nations (Pieters et al. 2013; Yadav et al. 2024a). A Comprehensive National Nutrition Survey (CNNS) conducted with 112,316 children under 16 years of age in 30 states of India from 2016 to 2018 revealed that more than 15% of the children aged 1–4, 5–9, and 9–16 had vitamin B_{12}, folate, vitamin D, vitamin A, and Zn contents insufficiency. It is estimated that 15.3% of India's population is undernourished, with 208.6 million individuals experiencing malnutrition, representing more than 30% of the global malnourished population (683.9 million) (FAO, IFAD,

UNICEF, WFP, and WHO 2021). Besides the scarcity of food, the diversity of diet has also declined over time. For example, while India's per capita availability of cereals rose from 334.2 grams per day in 1951 to 458.0 grams per day in 2020, simultaneously the availability of pulses fell from 60.7 to 43.8 grams per day during the same period, leading to a reduction in dietary diversity (DES 2020). Although establishing a direct link between crop diversification and its impact on food and nutrition security is challenging (Mazunda et al. 2015), there are indirect pathways that suggest crop diversification could positively affect food security.

1.5.1.1 Implications of Diversified Production on Food Availability

Crop diversification has led to a shift to non–food grain crops from food grains, especially cash crops. This change has led to the production of nuts, fruits, vegetables, and animal products, which are sources of organic nutrients. Crop diversification impacts the supply side of food (or food availability), potentially improving the nutritional outcomes of households. Many researchers have reported that crop diversification promotes crop production with positively impacting food availability and nutrition security in India's under varied ecological regions (Rathore et al. 2022; Walia et al. 2022; Das et al. 2022a).

1.5.1.2 Implications of Enhanced Purchasing Power to Improve Food Affordability

Growing profitable crops can increase farmers' incomes and improve their purchasing. Product diversity therefore affects the affordability (or demand side) of food. The Green Revolution in Asia demonstrates that diversification of crops can increase farmers' income and positively affect food security and poverty reduction (Ibrahim et al. 2009). Agricultural diversification targeting high-yielding crops helps in family nutrition (Babu et al. 2020).

1.5.1.3 Implications on Usability of Food

The usability of food refers to the ability to biologically utilize the consumed food. Many factors, like nutritional knowledge, practices, and stable hygiene and environmental conditions, determine the ability to absorb food and health status (Paul et al. 2016).

1.5.2 Access to a Variety of Foods

Crop diversification increases access to a variety of foods by increasing the number of crops grown. This practice not only improves dietary diversity but also makes agriculture more resilient to pests and climate-related challenges (Fig. 1.7). There are many ways to focus on solving macroeconomic and national food security problems. For example, importance is given to the availability of overall food products, including imports, exports, and local production, especially focusing on staple foods. While this strategy helps secure adequate national food supplies, it does not necessarily address nutrition quality and safety, processing and marketing, postharvest losses, dietary diversity, or equitable access to food for all individuals and families (FAO 1992). To solve these problems, the concept of Smart Food mainstreaming has been introduced, aiming to reintegrate "traditional" Smart Foods as staples across Africa and Asia. Smart Foods is defined as food that meets standards that are beneficial to human health, provides environmental sustainability, and supports farmers. Sorghum and millet are initially identified as Smart Foods, and a participatory approach was employed to develop culturally acceptable products, encourage behavioral change, and raise awareness to improve adoption, nutritional status, and

Fig. 1.7 Aspect to dietary diversity

dietary diversity (Diama et al. 2020). In arid regions, Smart Food approach is innovating nutritious crops (Barikmo et al. 2007; Chande et al. 2021) that can be better consumed by improving the local dishes; this can significantly improve the household's daily energy intake, particularly for women and children (Douyon et al. 2022). In rural remote areas, where household access to food heavily depends on production, crop diversification provides farmers with a variety of crops, which they might otherwise be unable to access due to costs or lack of infrastructure.

1.5.3 Resilience Against Pests and Climate Change

The adverse effects of climate change on agriculture have led to efforts to incorporate resilience into farming systems. An efficient and cost-effective way to achieve this could involve the increased implementation of agricultural crop diversification. This practice provides resilience against extreme and highly variable weather conditions caused by climate change, providing benefits such as mitigation of greenhouse gas emissions and increasing carbon sequestration in the soil. Crop diversification can contribute to resilience by improving the ability to suppress pest outbreaks and reducing the spread of disease and crop losses due to extreme events and greater climatic variability. These findings indicate that adopting crop diversification positively impacts resilience improvement.

It improves the ability of agroecosystems to withstand ecological stresses, enhancing resilience to pests, diseases, heat, and drought while reducing pollution circulation and helping to preserve natural resources (Clements et al. 2011; Degani et al. 2019). In the face of risk and uncertainty associated with climate change, crop diversification is becoming a key pillar of farmers' strategies to ensure income stability (Kurdyś-Kujawska et al. 2021). The aim of crop diversification in agriculture is to assist farmers in adapting to future climate changes (Myeni and Moeletsi 2020; Aribi and Sghaier 2020). By diversifying crops, small farms are more easily adapted to environmental changes and less vulnerable to production losses (Kurdyś-Kujawska and Sompolska-Rzechuła 2019).

1.5.4 Crop Diversification and Food Security

Mango et al. (2018) found a positive and significant relationship between crop diversification and food security of smallholder farm households in central Malawi. They concluded that diversifying crops is an effective strategy for building resilient agricultural systems that enhance food security for farmers. Adjimoti and Kwadzo (2018) found that in the Collines region of Benin, crop diversification had an encouraging impact on food security in rural households, with various farmer-specific and institutional variables influencing food security status. Similarly, Goshu et al. (2012) in central and eastern Ethiopia showed that crop diversification serves as a positive determinant of household food security. Herforth (2010) identified a positive relationship between the number of crops cultivated and food security among farmers in Kenya and Tanzania. Further studies supporting the positive relationship between crop diversification and food security include those by Mahbub and Shirazy (2016), Pandey et al. (2016), Mulmi et al. (2017), Deb and Bayes (2018), and Mulwa and Visser (2020).

1.6 Soil Health Improvement Through Crop Diversification

1.6.1 Reduced Soil Erosion and Degradation

Agricultural practices such as monocropping, synthetic fertilizer use, pesticide application, and mechanical tillage are recognized contributors to soil degradation. Monocropping, involving the continuous cultivation of the same crop on a piece of land, depletes soil nutrients, reduces organic matter, and leads to erosion. Indiscriminate use of synthetic fertilizers compromises soil health, while pesticide residues accumulate in the soil, negatively affecting microbial biodiversity and soil microbiology. Mechanical tillage causes soil compaction, erosion, and poor water absorption, impacting root growth and reducing yields. To address soil health issues, sustainable and regenerative diversification approaches are essential in agriculture. Agricultural diversification, often misconstrued as merely crop rotation, offers significant ecosystem benefits and can effectively combat soil degradation. Plant diversity positively influences soil conditions, benefiting belowground communities and processes and replacing the need for expensive agricultural inputs. In turn, soil biodiversity plays a vital role in ecosystem services and functions crucially in terms of plant growth and agricultural productivity. Soil health determines the performance and quality of soil-based agricultural functions (Gregory and Pj 2015). After crop rotation, wide heterogeneity of the crop production will be maintained, or improved soil function is achieved by increasing yield and diverse root systems, as well as microbial activity expansion (Studdert and Echeverría 2000). ‘O’Dea et al. (2015) and Blanco-Canqui et al. (2013) found that a 4-year wheat-fallow field pea (*Pisum sativum*) rotation provided enough cover to prevent soil deterioration by conserving the land’s physical structure.

1.6.2 Nutrient Cycling and Soil Fertility

Because of the distinct structure, function, and relationship of plant community with soil in diversified crop rotation (DCR), it contributes to the long-term soil health development by decreasing the incidence of insect, weed, and disease and increasing the physical and chemical structure of the soil (Shah et al. 2021). Plant nutrients are restored when certain plant species are included in crop rotation, requiring less chemical fertilizer. Soil mesofauna comprise key predators in soil food webs and are critical for recycling nutrients. They are generally thought to be more sensitive to agroecosystem management practices than soil microorganisms (Behan-Pelletier 1999; Wardle et al. 1995). The study of Wyland et al. (1996) discovered that cover crops like rye and phacelia (*Phacelia tanacetifolia* L.) were successful in absorbing nitrogen and water, reducing nitrate leaching in irrigated broccoli (*Brassica oleracea* var. *italica*) by 70%, and reducing fertilizer consumption in the soil.

The introduction of nitrogen-fixing crops, such as legumes, within a traditional cropping system, can improve the status of the soil, making atmospheric nitrogen available to other plants, thereby reducing the need for mineral fertilizers with their associated high energy costs and use of nonrenewable resources (Clements et al. 2011; Isbell et al. 2017). Generally, it seems that applying a cover crop species mixture may contribute to controlling soilborne diseases like nematodes (Hajjar et al. 2008). Next to plant parasitic nematodes, other nematodes thrive in the soil. They are mostly beneficial, as they participate in improving soil fertility, disease suppression, and soil structure.

1.6.3 Impact on Soil Microbial Diversity

Despite the global concern over biodiversity losses caused by intensive agriculture, the functional significance of changes in soil biological communities due to crop rotation and diversification remains poorly understood. Studies suggest that increasing temporal plant diversity can change soil microbial communities and enhance crop productivity through positive plant–soil feedback mechanisms mediated by soil biota (Zhou et al. 2017). The increase in microbial biomass carbon and nitrogen persistently observed in response to manure amendments (Kallenbach and Grandy 2011) and diversified crop rotations (McDaniel et al. 2014; McDaniel and Grandy 2016) indicate the potential benefits of these practices. Since C is thought to be limiting for soil microorganisms (Ekblad and Nordgren 2002; Soong et al. 2020), quality of C additions through manure and plant inputs with narrow C-to-N ratio likely increased microbial biomass (Campbell et al. 1991; Lynch 1984; McDaniel and Grandy 2016). Other mechanisms explaining the large increase in microbial biomass under diversified cropping systems include greater soil water content (Skopp et al. 1990) and near-neutral pH (Aciego Pietri and Brookes 2009; Geisseler and Scow 2014). Improvements in the soil environment and resource availability not only increase microbial abundance but also alter microbial community composition (Kravchenko et al. 2019; Zhalnina et al. 2015) and enhance microbial metabolism (McGill et al. 1986; Zhalnina et al. 2015). The influence of soil biodiversity is not limited to crops; evidence suggests that aboveground biodiversity as well as belowground communities can positively affect soil conditions (Tiemann et al. 2015). High diversity rotations, as well as intercropping or agroforestry systems, can sustain more diverse soil communities by increasing the quantity, quality, and chemical diversity of plant residues and root exudates, with positive effects on soil organic matter and soil fertility.

1.6.4 Improving Soil Structure and Water Retention

Crop rotation, especially in diversified systems, plays a crucial role in improving soil structure and water retention. Planting rotations that involve cover crops can produce a large root system that holds the soil together by producing a new source of fresh organic matter after the decay of roots. This organic matter provides a habitat suitable for earthworms, which form tunnels in the soil for large soil pores, subsequently enhancing water infiltration (Table 1.4). Organic matter also helps form new soil aggregates, which is important because it holds the soil together and therefore prevents the risk of soil loss due to erosion during high-intensity rainstorm events. According to certain recorded studies by Gan et al. (2015), diversifying

Table 1.4 Crop diversification for soil health improvement

Crop diversification	Location	Soil health benefit	References
Olive orchards being diversified with *Crocus sativus*	Cordoba, Spain	Increases in soil organic carbon stock by 5.88% at 0–10 cm and 14.47% at 10–30 cm soil depth	González-Rosado et al. (2022)
Maize-soybean-oat + alfalfa -alfalfa	Minnesota, USA	8% reduction in soil resistance, 16% increase in cation exchange capacity, 157% increase in soil carbon, and 62% increase in soil microbial biomass over control	Baldwin-Kordick et al. (2022)
Soybean–soybean/maize–maize	Northeast China	Decline in soil organic C when rotation soybean substituted with maize monocropping and increased hydrogenase activity in soybean system	Tong et al. (2017)
Dhaincha (*Sesbania aculeata*) as green manure in rice	West Bengal, India	Improved microbial population and SOC	Roy et al. (2017)
Sugarcane–ratoon–wheat (residue incorporation + *Trichoderma viride* application)	Lucknow, India	The bulk density of soil being lowered down; increased porosity (>48%) and SOC (20%) at 0–15 cm soil depth	Shukla et al. (2017)
Bt cotton + greengram–groundnut Fodder maize–lucerne Fodder sorghum + fodder cowpea–horse gram–sun hemp	Hyderabad, India	Improved soil N availability Enhanced soil phosphorus availability Improved soil potassium and sulfur availability	Nthebere et al. (2020)

cropping systems with crop rotation can benefit soil water conservation and increase soil nitrogen supply.

Farmers can improve the management of weed, insect, and disease; reduce soil depletion; and promote SOC and nitrogen fixation in agricultural production processes by farming different crops in the same location (crop rotation). For example, in two current long-term rotational tests of consistent corn, consistent soybean, corn–soybean, and corn–soybean–wheat rotation, the most diversified rotation of corn–soybean–wheat rotation increased the organic carbon capacity of the soil by 7% compared to consistent soybean, and rotational and tilling interactions with soil organic carbon were 7% greater than standard farming (Zuber et al. 2018). The adoption of a wider variety of crops can enhance soil organic carbon sequestration, soil composition, and texture while simultaneously improving production and reducing soil and environmental damage (Singh et al. 2021; Babu et al. 2023; Das et al. 2022a; Ansari et al. 2023).

1.7 Strategies for Promoting Crop Diversification

1.7.1 Crop Rotation

In contemporary agriculture, it is common for farms to cultivate a single crop repeatedly, leaving fields fallow or tilled for extended periods. Regenerative agriculture, however, embraces practices involving diverse cover crops and crop rotations. Crop rotation entails cultivating different sets of crops across fields instead of planting the same crop year after year. This approach contributes to the enhancement of soil health, promotes carbon sequestration, and ensures the presence of microbes in the soil throughout the year. The implementation of these practices has positive effects such as reducing soil erosion, enhancing soil water retention, and mitigating issues of compaction and nutrient leaching.

1.7.2 Fertilization Management

Fertilizers are a key component in farming, directly influencing crop productivity and environmental quality. Therefore, adopting a balanced approach to fertilizer application is crucial. Overapplication adversely affects soil health, creating a dependency on additional fertilizers to compensate for reduced fertility. Farmers should aim for a balanced approach, ensuring adequate nutrients based on crop requirements and the natural nutrient-providing capacity of the soil.

1.7.3 Non-Farm Diversification

The concept of establishing alternative revenue streams for the farming community aims to shield them from income fluctuations that result from adverse events. Shifting from exclusive agricultural activities to diversified ventures like dairy farming, poultry, livestock, fisheries, and horticulture can address employment challenges, supplement incomes, and reduce vulnerability to external circumstances.

1.7.4 Diversification Toward High-Value Crops

High-value crops (HVCs) refer to crops that yield higher net returns per hectare compared to staple commodities. Encouraging diversification toward high-value crops reflects a pro–small farm holder perspective. Given the significant role of smallholders in agriculture, diversifying cropping patterns to include high-value crops can boost farmer incomes and contribute to environmental conservation. High-value crops impose a lesser burden on soil systems and align with long-term sustainability goals in agriculture.

1.8 Government Policies and Subsidies in India

Despite the existing challenges, crop diversification offers a significant opportunity to double farmers' income and ensure food security for the nation. Consequently, the government should actively promote crop diversification by procuring crops other than wheat and rice at Minimum Support Price (MSP). Key developmental initiatives have been launched to support farmers and related stakeholders, including the Pradhan Mantri Krishi Sinchai Yojana (PMKSY), Paramparagat Krishi Vikas Yojana (PKVY), Pradhan Mantri Fasal Bima Yojana (PMFBY), and the soil health card mission. These programs aim to strengthen irrigation, traditional agriculture, crop insurance, and soil testing, respectively. The Department of Agriculture & Farmers Welfare (DA&FW) has been implementing the Crop Diversification Programme (CDP) since 2013–14, a sub-scheme of the Rashtriya Krishi Vikas Yojana (RKVY), focusing on the original Green Revolution states such as Haryana, Punjab, and western Uttar Pradesh. The goal is to shift the area of water-intensive paddy crops to alternative crops like pulses, oilseeds, coarse cereals, nutri-cereals, and cotton. The government of India is further supporting state governments in promoting diversified crop production through initiatives like the National Food Security Mission (NFSM) for pulses, coarse cereals, nutri-cereals, cotton, and oilseeds, as well as the Mission for Integrated Development of Horticulture (MIDH) for horticultural crops. Flexibility is provided to states under RKVY for addressing state-specific needs and priorities, subject to approval by the State Level Sanctioning Committee (SLSC) headed by the chief secretary of the state.

1.9 Education and Awareness Programs

The educational level within households significantly influences food security status. Therefore, interventions aimed at supporting rural households, whether by governmental or nongovernmental organizations, should prioritize capacity building through training and informal or alternative basic education. Informal education and training create opportunities for acquiring knowledge about nutritious food production and consumption, fostering the sharing of experiences among farmers. Empowering farmers through sufficient training and informal education has the potential to enhance household food security conditions. According to Mengistu et al. (2021), efforts to improve household food security should encompass farmer empowerment through adequate training, informal education, crop diversification, and the strengthening of rural infrastructure development, including roads, market centers, and cooperatives.

1.9.1 Support from Agricultural Extension Services

In the context of agricultural extension services, the linkage between research, extension services, and farmers plays a pivotal role in the success of adopted diversified cropping systems. Access to extension services facilitates crop diversification without affecting the intensity of diversification, underscoring the crucial role of extension services in encouraging farmers to engage in selected crop types suitable for their respective areas. Bizikova et al. (2020) suggested that formally educated respondents and household heads actively participating in social and farmer groups are more likely to generate food security benefits from interventions. However, the majority of global agrobiodiversity is produced in smallholder food-growing systems (Zimmerer and Vanek 2016). Hence, there could be a need for governments to provide farmers with additional incentives to conserve soil capital as a way to increase profits and reduce risks while promoting sustainable agriculture (Cong et al. 2014). Global initiatives like the Global Soil Partnership, Global Soil Biodiversity Initiative, and a planned Global Soil Biodiversity Assessment within the UN and FAO represent avenues for disseminating expert-based knowledge (Lloret et al. 2020).

1.9.2 Collaborations Between Different Stakeholders

To ensure adequate remunerative price to small and marginal farmers from farming is the cricual challanges for sustainble development of agriculture in India. Hence, there is a need to build partnerships among researchers, policymakers, and development workers and input suppliers to design the comprehensive approaches for sustaining the agriculture in the country. Stakeholder involvement and participatory research are often very useful tools in developing adaptation options that will be adopted by a local community because these methods recognize that knowledge often lies with the farmers in the field and that local considerations should be integrated into long-term planning (Rivington et al. 2007). Participatory approaches ensure the effectiveness, accuracy, and efficiency of the research process and its outputs. Participatory processes also support farmers' ability to strengthen social organization, experiment with different crops, cultivars, and management practices and find information (Clements et al. 2011).

1.10 Barriers in Adoption

However, despite the potential benefits, several barriers impede the widespread adoption of integrated, diversified farming systems, including farmer "path dependency" shaped by government policies, limited markets for diverse agricultural products, and the need for substantial investments in specialized equipment. Farmers often lack crucial future marketing information, hindering their ability to make informed decisions about crop diversification and impacting profit maximization. Resistance to adopting integrated systems is also influenced by concerns over complexity, infrastructure deficits, and apprehensions regarding equipment costs (Bowman and Zilberman 2013; Roesch-McNally et al. 2018; Weisberger et al. 2021).

Due to insufficient investments in machinery, infrastructure, and expertise, coupled with a lack of research evidence, practices related to diversified crop rotation (DCR) are rarely implemented. In recent decades, economic incentives favoring monocropping and biotechnological strategies, coupled with the perception of higher productivity in monocultures, have diverted farmers from diversified systems. Barriers like short-term land rental contracts, limited financing options, and insufficient technological support impede the adoption of diversified crop rotation. Additionally, in Mediterranean climate regions, farmers are hesitant to embrace intercropping and agroforestry due to concerns about the availability of water for the main cash crop. Traditional orchard also exhibits a preference for alleys without foliage growth, leading to cover crop removal and intensive tillage, perpetuating a historical view of vegetated alleys as undesirable. Short-term land rental contracts prevalent among farmers obstruct long-term planning and production or soil development incentives. Besides, the absence of experience, inadequate financing access for expenditure, and deficient technical support and information among farmers and stakeholders pose substantial barriers to the adoption of crop diversification (Roesch-McNally et al. 2018; Karki and Shrestha 2014). Moreover, the scarcity of markets poses challenges to extending crop rotations, with some farmers contributing to regulated biomass markets (Roesch-McNally et al. 2018). The relatively new concept of DCR may face resistance among farming communities, particularly those concerned about economic losses compared to their familiar, traditional rotations. Vulnerable farmers, marked by poor social status, land access, and education, need additional support systems and capacity building to benefit from interventions in crop diversification and food security (Bizikova et al. 2020).

1.11 Conclusion

The global challenge in crop production involves the need to enhance yields while concurrently minimizing adverse environmental impacts. Crop diversification presents an opportunity for growers to access international and national markets through advanced food stuffs, medicinal plants, and crop products. The shift away from traditional monoculture of staples benefited significant nutritional advantages for farmers in the emerging economies of developing nations and contributed to enhancing a country's self-reliance in food production. Diversified crop rotation promotes positive interactions among soil microbes, disrupts disease cycles, and reduces weed populations. This practice increases crop production and land use and improves the chemical and physical properties of the soil. Improving ecosystem services can replace expensive inputs such as energy, inorganic fertilizers, and pesticides. Additionally, diversification helps manage price risk as not all products will be exposed to low market prices at the same time, thus benefiting the rural community. Crop diversification has the potential to be an important tool for managing sustainable production and ecosystem services and represents an important management approach for soil sustainability and food safety. Scientific evidence shows that crop diversification can reduce agricultural risk and improve the future of farming (Cong et al. 2014), increase stress resistance, and promote more resilient systems (Lin 2011; Degani et al. 2019). Diversifying agricultural practices may play a crucial role in managing ecosystem services related to croplands and crop production and pasture and forest production and will become a vital management strategy for soil sustainability and food security (Isbell et al. 2017).

1.12 Future Prospects

Crop diversification, promotes increased production and sustainability by fostering plant diversity, efficient nutrient utilization, and mitigating pollution risks associated with synthetic inputs. Government intervention is crucial for policy implementation, investment in research, infrastructure, and support services. Future research should focus on managing low-input crop productivity, understanding interactions between crop diversity and belowground biodiversity, and adapting diversified rotation practices to changing climates. Additionally, there is a need for technology development, knowledge transfer, and monitoring mechanisms to ensure agricultural sustainability.

References

Aciego Pietri JC, Brookes PC (2009) Substrate inputs and pH as factors controlling microbial biomass, activity and community structure in an arable soil. Soil Biol Biochem 41(7):1396–1405. https://doi.org/10.1016/j.soilbio.2009.03.017

Adjimoti GO, Kwadzo GTM (2018) Crop diversification and household food security status: evidence from rural Benin. Agric Food Secur 7(1):1–12. https://doi.org/10.1186/s40066-018-0233-x

Ansari MA, Ravisankar N, Ansari MH, Subhash B, Layek J, Panwar AS (2023) Integrating conservation agriculture with intensive crop diversification in the maize-based organic system: impact on sustaining food and nutritional security. Front Nutr 10:1137247. https://doi.org/10.3389/fnut.2023.1137247

Aribi F, Sghaier M (2020) Determinants and strategies of farmers' adaptation to climate change: the case of Medenine governorate, Tunisia. Agrofor Int J 5:122–129. https://doi.org/10.7251/AGRENG2002124A

Babu S, Singh R, Avasthe RK, Yadav GS, Rajkhowa DJ (2016) Intensification of maize (*Zea mays*)–based cropping sequence in rainfed ecosystem of Sikkim Himalayas for improving system productivity, profitability, employment generation and energy-use efficiency under organic management condition. Indian J Agric Sci 86(6):778–784. https://doi.org/10.56093/ijas.v86i6.58980

Babu S, Singh R, Avasthe RK, Yadav GS, Mohapatra KP, Selvan T, Das A, Singh VK, Valente D, Petrosillo I (2020) Soil carbon dynamics in Indian Himalayan intensified organic rice-based cropping sequences. Ecol Indic 114:106292. https://doi.org/10.1016/j.ecolind.2020.106292

Babu S, Mohapatra KP, Das A, Yadav GS, Tahasildar M, Singh R, Panwar AS, Yadav V, Chandra P (2020b) Designing energy-efficient, economically sustainable and environmentally safe cropping system for the rainfed maize–fallow land of the Eastern Himalayas. Sci Total Environ. https://doi.org/10.1016/j.ecolind.2020.106292

Babu S, Mohapatra KP, Yadav GS, Lal R, Singh R, Avasthe RK, Das A, Chandra P, Gudade BA, Kumar A (2020c) Soil carbon dynamics in diverse organic land use systems in North Eastern Himalayan ecosystem of India. Catena 194:104785. https://doi.org/10.1016/j.catena.2020.104785

Babu S, Singh R, Avasthe RK, Kumar S, Rathore SS, Singh VK, Ansari MA, Valente D, Petrosillo I (2023) Soil carbon dynamics under organic farming: impact of tillage and cropping diversity. Ecol Indic 147:109940. https://doi.org/10.1016/j.ecolind.2023.109940

Baldwin-Kordick R, De M, Lopez MD, Liebman M, Lauter N, Marino J, McDaniel MD (2022) Comprehensive impacts of diversified cropping on soil health and sustainability. Agroecol Sustain Food Syst 46(3):331–363. https://doi.org/10.1080/21683565.2021.2019167

Barikmo I, Ouattara F, Oshaug A (2007) Differences in micronutrients content found in cereals from various parts of Mali. J Food Compos Anal 20(8):681–687. https://doi.org/10.1016/j.jfca.2007.04.002

Behan-Pelletier VM (1999) Oribatid mite biodiversity in agroecosystems: role for bioindication. Agric Ecosyst Environ 74(1–3):411–423. https://doi.org/10.1016/S0167-8809(99)00046-8

Behera UK, Jha KP (1999) Agro-economic assessment of different crops and cropping systems for the drought prone uplands of Kalahandi, Orissa. J Andaman Sci Assoc 15(1):30–34

Berg MM, Hengsdijk H, Wolf J, Van Ittersum MK, Guanghuo W, Roetter RP (2007) The impact of increasing farm size and mechanization on rural income and rice production in Zhejiang province, China. Agric Syst 94(3):841–850. https://doi.org/10.1016/j.agsy.2006.11.010

Bizikova L, Nkonya E, Minah M, Hanisch M, Turaga RMR, Speranza CI, Timmers B (2020) A scoping review of the contributions of farmers' organizations to smallholder agriculture. Nat Food 1(10):620–630. https://doi.org/10.1038/s43016-020-00164-x

Blanco-Canqui H, Holman JD, Schlegel AJ, Tatarko J, Shaver TM (2013) Replacing fallow with cover crops in a semiarid soil: effects on soil properties. Soil Sci Soc Am J 77(3):1026–1034. https://doi.org/10.2136/sssaj2013.01.0006

Bowles TM, Mooshammer M, Socolar Y, Calderón F, Cavigelli MA, Culman SW, Deen W, Drury CF, Garcia AG, Gaudin ACM (2020) Long-term evidence shows that crop-rotation diversification increases agricultural resilience to adverse growing conditions in North America. One Earth 2(3):284–293. https://doi.org/10.1016/j.oneear.2020.02.007

Bowman MS, Zilberman D (2013) Economic factors affecting diversified farming systems. Ecol Soc 18(1):33. https://doi.org/10.5751/ES-05574-180133

Bravo-Ureta B, Cocchi H, Solis D (2006) Output diversifiication among small-scale hillside farmers in El Salvador, research papers in economics (RePEc). Inter-American Development Bank, Washington, DC. https://doi.org/10.18235/0011177

Campbell CA, Biederbeck VO, Zentner RP, Lafond GP (1991) Effect of crop rotations and cultural practices on soil organic matter, microbial biomass and respiration in a thin Black Chernozem. Can J Soil Sci 71(3):363–376. https://doi.org/10.4141/cjss91-035

Chande M, Muhimbula H, Mremi R, Muzanila YC, Kumwenda NC, Msuya J (2021) Drivers of millet consumption among school aged children in Central Tanzania. Front Sustain Food Syst 5:694160. https://doi.org/10.3389/fsufs.2021.694160

Chaudhary S, Chaturvedi S, Dhyani VC (2020) Effect of crop diversification and residue management techniques on yield attributes, yield and soil nutrient. Int J Chem Stud 8(2):1107–1111. https://doi.org/10.22271/chemi.2020.v8.i2q.8915

Clements R, Haggar J, Quezada A, Torres (2011) Technologies for climate change adaptation – agriculture sector, ed. Zhu X. UNEP Risø Centre, Roskilde. http://www.risoe.dtu.dk/rispubl/NEI/NEI-DK-5548.pdf; OSTI as DE01026421

Cong RG, Hedlund K, Andersson H, Brady M (2014) Managing soil natural capital: an effective strategy for mitigating future agricultural risks? Agric Syst 129:30–39. https://doi.org/10.1016/j.agsy.2014.05.003

Das A, Babu S, Datta M, Kumar S, Singh R, Avasthe RK, Rathore SS, Yadav SK, Singh VK (2022a) Restoring soil carbon in marginal land of Indian Himalayas: impact of crop intensification and conservation tillage. J Environ Manag. https://doi.org/10.1016/j.jenvman.2022.115603

Das BS, Wani SP, Benbi DK, Muddu S, Bhattacharya T, Mandal B, Santra P, Chakraborthy D, Bhattacharya R, Basak N, Reddy NN (2022b) Soil health and its relationship with food security and human health to meet the sustainable development goals in India. Soil Secur:100071. https://doi.org/10.1016/j.soisec.2022.100071

Deb U, Bayes A (2018) Crop diversity, dietary diversity and nutritional outcome in rural Bangladesh: evidences from VDSA panel household surveys. https://opendocs.ids.ac.uk/opendocs/handle/20.500.12413/13873

Degani E, Leigh SG, Barber HM, Jones HE, Lukac M, Sutton P, Potts SG (2019) Crop rotations in a climate change scenario: short-term effects of crop diversity on resilience and ecosystem service provision under drought. Agric Ecosyst Environ 285(December):106625. https://doi.org/10.1016/j.agee.2019.106625

DES (2020) Directorate of Economics and Statistics. Department of Agriculture and Farmers Welfare, Ministry of Agriculture and Farmers' Welfare, Government of India, New Delhi

Diama A, Anitha S, Kane-Potaka J, Htut TT, Jalagam A, Kumar P (2020) How the smart food concept can lead to the transformation of food systems and combat malnutrition: different approaches in Africa, globally, and a case study from Myanmar with lessons learnt for creating behaviour change in diets. World Rev Nutr Diet 121:149–158. https://doi.org/10.1159/000507494

Doran JW (1996) Soil health and global sustainability. In: MacEwan RJ, Carter MR (eds) Soil quality is in the hands of the land manager, vol 45. CRC, Ballarat, p 46

Douyon A, Worou ON, Diama A, Badolo F, Denou RK, Toure S, Sidibe A, Nebie B, Tabo R (2022) Impact of crop diversification on household food and nutrition security in southern and Central Mali. Front Sustain Food Syst 5:751349. https://doi.org/10.3389/fsufs.2021.751349

Ekblad A, Nordgren A (2002) Is growth of soil microorganisms in boreal forests limited by carbon or nitrogen availability? Plant Soil 242(1):115–122. https://doi.org/10.1023/A:1019698108838

FAO (1992) Food, nutrition and agriculture – 04 – food security. FAO Repository. Available online at: https://www.fao.org/3/U8050t/u8050t01.htm. Accessed 13 Oct 2021

FAO (1996) The Rome declaration on World Food Security and plan of action of the World Food Summit. Available online at: https://www.fao.org/3/w3613f/w3613f00.htm. Accessed 7 July 2021

FAO (2015) Soil ecosystem services [WWW Document]. Food and Agriculture Organization, Rome. http://www.fao.org/resources/infographics/infographics-details/en/c/284478

FAO, IFAD, UNICEF, WFP, WHO (2021) The State of Food Security and Nutrition in the World 2021. Transforming food systems for food security, improved nutrition and affordable healthy diets for all. FAO, Rome. https://doi.org/10.4060/cb4474en

Frankenberger S, Maxwell TR (1992) Household food security: concepts, indicators and measurement. UNICEF/IFAD, New York/Rome

Gan Y, Hamel C, O'Donovan JT (2015) Diversifying crop rotations with pulses enhances system productivity. Sci Rep 5(1):14625. https://doi.org/10.1038/srep14625; https://doi.org/10.4060/cb4474en

Garbach K, Milder JC, DeClerck FA, Montenegro de Wit M, Driscoll L, Gemmill-Herren B (2017) Examining multifunctionality for crop yield and ecosystem services in five systems of agroecological intensification. Int J Agric Sustain 15(1):11–28. https://doi.org/10.1080/14735903.2016.1174810

Geisseler D, Scow KM (2014) Long-term effects of mineral fertilizers on soil micro-organisms – a review. Soil Biol Biochem 75:54–63. https://doi.org/10.1016/j.soilbio.2014.03.023

González-Rosado M, Parras-Alcántara L, Aguilera-Huertas J, Lozano-García B (2022) Crop diversification effects on soil aggregation and aggregate-associated carbon and nitrogen in short-term rainfed olive groves under semiarid mediterranean conditions. Horticulturae 8(7):618. https://doi.org/10.3390/horticulturae8070618

Goshu D, Kassa B, Ketema M (2012) Does crop diversification enhance household food security? Evidence from rural Ethiopia. Adv Agric Sci Eng Res 2(11):503–515. https://www.researchgate.net/publication/275031635

Gregory MA, Pj G (2015) Soil, food security and human health: a review. Eur J Soil Sci 66(2):257–276. https://doi.org/10.1111/ejss.12216

Hajjar R, Jarvis DI, Gemmill-Herren B (2008) The utility of crop genetic diversity in maintaining ecosystem services. Agric Ecosyst Environ 123(4):261–270. https://doi.org/10.1016/j.agee.2007.08.003

Hansen J, Hellin J, Rosenstock T (2019) Climate risk management and rural poverty reduction. Agric Syst 172:28–46. https://doi.org/10.1016/j.agsy.2018.01.019

Hedley DD (1987) Diversification: concepts and directions in Indonesian agricultural policy. Soybean Res Dev Indonesia 10:17

Herforth A (2010) Promotion of traditional African vegetables in Kenya and Tanzania: a case study of an intervention representing emerging imperatives in global nutrition. PhD thesis, Cornell University, Ithaca. https://hdl.handle.net/1813/17139

Ibrahim H, Bello M, Ibrahim H (2009) Food security and resource allocation among farming households in North Central Nigeria. Pak J Nutr 8(8):1235–1239. https://doi.org/10.3923/pjn.2009.1235.1239

Iocola I, Angevin F, Bockstaller C, Catarino R, Curran M, Messéan A, Schader C, Stilmant D, VanStappen F, Vahove P (2020) An actor-oriented multi-criteria assessment framework to support a transition towards sustainable agricultural systems based on crop diversification. Sustain For 12(13):5434. https://doi.org/10.3390/su12135434

Isbell F, Adler PR, Eisenhauer N, Fornara D, Kimmel K, Kremen C, Letourneau DK (2017) Benefits of increasing plant diversity in sustainable agroecosystems. J Ecol 105(4):871–879. https://doi.org/10.1111/1365-2745.12789

Jat RA, Dungrani RA, Arvadia MK, Sahrawat KL (2012) Diversification of rice (*Oryza sativa* L.)-based cropping systems for higher productivity, resource-use efficiency and economic returns in South Gujarat, India. Arch Agron Soil Sci 58(6):561–572. https://doi.org/10.1080/03650340.2010.533172

Kallenbach C, Grandy AS (2011) Controls over soil microbial biomass responses to carbon amendments in agricultural systems: a meta-analysis. Agric Ecosyst Environ 144(1):241–252. https://doi.org/10.1016/j.agee.2011.08.020

Karki TB, Shrestha J (2014) Conservation agriculture: significance, challenges and opportunities in Nepal. Adv Plants Agric Res 1(5). https://doi.org/10.15406/apar.2014.01.00029

Kravchenko AN, Guber AK, Razavi BS, Koestel J, Quigley MY, Robertson GP, Kuzyakov Y (2019) Microbial spatial footprint as a driver of soil carbon stabilization. Nat Commun 10:1–10. https://doi.org/10.1038/s41467-019-11057-4

Kurdyś-Kujawska A, Sompolska-Rzechuła A (2019) Public support for agriculture of EU countries under the CAP. Scale, dynamics and trends of changes. In Subsidies versus economics, finance and income of farms (4); Monographs of multi-annual programme 2015–2019, 77.1; Soliwoda M, ed. IAFE-NRI, Warsaw 2018

Kurdyś-Kujawska A, Strzelecka A, Zawadzka D (2021) The impact of crop diversification on the economic efficiency of small farms in Poland. Agriculture 11(3):250. https://doi.org/10.3390/agriculture11030250

Lin BB (2011) Resilience in agriculture through crop diversification: adaptive management for environmental change. Bioscience 61(3):183–193. https://doi.org/10.1525/bio.2011.61.3.4

Lloret E, Waeyenberge L, Schrader S, Fernandez JA, Zornoza R (2020) Crop diversification and soil biodiversity. In Interactions between agricultural management and soil biodiversity: an overview of current knowledge, pp 26–34. http://soildiveragro.webs.uvigo.es/knowledge-center/publications/

Lynch JM (1984) Interactions between biological processes, cultivation and soil structure. Plant Soil 76:307–318. https://doi.org/10.1007/bf02205589

Mahbub MM, Shirazy BJ (2016) Evaluation of genetic diversity in different genotypes of soybean (*Glycine max* (L.) Merrill). Am J Plant Biol 1(1):24–29. https://doi.org/10.11648/j.ajpb.20160101.14

Makate C, Wang R, Makate M, Mango N (2016) Crop diversification and livelihoods of smallholder farmers in Zimbabwe: adaptive management for environmental change. Springerplus 5(1):1135. https://doi.org/10.1186/s40064-016-2802-4

Mango N, Makate C, Mapemba L, Sopo M (2018) The role of crop diversification in improving household food security in Central Malawi. Agric Food Secur 7(7):1–10. https://doi.org/10.1186/s40066-018-0160-x

Manjunatha AV, Anik AR, Speelman S, Nuppenau EA (2013) Impact of land fragmentation, farm size, land ownership and crop diversity on profit and efficiency of irrigated farms in India. Land Use Policy 31:397–405. https://doi.org/10.1016/j.landusepol.2012.08.005

Mazunda J, Kankwamba H, Pauw K (2015) Food and nutrition security implications of crop diversification in Malawi's farm households. In: Aberman N-L, Meerman J, Benson T (eds) Mapping the linkages between agriculture, food security and nutrition in Malawi, pp 44–49. https://doi.org/10.2499/9780896292864_05

McDaniel MD, Grandy AS (2016) Soil microbial biomass and function are altered by 12 years of crop rotation. Soil 2(4):583–599. https://doi.org/10.5194/soil-2-583-2016

McDaniel MD, Grandy AS, Tiemann LK, Weintraub MN (2014) Crop rotation complexity regulates the decomposition of high- and low-quality residues. Soil Biol Biochem 78:243–254. https://doi.org/10.1016/j.soilbio.2014.07.027

McGill WB, Cannon KR, Robertson JA, Cook FD (1986) Dynamics of soil microbial biomass and water-soluble organic C in Breton L after 50 years of cropping to two rotations. Can J Soil Sci 66(1):1–19. https://doi.org/10.4141/cjss86-001

Mengistu DD, Degaga DT, Tsehay AS (2021) Analyzing the contribution of crop diversification in improving household food security among wheat dominated rural households in Sinana District, Bale Zone, Ethiopia. Agric Food Secur 10(1):1–15. https://doi.org/10.1186/s40066-020-00280-8

Michlera JD, Josephson AL (2017) To specialize or diversify: agricultural diversity and poverty dynamics in Ethiopia. World Dev 89:214–226. https://doi.org/10.1016/j.worlddev.2016.08.011

Mulmi P, Masters WA, Ghosh S, Namirembe G, Rajbhandary R, Manohar S, Webb P (2017) Household food production is positively associated with dietary diversity and intake of nutrient-dense foods for older preschool children in poorer families: results from a nationally-representative survey in Nepal. PLoS One 12(11):e0186765. https://doi.org/10.1371/journal.pone.0186765

Mulwa CK, Visser M (2020) Farm diversification as an adaptation strategy to climatic shocks and implications for food security in northern Namibia. World Dev 129(1):104906. https://doi.org/10.1016/j.worlddev.2020.104906

Myeni L, Moeletsi ME (2020) Factors determining the adoption of strategies used by smallholder farmers to cope with climate variability in the Eastern Free State, South Africa. Agriculture 10(9):410. https://doi.org/10.3390/agriculture10090410

Nishan MA (2014) Crop diversification for sustainability. Popular Kheti 2(2):20–24

Nthebere K, Sharma SHK, Kumari CP, Qureshi AA (2020) Different cropping system effects of depth-wise distribution of available NPKS in an inceptisol of Southern Telangana Zone. Int Res J Pure Appl Chem 21(17):32–39. https://doi.org/10.9734/IRJPAC/2020/v21i1730263

O'Dea JK, Jones CA, Zabinski CA, Miller PR, Keren IN (2015) Legume, cropping intensity, and N-fertilization effects on soil attributes and processes from an eight-year-old semiarid wheat system. Nutr Cycl Agroecosyst 102(2):179–194. https://doi.org/10.1007/s10705-015-9687-4

Pandey VL, Dev SM, Jayachandran U (2016) Impact of agricultural interventions on the nutritional status in South Asia: a review. Food Policy 62:28–40. https://doi.org/10.1016/j.foodpol.2016.05.002

Paul S, Laha A, Kuri PK (2016) Crop diversification and its implications to food security: a case study of India with special reference to West Bengal. Asia-Paciöc J Rural Dev XXVI(1):61–84. https://doi.org/10.1177/1018529120160102

Perz SG (2004) Are agricultural production and forest conservation compatible? Agricultural diversity, agricultural incomes and primary forest cover among small farm colonists in the Amazon. World Dev 32(6):957–977. https://doi.org/10.1016/j.worlddev.2003.10.012

Pieters H, Guariso A, Vandeplas A (2013) Conceptual framework for the analysis of the determinants of food and nutrition security, FOODSECURE working paper no 13. LEI Wageningen UR, The Hague. https://doi.org/10.22004/ag.econ.285141

Rathore SS, Subhash B, Shekhawat K, Singh R, Yadav SK, Singh VK, Singh C (2022) Designing energy cum carbon-efficient environmentally clean production system for achieving green economy in agriculture. Sustain Energy Technol Assess 52(34):102190. https://doi.org/10.1016/j.seta.2022.102190

Rivington M, Matthews KB, Bellocchi G, Buchan K, Stockle CO, Donatelli M (2007) An integrated assessment approach to conduct analyses of climate change impacts on whole-farm systems. Environ Model Softw 22(2):202–210. https://doi.org/10.1016/j.envsoft.2005.07.018

Roesch-McNally GE, Arbuckle JG, Tyndall JC (2018) Barriers to implementing climate resilient agricultural strategies: the case of crop diversification in the U.S. Corn Belt. Glob Environ Change 48:206–215. https://doi.org/10.1016/j.gloenvcha.2017.12.002

Roy MD, Sarkar GK, Das I, Karmakar R, Saha T (2017) Integrated use of inorganic, biological and organic manures on rice productivity, nitrogen uptake and soil health in gangetic alluvial soils of West Bengal. J Indian Soc Soil Sci 65(1):72–79. https://doi.org/10.5958/0974-0228.2017.00010.X

Sen AK (1981) Poverty and famines: an essay on entitlement and deprivation. Clarendon Press, Oxford. https://doi.org/10.2307/1240373

Shah KK, Modi B, Pandey HP, Subedi A, Aryal G, Pandey M, Shrestha J (2021) Diversified crop rotation: an approach for sustainable agriculture production. Adv Agric. Article ID 8924087. https://doi.org/10.1155/2021/8924087

Shukla SK, Shee S, Maity SK, Solomon S, Awasthi SK, Gaur A, Pathak AD, Jaiswal VP (2017) Soil carbon sequestration and crop yields in rice – wheat and sugarcane – ratoon – wheat cropping systems through crop residue management and inoculation of Trichoderma viride in subtropical India. Sugar Tech 19(4):347–358. https://doi.org/10.1007/s12355-016-0470-x

Singh R, Babu S, Avasthe RK, Meena RS, Yadav GS, Das A, Mohapatra KP, Rathore SS, Kumar A, Singh C (2021) Conservation tillage and organic nutrients management improve soil properties, productivity, and economics of the maize-vegetable pea system in the Eastern Himalayas. Land Degrad Dev 32(6):1–18. https://doi.org/10.1002/ldr.4066

Singh R, Kumar A, Babu S, Avasthe R, Das A, Rathore SS, Kumar S, Singh C, Sharma V, Bhupenchandra I (2023) Development of organic nutrients management system for profitable and soil-supportive French bean (*Phaseolus vulgaris* L.) farming in North Eastern Himalayas, India. Front Sustain Food Syst 7:1115521. https://doi.org/10.3389/fsufs.2023.1115521

Skopp J, Jawson MD, Doran JW (1990) Steady-state aerobic microbial activity as a function of soil water content. Soil Sci Soc Am J 564:1619–1625. https://doi.org/10.2136/sssaj1990.03615995005400060018x

Soong JL, Fuchslueger L, Marañon-Jimenez S, Torn MS, Janssens IA, Penuelas J, Richter A (2020) Microbial carbon limitation: the need for integrating microorganisms into our understanding of ecosystem carbon cycling. Glob Chang Biol 26(4):1953–1961. https://doi.org/10.1111/gcb.14962

Stanger TF, Lauer JG (2008) Corn grain yield response to crop rotation and nitrogen over 35 years. Agron J 100(3):643–650. https://doi.org/10.2134/agronj2007.0280

Studdert GA, Echeverría HE (2000) Crop rotations and nitrogen fertilization to manage soil organic carbon dynamics. Soil Sci Soc Am J 64(4):1496–1503. https://doi.org/10.2136/sssaj2000.6441496x

Tiemann LK, Grandy AS, Atkinson EE, Marin-Spiotta E, Mcdaniel MD (2015) Crop rotational diversity enhances belowground communities and functions in an agroecosystem. Ecol Lett 18(8):761–771. https://doi.org/10.1111/ele.12453

Tong Y, Liu J, Li X, Sun J, Herzberger A, Wei D, Zhang W, Dou Z, Zhang F (2017) Cropping system conversion led to organic carbon change in China's mollisols regions. Sci Rep 7(1):18064. https://doi.org/10.1038/s41598-017-18270-5

Walia SS, Babu S, Gill RS, Kaur T, Kohima N, Panwar AS, Yadav DK, Ansari MA, Ravishankar N, Kumar S, Kaur K, Ansari MH (2022) Designing resource-efficient and environmentally safe cropping systems for sustainable energy use and economic returns in Indo-Gangetic Plains, India. Sustainability 14:14636. https://doi.org/10.3390/su142114636

Wardle DA, Yeates GW, Watson RN, Nicholson KS (1995) The detritus food-web and the diversity of soil fauna as indicators of disturbance regimes in agro-ecosystems. Plant Soil 170(1):35–43. https://doi.org/10.1007/BF02183053

Weisberger DA, McDaniel MD, Arbuckle JG, Liebman M (2021) Farmer perspectives on benefits of and barriers to extended crop rotations in Iowa, USA. Agric Environ Lett 6(2):e20049. https://doi.org/10.1002/ael2.20049

Wyland U, Jackson LE, Chaney WE, Klonsky K, Koike ST, Kimple B (1996) Winter cover crops in a vegetable cropping system: impacts on nitrate leaching, soil water, crop yield, pests and management costs. Agric Ecosyst Environ 59:1–17. https://doi.org/10.1016/0167-8809(96)01048-1

Yadav A, Babu S, Krishnan P, Kaur B, Bana RS, Chakraborty D, Kumar V, Joshi B, Lal SK (2024a) Zinc oxide and ferric oxide nanoparticles combination increase plant growth, yield, and quality of soybean under semiarid region. Chemosphere 352(3):141432. https://doi.org/10.1016/j.chemosphere.2024.141432

Yadav D, Singh D, Babu S, Madegowda M, Singh DV, Mandal D, Rathore AC, Sharma VK, Singhal V, Kumawat A, Yadav DK, Yadav RK, Kumar S (2023) Intensified cropping reduces soil erosion and improves rainfall partitioning and soil properties in the marginal land of the Indian Himalayas. Int Soil Water Conserv Res. https://doi.org/10.1016/j.iswcr.2023.10.002

Yadav D, Singh D, Madhu M, Singh DV, Islam S, Kumar G, Gupta AK, Mandal D, Babu S, Barh A, Kumawat A, Dubey R, Pati K, Bhargavi B, Yadav DK (2024b) Interplanted crops and rainfall variability regulate soil erosion in rainfed maize of the Indian Himalayas. Land Degrad Dev. https://doi.org/10.1002/ldr.5221

Yadav GS, Das A, Kandpal BK, Babu S, Lal R, Datta M, Das B, Singh R, Singh VK, Mohapatra KP, Chakraborty M (2021) The food-energy-water-carbon nexus in a maizemaize-mustard cropping sequence of the Indian Himalayas: an impact of tillage-cum-live mulching. Renew Sust Energ Rev 151:111602. https://doi.org/10.1016/j.rser.2021.111602

Zhalnina K, Dias R, de Quadros PD, Davis-Richardson A, Camargo FAO, Clark IM, McGrath SP, Hirsch PR, Triplett EW (2015) Soil pH determines microbial diversity and composition in the park grass experiment. Microb Ecol 69(2):395–406. https://doi.org/10.1007/s00248-014-0530-2

Zhou X, Liu J, Wu F (2017) Soil microbial communities in cucumber monoculture and rotation systems and their feedback effects on cucumber seedling growth. Plant Soil 415(1–2):507–520. https://doi.org/10.1007/s11104-017-3181-5

Zimmerer KS, Vanek SJ (2016) Toward the integrated framework analysis of linkages among agrobiodiversity, livelihood diversification, ecological systems, and sustainability amid global change. Land 5(2):1–28. https://doi.org/10.3390/land5020010

Zuber SM, Behnke GD, Nafziger ED, Villamil MB (2018) Carbon and nitrogen content of soil organic matter and microbial biomass under long-term crop rotation and tillage in Illinois, USA. Agriculture 8(3):37. https://doi.org/10.3390/agriculture8030037

Chapter 2
Approaches for Capturing Soil Carbon for Sustained Food and Nutritional Security

Roopa Patel, Sanjeev Kumar, Alok Kumar Pandey, Abha Yadav, and Venkatesh Padimi

Abstract Soils of the world's agroecosystems (croplands, grazing lands, rangelands) are depleted of their organic carbon (SOC) by 25–75% depending on climate, soil type, and historic management. The extent of drain may be of 10–50 Mg carbon/ha. Soils with acute depletion of their SOC have low agronomic yield and use the efficiency of supplemental inputs. Adopting recommended management practices and changing to a restorative land use can improve soil quality, boost agronomic productivity, improve soil resilience to withstand extreme weather events, improve global food security, enrich the SOC sink, and mitigate climate change by offsetting emissions from fossil fuels. Agroecosystems soils have the technical capacity to sequester ~1.2 and 3.1 billion tons of carbon dioxide (CO_2) annually. Improved soil quality can boost food production in underdeveloped nations by 24–32 million tons. The global agriculture has the capacity to sequester around 0.75 ± 1.0 Pg of carbon each year. Thus, in addition to reducing the possibility of global warming, carbon sequestration helps in increasing food security in a sustainable way.

Keywords Agricultural sustainability · Carbon sequestration · Food security · Nutritional security · Soil health

---

R. Patel
Department of Soil Science & Agricultural Chemistry, Institute of Agricultural Sciences, Banaras Hindu University, Varanasi, Uttar Pradesh, India

S. Kumar (✉)
Department of Soil Science & Agricultural Chemistry, College of Agriculture, Sardar Vallabhbhai Patel University of Agriculture and Technology, Meerut, Uttar Pradesh, India

A. K. Pandey · A. Yadav
Department of Soil Science and Agricultural Chemistry, College of Agriculture, Acharya Narendra Deva University of Agriculture and Technology, Ayodhya, Uttar Pradesh, India

V. Padimi
Department of Agricultural Biology, Colorado State University, Fort Collins, CO, USA

S. Babu et al. (eds.), *Agricultural Diversification for Sustainable Food Production*, Sustainability Sciences in Asia and Africa,
https://doi.org/10.1007/978-981-97-7517-0_2

2.1 Introduction

Unsustainable intensive agricultural activities depleted soil organic carbon (SOC) from active carbon pool, which is detrimental to crop production activities (Yadav et al. 2021a; Babu et al. 2020a), whereas better SOC stocks and capturing potential play a pivotal role in sustainable systems (Das et al. 2022), contributing to climate change mitigation, enhanced agricultural resilience, and ensuring improved land quality through better soil microbial biodiversity and biomass carbon (Rathore et al. 2022), which are being considered as the major indicators of land quality (Saharawat et al. 2010; Kalaiselvi et al. 2023). Moreover, climate change has become a worldwide emerging issue as it directly or indirectly controls greenhouse gas (GHG) emission, global temperature rise, changes in weather patterns, and other extreme weather events (Yadav et al. 2021b). Along with being a major factor in regulating the climate and moving towards a sustainable agricultural system, soil can either be a source or a sink of atmospheric CO_2. Elevated levels of soil C improve plant nutrient availability, which has a beneficial effect on plant growth and nutritional value. Increasing the SOC pool by 1 Mg C/ha/year in the root zone can boost food output in underdeveloped nations by 24–32 million Mg of cereal grains and 6–10 million Mg of roots and tubers (Lal 2011; Yadav et al. 2020a). It is now broadly recognized that to fight with the climate change, it will be necessary to adopt negative emission technologies in addition to substantially reducing fossil fuel emissions. One of the most important and affordable strategies for mitigating climate change is SOC sequestration, which has the added advantages of enhancing soil fertility and providing other ecosystem services (Rathore et al. 2022; Babu et al. 2023a). Sequestering organic C in soil may potentially, and in a technically profitable manner, capture ~0.79 and 1.54 Giga tons C/year from the atmosphere (Fuss et al. 2018). The stakeholders in the United Nations Framework on Climate Change (UNFCC) have voluntarily committed to promoting soil carbon enhancing agricultural practices at the Conference of Parties (COP 21) in Paris (Chambers et al. 2016). Maintaining or enhancing SOC stocks was recognized as a part of Sustainable Development Goal (UNCCD 2016). Agriculture constitutes one-third of the world arable land (World Bank 2015). Hence, it is vital to devise appropriate soil and land management practices to increase SOC stocks by utilizing soil as a sink (FAO ITPS 2015). Additional benefits of soil C sequestration include advancing food and nutritional security, increasing renewability and quality of water, improving biodiversity, and strengthening elemental recycling. Threshold level of SOC in the root zone is 1.5–2.0% (Babu et al. 2023a). SOC is influenced by land use, soil management, and farming systems. More than 50% of the total C pool is contained between 0.3 and 1 m layer of surface soils (Lorenz and Lal 2018). Restoring soil quality necessitates increasing SOC concentration by adopting best management practices that create a positive C budget. Therefore, it is important to address the potential of SOC sequestration in agricultural ecosystems to mitigate climate change, improve soil quality, and advance food and nutritional security (Fig. 2.1).

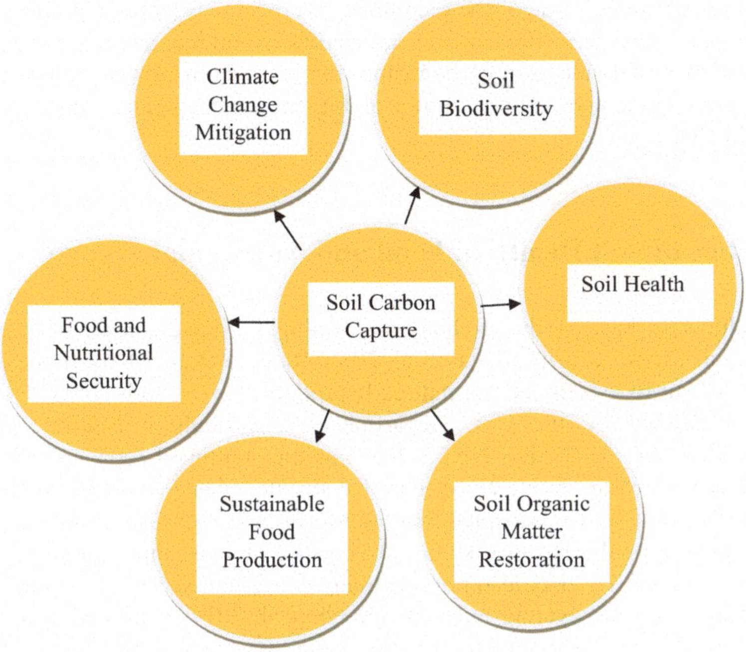

Fig. 2.1 Benefits of soil carbon capture for sustainable agroecosystems

2.2 Sustainable Development Goals and Soil Carbon Capturing

One of the 17 sustainable development goals (SDGs) is to maintain the balance between the inflow and outflow of soil C through sustainable agricultural management practices. Limiting the increase in the global mean temperature to 2 °C above preindustrial levels in order to mitigate climate change through carbon sequestration is the primary goal of the climate change summit and policies (Wei et al. 2021). Since practicing agriculture is the most common land use system worldwide, capturing atmospheric C into the terrestrial pool through improved agricultural management practices lowers the atmospheric CO_2 concentration, improves soil quality, and ensures sustainability (Kumar et al. 2022). The UN's Sustainable Development Goals (SDGs) aim to establish a framework of principles for a sustainable future. Responsible production and consumption, one of the SDGs, emphasizes the application of scientifically derived, data-driven, and sustainable production practices on farmlands. Thus, carbon capture and storage can support the goals of sustainable development. It is essential to achieve goal 13, which is to fight climate change (Mikunda et al. 2021). Several types of soil C capture help restore ecosystems that have been damaged by coal mine waste, soil erosion, and degradation (Ghosh and Maiti 2021). This is done to help reach Goals 13 (climate action) and 15 (biodiversity protection) by storing more carbon and promoting ecosystem growth. Hence, carbon sequestration is one of the recommended processes by the UN SDGs, since the method is extremely effective and contributes to building a better tomorrow for agriculturists worldwide. Soil carbon capturing is therefore a sustainable option to combat climate change and does not prohibit the achievement of any other SDG (Raj et al. 2022).

2.3 Role of Soil Health in Food and Nutritional Security

Soil health is the base for the production of healthy food and consequently contributes to global food security (Rojas et al. 2016; Rathore et al. 2020). It is estimated that 95% of our food is directly or indirectly produced on our soils (FAO 2015). Soil health is defined by the U.S. Department of Agriculture Natural Resource Conservation Service (USDA-NRCS 2012) as the continued capacity of a soil to function as a necessary living ecosystem that sustains plants, animals, and humans. From this definition, it is accepted that long-term food security cannot be attained without healthy soils. Healthy soils are versatile, that is, they not only provide humans with food, but they also provide many ecosystem services, such as the distillation of water and cycling nutrients (Koch et al. 2014; Dominati et al. 2010). Healthy soils supply the essential nutrients, water, oxygen, and root support that our food-producing plants need to grow and flourish. There are 14 elements that are essential for plant growth that come from the soil, and many of these elements, such as calcium (Ca), iron (Fe), potassium (K), and others, are also essential for human health. Essential soil elements that end up in the human diet are supplied through food from either plants that took the elements up from the soil during growth or animal products after the animal obtained those necessary elements from plants. Because plants rely on the soil for their nutritional needs, and all higher animals as well as humans depend directly or indirectly on plants for their nutrition, plants form the base of the food chain and, thus, a major portion of the nutrients needed for human health emerge with the soil.

Climate change affects soil health mainly by altering the C and water cycles. Soils act as C sinks through C sequestration; the absorption of CO_2 positively affects agronomic productivity and food security by increasing crop yields (Lal 2004; Singh et al. 2021). Soil also contributes to the production of crops for biofuel and fibers. Besides producing our food, fuel, and clothing materials, soil provides numerous ecosystem services to society, including filtering water, wildlife habitat, biodiversity, and carbon storage (Koch et al. 2014; McBratney et al. 2013). Nutrient-poor soils contribute to systemic food and nutritional security issues. Over two billion people suffer from micronutrient deficiencies (Ritchie and Roser 2024), which are a significant cause of morbidity and mortality (Black 2003; Shetty 2009). Indeed, human (protein-energy) malnutrition is related to food scarcity and to the consumption of crops with low tissue concentrations of trace elements, which are directly attributable to nutrient-impoverished soils (St. Clair and Lynch 2010). Soil is one of the fundamental production factors determining food production stability, food nutrient quality, and yield quantity (FAO 2008). Additionally, soil health encompasses physical, chemical, and biological components of soil and integrates ecosystem functioning and services provisioning by soils, including the capacity of soils to encourage and sustain agricultural productivity (Kibblewhite et al. 2008), sequester C (Peichl et al. 2006; Lal 2016), reduce nutrient leaching (Abrahams 2002; Khan et al. 2017), and protect biodiversity (Thevathasan and Gordon 2004). Therefore, improving soil health is often a primary goal in the development of sustainable agricultural practices (Singh et al. 2023). Healthy soils are the foundation for high yields, high-quality crops, and long-term cropping potential. Healthy soils also have the capability to sustain the environment (Wang et al. 2022; Wani et al. 2023b). Soil holds more than three times the amount of carbon in the Earth's atmosphere—and unhealthy soils liberate that carbon, contributing to greenhouse gas emissions. Hence, soil health directly affects crop quality and nutrient availability; it is essential to food and nutritional security. Furthermore, resilient soils are essential for reducing the effects of environmental stressors, which supports stable agricultural yields and high-quality nutrition. Therefore, adopting sustainable soil management techniques is essential to tackling the world's nutritional challenges and advancing long-term food security.

2.4 Main Sources of Agricultural GHG Emissions

Agriculture has a significant potential to reduce greenhouse gas (GHG) emissions, as it is one of the primary contributors to these emissions. The Food and Agriculture Organization (FAO) states that forestry, agriculture, and other land-use practices contribute to 24% of the world's GHG emissions and 7.1 billion tons (Gt) of CO_2-equivalent emissions from livestock annually, which is equivalent to 14.5% of all anthropogenic GHG emissions. China, the United States, India, Russia, Japan, Africa, Germany, South Korea, Iran, and Saudi Arabia are the top 10 countries in the world for CO_2 emissions. Due to their growing needs for fossil fuels, which are rising at rates of 3.5% and 3.9% annually, respectively, China and India are the top two countries in terms of CO_2 emissions (Nanda et al. 2015). Increased use of production inputs, such as inorganic fertilizer, has made Indian agriculture more GHG emitters, currently accounting for 18% of total GHG emissions in India (INCCA 2010). Agriculture is also a significant contributor to global emissions of NH_3, CO_X, and NO. The CO_2 emissions rose by 1.1% in 2023, and global CO_2 emissions were 37.4 billion tons (IEA 2023). These emissions have increased by 48% in the last 20 years, and if mitigation techniques are not implemented, they are expected to increase by 25–90% by 2030 (Zahed et al. 2021). Hence, the sources of agricultural emissions are diverse and significant, contributing to the global challenge of climate change (Fig. 2.2). Livestock production, particularly enteric fermentation and manure left on pasture, is an important source of agricultural greenhouse gas emissions. Addressing this issue requires promoting efficient feed conversion, adopting improved manure management systems, and exploring alternative sources.

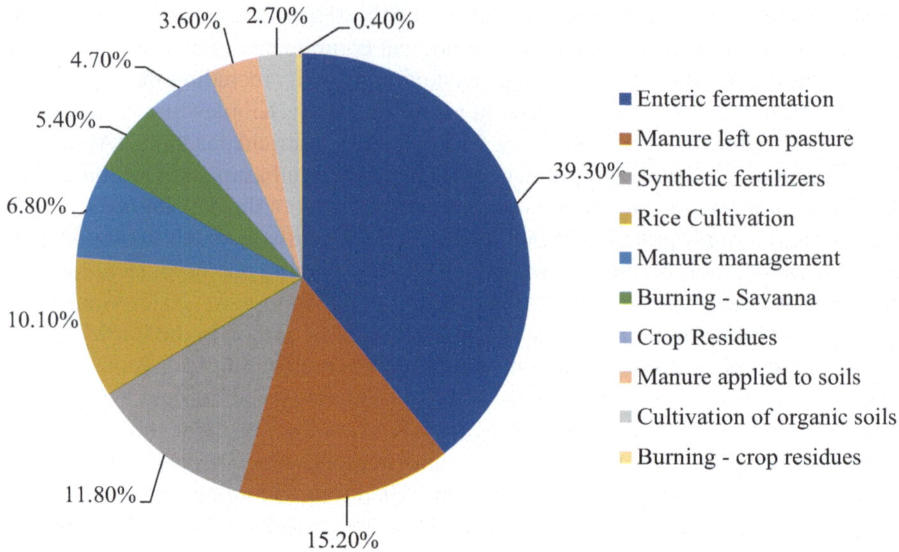

Fig. 2.2 Sources of GHG emissions in the agricultural sector (https://www.wri.org/insights/5-questions-about-agricultural-emissions-answered) accessed on 20 April 2024

2.5 Need for Soil Carbon Restoration for Enhanced Crop Resilience and Soil Health

Carbon levels in most agricultural land have decreased over the last 100 years. In a global analysis, the largest SOC losses occur simultaneously with cropping regions, but grazing, especially in arid and semiarid regions that are worldwide more extensive, was responsible for at least half of the total SOC loss (Sanderman et al. 2017). The total potential of C sequestration by the world cropland is about 0.75 ± 1.0 Pg/yr. Each 1 Mg/ha rise in SOC pool in the root zone under the soil would enhance yields of crops by 20–70, 10–50, and 30–300 kg/ha, respectively, for wheat, rice, and corn, augmenting production of cereals and legumes in the developing countries by 32 and 11 million tons/year, respectively (Kaur et al. 2023). Contrary to 0.15–0.20% in uncultivated soils, the SOC concentration in cultivated soils is less than 0.05% (Sahoo et al. 2019). Consequently, there is an abundance of opportunity to raise the SOC pool in cultivated soils. One of the essential aspects of sustainability of agricultural systems is the maintenance or improvement of SOC. A healthy soil is productive, sustainable, and resilient, but rising atmospheric concentrations of GHGs pose a threat to the ecosystem and the global climate system. Processes that lead to depletion of the SOC pool are deforestation, biomass burning, and the attendant soil erosion, which create a large C debt (Fargione et al. 2008). Indiscriminate and inappropriate land use and management practices, and drainage and cultivation of peat lands may also exacerbate gaseous emissions even from biofuel/energy plantations (Searchinger et al. 2008). Therefore, C sequestration plays a vital role in enhancing crop resilience through its profound effects on soil health and ecosystem functioning. Increased SOC content, facilitated by carbon sequestration practices, such as conservation tillage, cover cropping, and crop rotation, has been extensively documented to improve soil structure, water retention, and nutrient availability, thus creating a more favorable environment for crop growth and resilience (Six et al. 2000; Lal 2004; Yadav et al. 2020a). Moreover, carbon-enriched soils support a diverse array of microbial communities, which contribute to disease suppression, nutrient cycling, and the enhancement of soil fertility, all essential components of crop resilience (Lehmann et al. 2015). The soil carbon capture approaches aim to reduce losses of carbon and other nutrients from the soil, build soil organic matter, and promote healthy and sustainable nutritional security (Fig. 2.3).

2.6 Approaches of Soil Carbon Capture for Sustainable Production in Agroecosystems

Meeting the food demands of a population that is rising quickly without appreciably increasing the amount of land utilized for agriculture or negatively impacting the environment is one of the main issues facing the agriculture sector (Yadav et al. 2020b). Agricultural and partially degraded soils can both significantly reduce the atmospheric concentrations of increased CO_2. Currently, 33% of the global soils have been degraded and have lost much of their SOC through the historical expansion of agriculture and pastoralism and subsequent land-use conversion from native ecosystems (e.g., peatlands, forests, grasslands) to arable land (FAO 2019). This has resulted in a decline in soil structural stability, increased erosion risks, and reduced

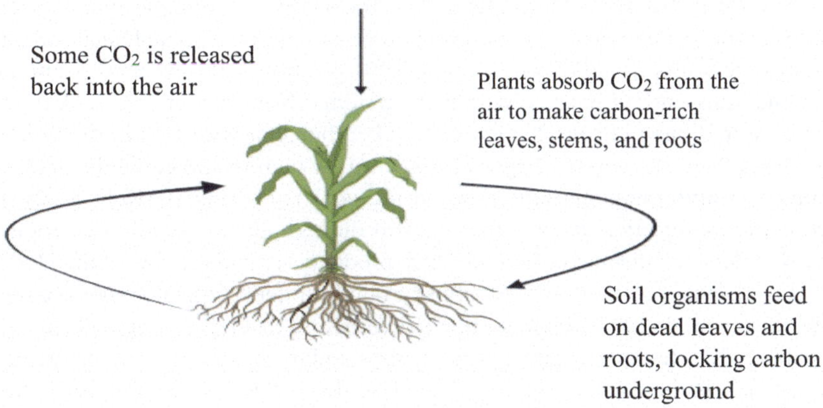

Fig. 2.3 Representative mechanism of organic carbon storage in soil

water storage and nutrient supplies. Soil degradation has thus become a major threat to food security, especially in developing countries. Therefore, technically sound and financially feasible solutions are needed to reduce the impact and amount of CO_2 in the Earth's atmosphere while also enhancing soil health and quality. Apart from improving the quality of the soil for better plant development, sequestering carbon is a workable way to lower the atmospheric concentration of CO_2 and fight climate change (Sanaullah et al. 2019). As a result, there are numerous soil carbon capture strategies that have been categorized differently by different individuals. There are a number of potential classification criteria, including techniques for capturing carbon, types of species that capture greenhouse gases, potential carbon sinks, and many more. Natural, biological, traditional, and modern innovative agricultural carbon capture tactics form the basis of the main classification system.

2.6.1 Natural Mechanism of Carbon Capturing

An integrated strategy based on the improvement of natural biogeochemical cycles in agroecosystems that promote carbon capture and storage while boosting resilience and long-term productivity could be nature's method of capturing carbon from soil. The process combines photosynthesis and plant roots with forests and improved weathering to accelerate carbon capture.

2.6.1.1 Plants Roots and Photosynthesis

Sequestering carbon is one important way that plant roots function. During photosynthesis, plants take in CO_2 and release some of it into the soil through root exudates, which helps stable organic matter to develop. Rui et al. (2016) emphasize the significance of plant-derived carbon inputs in enhancing SOC. Additionally, microbial activity in the soil contributes to the breakdown of organic matter, forming stable carbon compounds. These natural processes collectively enhance soil carbon sequestration, emphasizing the vital role of vegetation and microbial communities in maintaining carbon balance within ecosystems (Kumar et al. 2023). A few optimistic estimations indicate that enough soil carbon can be stored in the ground by improving the root growth of agricultural crops, which would equal the anticipated quantity of anthropogenic emissions in the next two decades (Kell 2012). Plant carbon eventually finds its way into the soil through aboveground litter, roots, and the rhizo-deposits they leave behind. Plant productivity and the size of the SOC pool are related; methods that increase aboveground biomass typically also accumulate SOC (Smith et al. 2007). Plant carbon stored below ground is effectively converted to stable SOC. Remaining aboveground crops and trees are pyrolyzed to create biochar, which is then put into the soil to slow down the release of carbon through decomposition (Babu et al. 2023c; Wani et al. 2023a).

2.6.1.2 Forests

By storing organic materials underground or in their biomass, forests and soils absorb CO_2 from the atmosphere. While the majority of carbon stored on land is stored in soils and forests, the majority of carbon in the world is stored in the oceans. Globally, trees store around 47% of all C emissions (Malhi et al. 2002). More than half of the dry biomass of trees in forests is C. The C can be found in biomass, which is any portion of the living or nonliving tissue of the tree, such as the trunk, branches, leaves, or roots. C is also present in plant cells, which store carbon within the cell walls made of cellulose and/or lignin fibers. Plant cells, in contrast to animal cells, have cell walls that provide the organism with support and structure; carbon is consequently used in the synthesis of these substances. C is found in biomass, both aboveground and belowground biomass, dead wood, litter, and soil organic carbon (Yadav et al. 2019). Owing to their lengthy lifespans and possibly substantial bulk, forests can—and frequently do—store significant amounts of carbon in their cells. Since net biological growth increases the amount of carbon stored in forests, these ecosystems essentially function as carbon silos. Decomposition or fire that kills biological material releases carbon, usually back into the atmosphere, and causes biological decline or destruction. Decomposition and fires have the ability to release significant amounts of carbon stored in forests (Sedjo and Sohngen 2012).

Planting trees contributes significantly to enhancing both the ecological elements of our planet and the well-being of humankind by expanding the number of trees that are present in appropriate locations worldwide (Zhang and Qian 2024). However, planting trees is not a sufficient approach to reduce GHG emissions; other activities with larger potential are required (Brancalion and Holl 2020). Reducing the amount of CO_2 in the atmosphere is frequently advocated via promoting carbon sequestration and storage in forest ecosystems (Ameray et al. 2021). Because they serve as primary habitats for a variety of species, provide wood, support biodiversity conservation and maintenance, and are essential to both indigenous societies and the general maintenance of human health, forest ecosystems provide societies with a wide range of important and diverse services (Nelson et al. 2009). As a result, the forest stores more carbon in the biomass of its trees as it grows. When trees are still young and growing, they absorb the most carbon in forests.

2.6.1.3 Enhanced Weathering

Enhanced weathering (EW) sick's provide a natural sink for CO_2 through the addition of powdered silicate rock with high reactive surface area into agricultural soils (Taylor et al. 2017). The mechanism via which CO_2 is removed from the atmosphere by enhanced weathering is the disintegration of silicate minerals on the surface of the land. With their abundance, the silicate resources have the potential to absorb 430 billion tons of CO_2. Basic rocks, which have a 0.3 tons CO_2 per tons rock carbon capture capacity, contain the majority of this resource (Renforth et al. 2013). A further tactic to combat climate change is enhanced rock weathering, which involves removing carbon from the atmosphere and storing it in rocks. It is one of many "carbon removal" methods aimed at removing CO_2, the main GHG that humans have been contributing to climate change. Enhanced rock weathering is an extension of our climate system's inherent processes. When rocks are "weathered," or eroded by rain, minerals like calcium and magnesium are released. In the meantime, other natural processes convert CO_2 in the atmosphere to bicarbonate, which is found in the ocean, or carbonic acid, which is found in rain. These various substances combine to create new rocks, such as limestone or calcium carbonate. Rocks assist in removing CO_2 from the atmosphere in this way. Rocks assist in removing CO_2 from the atmosphere in this way (Beerling et al. 2020). Furthermore, a suggested CO_2 removal (CDR) technique called enhanced rock weathering (ERW) entails quickening the natural carbon sequestration processes caused by the chemical weathering of silicate rocks.

2.6.2 *Biological Carbon Capturing*

Biological carbon capture stores C in vegetation, soils, woody products, and aquatic habitats, and it necessitates the implementation of advanced management strategies that increase the quantity of carbon stored by vegetation (cropland, grassland, and forest), soil, ocean, and microorganisms. However, biological C capture not only mitigates climate change by reducing atmospheric CO_2 levels but also enhances biodiversity, improves soil fertility, and provides ecosystem services crucial for sustainable development. Following are the important approaches that are helpful in carbon capture and storage through biological processes.

2.6.2.1 Microalgae-Based Carbon Capturing

Biological capture of CO_2 using microalgae is considered an appropriate medium for recycling the excess CO_2 produced from power plants, vehicles, volcanic eruptions, breakdown of organic matter, and forest fires (Onyeaka et al. 2021). Microalgae farming appears to be a novel approach to lowering the atmospheric concentration of CO_2. Furthermore, through microalgae, CO_2 can be captured and reused into biomass, which in turn could be utilized as a carbon source to build lipids for the production of bioenergy and other value-added products. In comparison to terrestrial plants, microalgae are rapidly developing microorganisms that biofix CO_2 at a comparatively high pace. Microalgae are unicellular, quickly growing microorganisms that can split their cells in 3–4 hours, but under ideal growing conditions, they typically divide every 12 days (Williams and Laurens 2010). Microalgae cells absorb CO_2 through the process of photosynthesis, which then turns C into carbohydrates. These carbohydrates are then used by the cells to make proteins, lipids, and nucleic acids (Beer et al. 2009). Owing to their rapid growth rate and straightforward cell structure, microalgae are predicted to have 10–50 times better CO_2 bio-fixation efficiency than terrestrial plants (Khan et al. 2009).

The microalgae utilize light energy from the sun, water from their environment, and CO_2 from the atmosphere as a carbon source to produce oxygen and energy, which is accumulated inside the cell as fatty acids and starch (Fig. 2.4). However, the fatty acids from microalgae can potentially be separated and trans-esterified to create biofuels, bioethanol, biodiesel, and gas that are all carbon neutral (Mondal et al. 2016). Since extracting CO_2 from the atmosphere and flue gas presents a number of safety and environmental concerns, research efforts have shifted to focus on using CO_2 as a carbon source for the cultivation of microalgae.

2.6.2.2 Biochar

Biochar production is a process through which carbon from certain biomass sources is transformed into stable carbon that can be absorbed into the soil (Babu et al. 2023c). Besides its role in long-term carbon sequestration, biochar is also beneficial to soil performance as it enhances the retention and diffusion of water and nutrients in soil (Wani et al. 2023a). Biochar is a highly resistant soil additive produced by burning or gasifying biomass. After waste (manure, agricultural, or forestry residues) has been processed in high-temperature furnaces in an oxygen-limited environment, it is transformed into a solid, porous, carbon-rich, and stable material (Fang et al. 2014). Municipal sewage sludge and agricultural wastes are two examples of the many organic waste feedstocks used to create biochar, a carbon-rich solid. The removal of CO_2 from the atmosphere is accomplished in a biochar system by the sequestration of C by photosynthetic organisms (Terlouw et al. 2021). Increased interests in biochar are mainly due to its unique properties, which include a large specific surface area, high carbon content, cation exchange capacity, and stable structure. Utilizing biochar, an indispensable tool in the fight against GHG

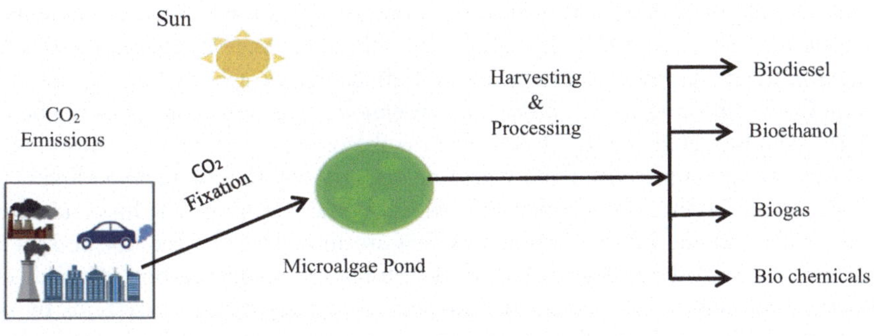

Fig. 2.4 Schematic diagrams of microalgae-based carbon capturing for biofuels (Onyeaka et al. 2021)

emissions, one can obtain increased C sequestration, improved soil fertility, agricultural waste recycling, and remediation (Ali et al. 2021; Zahed et al. 2021).

When added to soil, biochar improves the soil faunal population, plant root development, and accessible water holding capacity (Major et al. 2010). The micropores in biochar allow water and air to permeate, reducing the bulk density of the soil as a whole (Babu et al. 2023c). Biochar is superior at adsorbing soil nutrients and humus because it has a higher internal surface area than soil. Biochar allows plants to absorb the soil's mineral content through their roots by holding onto large amounts of it (Idapuganti et al. 2024).

2.6.2.3 Dedicated Energy Crop

A nonfood plant variety produced primarily for energy or biofuel generation is called a dedicated energy crop. Since they require less upkeep during growing, the majority of energy crops are lignocellulosic in nature. Biofuels, including bioethanol, biobutanol, bio-oil, biodiesel, and jet fuels, are produced from energy crops, which may also be burned to produce heat and electricity. The energy crops provide enormous potential for carbon sequestration in addition to producing ethanol (Nanda et al. 2016). The energy crops can come from herbaceous (elephant grass, switch grass, and timothy grass) or woody (willow and poplar, for example). These crops can be genetically altered to grow more quickly, produce more biomass, use less water, and require fewer nutrients.

2.6.2.4 Blue Carbon

There are many natural methods capable of reducing the atmospheric CO_2. Traditional blue carbon plants are responsible for more than half of the carbon sequestered in ocean sediments, although they make up less than 2% of the world's ocean area (Chen and Xu 2020). The phrase "blue carbon" was first used 20 years ago to highlight the noteworthy ability of coastal vegetation ecosystems to sequester carbon (Santos et al. 2021). In light of the climate change dilemma, the word has aided in the development of laws and initiatives that protect and restore these vulnerable coastal ecosystems and reduce greenhouse gas emissions (Macreadie et al. 2019; Wedding et al. 2021). Blue carbon is the organic carbon stored for centuries or even millennia in coastal and maritime ecosystems. Keeping and adding to blue carbon is one of the simplest ways to reduce global warming. The loss and deterioration of vegetated marine habitats, which are thought to be the hotspots of blue carbon, have resulted in a decline in the stores of organic carbon (Geraldi et al. 2019). Therefore, conserving and restoring coastal vegetated ecosystems by some activities (blue carbon strategies) are necessary to prevent and mitigate CO_2 emissions.

2.6.2.5 Nitrogen Fixation

Compared to non-nitrogen-fixing plants, nitrogen-fixing plants have a greater tendency to accumulate carbon in the soil (Binkly and Giardina 1998) and have a greater effect on soil organic biomass (McBratney and Stockmann 2011). By adding fixed nitrogen to the ecosystem's nutrient cycle, soil carbon storage is increased and soil carbon breakdown is further reduced. Nitrogen fixation process favors litter fall and mycorrhizal production, thereby increasing the carbon input of soil.

2.6.2.6 Microbial Carbon Capturing

Magnificent heterotrophic organisms, microbes, can affordably reduce GHG emissions, especially CO_2, while producing large quantities of chemicals from renewable resources (Hu et al. 2019). The primary factors influencing the microbial process's ability to sequester carbon in soil are the dynamics of microbial communities and the equilibrium of microbial by-product synthesis and degradation (Cotrufo et al. 2012). Soil bacteria improve soil aggregation, which affects carbon storage. Understanding the various ways that microbial diversity responds to climate change is one of the major difficulties faced by the diverse communities of microbes. Many organisms and species are disappearing globally at previously unheard of rates as a result of climate change, leading to an overall reduction in biodiversity. The activities of soil microbes and aboveground populations, including parasites, pathogens, and herbivores, are intrinsically linked. Thus, interactions among by-products, community structure, microbial biomass, and soil features such as texture, pore-size distribution, clay mineralogy, and aggregate dynamics determine the microbial contribution to carbon sequestration (Das et al. 2021). The characteristics of reactive clays influence the soil's ability to retain microbial biomass and microbial-generated organic matter (Mikutta et al. 2019). In addition to being the most prevalent species on Earth, heterotrophic microbes control the majority of oceanic biogeochemical cycles, including the carbon cycle (Kirchman et al. 2009). Accordingly, understanding how climate change affects carbon dynamics requires paying close attention to the feedbacks that arise between aboveground and belowground populations as well as how they react to it (Bardgett et al. 2013). Soil microbes play a major role in determining the plant community, variety, efficiency, and amount and quality of carbon input to the soil. Dutta and Dutta (2016) estimate that soil contains about 70% of all terrestrial carbon. But since soil microbial decomposition generates eight times more CO_2 annually than human activity does, it is a significant contributor to CO_2 emissions. Climate change is impacted by both direct and indirect effects of soil microbial activity. Microbes are more likely to be released from the soil into the atmosphere when the weather turns mild. Meanwhile, changes in plant productivity and diversity brought on by climate change alter the physicochemical characteristics of soil indirectly. The amount of carbon released from soil depends on the makeup and activity of the microbial communities involved in the decomposition processes. Changing the physicochemical conditions of the soil to encourage microbial activity is one potential substitute for reducing SOC turnover and enhancing sequestration (Prommer et al. 2020).

2.6.3 Traditional Farming Practices and Soil Carbon Enhancement

The potential impact of global climate change on agricultural productivity has alarmed scientists because it has the potential to jeopardize local and global food security by altering critical climate factors. It is now known to researchers that many small farmers use a variety of agro-ecological techniques to minimize crop failure in order to adapt and even anticipate the effects of climate change. It argued that the only practical and reliable way to boost the productivity, sustainability, and resilience of peasant-based agricultural production under projected climate conditions is to salvage traditional management systems and apply agroecologically based management strategies. Traditional farming techniques, which are still commonly employed, are based on the wisdom and experience that indigenous people have amassed throughout the millennia. Common traditional farming practices include agroforestry, intercropping, crop rotation, traditional organic composting, and integrated crop-animal farming, among others. Although there are many benefits involved with these practices, such as improved soil fertility, carbon capturing, resource utilization, biodiversity maintenance, sustainability, and environment protection. Traditional farming is getting world interaction for being a source of sustainable food production in times of environmental degradation and the need for safe food production (Hamadani et al. 2021).

Incorporating traditional methods into high-diversity landscapes and increasing farm diversification may also greatly cut GHG emissions, perhaps aiding in the fight against global warming (Tscharntke et al. 2005). Therefore, traditional farmers who diversify their traditional farms and rural landscapes instead of using agrochemical inputs could significantly reduce the greenhouse gas emissions that come from the manufacture of fertilizer and pesticides. Traditional bio-diverse farms are able to prevent emissions since they require less energy, fertilizer, and pesticides (Niggli et al. 2008). The notions of food sovereignty and agro-ecologically based production systems have received a lot of attention globally in the last 20 years as a result of the realization of the contribution of indigenous and peasant agriculture to food security in the face of scenarios of climate change, economic crisis, and energy crisis (Altieri and Toledo 2005).

It is well known that traditional farming methods have a significant impact on the input and turnover rates of SOM, which in turn allows for strong control over the concentration of SOC. It affects environmental policy, especially in relation to soil carbon sequestration, whether specific actions result in a rise or fall in SOM content. Through increased SOM levels, organic farming is thought to increase soil fertility. The term "organic" farming refers to a set of practices that go into farming.

High-quality food produced in a sustainable manner with the least possible negative impact on the environment is the aim of organic farming. Important components of organic farming in terms of soil health include varied crop rotations, planting legumes to provide nitrogen to the farm, and minimal fertilizer inputs from outside sources (Leifeld and Fuhrer 2010). Legume-based systems had the highest carbon sequestration under integrated nutrient management (17.3%), no-tillage (15.04%), residue retention (17.94%), and organic amendment (40.03%). Under the maize–wheat system, integrated nutrient management (18.71%) and no-tillage (14.43%) positively affected soil C stock. Implementation of crop rotation had increased additional C sequestration by 18.01%. The organic amendment and crop rotation under these groups had increased C sequestration by 56.11% and 26.28% (Kumara et al. 2023). The carbon sequestration potential of various traditional farming practices is shown in Table 2.1. Adopting traditional agricultural practices that conserve SOC can increase crop productivity of soybean, maize, and wheat by 17.6 megatons per year (de Moraes Sá et al. 2017).

Although traditional agricultural practices do reduce GHGs or better store carbon compared to industrial agriculture, carbon offset credits are unlikely to be suitable to support the mitigation potential of traditional agriculture. Traditional farming systems have an important role to play in mitigation of climate change, especially through carbon sequestration, because of the carbon storage potential in their multiple plant species and soil (Mutuo et al. 2005).

2.6.4 Modern Agricultural Innovations for Soil Carbon Capturing

Improving food security, environmental conservation, and enhancing livelihood should be the main focus of the innovators of today's farming systems. Unfortunately, current agricultural practices are optimized primarily for yield and use large amounts of fossil energy to produce fertilizers, chemicals, and mechanical energy (Searchinger et al. 2019). When SOC accumulation exceeds emissions, the sector will be among the few to achieve net negative emissions and lead in the climate change solution. Modern agricultural technologies such as no-till and reduced tillage, cover cropping, and regenerative agriculture are widely regarded as crucial solutions for mitigating the impacts of climate change. It has a unique potential to provide beneficial contributions to the global carbon budget because it involves several modern approaches that remove CO_2 from the atmosphere. Soil carbon capturing through new technologies and the adoption of methods builds soil organic carbon stocks. There are various ways of modern agricultural innovation for soil carbon capturing, described hereunder.

Table 2.1 The carbon sequestration potential of various farming practices

Farming practices	Sequestration potential (Mg C/ha/year)
Crop lands	0.10–0.20
Afforestation, grasslands, and pastures	0.05–0.1
Elimination of bare fallow	0.1–0.2
Silvopasture	0.25–0.9
Integrated nutrient management (INM)	0.1–0.2
Mulch farming (4–6 Mg/ha/year)	0.1–0.2
Intensified agriculture	0.1–0.3

Source: Modified from Lal et al. (2018)

2.6.4.1 Conservation Agriculture (CA)

CA is an approach to farming that aims to improve food security, reduce poverty, preserve biodiversity, and protect ecosystem services. According to the FAO definition, CA is a farming system that encourages maintenance of (1) minimum soil disturbance to prevent soil inversion (i.e., no-tillage or minimum tillage), (2) a permanent soil cover with crop residues and/or cover crops, and (3) diversification through varied crop series (FAO 2017). Moreover, CA practice can help strengthen agricultural systems' resistance to climate change. It has been demonstrated that CA frequently lowers the GHG emissions of farming systems while improving their capacity as C sinks. Meta-data analyses revealed C sequestration rate of 0.4% per year is achievable under CA (Valkama et al. 2020). Through the mitigation of climate change and the maintenance of biodiversity, CA systems improve agricultural sustainability and soil health while having an impact on several ecosystem services in a variety of locales (Raj et al. 2022).

No-Till and Reduced Tillage

No-till and reduced tillage practices as components of CA play a crucial role in capturing carbon in soils. By minimizing soil disturbance, these techniques help preserve organic matter and promote the sequestration of carbon. Research suggests that adopting no-till or reduced tillage can lead to increased soil carbon storage over time (Pittelkow et al. 2015; Yadav et al. 2021a; Singh et al. 2021). No-till or reduced-tillage farming techniques reduce soil disturbance, which aids in the soil's ability to store carbon. Lessening or stopping tilling can be a successful carbon capture approach since it prevents the release of carbon captured into the atmosphere (Babu et al. 2023b). Farmers typically favor tillage techniques, which include physically disturbing the soil to improve soil respiration and break down the soil's organic pool. This practice increases turnover of soil aggregates and results in a reduction in the carbon content of soil. These days, "no-till" farming methods—growing crops year after year without disturbing the soil with tillage—are used to sequester or retain atmospheric carbon in the soil system. Because the soil is not thoroughly mixed and retains greater fertility from the retention of past crop residue, which raises the soil's carbon content, no-till farming practices encourage the least amount of soil loss. This preservation method lessens soil erosion, improves infiltration and the soil's ability to retain water, and encourages the development of soil aggregates. Therefore, no-till methods aid in the sequestration process by either directly or indirectly retaining soil carbon. Reducing soil disturbance is one way to improve SOC supplies in croplands. In contrast, irresponsible plowing or tilling causes the SOC stock to drop and releases the stored CO_2 back into the sky. Expanding this method to a worldwide scale will result in higher greenhouse gas emissions and more infertile soil.

Cover Cropping

A promising alternative to sequester carbon in agricultural soils is the practice of cover cropping in cropping systems. Cover crops, also known as inter-crops or catch crops, are crops that replace bare fallow before the next major crop is sown (Yadav et al. 2023). Compared to soils with natural vegetation, cropland soils have less SOC. Thirty to forty percent less SOC loss during cultivation than in seminatural or natural vegetation has been reported (Don et al. 2011; Poeplau et al. 2011). Moreover, cover crops can increase C concentration and stocks, potentially offsetting residue removal-induced losses to SOC and harm to other soil properties (Ruis and Blanco-Canqui 2017; Yadav et al. 2023). Legumes or certain grasses can be used as cover crops to trap atmospheric carbon and deposit it in the soil (Quintarelli et al. 2022). These crops reduce erosion while promoting soil health and organic matter content (Yadav et al. 2023). Cover crops used as green manure are an important management option to increase stocks. Such techniques could bring soil carbon sequestration of up to 17 Gt worldwide (Poeplau and Don 2015). Table 2.2 provides the rate of C sequestration potential by various cover crops in different cropping systems. When utilized as mulch cover in water-limited systems, cover crops have been demonstrated to enhance biodiversity (Lal 2004) and lessen soil erosion and drought stress for the succeeding crop (Frye et al. 1988), in addition to increasing carbon input. When it comes to managing SOC, cover crops have an advantage over other methods because they do not reduce yields as extensification does or contribute to carbon losses in other systems like applying organic manure (Babu et al. 2020b). Cover crops sequestrated carbon 3.55 Mg C/ha/ year (Peng et al. 2023).

2.6.4.2 Regenerative Agriculture

Regenerative agriculture has been suggested as an alternative means of producing food that may have fewer negative effects on the environment and society overall (Rhodes 2017). Regenerative agriculture and other sustainable agricultural systems are leveraging the potential for soil carbon sequestration as a benefit of innovative land management techniques. Regenerative agriculture is an approach to farming

Table 2.2 Carbon sequestration potential of different cover crops

Cover crops	Potential of C sequestration (Mg/C/year)	References
Ryegrass (*Lolium perenne*)	0.36 + 0.40	Zhang and Fang (2007)
Sesbania in rice–wheat rotation	0.09 + 0.03	Singh et al. (2004)
Lablab (*Lablab purpureum* L. Sweet)	0.83	Diekow et al. (2005)
Oat (*Avena strigosa* Scherb)	0.82	Bayer et al. (2000)

that emphasizes the health of the soil by boosting biodiversity and using soil-focused techniques (McCauley and Barlow 2023). Regenerative annul cropping has the potential to reduce or sequester 14.5–22 gigatons of CO_2 by 2050, according to Project Drawdown, which claims that "regenerative agriculture enhances and sustains the health of the soil by restoring its carbon content, which in turn improves productivity-just opposite to the conventional agriculture" (Project Drawdown 2020). While there is not a single definition for regenerative agriculture, the term is frequently used to refer to methods that restore the organic carbon content of the soil in order to promote soil health. As a natural "carbon sinks," the world's soils store several times as much carbon as the atmosphere does. However, due to several factors like overgrazing and the conversion of natural landscapes to croplands, soil carbon stores have been decreasing globally. Utilizing part of the carbon that plants have taken up from the environment to aid in the restoration of soil carbon is one objective of regenerative techniques (Ranganathan 2020).

Regenerative principles focus on building soil organic matter and promoting biodiversity through management practices that include reduced tillage and synthetic inputs, eliminating bare soil and fallow events, the implementation of multispecies cover crops in rotations to foster plant diversity, and the use of rotational grazing or animal manures (Lal 2004).

2.6.4.3 Perennial Crops

A way for managing SOC is the introduction of perennial crop species, in addition to the use of the no-till farming method and the use of biochar. This technique helps with long-term carbon storage by raising the soil's organic pool. Because of their expanded root system, a larger amount of carbon can be stored in the root biomass of perennial crops (Lal 2008). Table 2.3 shows the different perennial crops and their potential for C sequestration. Furthermore, perennial species react favorably to agricultural practices that cause the least amount of soil disturbance; for instance, extending the growing season is seen to be a higher-yielding and better farming alternative than cultivating annual crops (Luo et al. 2010).

Table 2.3 Depicted different perennial crops and their potential for C sequestration

Perennial crops	Potential of C sequestration (Mg C/ha/year)	References
Tree crops-temperate	2.1	Lasco et al. (2006)
Bamboo plantations	6.0–13.0	Nath et al. (2015)
Vitis vinifera	0.3–0.8	Vicente-Vicente et al. (2016)
Giant biomass grasses	1	Lorenz and Lal (2018)
Perennial grains	0.3–0.5	Olsson and Barbosa (2019)
Bactris gasipaes	5.1	Schroth et al. (2002)
Olea europaea	2.6	Sofo et al. (2005)

2.7 Conclusion and Future Perspectives

Carbon sequestration is one of the practical strategies to reduce the rising GHG emissions that contribute to climate change and global warming. However, there are multiple methods currently available and many more being developed; there is still no guarantee of sustainability or long-term consequences when it comes to sequestering atmospheric carbon. The approaches for carbon capture that are currently accessible for ensuring food and nutrition security can be divided into four categories: natural, biological, traditional farming practices, and modern innovative agricultural practices. The enhanced weathering process, photosynthesis, plant roots, and forests are examples of the natural carbon-capturing mechanisms. The biological strategy, on the other hand, takes a longer-term approach and uses blue carbon, energy crops, microalgae, microorganisms, and biochar for carbon fixation. Likewise, modern innovative and sustainable agricultural practices like conservation agriculture, regenerative agriculture, cover crops, perennial crops, no-till and reduced tillage, precision agriculture, and others help restore and capture carbon in the soil's organic pool. Even though all of these strategies appear to have potential, commercial applications are necessary before they can be put into practice for soil carbon capture. Currently, the most contentious problem that needs to be examined, debated, and, most importantly, resolved is the rise in atmospheric carbon levels due to environmental factors. Because of the numerous disadvantages connected with each and every option, it is not possible to solve the issue of rising carbon levels by employing a single approach. Therefore, it can be said that in order to attain sustainability through effective carbon capturing, multiple strategies must be employed in an integrated manner involving multiple agencies and people's participation. Such an integrated approach would be helpful in balancing the atmospheric carbon level in addition to soil carbon-capturing strategies. Although extensive research is going on in this area, still more research is the need of the hour. Long-term resilience and increased food security can be achieved by governmental policy supporting research and technology that improve soil health. Investments and policy focus should be targeted at encouraging soil carbon-capturing and maintaining techniques since these will improve the overall resilience of agricultural systems and move us closer to SDGs.

References

Abrahams PW (2002) Soils: their implications to human health. Sci Total Environ 291(1–3):1–32. https://doi.org/10.1016/S0048-9697(01)01102-0

Ali EF, Al-Yasi HM, Kheir AM, Eissa MA (2021) Effect of biochar on CO_2 sequestration and productivity of pearl millet plants grown in saline sodic soils. J Soil Sci Plant Nutr 21:897–907. https://doi.org/10.1007/s42729-021-00409-z

Altieri MA, Toledo VM (2005) Natural resource management among small-scale farmers in semi-arid lands: building on traditional knowledge and agroecology. Ann Arid Zone 44(3/4):365. https://epubs.icar.org.in/index.php/AAZ/article/view/66142

Ameray A, Bergeron Y, Valeria O, Montoro GM, Cavard X (2021) Forest carbon management: a review of silvicultural practices and management strategies across boreal, temperate and tropical forests. Curr For Rep:1–22. https://doi.org/10.1007/s40725-021-00151-w

Babu S, Mohapatra KP, Das A, Yadav GS, Tahasildar M, Singh R, Panwar AS, Yadav V, Chandra P (2020a) Designing energy-efficient, economically sustainable and environmentally safe cropping system for the rainfed maize–fallow land of the Eastern Himalayas. Sci Total Environ. https://doi.org/10.1016/j.scitotenv.2020.137874

Babu S, Singh R, Avasthe RK, Yadav GS, Mohapatra KP, Selvan T, Das A, Singh VK, Valente D, Petrosillo I (2020b) Soil carbon dynamics in Indian Himalayan intensified organic rice-based cropping sequences. Ecol Indic. https://doi.org/10.1016/j.ecolind.2020.106292

Babu S, Singh R, Avasthe RK, Kumar S, Rathore SS, Singh VK, Ansari MA, Valente D, Petrosillo I (2023a) Soil carbon dynamics under organic farming: impact of tillage and cropping diversity. Ecol Indic. https://doi.org/10.1016/j.ecolind.2023.109940

Babu S, Singh R, Avasthe R, Rathore SS, Kumar S, Das A, Layek J, Sharma V, Wani OA, Singh VK (2023b) Conservation tillage and diversified cropping enhance system productivity and eco-efficiency and reduce greenhouse gas intensity in organic farming. Front Sustain Food Syst 7:1114617. https://doi.org/10.3389/fsufs.2023.1114617

Babu S, Singh R, Kumar S, Rathore SS, Yadav D, Yadav SK, Yadav V, Ansari MA, Das A, Rajanna GA, Wani OA, Raj R, Yadav DK, Singh VK (2023c) Biochar implications in cleaner agricultural production and environmental sustainability. Environ Sci Adv 2(8):1042–1059. https://doi.org/10.1039/D2VA00324D

Bardgett RD, Manning P, Morriën E, De Vries FT (2013) Hierarchical responses of plant–soil interactions to climate change: consequences for the global carbon cycle. J Ecol 101(2):334–343. https://doi.org/10.1111/1365-2745.12043

Bayer C, Mielniczuk J, Amado TJC, Martin-Neto L, Fernandes SV (2000) Organic matter storage in a sandy clay loam Acrisol affected by tillage and cropping systems in southern Brazil. Soil Tillage Res 54:101–109. https://doi.org/10.1016/S0167-1987(00)00090-8

Beer LL, Boyd ES, Peters JW, Posewitz MC (2009) Engineering algae for biohydrogen and biofuel production. Curr Opin Biotechnol 20:264–271. https://doi.org/10.1016/j.copbio.2009.06.002

Beerling DJ, Kantzas EP, Lomas MR, Wade P, Eufrasio RM, Renforth P, Banwar SA (2020) Potential for large-scale CO_2 removal via enhanced rock weathering with croplands. Nature 583(7815):242–248. https://doi.org/10.1038/s41586-020-2448-9

Binkly D, Giardina C (1998) Why do tree species affect soils? The warp and woof of trees-soil interactions. Biogeochemistry 42:89–106. https://doi.org/10.1007/978-94-017-2691-7_5

Black R (2003) Micronutrient deficiency – an underlying cause of morbidity and mortality. Bull World Health Organ 81(2):79

Brancalion PH, Holl KD (2020) Guidance for successful tree planting initiatives. J Appl Ecol 57(12):2349–2361. https://doi.org/10.1111/1365-2664.13725

Chambers A, Lal R, Paustian K (2016) Soil carbon sequestration potential of US croplands and grasslands: implementing the 4 per Thousand Initiative. J Soil Water Conserv 71:68A–74A. https://doi.org/10.2489/jswc.71.3.68A

Chen Y, Xu C (2020) Exploring new blue carbon plants for sustainable ecosystems. Trends Plant Sci 25(11):1067–1070. https://doi.org/10.1016/j.tplants.2020.07.016

Cotrufo MF, Wallenstein MD, Boot CM, Denef K, Paul E (2012) The Microbial Efficiency-Matrix Stabilization (MEMS) framework integrates plant litter decomposition with soil organic matter stabilization: do labile plant inputs form stable soil organic matter? Glob Chang Biol 19:988–995. https://doi.org/10.1111/gcb.12113

Das A, Krishnappa R, Basavaraj S, Dey U, Haloi M, Layek J, Ramkrushna GI, Lal R, Deshmukh NA, Yadav GS, Babu S, Ngachan SV (2021) Conservation tillage and nutrient management practices in summer rice (*Oryza sativa* L.) favoured root growth and phenotypic plasticity of succeeding winter pea (*Pisum sativum* L.) under eastern Himalayas, India. Heliyon 7(5). https://doi.org/10.1016/j.heliyon.2021.e07078

Das A, Babu S, Datta M, Kumar S, Singh R, Avasthe R, Rathore SS, Yadav SK, Singh VK (2022) Restoring soil carbon in marginal land of Indian Himalayas: impact of crop intensi-

fication and conservation tillage. J Environ Manag 318:115603. https://doi.org/10.1016/j.jenvman.2022.115603
de Moraes Sá JC, Lal R, Cerri CC, Lorenz K, Hungria M, de Faccio Carvalho PC (2017) Low-carbon agriculture in South America to mitigate global climate change and advance food security. Environ Int 98:102–112. https://doi.org/10.1016/j.envint.2016.10.020
Diekow J, Mielniczuk J, Knicker H, Bayer C, Dick DP, Kogel-Knabner I (2005) Soil C and N stocks as affected by cropping systems and nitrogen fertilization in a Southern Brazil Acrisol managed under no-tillage for 17 years. Soil Tillage Res 81:87–95. https://doi.org/10.1016/j.still.2004.05.003
Dominati E, Patterson M, Mackay A (2010) A framework for classifying and quantifying the natural capital and ecosystem services of soils. Ecol Econ J 69(9):1858–1868. https://doi.org/10.1016/j.ecolecon.2010.05.002
Don A, Schumacher J, Freibauer A (2011) Impact of tropical land-use change on soil organic carbon stocks–a meta-analysis. Glob Chang Biol 17(4):1658–1670. https://doi.org/10.1111/j.1365-2486.2010.02336.x
Dutta H, Dutta A (2016) The microbial aspect of climate change. Energy Ecol Environ 1:209–232. https://doi.org/10.1007/s40974-016-0034-7
Fang Y, Singh B, Singh BP, Krull E (2014) Biochar carbon stability in four contrasting soils. Eur J Soil Sci 65(1):60–71. https://doi.org/10.1111/ejss.12094
FAO (2008) Food and Agriculture Organization of the United Nations. FAO statistical yearbook 2007–2008. FAO, Rome
FAO (2015) Healthy soils are the basis for healthy food production. Food and Agriculture Organization of United Nations, Rome. http://www.fao.org/soils-2015/news/news-detail/en/c/277682/
FAO (2017) Conservation agriculture. 2p. http://www.fao.org/documents/card/en/c/981ab2a0-f3c6-4de3-a058-f0df6658e69f/
FAO (2019) Recarbonization of global soils – a dynamic response to offset global emissions. FAO. http://www.fao.org/3/i7235en/I7235EN.pdf
FAO ITPS (2015) Status of the World's soil resources (SWSR) – main report. Food and Agriculture Organization of the United Nations and Intergovernmental Technical Panel on Soils, Rome. https://openknowledge.fao.org/server/api/core/bitstreams/6ec24d75-19bd-4f1f-b1c5-5becf50d0871/content
Fargione J, Hill J, Tilman D, Polasky S, Hawthorne P (2008) Land clearing and the biofuel carbon debt. Science 319(5867):1235–1238. https://doi.org/10.1126/science.1152747
Frye WW, Blevins RL, Smith MS, Corak SJ, Varco JJ (1988) Role of annual legume cover crops in efficient use of water and nitrogen. In: Cropping strategies for efficient use of water and nitrogen, vol 51, pp 129–154. https://doi.org/10.2134/asaspecpub51.c8
Fuss S, Lamb WF, Callaghan MW, Hilaire J, Creutzig F, Amann T et al (2018) Negative emissions – part 2: costs, potentials and side effects. Environ Res Lett 13(6):063002. https://doi.org/10.1088/1748-9326/aabf9f
Geraldi NR, Ortega A, Serrano O, Macreadie PI, Lovelock CE, Krause-Jensen D, Kennedy H, Lavery PS, Pace ML, Kaal J, Duarte CM (2019) Fingerprinting blue carbon: rationale and tools to determine the source of organic carbon in marine depositional environments. Front Mar Sci:263. https://doi.org/10.3389/fmars.2019.00263
Ghosh D, Maiti SK (2021) Biochar-assisted eco-restoration of coal mine degraded land to meet United Nation Sustainable Development Goals. Land Degrad Dev 32(16):4494–4508. https://doi.org/10.1002/ldr.4055
Hamadani H, Rashid SM, Parrah JD, Khan AA, Dar KA, Ganie AA, … Ali A (2021) Traditional farming practices and its consequences. In: Microbiota and biofertilizers, vol 2. Ecofriendly tools for reclamation of degraded soil environs, pp 119–128. https://doi.org/10.1007/978-3-030-61010-4_6
Hu G, Li Y, Ye C, Liu L, Chen X (2019) Engineering microorganisms for enhanced CO_2 sequestration. Trends Biotechnol 37(5):532–547. https://doi.org/10.1016/j.tibtech.2018.10.008

Idapuganti RG, Das A, Sungoh H, Layek J, Mandal S, Verma BC, Lal R, Rangappa K, Babu S, Hazarika S (2024) Can biochar conserve soil moisture and improve soil properties for sustainable intensification of acid soils in the Eastern Indian Himalayas? Land Degrad Dev 35(3):1192–1208. https://doi.org/10.1002/ldr.4981

IEA (2023) World Energy Outlook 2023. IEA, Paris. https://www.iea.org/reports/world-energy-outlook-2023, Licence: CC BY 4.0 (report); CC BY NC SA 4.0 (Annex A)

Indian Network for Climate Change Assessment, and India, Ministry of Environment. (2010). Climate change and India: a 4 × 4 assessment, a sectoral and regional analysis for 2030s, 2. Ministry of Environment & Forests, Government of India

Kalaiselvi B, Kumari S, Sathya S, Dharumarajan S, Kumar KA, Hegde R (2023) Crop management practices for carbon sequestration. In: Agricultural soil sustainability and carbon management. Academic Press, pp 27–68. https://doi.org/10.1016/B978-0-323-95911-7.00008-6

Kaur R, Kaur N, Kumar S, Dass A, Singh T (2023) Carbon capture and sequestration for sustainable land use–a review. Indian J Agric Sci 93(1):11–18. https://doi.org/10.56093/ijas.v93i1.124838

Kell DB (2012) Large-scale sequestration of atmospheric carbon via plant roots in natural and agricultural ecosystems: why and how. Philos Trans R Soc B Biol Sci 367(1595):1589–1597. https://doi.org/10.1098/rstb.2011.0244

Khan S, Khan MA, Hanjra MA, Mu J (2009) Pathways to reduce the environmental footprints of water and energy inputs in food production. Food Policy 34(2):141–149. https://doi.org/10.1016/j.foodpol.2008.11.002

Khan S, Shah A, Nawaz M, Khan M (2017) Impact of different tillage practices on soil physical properties, nitrate leaching and yield attributes of maize (*Zea mays* L.). J Soil Sci Plant Nutr 17(1):240–252. https://doi.org/10.4067/S0718-95162017005000019

Kibblewhite MG, Ritz K, Swift MJ (2008) Soil health in agricultural systems. Philos Trans R Soc B Biol Sci 363(1492):685–701. https://doi.org/10.1098/rstb.2007.2178

Kirchman DL, Morán XAG, Ducklow H (2009) Microbial growth in the polar oceans – role of temperature and potential impact of climate change. Nat Rev Microbiol 7(6):451–459. https://doi.org/10.1038/nrmicro2115

Koch A, McBratney AB, Adams M, Field DJ, Hill R, Lal R, Abbott L, Angers D, Baldock J, Barbier E, Bird M, Bouma J, Chenu C, Crawford J, Flora CB, Goulding K, Grunwald S, Jastrow J, Lehmann J et al (2014) Soil security: solving the global soil crisis. Glob Policy J 4(4):434–441. https://doi.org/10.1111/1758-5899.12096

Kumar SS, Mir SA, Wani OA, Babu S, Yeasin M, Bhat MA, Hussain N, Wani AA, Kumar R, Devideen Y, Dar SR (2022) Land-use systems regulate carbon geochemistry in the temperate Himalayas, India. J Environ Manag 320:115811. https://doi.org/10.1016/j.jenvman.2022.115811

Kumar A, Saha S, Layek J, Babu S, Kumar R, Choudhary AK, Das A (2023) Organic farming in Indian Himalayan region: innovations for sustainability. Indian J Agron 68. (XXII Biennial National Symposium Special issue):S36–S50

Kumara K, Pal S, Chand P, Kandpal A (2023) Carbon sequestration potential of sustainable agricultural practices to mitigate climate change in Indian agriculture: a meta-analysis. Sustain Prod Consum 35:697–708. https://doi.org/10.1016/j.spc.2022.12.015

Lal R (2004) Soil carbon sequestration impacts on global climate change and food security. Science 304(5677):1623–1627. https://doi.org/10.1126/science.1097396

Lal R (2008) Carbon sequestration. Philos Trans R Soc B Biol Sci 363(1492):815–830. https://doi.org/10.1098/rstb.2007.2185

Lal R (2011) Sequestering carbon in soils of agro-ecosystems. Food Policy 36:S33–S39. https://doi.org/10.1016/j.foodpol.2010.12.001

Lal R (2016) Soil health and carbon management. Food Energy Secur 5(4):212–222. https://doi.org/10.1002/fes3.96

Lal R, Smith P, Jungkunst HF, Mitsch WJ, Lehmann J, Nair PR, … Ravindranath NH. (2018). The carbon sequestration potential of terrestrial ecosystems. J Soil Water Conserv 73(6):145A–152A. https://doi.org/10.2489/jswc.73.6.145A

Lasco R, Ogle S, Raison J, Verchot L, Wassmann R, Yagi K (2006) IPCC guidelines for national greenhouse gas inventories: vol 4. Agriculture, forestry and other land use. Institute for Global Environmental Strategies, Hayama

Lehmann J, Kuzyakov Y, Pan G, Ok YS (2015) Biochars and the plant-soil interface. Plant Soil 395:1–5. https://doi.org/10.1007/s11104-015-2658-3

Leifeld J, Fuhrer J (2010) Organic farming and soil carbon sequestration: what do we really know about the benefits? Ambio 39(8):585–599. https://doi.org/10.1007/s13280-010-0082-8

Lorenz K, Lal R (2018) Carbon sequestration in agricultural ecosystems. Springer, Dordrecht

Luo Z, Wang E, Sun OJ (2010) Soil carbon change and its responses to agricultural practices in Australian agro-ecosystems: a review and synthesis. Geoderma 155:211–223. https://doi.org/10.1016/j.geoderma.2009.12.012

Macreadie PI, Anton A, Raven JA, Beaumont N, Connolly RM, Friess DA, … Duarte CM (2019) The future of blue carbon science. Nat Commun 10(1):3998. https://doi.org/10.1038/s41467-019-11693-w

Major J, Lehmann J, Rondon M, Goodale C (2010) Fate of soil-applied black carbon: downward migration, leaching and soil respiration. Glob Change Biol 16:1366–1379. https://doi.org/10.1111/j.1365-2486.2009.02044.x

Malhi Y, Meir P, Brown S (2002) Forests, carbon and global climate. Philos Trans R Soc London, Ser A 360(1797):1567–1591. https://doi.org/10.1098/rsta.2002.1020

McBratney, A., Stockmann, U. (2011). Some Key Management Options for increasing Soil Organic Carbon: A primer of practical management for those interested in managing soil carbon. Soil carbon summit. 31 January to 3 February 2011. Sydney, Australia

McBratney AB, Field DJ, Koch A (2013) The dimensions of soil security. Geoderma 213:203–213. https://doi.org/10.1016/j.geoderma.08.013

McCauley K, Barlow K (2023) Regenerative agriculture: increasing plant diversity and soil carbon sequestration on agricultural landscapes. Surg J 15(1). https://doi.org/10.21083/surg.v15i1.7196

Mikunda T, Brunner L, Skylogianni E, Monteiro J, Rycroft L, Kemper J (2021) Carbon capture and storage and the sustainable development goals. Int J Greenhouse Gas Control 108:103318. https://doi.org/10.1016/j.ijggc.2021.103318

Mikutta R, Turner S, Schippers A, Gentsch N, Meyer-Stüve S, Condron LM, Peltzer AD, Richardson SJ, Eger A, Hempel G, Kaiser K, Klotzbücher T, Guggenberger G (2019) Microbial and abiotic controls on mineral-associated organic matter in soil profiles along an ecosystem gradient. Sci Rep 9(1):10294. https://doi.org/10.1038/s41598-019-46501-4

Zhang M-K, Fang L-p (2007) Effect of tillage, fertilizer and green manure cropping on soil quality at an abandoned brick making site. Soil Tillage Res 93:87–93. https://doi.org/10.1016/j.still.2006.03.016s

Mondal M, Khanra S, Tiwari ON, Gayen K, Halder GN (2016) Role of carbonic anhydrase on the way to biological carbon capture through microalgae – a mini review. Environ Prog Sustain Energy 35(6):1605–1615. https://doi.org/10.1002/ep.12394

Mutuo PK, Cadisch G, Albrecht A, Palm CA, Verchot L (2005) Potential of agro forestry for carbon sequestration and mitigation of greenhouse gas emissions from soils in the tropics. Nutr Cycl Agroecosyst 71:43–54. https://doi.org/10.1007/s10705-004-5285-6

Nanda S, Azargohar R, Dalai AK, Kozinski JA (2015) An assessment on the sustainability of lignocellulosic biomass for biorefining. Renew Sust Energ Rev 50:925–941. https://doi.org/10.1016/j.rser.2015.05.058

Nanda S, Reddy SN, Mitra SK, Kozinski JA (2016) The progressive routes for carbon capture and sequestration. Energy Sci Eng 4(2):99–122. https://doi.org/10.1002/ese3.117

Nath AJ, Lal R, Das AK (2015) Managing woody bamboos for carbon farming and carbon trading. Glob Ecol Conserv 3:654–663. https://doi.org/10.1016/j.gecco.2015.03.002

Nelson E, Mendoza G, Regetz J, Polasky S, Tallis H, Cameron D, Chan MA, Daily GC, Goldstein J, Kareiva PM, Lonsdorf E, Naidoo R, Ricketts TH, Shaw M (2009) Modeling multiple ecosystem services, biodiversity conservation, commodity production, and tradeoffs at landscape scales. Front Ecol Environ 7(1):4–11. https://doi.org/10.1890/080023

Niggli U, Slabe A, Schmid O, Halberg N, Schlüter M (2008) Vision for an organic food and farming research agenda 2025. Organic knowledge for the future. Food Fairness Ecol. http://www.organic-research.org/

Olsson L, Barbosa H (2019) Land degradation. In: IPCC. Climate and change and land: a special report of the Intergovernmental Panel on Climate Change. https://www.ipcc.ch/srccl-report-downloadpage/

Onyeaka H, Miri T, Obileke K, Hart A, Anumudu C, Al-Sharify ZT (2021) Minimizing carbon footprint via microalgae as a biological capture. Carbon Capture Sci Technol 1:100007. https://doi.org/10.1016/j.ccst.2021.100007

Peichl M, Thevathasan NV, Gordon AM, Huss J, Abohassan RA (2006) Carbon sequestration potentials in temperate tree-based intercropping systems, southern Ontario, Canada. Agrofor Syst 66:243–257. https://doi.org/10.1007/s10457-005-0361-8

Peng Y, Rieke EL, Chahal I, Norris CE, Janovicek K, Mitchell P, … Van Eerd LL (2023) Maximizing soil organic carbon stocks under cover cropping: insights from long-term agricultural experiments in North America. Agric Ecosyst Environ 356:108599. https://doi.org/10.1016/j.agee.2023.108599

Pittelkow CM, Linquist BA, Lundy ME, Liang X, Van Groenigen KJ, Lee J, Gestel VN, Six J, Venterea RT, Van Kessel C (2015) When does no-till yield more? A global meta-analysis. Field Crop Res 183:156–168. https://doi.org/10.1016/j.fcr.2015.07.020

Poeplau C, Don A (2015) Carbon sequestration in agricultural soils via cultivation of cover crops–a meta-analysis. Agric Ecosyst Environ 200:33–41. https://doi.org/10.1016/j.agee.2014.10.024

Poeplau C, Don A, Vesterdal L, Leifeld J, Van Wesemael BAS, Schumacher J, Gensior A (2011) Temporal dynamics of soil organic carbon after land-use change in the temperate zone–carbon response functions as a model approach. Glob Chang Biol 17(7):2415–2427. https://doi.org/10.1111/j.1365-2486.2011.02408.x

Project Drawdown (2020) Regenerative annual cropping. Available online at: https://www.drawdown.org/solutions/regenerative-annual-cropping. Accessed 19 Apr 2024

Prommer J, Walker TW, Wanek W, Braun J, Zezula D, Hu Y, Hofhansl F, Richter A (2020) Increased microbial growth, biomass, and turnover drive soil organic carbon accumulation at higher plant diversity. Glob Chang Biol 26(2):669–681. https://doi.org/10.1111/gcb.14777

Quintarelli V, Radicetti E, Allevato E, Stazi SR, Haider G, Abideen Z, Bibi S, Jamal A, Mancinelli R (2022) Cover crops for sustainable cropping systems: a review. Agriculture 12(12):2076. https://doi.org/10.3390/agriculture12122076

Raj R, Das TK, Chakraborty D, Bhattacharyya R, Babu S, Govindasamy P, Kumar V, Ekka U, Sen S, Ghosh S, Roy A, Sharma T (2022) Soil physical environment and active carbon pool in rice–wheat system of South Asia: impact of long-term conservation agriculture practices. Environ Technol Innov 29:102966. https://doi.org/10.1016/j.eti.2022.102966

Ranganathan T (2020) Risk and risk management in agriculture. In: Contemporary issues in sustainable development. Routledge, New Delhi, pp 9–29

Rathore SS, Shekhawat K, Babu S, Singh VK (2020) Mitigating moisture stress in Brassica juncea through deficit irrigation scheduling and hydrogel in ustocherpts soils of semi-arid India. Heliyon 6(12). https://doi.org/10.1016/j.heliyon.2020.e05786

Rathore SS, Babu S, Shekhawat K, Singh R, Yadav SK, Singh VK, Singh C (2022) Designing energy cum carbon-efficient environmentally clean production system for achieving green economy in agriculture. Sustain Energy Technol Assess 52:102190. https://doi.org/10.1016/j.seta.2022.102190

Renforth P, Jenkins BG, Kruger T (2013) Engineering challenges of ocean liming. Energy 60:442–452. https://doi.org/10.1016/j.energy.2013.08.006

Rhodes CJ (2017) The imperative for regenerative agriculture. Sci Prog 100(1):80–129. https://doi.org/10.3184/003685017X14876775256165

Ritchie H, Roser M (2024) Micronutrient deficiency. Our World in Data. https://ourworldindata.org/micronutrient-deficiency

Rojas RV, Achouri M, Maroulis J, Caon L (2016) Healthy soils: a prerequisite for sustainable food security. Environ Earth Sci 75:1–10. https://doi.org/10.1007/s12665-015-5099-7
Rui Y, Murphy DV, Wang X, Hoyle FC (2016) Microbial respiration, but not biomass, responded linearly to increasing light fraction organic matter input: consequences for carbon sequestration. Sci Rep 6(1):35496. https://doi.org/10.1038/srep35496
Ruis SJ, Blanco-Canqui H (2017) Cover crops could offset crop residue removal effects on soil carbon and other properties: a review. Agron J 109(5):1785–1805. https://doi.org/10.2134/agronj2016.12.0735
Saharawat YS, Singh B, Malik RK, Ladha JK, Gathala M, Jat ML, Kumar V (2010) Evaluation of alternative tillage and crop establishment methods in a rice–wheat rotation in North Western IGP. Field Crop Res 116(3):260–267. https://doi.org/10.1016/j.fcr.2010.01.003
Sahoo UK, Singh SL, Gogoi A, Kenye A, Sahoo SS (2019) Active and passive soil organic carbon pools as affected by different land use types in Mizoram, Northeast India. PloS One 14(7):e0219969. https://doi.org/10.1371/journal.pone.0219969
Sanaullah M, Afzal T, Shahzad T, Wakeel A (2019) Carbon sequestration for sustainable agriculture. Innov Sustain Agric:469–500. https://doi.org/10.1007/978-3-030-23169-9_15
Sanderman J, Hengl T, Fiske GJ (2017) Soil carbon debt of 12,000 years of human land use. Proc Natl Acad Sci U S A 114:9575–9580. https://doi.org/10.1073/pnas.1706103114
Santos IR, Burdige DJ, Jennerjahn TC, Bouillon S, Cabral A, Serrano O, Wernberg T, Dexter FK, Guimond JA, Tamborski JJ (2021) The renaissance of Odum's outwelling hypothesis in 'Blue Carbon' science. Estuar Coast Shelf Sci 255:107361. https://doi.org/10.1016/j.ecss.2021.107361
Schroth G, D'Angelo S, Teixeria W, Haag D, Leiberei R (2002) Conversion of secondary forest into agroforestry and monoculture plantations in Amazonia: consequences for biomass, litter and soil carbon stocks after 7 years. For Ecol Manag 163(1):131–150. https://doi.org/10.1016/S0378-1127(01)00537-0
Searchinger T, Heimlich R, Houghton RA, Dong F, Elobeid A, Fabiosa J, Tokgoz S, Hayes D, Yu TH (2008) Use of US croplands for biofuels increases greenhouse gases through emissions from land-use change. Science 319(5867):1238–1240. https://doi.org/10.1126/science.1151861
Searchinger T, Waite R, Hanson C, Ranganathan J, Dumas P, Matthews E, Klirs C (2019) Creating a sustainable food future: a menu of solutions to feed nearly 10 billion people by 2050. Final report https://agritrop.cirad.fr/593176/1/WRR_Food_Full_Report_0.pdf
Sedjo R, Sohngen B (2012) Carbon sequestration in forests and soils. Ann Rev Resour Econ 4(1):127–144. https://doi.org/10.1146/annurev-resource-083110-115941
Shetty P (2009) Incorporating nutritional considerations when addressing food insecurity. Food Secur 1:431–440. https://doi.org/10.1007/s12571-009-0039-6
Singh Y, Singh B, Ladha JK, Khind CS, Khera TS, Bueno CS (2004) Effects of residue decomposition on productivity and soil fertility in rice-wheat rotation. Soil Sci Soc Am J 68:854–864. https://doi.org/10.2136/sssaj2004.8540
Singh R, Babu S, Avasthe RK, Yadav GS, Das A, Mohapatra KP, Kumar A, Singh VK, Chandra P (2021) Crop productivity, soil health, and energy dynamics of Indian Himalayan intensified organic maize-based systems. Int Soil Water Conserv Res 9(2):260–270. https://doi.org/10.1016/j.iswcr.2020.11.003
Singh R, Kumar A, Babu S, Avasthe R, Das A, Rathore SS, Kumar S, Singh C, Sharma V, Bhupenchandra I (2023) Development of organic nutrients management system for profitable and soil-supportive French bean (*Phaseolus vulgaris* L.) farming in North Eastern Himalayas, India. Front Sustain Food Syst 7:1115521. https://doi.org/10.3389/fsufs.2023.1115521
Six JAET, Elliott ET, Paustian K (2000) Soil macroaggregate turnover and microaggregate formation: a mechanism for C sequestration under no-tillage agriculture. Soil Biol Biochem 32(14):2099–2103. https://doi.org/10.1016/S0038-0717(00)00179-6
Smith P, Martino D, Cai Z, Gwary D, Janzen H, Kumar P, McCarl B, Ogle S, O'Mara F, Rice C, Scholes B, Sirotenko O, Howden M, McAllister T, Pan G, Romanenkov V, Schneider U, Towprayoon S, Wattenbach M, Smith P (2007) Greenhouse gas mitigation in agriculture. Philos Trans R Soc B 363(1492):789–813. https://doi.org/10.1098/rstb.2007.2184

Sofo A, Nuzzo V, Palese A, Xiloyannis C, Celano G, Zukowskyj P (2005) Net CO_2 storage in Mediterranean olive and peach orchards. Sci Hortic 107(1):17–24. https://doi.org/10.1016/j.scienta.2005.06.001

St. Clair SB, Lynch JP (2010) The opening of Pandora's Box: climate change impacts on soil fertility and crop nutrition in developing countries. Plant Soil 335:101–115. https://doi.org/10.1007/s11104-010-0328-z

Taylor LL, Beerling DJ, Quegan S, Banwart SA (2017) Simulating carbon capture by enhanced weathering with croplands: an overview of key processes highlighting areas of future model development. Biol Lett 13(4):20160868. https://doi.org/10.1098/rsbl.2016.0868

Terlouw T, Treyer K, Bauer C, Mazzotti M (2021) Life cycle assessment of direct air carbon capture and storage with low-carbon energy sources. Environ Sci Technol 55(16):11397–11411. https://doi.org/10.1021/acs.est.1c03263

Thevathasan NV, Gordon AM (2004) Ecology of tree intercropping systems in the North temperate region: experiences from southern Ontario, Canada. In: New vistas in agroforestry: a compendium for 1st World Congress of Agroforestry, 2004. Springer, Netherlands, pp 257–268. https://doi.org/10.1007/978-94-017-2424-1_18

Tscharntke T, Klein AM, Kruess A, Steffan-Dewenter I, Thies C (2005) Landscape perspectives on agricultural intensification and biodiversity–ecosystem service management. Ecol Lett 8(8):857–874. https://doi.org/10.1111/j.1461-0248.2005.00782.x

U.S. Department of Agriculture, Natural Resources Conservation Service (2012) Soil health: unlock the secrets of the soil. Available at: http://www.nrcs.usda.gov/wps/portal/nrcs/main/national/soils/health/. Accessed 24 Apr 2024

United Nations Convention to Combat Desertification (UNCCD) (2016) Report of the Conference of the Parties on its twelfth session, held in Ankara from 12 to 23 October 2015

Valkama E, Kunypiyaeva G, Zhapayev R, Karabayev M, Zhusupbekov E, Perego A, Schillaci C, Sacco D, Moretti B, Grignani C, Acutis M (2020) Can conservation agriculture increase soil carbon sequestration? A modelling approach. Geoderma 369:114298. https://doi.org/10.1016/j.geoderma.2020.114298

Vicente-Vicente J, García-Ruiz R, Francaviglia R, Aguilera E, Smith P (2016) Soil carbon sequestration rate under Mediterranean woody crops using recommended management practices: a meta-analysis. Agric Ecosyst Environ 235:204–214. https://doi.org/10.1016/j.agee.2016.10.024

Wang J, Lu P, Valente D, Petrosillo I, Babu S, Xu S, Li C, Huang D, Liu M (2022) Analysis of soil erosion characteristics in small watershed of the loess tableland Plateau of China. Ecol Indic 137:108765. https://doi.org/10.1016/j.ecolind.2022.108765

Wani OA, Akhter F, Kumar SS, Babu S, Kanth RH, Mir SA, Mahdi SS, Malik AR, Bangroo S, Gaafar ARZ, Popescu SM, Rathore SS (2023a) Mitigating soil erosion through biomass-derived biochar: exploring the influence of feedstock types and pyrolysis temperature. Land 12(12):2111. https://doi.org/10.3390/land12122111

Wani OA, Sharma V, Kumar SS, Babu S, Sharma KR, Rathore SS, Ganai NA, Dar SR, Yeasin M, Singh R, Tomar J (2023b) Climate plays a dominant role over land management in governing soil carbon dynamics in North Western Himalayas. J Environ Manag 338:117740. https://doi.org/10.1016/j.jenvman.2023.117740

Wedding LM, Moritsch M, Verutes G, Arkema K, Hartge E, Reiblich J, Strong AL (2021) Incorporating blue carbon sequestration benefits into sub-national climate policies. Glob Environ Chang 69:102206. https://doi.org/10.1016/j.gloenvcha.2020.102206

Wei YM, Liu LJ, Liang QM, Yu BY, Liu LC, Yao YF, Zhang K, Wang C, Han R, Kang NK, Liao H (2021) Pathway comparison of limiting global warming to 2°C. Energy Climate Change 2:100063. https://doi.org/10.1016/j.egycc.2021.100063

Williams PJLB, Laurens LM (2010) Microalgae as biodiesel & biomass feedstocks: review & analysis of the biochemistry, energetics & economics. Energy Environ Sci 3(5):554–590. https://doi.org/10.1039/B924978H

World Bank (2015). Available at: https://data.worldbank.org/indicator/AG.LND.AGRI.ZS

Yadav GS, Babu S, Das A, Bhowmik SN, Datta M, Singh R (2019) Soil carbon dynamics and productivity of rice–rice system under conservation tillage in submerged and unsubmerged ecologies of Eastern Indian Himalaya. Carbon Manag 10(1):51–62. https://doi.org/10.1080/17583004.2018.1545518

Yadav GS, Babu S, Das A, Mohapatra KP, Singh R, Avasthe RK, Roy S (2020a) No-till and mulching enhance energy use efficiency and reduce carbon footprint of a direct-seeded upland rice production system. J Clean Prod 271:122700. https://doi.org/10.1016/j.jclepro.122700

Yadav GS, Das A, Babu S, Mohapatra KP, Lal R, Rajkhowa D (2020b) Potential of conservation tillage and altered land configuration to improve soil properties, carbon sequestration and productivity of maize based cropping system in Eastern Himalayas, India. Int Soil Water Conserv Res, 9(2), 279–290. hhttps://doi.org/10.1016/j.iswcr.2020.12.003

Yadav GS, Babu S, Das A, Datta M, Mohapatra KP, Singh R, Singh VK, Rathore SS, Chakraborty M (2021a) Productivity, soil health, and carbon management index of Indian Himalayan intensified maize-based cropping systems under live mulch-based conservation tillage practices. Field Crop Res 264:108080. https://doi.org/10.1016/j.fcr.2021.108080

Yadav GS, Kandpal BK, Das A, Babu S, Mohapatra KP, Devi AG, Devi HL, Chandra P, Singh R, Barman KK (2021b) Impact of 28-year-old agroforestry systems on soil carbon dynamics in Eastern Himalayas. J Environ Manag 283:111978. https://doi.org/10.1016/j.jenvman.2021.111978

Yadav D, Singh D, Babu S, Madegowda M, Singh D, Mandal D, Rathore AC, Sharma VK, Singhal V, Kumawat A, Yadav DK, Yadav RK, Kumar S (2023) Intensified cropping reduces soil erosion and improves rainfall partitioning and soil properties in the marginal land of the Indian Himalayas. Int Soil Water Conserv Res. https://doi.org/10.1016/j.iswcr.2023.10.002

Zahed MA, Movahed E, Khodayari A, Zanganeh S, Badamaki M (2021) Biotechnology for carbon capture and fixation: critical review and future directions. J Environ Manag 293:112830. https://doi.org/10.1016/j.jenvman.2021.112830

Zhang F, Qian H (2024) A comprehensive review of the environmental benefits of urban green spaces. Environ Res 252:118837. https://doi.org/10.1016/j.envres.2024.118837

Chapter 3
Biochar for Food Security and Environmental Sustainability Under Current Climate Change Scenario

Shamal Shasang Kumar, Owais Ali Wani, Ab Raouf Malik, Sanjeev Kumar, and Roopa Patel

Abstract Biochar properties have diverse influence on ecological systems and are significantly altered by various attributes, which include the type of biomass utilized, the length of time, and the specific conditions employed during the pyrolysis procedure. Biochar has proven potential as an efficient adsorbent that eliminates organic as well as inorganic pollutants from soil. Biochar incorporation into the soil is well acknowledged for its capacity to enhance soil quality by promoting soil microbiome and enzyme activities, increasing the soil organic carbon (SOC) content, and improving nutrient availability and retention. The use of biochar as a soil amendment has produced encouraging improvements in the overall state of the soil. Moreover, biochar exhibits significant potential for nutrient capture and carbon (C) sequestration within soil systems. The physicochemical properties of biochar vary depending on the feedstock raw material and the pyrolysis conditions employed, such as temperature, duration, and activation methods. These factors play a crucial role in deterring the final characteristics of the produced biochar. This study sum-

---

S. S. Kumar (✉)
Crop Research Division, Ministry of Agriculture & Waterways (MOA & W), Rakiraki, Fiji

O. A. Wani
Division of Soil Science and Agricultural Chemistry, Faculty of Agriculture, Sher-e-Kashmir University of Agricultural Sciences and Technology of Kashmir, Srinagar, Jammu and Kashmir, India

A. R. Malik
Division of Fruit Science, Faculty of Horticulture, Sher-e-Kashmir University of Agricultural Sciences and Technology of Kashmir, Srinagar, Jammu and Kashmir, India

S. Kumar
Department of Soil Science & Agricultural Chemistry, Sardar Vallabhbhai Patel University of Agriculture and Technology, Meerut, Uttar Pradesh, India

R. Patel
Department of Soil Science & Agricultural Chemistry, Banaras Hindu University, Varanasi, Uttar Pradesh, India

S. Babu et al. (eds.), *Agricultural Diversification for Sustainable Food Production*, Sustainability Sciences in Asia and Africa,
https://doi.org/10.1007/978-981-97-7517-0_3

marizes recent research findings regarding the role of biochar toward food security amidst the changing climate and composition of biochar, focusing on its influence on C dynamics within soil. It also explores the contribution of biochar in enhancing soil productivity and its capacity for C storage, which have received limited attention in prior research endeavors. Our research aims to shed light on these aspects and provide an in-depth knowledge of the advantages and potential applications of biochar in agricultural and environmental contexts. The primary finding of our study indicates that elevated pyrolytic temperatures used during the production of biochar can potentially have negative effects on the availability of essential nutrients for plants. The release of nutrients in the soil by biochar is impacted by the specific selection of feedstock and the particular pyrolysis techniques employed. This emphasizes the significance of meticulous evaluation of feedstock and pyrolysis parameters to maximize the nutrient release potential of biochar in agricultural contexts. To ensure the economic feasibility of biochar manufacturing, it is essential to devise novel methodologies for producing customized biochar. These techniques should involve controlled pyrolysis processes and the utilization of C-based materials that are specifically designed for meeting sustainable agriculture. By tailoring the characteristics of biochar to align with specific agricultural needs, we can improve its efficacy as a soil enhancer and encourage its widespread use in a manner that is both cost-effective and sustainable. It is crucial for future research efforts to concentrate on optimizing production methodologies and assessing the suitability of biochar in local farming communities, thereby promoting the adoption of sustainable agricultural practices. By developing a more comprehensive comprehension of the intricate interplays between the attributes of biochar, soil health, and environmental systems, we can achieve substantial advancements in harnessing the potential of biochar as a valuable instrument for mitigating environmental pollution and improving soil productivity. Through continued research and exploration, we can unlock the full potential of biochar as a sustainable solution for addressing environmental challenges while simultaneously promoting agricultural sustainability and resilience.

Keywords Biochar · C-sequestration · Amendments · Temperature · Sustainability · Productivity

3.1 Introduction

Biochar is an emerging subject of interest and extensive scientific investigation due to its versatile applications in soil improvement and its potential as an effective adsorbent (Zhang et al. 2019; He et al. 2022). The generation of biochar is widely acknowledged as a method that promotes ecological sustainability (Goswami et al. 2022; Wani et al. 2023a, b). Nonetheless, it is vital to recognize that employing a range of biomass sources for the production of biochar can introduce significant disparities in the physical and chemical attributes of the resulting biochar materials

(Biederman and Harpole 2013; Janu et al. 2021). The characteristics of biochar, including its adsorption properties, surface area, porosity, alkalinity, and C content, are influenced by both the choice of biomass feedstock and the specific conditions employed during the production process. These elements play a substantial role in determining the functional properties of biochar (Inyang and Dickenson 2015; Murtaza et al. 2022; Layek et al. 2022). Biochar chemical structure displays the presence of chemically bonded heteroatoms that are linked to the hydro-C skeleton on its surface. Furthermore, the foundational structure of biochar comprises unsaturated C atoms that possess elevated concentrations of electron pairs, thus exerting a substantial influence on the chemisorption of oxygen (O) atoms. The complex configuration of chemically bonded heteroatoms on the biochar surface has a significant impact on the chemical surface quality of the biochar. The presence and distribution of these heteroatoms influence the biochar's reactivity, adsorption properties, and interactions with the surrounding environment. The specific arrangement of these atoms can determine the biochar's affinity for certain contaminants, nutrient availability, and its ability to support microbial activity in soil. Therefore, understanding and characterizing the chemical surface of biochar is crucial for assessing its potential applications and effects in various environmental and agricultural systems (Yang et al. 2018; Panahi et al. 2020).

The presence of acid and base-forming radicals characterizes the surface of biochar. Acidic radicals are linked to O-rich surfaces and possess an anion exchange capability. These acidic functional groups can interact with negatively charged ions, promoting adsorption and retention of anions within the matrix of biochar (Kemmou et al. 2018; Enaime et al. 2020). The positive attributes of biochar contribute to its effectiveness as a soil amendment for improving polluted soils and promoting soil health. These characteristics include a larger surface area, greater micropore volume, enhanced cation exchange capacity (CEC), and the presence of surface functional groups. These attributes collectively enhance biochar's efficacy in addressing soil contamination issues and promoting soil health (Uras et al. 2012; Novak et al. 2016). The utilization of biochar has the potential to bring about significant changes in various soil characteristics. This includes increasing soil pH and CEC, which subsequently affects the availability and retention of nutrients. Furthermore, the application of biochar can result in modifications to the composition of soil organic matter (SOM), potentially leading to positive priming effects that stimulate the mineralization of organic matter (OM) and impact nutrient dynamics in the soil. It is crucial to acknowledge that biochar application can also facilitate the leaching of SOC. These effects highlight the intricate interactions between biochar and soil properties, emphasizing the importance of carefully considering biochar application techniques and conducting long-term monitoring of soil processes (Lorenz and Lal 2014; Han et al. 2020; Rasul et al. 2022). To address the challenges posed by climate change (CC), various management strategies have been employed, including utilization of animal manure, composts, mulches, and sewage sludge amendments. These organic amendments can improve soil fertility, enhance nutrient availability, and promote soil structure and water retention. In addition to organic additives, the integration of inorganic chemicals, fertilizers, nanomaterials, and clay minerals has also been explored. Inorganic chemicals and fertilizers provide essential nutrients to plants, while nanomaterials offer potential benefits such as enhanced nutrient uptake efficiency and targeted delivery of agrochemicals. By combining these diverse management strategies, farmers and researchers aim to optimize soil health, promote sustainable agriculture, and mitigate CC impacts. Nevertheless, it is essential to consider the possible environmental repercussions and long-lasting effects of these additives, as well as their interactions with soil and ecosystems. Regular monitoring and adherence to best management practices are crucial to ensure the effectiveness and sustainability of these strategies (Sujatha et al. 2021; Rogiers et al. 2022). Nevertheless, these approaches frequently exhibit limited efficacy in terms of soil C storage due to the rapid decomposition of SOC, resulting in carbon dioxide (CO_2) emission and an imbalance in the C cycle (Wang et al. 2022; Wani et al. 2023a, b). It is essential to recognize that when composts and sewage sludge are utilized as soil amendments, they have the potential to introduce phytopathogens, potentially hazardous metals, and even residues of drugs or antibiotics into the soil (Kacprzak et al. 2022). Furthermore, it is important to note that when composts and manures are used as soil amendments, there is a possibility of increased greenhouse gas (GHG) emissions, e.g., NH_3, methane (CH_4), and nitrous oxide (N_2O). The release of these GHGs has the capacity to pollute both groundwater and surface water while also exacerbating CC by adding to the overall GHG emissions (Elbasiouny et al. 2020; Sohil and Kichloo 2023).

The utilization of a novel category of biochar-based composites commonly referred to as engineered or modified biochar holds significant promise for environmental management and the mitigation of pollution in multi-metal soils (Chen et al. 2023). Multiple studies showed evidence of the potential of these composites to regulate soil pH, sequester C, enhance soil aggregation, and facilitate long-term stabilization of heavy metals in such soils. Engineered biochar, like conventional biochar, is a porous and C-rich material made by controlled pyrolysis of diverse biomass sources and C-based materials (Song et al. 2022). Several factors determine the chemical composition of engineered biochar, namely pyrolysis conditions, biomass feedstock composition, reaction atmosphere, inclusion of metal oxides, and nanometals. These factors collectively influence the chemical properties and potential applications of engineered biochar (Bolan et al. 2022). Engineered biochar exhibits superior qualities compared to regular biochar, such as a greater surface area, more efficient functional groups, and enhanced adsorption capacity and reactivity. These superior attributes contribute to the increased performance and versatility of engineered biochar in various applications (Zhang et al. 2022). Consequently, the development of enhanced supported metal oxide or nanometal systems can effectively address significant limitations and potentially improve soil properties. The unique physicochemical properties of biochar-based composites have been extensively investigated for their potential as efficient bioadsorbents, with reactivity and sorption capacity being contingent upon the fractions of carbonized or non-carbonized components (Pal et al. 2021).

Biochar has garnered significant attention for its potential to enhance food security by improving soil health and agricultural productivity. When applied to soils, biochar can significantly enhance soil fertility by increasing nutrient retention, water holding capacity, and promoting beneficial microbial activity, which are all critical factors for healthy crop growth (Khan et al. 2024). Its porous structure allows it to hold nutrients and water more effectively than typical soils, reducing the need for chemical fertilizers and irrigation (Ayaz et al. 2021). Additionally, biochar's long-term C-sequestration capabilities play a crucial role in mitigating CC, which indirectly supports food security by stabilizing agricultural ecosystems and reducing the vulnerability of crops to extreme weather conditions (Nyambo et al. 2020). In regions with degraded soils, biochar amendments have been particularly effective in restoring soil health and increasing crop yields. For instance, studies have shown that biochar can improve soil pH, enhance CEC, and reduce nutrient leaching, leading to more sustainable and productive agricultural practices (Troy et al. 2014). These improvements are especially beneficial in tropical and subtropical regions, where soil degradation is a major challenge. By enhancing soil structure and fertility, biochar helps ensure a more stable and resilient food supply, which is crucial for food security in vulnerable regions (Kundu and Kumar 2024). The use of biochar aligns with sustainable agricultural practices by promoting a circular economy. The production of biochar from agricultural waste not only reduces waste but also creates a valuable soil amendment that can improve crop yields and enhance soil health (Agegnehu et al. 2017). This sustainable approach helps smallholder farmers, who are often the most affected by food insecurity, to achieve better yields and more reliable food production, thereby contributing to local and global food security (Mekuria and Noble 2013).

Recent research studies have generated conflicting findings regarding the impact of temperature and feedstock selection for biochar production on nutrient phytoavailability and crop productivity (Tsai and Chang 2019; Karim et al. 2019). For instance, Hussain et al. (2016) reported a diverse range of variations in crop yield index (−36% to 31%) and soil properties index (−21% to 101%) subsequent to biochar application. The study by Jeffery et al. (2017) indicated that these variances could be ascribed to the specific characteristics of the feedstock employed during the biochar production process. Moreover, several studies have reported negligible effects on nitrogen (N) levels, plant productivity, or nutrient uptake following the application of biochar to soil (Rajkovich et al. 2012). Based on their findings, Hussain et al. (2016) concluded that the utilization of biochar could lead to better crop yields in severely degraded lands characterized by low nutrient content. However, its effectiveness might be limited in already fertile and healthy soils. As a result, biochar serves as an effective bioremediation agent for degraded and infertile soils. To achieve successful management strategies aimed at enhancing crop productivity across diverse soil types, an integrated approach is needed that involves the combined application of biochar with the inclusion of chemical fertilizers (Hou et al. 2020; Gurmessa 2021; Al-Rumaihi et al. 2022; Gudade et al. 2022). Biochar application increases the yields of different crops. For example, maize yield increased by 90–150%, while water use efficiency (WUE) increased by 90–140% (Fan et al., 2017). The yield of groundnut increased by 23–24% and of wheat by 9–10% when the biochar compost was used (Agegnehu et al., 2015; Babu et al.

2023). Long-term (10 years) application of biochar at the rate of 50 Mg/ha increases maize yield by 1.2 Mg/ha over control (Kätterer et al. 2019). This indicates that biochar has a sustainable beneficial impact on crop yields and soil fertility. Further research is essential to obtain a comprehensive understanding of the outcomes associated with the implementation of this integrated approach (Brantley et al. 2016; Ali et al. 2016). Nevertheless, it is worth noting that there can be variations in results. For instance, a study conducted by Tammeorg et al. (2014) demonstrated no significant improvement in wheat productivity when a combination of meat bone meal biochar and chemical fertilizer was applied. Their findings indicate that under the specific experimental conditions of their study, the combined application of these substances may not result in enhanced wheat yields.

The specific characteristics of the feedstock used and the production conditions directly influence the unique properties and functions of biochar. The selection of feedstock, which includes the biomass or organic material used, plays a crucial role in defining the chemical composition, elemental content, and physical structure of the resultant biochar. Likewise, parameters like pyrolysis temperature, residence time, and heating rate during the production process greatly influence the porosity, surface area, and stability of the biochar.

Hence, it is essential to comprehend and consider the particular feedstock and production conditions to customize the properties of biochar in order to achieve desired objectives and maximize its efficacy as a soil amendment or environmental remediation tool (Zhao et al. 2013; Steiner et al. 2016; Mohammadi 2021; Pooja et al. 2024). However, the influence of biochar composition on soil quality has not been extensively investigated in the available literature. The specific influence of biochar composition, encompassing its chemical and physical properties, on various aspects of soil quality remains an understudied area. Therefore, there is a need for further research to comprehensively examine and understand the connection between biochar composition and its influence on soil health, nutrient availability, microbial activity, and other relevant soil properties. Conducting such investigations will provide valuable insights into the potential advantages and limitations of various biochar compositions in improving soil quality and fostering sustainable agricultural practices. Therefore, this review aims to address this research gap by examining the composition-dependent effects of biochar in key domains, including nutrient content and release, soil structure, and potential environmental consequences.

3.2 Current Climate Change Scenario and Role of Biochar

The current CC scenario is characterized by increasing global temperatures, rising sea levels, and more frequent and severe weather events, all largely driven by anthropogenic GHG emissions. According to the Intergovernmental Panel on Climate Change (IPCC), the Earth's average surface temperature has risen by approximately 1.2 °C since the preindustrial period, primarily due to CO_2 emissions from fossil fuel combustion, deforestation, and other human activities (Kumar et al. 2020). This warming trend is expected to continue, with potentially catastrophic impacts on ecosystems, biodiversity, and human societies if significant mitigation efforts are not implemented. One promising mitigation strategy is the use of biochar, a stable, C-rich material produced through the pyrolysis of organic biomass. Biochar has the potential to address CC in multiple ways. First, it sequesters C in a stable form that can persist in soils for hundreds to thousands of years, thus reducing the amount of CO_2 in the atmosphere (Arif et al. 2020). This long-term C-sequestration makes biochar a valuable tool in efforts to achieve negative emissions, which are essential for limiting global warming to below 2 °C, as outlined in the Paris Agreement (Rogelj et al. 2018). Moreover, biochar application to soils can improve soil health and fertility, which can enhance crop productivity and resilience to CC. Biochar improves soil structure, increases water retention, and enhances nutrient availability, all of which are crucial for maintaining agricultural productivity in the face of changing climate conditions (Diatta et al. 2020). For instance, in drought-prone areas, biochar's ability to retain water can help crops survive periods of water scarcity, thereby stabilizing food production and reducing vulnerability to climate-induced food insecurity (Biederman and Harpole 2013). Biochar also contributes to CC mitigation by reducing GHG emissions from soils. Studies have shown that biochar can lower emissions of N_2O and CH_4, two potent GHGs, by altering soil microbial activity and nutrient cycling (Cayuela et al. 2014). This reduction in soil emissions can significantly enhance the overall GHG mitigation potential of biochar. Additionally, biochar production can be integrated into sustainable waste management systems. By converting agricultural residues, forestry waste, and other biomass materials into biochar, we can reduce the amount of organic waste that decomposes anaerobically to produce CH_4, a potent GHG. This process not only reduces methane emissions but also produces a valuable soil amendment that can improve soil health and agricultural productivity (Woolf et al. 2010). Furthermore, biochar's role in CC mitigation is not limited to terrestrial ecosystems. Research is exploring the potential for biochar to enhance C-sequestration in aquatic systems and to be used in the remediation of polluted water bodies, thus contributing to broader environmental benefits (Liu et al. 2016). These findings collectively underscore biochar's multifaceted benefits and its essential role in climate change mitigation and sustainable agriculture. To fully leverage the potential of biochar, further research and large-scale implementation are essential. Policymakers should consider supporting biochar initiatives through subsidies and incentives, while farmers and agricultural stakeholders need to be educated on its benefits and application techniques. Additionally, integrating biochar production with sustainable waste management systems can enhance its environmental impact. By prioritizing biochar, we can make significant strides toward mitigating climate change, enhancing soil health, and securing global food systems for future generations.

3.3 Enhancing Soil Fertility: The Impact of Biochar Amendment

The unique properties of biochar stand out in contrast to its original materials, and these differences are influenced by various factors during the pyrolysis process. These factors include the levels of O, the maximum temperature reached, the pressure involved, the duration of heating, and the rate at which the heating process occurs (Crombie et al. 2013; Luo et al. 2015). When Douglas fir bark is subjected to pyrolysis at 623 °C, the resulting biochar shows the presence of amorphous aromatic compounds and graphene sheets. These aromatic rings contain elements such as O, nitrogen (N), phosphorus (P), and sulfur (S), and their relative concentrations appear to be influenced by both the pyrolysis temperature and the specific biomass source utilized (McElligott et al. 2011; Tripathi et al. 2016). The heterogeneous and reactive nature of biochar arises from the inclusion of heteroatoms within its aromatic rings, which in turn influences its surface properties (Zhang et al. 2023a). Cayuela et al. (2015) revealed that biochar possesses a lower hydrogen (H)-to-C ratio compared to its feedstock, suggesting polymerization and potential recalcitrance. The use of biochar as a soil amendment shows potential for improving various soil characteristics, including soil aeration, soil structure, water retention capacity, microbial biomass C, enzyme activity, and SOC content (Irfan et al. 2019; Tauqeer et al. 2022; Bogale et al. 2023), along with nutrient retention and availability (Goldschmidt and Buffam 2023; Bilias et al. 2023). The utilization of biochar holds promise in reducing nutrient runoff from the soil's surface, thereby averting the downward migration of essential nutrients into groundwater sources (Lei et al. 2024). Table 3.1 provides a comprehensive summary of the effects of incorporating biochar into soil. Multiple research studies have emphasized the effectiveness of biochar as a soil amendment, as indicated in Table 3.1. Nevertheless, Schmidt et al. (2015) reported a decline in crop productivity following the application of biochar, which might be attributed to reduced soil C mineralization or diminished nutrient absorption by plants, as suggested by Sarma et al. (2017). The contradictory results observed regarding crop yields in soils amended with biochar are likely due to differences in biochar characteristics and soil properties. For example, elevated pyrolytic temperatures (500–800 °C) can decrease nutrient accessibility to plants by encouraging binding mechanisms and restricting improvements in adsorption capabilities (Chatterjee et al. 2020; Murtaza et al. 2023).

Table 3.1 Effect of biochar application on soil properties

Feedstock	Pyrolysis temperature (°C)	Soil type	Effect on soil properties	References
Pinus halepensis	450	Red slate	Promotes immobilization of potentially toxic elements and N biotransformation	Huang and He (2023)
Wheat straw	350–650	Clay texture	Increased water-stable aggregates and available water content up to 166%	Ghorbani et al. (2022)
Poultry litter	350–600	Sandy loam	Increased soil pH, SOM, and CEC	Masocha and Dikinya (2022)
Vine shoots	400–600	Calcisol and Cambisol	Enhanced WHC, K, Ca, and Mg contents by biochar addition	Videgain-Marco et al. (2020)
Solid municipal waste	350, 500, and 750	Sandy clay loam	Increased soil pH, improved EC, aggregate stability, water retention, and micronutrients	Randolph et al. (2017)
Weed biomass	400	Sandy clay loam	Contains 50% more SOC	Mandal et al. (2020)
Bamboo	600	Mineral soil	Affected the biogeochemical cycles such as release of CH_4 and N_2O from the soil	Liu et al. (2012)
Wheat straw	525	Planosol	Effectively improved soil aggregate stability	Burrell et al. (2016)
Birch	400	Silt loam	Increased soil porosity by 52.8%	Karhu et al. (2011)
Pine	630	Allophanic	Enhanced water retention capacity by 98.23%	Rahman et al. (2011)

3.4 Nutrient Dynamics in Biochar

Lignocellulosic biomass is composed of an intricate blend of naturally occurring carbohydrate polymers, primarily cellulose and hemicellulose, in addition to lignin, and it also contains minor quantities of other elements like extractives and ash (Eqbalpour et al. 2023; Zhang et al. 2023b). The makeup of biomass plays a pivotal role in shaping the outcome of product distribution in the pyrolysis process (Chen et al. 2019). Upon pyrolysis, each constituent component of the biomass undergoes degradation at specific temperatures, leading to the formation of distinct compounds. Biochar consists primarily of C and H, and it also contains essential plant nutrients like potassium (K), calcium (Ca), magnesium (Mg), and sodium (Na), as well as N.

The characteristics of biochar are significantly influenced by both the choice of feedstock and the specific conditions employed during the pyrolysis process, including factors such as temperature (Vijayaraghavan 2019; Ippolito et al. 2020). Many research studies have been conducted to investigate how the properties of biochar are influenced by pyrolysis temperature and the type of feedstock used, as depicted in Fig. 3.1. Research has indicated that higher pyrolysis temperatures result in a decrease in the H and O content of biochar, along with an increase in the concentration of aromatic C (Griffin et al. 2022). The application of biochar has been demonstrated to enhance the availability of essential nutrients like N, P, and K, resulting in substantial increases of up to 1.41-fold, 2.65-fold, and 2.60-fold, respectively (Karimi et al. 2020). Moreover, biochar exhibits impressive chemical stability and resilience against decomposition by both microbial and chemical processes within the soil. These properties can be attributed to its unique composition (Diatta et al. 2020). Additionally, with the escalation of pyrolysis temperature, the pH of biochar tends to rise. This increase in pH can be ascribed to the greater presence of alkaline elements, ash salts, and soluble/exchangeable cations within the biochar (Bilias et al. 2023). The distinctive porous framework of biochar, coupled with its extensive surface area, enables it to exhibit remarkable capabilities in absorbing both moisture and nutrients in the soil (Bolan et al. 2022). Conversely, elevated pyrolysis temperatures have been observed to have detrimental effects on the levels of acidic functional groups, volatile matter, biochar yield, and CEC (Ferjani et al. 2019; Balmuk et al. 2023). Moreover, the choice of feedstock significantly influences the concentrations of fixed C, total organic C (TOC), and mineral elements found in biochar (Das et al., 2021). Recent research efforts have concentrated on generating biochar from various sources and origins, including organic waste rich in nutrients, in order to investigate its potential uses in both environmental and agricultural contexts.

The moisture content (MC) of biomass plays a pivotal role in determining the outcome of pyrolysis, especially in relation to the yield of pyrolysis products. Increased humidity levels have been observed to decrease product yields (Hoang et al. 2021). The presence of moisture impacts temperature reduction within the reactor, resulting in a decrease in reaction rate and overall process efficiency. Energy is needed for the vaporization of the water contained within the biomass, which contributes to these outcomes (Yek et al. 2021; Thengane et al. 2022). In fast pyrolysis processes, it is generally preferable to utilize biomass with low MC, approximately 10%, to ensure that the heating rate is not hindered by the evaporation of

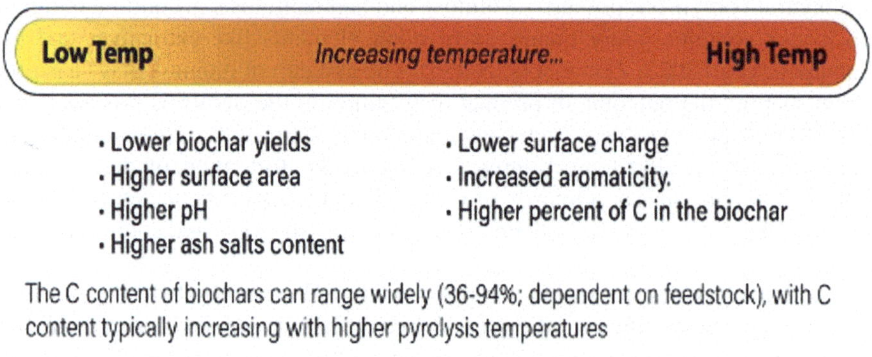

Fig. 3.1 Effect of biochar feedstock and pyrolysis temperature on biochar properties

water (Polin et al. 2019). Bulmău et al. (2010) reported higher MC in wood during pyrolysis results in increased gas fractions, leading to reduced production of liquid and solid fractions. The size of biomass particles is another crucial factor that influences the production of pyrolysis products, primarily because it offers a greater surface area for the process of thermal breakdown. This increased surface area facilitates the release of condensable gases and vapors, as discussed by Ryu et al. (2006), particularly when larger particle sizes are involved. Lower volatilization rates associated with larger particles allow for heat transfer between adjacent particles.

The equality of particle sizes is an additional factor related to particle size that significantly impacts the pyrolysis process. The presence of larger particles can create spaces between bed particles and result in the formation of preferential channels, leading to temperature fluctuations and reactor heterogeneity (Warnecke 2000). Therefore, ensuring a uniform particle size distribution is crucial in pyrolysis processes. In a study, Raj et al. (2013) used waste tires as biomass, and it was observed that smaller particle sizes, ranging from 0.3 to 1.18 mm, led to increased bio-oil yields and a decreased amount of solid waste (charcoal). Haydary et al. (2012) also observed that particle size significantly affected the pyrolysis time. When burning biomass with 5 mm particles at 550 °C, the decomposition occurred in less than 1 minute, whereas combustion of 50 mm particles took over 25 minutes. Larger particle sizes tend to generate more charcoal, so smaller particles are favored to maximize net yields (Sharma et al. 2019). The duration that biomass spends inside the reactor, known as residence time, is another factor that impacts pyrolysis yield. Studies have shown that shorter residence times yield higher results compared to longer residence times. This is due to shorter residence times preventing the formed bio-oil from remaining at high temperatures for an extended period, which could result in cracking and reduced bio-oil yield overall (Bridgwater 2012). Bulmău et al. (2010) explored residence times spanning from 5 to 30 minutes and noted a substantial inverse correlation between residence time and yield.

The temperature profile is a critical element in managing and refining pyrolysis processes. It plays a pivotal role in establishing key parameters like the heating rate, minimum and maximum temperatures, and residence time, all of which have a significant impact on the ultimate quality of the products (Muzyka et al. 2023). Precisely managing the temperature inside the reactor enables the customization of the pyrolysis process to attain specific objectives and enhance the production of particular products. The final pyrolysis temperature is a key factor significantly affecting the yields of bio-oil and biochar. Generally, higher temperatures tend to reduce the production of solid waste. Nonetheless, surpassing the typical maximum temperatures used in the pyrolysis process (>700 °C) can trigger additional decomposition reactions, leading to decreased net yields of the intended products (Torres-Herrador et al. 2021). Careful control of the pyrolysis temperature is essential to achieve optimal yields and desired product compositions.

Using pyrolysis to transform waste materials into biochar is widely recognized as an efficient method for recycling waste and increasing its value. During the pyrolysis process, waste materials undergo thermal decomposition at temperatures typically ranging from 300 °C to 800 °C, often in the absence of oxygen or with limited oxygen supply. As highlighted by Lehmann et al. (2015), this transformative process allows for the conversion of industrial or agricultural waste into multiple valuable forms, including solid biochar, gaseous fuels (such as CO, CO_2, H_2, CH_4, and other hydro-Cs), and liquid bio-oil, all of which can be used as valuable energy resources. Consequently, pyrolysis has emerged as a promising alternative for the treatment and utilization of industrial or agricultural waste. Furthermore, the applications of biochar go beyond waste management and encompass soil improvement, environmental cleanup, and even electricity generation (Patra et al. 2021; Mishra et al. 2023). The adaptability and potential advantages of biochar render it a valuable asset across various domains, playing a role in waste reduction, enhancing soil quality, and contributing to energy production.

3.5 Unraveling the Connection: Exploring the Nexus Between Biochar Composition and Nutrient Release

The advantageous effects of biochar on soil, plant, and microbial systems have been extensively substantiated, primarily ascribed to the nutrient content and the biochar's inherent ability to absorb and retain nutrients (Hou et al. 2022). However, it is important to consider the potential negative consequences of nutrient leaching, such as eutrophication in surface and groundwater systems. The nutrient composition of biochar exhibits considerable variability contingent upon the specific biomass source and the conditions of combustion (Mukherjee et al. 2011). Reports by Wu et al. (2011) suggest that approximately 15–20% of Ca, 10–60% of P, and less than 2% of other elements are easily soluble and leachable from mallee-wood biochar when subjected to distilled water. The rate of leaching varies according to temperature and the specific plant component under investigation. The interaction between biochar and soil type significantly influences the release of nutrients accessible to plants. As an illustration, when column leachate from Norfolk loamy soil was enriched with biochar produced from pecan shells at 700 °C, it demonstrated reduced levels of P, Ca, manganese (Mn), and zinc (Zn) in comparison to soil devoid of biochar, with P showing a decrease of approximately 35% (Yuan et al. 2016). This suggests that biochar may exchange multivalent cations for monovalent cations that are sorbed superficially. Hangs et al. (2016) reported N immobilization within stable biochar, which they attributed to the elevated C/N ratio of biochar and the confinement of N. Short-term experiments conducted by Nelson et al. (2011) furnish evidence supporting the application of N fertilization to soils amended with biochar to augment their N status.

The outcome from consecutive batch extractions of biochar, with the exception of those containing a high N content, consistently showed minimal N release, as detailed in Table 3.2. The composition of nutrients released from biochar into soil water displayed variations depending on the specific feedstocks employed. Among the leachates obtained from various biochar products, ammonium (NH_4) was the primary form of N, followed by organic N. The percentage contribution of nitrate-derived N ranged from 2% to 30%, while organic N content reached levels of up to

Table 3.2 Proportion of DOC, total N, and total P in biochar's (% nutrients lost)

Feedstock	Pyrolysis temperature (°C)	DOC	TN	P	Reference
Oak	250	0.9	1.5	151	Mukherjee and Zimmerman (2013)
Oak	650	0.1	0.5	22	
Grass	250	1.7	5.3	117	
Grass	650	0.4	3	39	

59%. In alignment with the results reported by Mukherjee and Zimmerman (2013), the concentrations of dissolved organic C, N, and P in soil solution water exhibited associations with the abundance of acid functional groups and the content of volatile matter within the biochar.

3.6 Exploring the Correlation Between Physicochemical Properties of Biochar and Nutrient Release

The capacity of particular biochar's to adsorb and potentially release organic matter (OM) or nutrients over time is likely impacted by variables like surface area, porosity, and ion exchange capacity. These variables play a pivotal role in shaping the interaction between biochar and its immediate surroundings, whether it be soil or other substrates (Nobaharan et al. 2021). Mukherjee et al. (2011) observed a wide range of CEC in biochars, varying from 0 to 70 cmol/kg, with higher values found in biochars developed at lower temperatures. Additionally, aged biochars have shown noteworthy levels of anion exchange capacity (AEC). As a result, fresh biochar's are anticipated to retain NH_4^+ while releasing exchangeable nitrate (NO_3^-) and phosphate (PO_4^{3-}). Wu et al. (2011) pointed out that there was no direct association between the BET surface area and catalyst activity. This observation implies that substantial alterations in the C structure have a more pronounced influence on catalyst activity. Nguyen et al. (2017) proposed that the biochar CEC is significantly correlated with NH_4^+-N adsorption, as supported by the BRT model. Changes in the composition of lignin, cellulose, and hemicellulose within biochar can alter its physical properties, thereby influencing the release of nutrients.

3.7 Utilization and Impact of Engineered Biochar

3.7.1 Influence on Soil Health and Carbon Sequestration Through Engineered Biochar Implementation

The occurrence of soil pollution has been linked to various incidents related to food safety, highlighting the importance of identifying reliable and safe soil remediation agents. Researchers, including Qadeer et al. (2017) and Dang et al. (2020), have suggested that biochar could potentially exert a positive influence on the transportability of specific inorganic and organic soil contaminants. Biochar exhibits noteworthy traits, such as its significant surface area, O-containing functional groups, and CECs, which empower it to proficiently capture, secure, and immobilize harmful heavy metals. Through the reduction of phytotoxicity and the availability of heavy metal ions in the soil, biochar plays a role in addressing soil contamination (Paz-Ferreiro et al. 2014). According to Beesley et al. (2011), the existence of water-soluble C and the highly alkaline pH linked to biochar can lead to the immobilization of particular elements. In a separate study, Rees et al. (2016) observed that the use of biochar primarily controls the movement of heavy metals in soil by elevating soil pH and enhancing intraparticle diffusion. Modified biochar variants display impressive adsorption capacities for both inorganic and organic contaminants. Nevertheless, the available literature on engineered biochar, which involves processing biochar under specific temperature and conditions to enhance its properties, is relatively scarce compared to pristine biochar (unmodified carbonaceous biochar), especially concerning its application for soil heavy metal remediation. The interplay between particulate organic matter and various soil types introduces disparities in nutrient accessibility and release. Nevertheless, despite the conduct of laboratory experiments, the translation and real-world implementation of these research discoveries remain constrained. Further inquiries are imperative to address aspects such as the production expenses of engineered biochar, its durability, collection, and regeneration procedures, alongside potential environmental hazards linked to its application. The resilience of OC within biochar is primarily contingent on the soil environment. For instance, biochar originating from wood displays greater durability in comparison to biochar derived from rice residues (Chowdhury et al. 2016). Bird et al. (2011) revealed that biochar produced from wood exhibits greater C stability, which can be attributed to its elevated lignin content when contrasted with biochar derived from crop residues. Kloss et al. (2012) have offered insights into the properties of biochar derived from biomass subjected to lower pyrolysis temperatures. It is worth noting that the stability of C in biochar is also influenced by the pyrolysis temperature, as it can impact the balance between aliphatic and aromatic C compounds, along with the quantity of condensed aromatic C present within the biochar structure (Du et al. 2018; Wang et al. 2021). Biochar made at elevated pyrolysis temperatures contains a higher concentration of aromatic C in comparison to biochar produced at lower pyrolysis temperatures. Consequently, biochar produced at lower pyrolysis temperatures displays decreased susceptibility to decomposition within the soil environment when compared to biochar generated at higher pyrolysis temperatures. The durability of biochar within the soil holds significant importance as it contributes to and sustains essential soil attributes necessary for optimal plant growth and productivity.

3.7.2 *Enhancing Soil Quality and Assessing Environmental Parameters Through Engineered Biochar Amendment*

The characteristics of engineered biochar are intricately linked to the pyrolysis procedure, the selection of feedstock, and the techniques employed for biochar transformation (Mishra et al. 2023). Engineered biochar is derived from different feedstocks obtained from the agricultural and forestry sectors. Nevertheless, there is an incomplete comprehension of how manufacturing conditions influence the attributes of engineered biochar, as highlighted by Mayer et al. (2014). A thorough understanding of the enduring stability of pyrolyzed biochar, as well as its reliance on production conditions and inherent traits, is imperative for the progression of engineered biochar (Sun et al. 2021). Henceforth, the adsorption properties of engineered biochar predominantly hinge on its unique surface area, pore arrangement, and surface functional groups. This awareness has prompted the development of diverse engineered biochar's tailored to meet various application needs. The performance of an engineered biochar adsorbent can be influenced by factors such as the attributes of the source materials (feedstocks), methods employed for biochar preparation, and specific process requisites, as indicated by Yaashikaa et al. (2019). By carefully selecting particular feedstocks or utilizing chemical, physical, or biological adjustments to the raw material, along with employing physiochemical or chemical–physical amalgamation techniques, it is feasible to regulate the pore structure of the carbonaceous adsorbent to some extent (Fig. 3.2).

The increasing enthusiasm for the creation and utilization of biochar has led to the emergence of multidisciplinary fields within the realms of environmental science and engineering. Biochar derived from various feedstock materials has demonstrated promising potential for treating contaminated soils, particularly in agricultural settings. Engineered biochar, subject to chemical oxidation to alter its characteristics, serves as a valuable tool in addressing environmental contamination and is widely applied in soil and air purification processes. Engineered biochars, as opposed to conventional biochars, demonstrate elevated levels of oxygen functional groups as a result of chemical oxidation (Fan et al. 2018). These functional groups, particularly those containing N or O atoms, enhance the reactivity of biochars. Moreover, engineered biochars contain reactive functional groups and nonaromatic carbon structures that form the labile fraction of biochar, rendering them susceptible to biotic decomposition in soil. Consequently, during the initial phases of aging, engineered biochars may display comparatively lower stability when compared to regular biochars due to these characteristics (Fig. 3.3). Duan et al. (2019) found that the effectiveness of engineered biochars in removing contaminants was notably improved when they were modified with Fe_3O_4. They also noted that the functions of biochars might decrease as they age. Moreover, Yaashikaa et al. (2020) underscored the activation of biochar as a promising strategy for the effective elimination of particular pollutants. They also emphasized the possibilities presented by closed-loop systems in biochar production, which can contribute to waste management and resource efficiency within the circular bioeconomy.

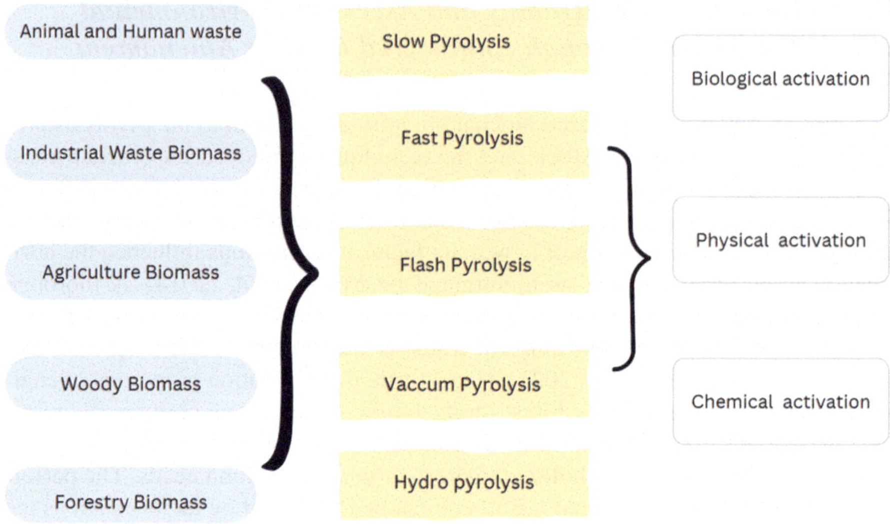

Fig. 3.2 Biochar-engineered production methods

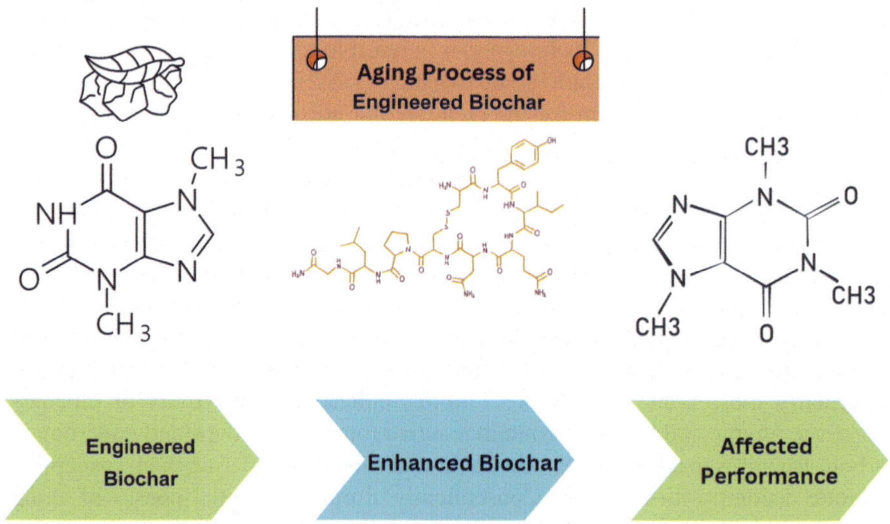

Fig. 3.3 The contaminant removal performance of EngBC

3.8 Constraints Associated with Biochar Utilization as a Soil Amendment: Considerations in Feedstock Selection, Production Conditions, and Biochar Stability

Numerous studies have highlighted various limitations associated with the implementation of biochars, despite their demonstrated beneficial aspects (Fig. 3.4). A notable drawback is the potential influence of biochar application on soil aging,

Fig. 3.4 Influence of biochar amendment in soil

which could impede natural soil aging processes. Therefore, periodic incorporation of alternative biomass into the soil may be required to maintain appropriate nutrient cycling. Anyanwu et al. (2018) have presented evidence indicating that biochar may have enduring adverse effects on soil microbiota and populations of earthworms. Similarly, aged biochar has been observed to restrict the underground root biomass in rice (*Oryza sativa*) and tomato (*Solanum lycopersicum*). Furthermore, biochar has the potential to reduce soil thermal conductivity due to its low thermal diffusivity (Zhao et al. 2016). However, it is crucial to emphasize that the positive impacts of biochar often hinge on the specific attributes of the soil to which it is applied. The efficacy of biochar as a soil amendment can fluctuate depending on the soil type (Siedt et al. 2021). Overapplication of biochar has been found to result in excessive weed growth, highlighting the need for careful dosage to avoid exacerbating weed control (Yang et al. 2019). The influence of biochar on crop productivity can also vary based on the particular crop species and the specific plant part being considered. For instance, although applying biochar at a certain rate improved the vegetative growth of tomato plants, it did not lead to a significant enhancement in overall productivity (Zhu et al. 2015). Additionally, the use of biochar has been linked to a delay in the flowering of plants (Hol et al. 2017). Biochar exhibits selectivity with its ability to adsorb pollutants, as evidenced by the fact that biochar supplementation did not impede the uptake of DDT pesticide in soil (Denyes et al. 2016). Biochar generated at higher temperatures typically contains greater ash content in comparison to biochar produced at lower temperatures. This elevated ash content could potentially have detrimental effects on crops cultivated in soils treated with high-temperature biochar (Al-Wabel et al. 2013). Furthermore, biochar's capacity to adsorb N and other essential nutrients, such as Fe, may negatively influence plant growth (Mohamed et al. 2017). Biochar can also engage in competition with soil for accessible nutrients, as observed when it is used in conjunction with P fertilizer in saline-sodic soil. This interaction can promote precipitation or sorption reactions of phosphate, leading to a reduction in the availability of P to plants (Enaime and Lübken 2021). The application of biochar to the soil influences the decomposition of SOM, resulting in a decrease in the populations of fungi, particularly *Ascomycota* and *Basidiomycota* (Jenkins et al. 2017). The origin of biochar-derived products is pivotal because the efficacy of these products can vary significantly depending on the specific type of biochar used. For example, coconut-husk-derived biochar greatly improves maize biomass, while orange-bagasse-derived biochar has a lesser effect at a similar dosage (Gonzaga et al. 2018). Contamination of biochar can likewise exert an adverse influence on plant growth (Jones and Quilliam 2014). Additionally, the availability of feedstock can influence the production costs of biochar. Shackley et al. (2011) estimated that the production of one ton of biochar, encompassing all supplementary expenses, could range between 148 and 389 British pounds. Additionally, regulatory demands and the necessity for testing biomass feedstock can contribute to cost considerations.

3.9 Recent Advancements of Biochar Production to Enhance Soil C-Sequestration and Soil Fertility

Within the realm of materials, biochar signifies the residue produced when OM undergoes incomplete combustion and is subsequently employed in soil-related applications. However, these materials have garnered noteworthy attention in recent years due to their potential benefits and the objectives attainable through their production and use. Black C, constituting a significant portion of the global SOC content, represents a long-lasting reservoir that withstands microbial decomposition and is deposited as a result of historical fires (Lorenz and Lal 2014). Researchers have directed their attention towards enhancing this almost universally present pool with additional stable C to enhance soil quality and enhance C-sequestration. This initiative has garnered support from prominent climate scientists and environmental champions. Biochar serves as a human-made form of stable C designed specifically for soil application, drawing inspiration from the techniques employed by precolonial native Amazonian societies. These practices effectively transformed infertile soils into highly productive ones that have sustained their fertility for centuries (Gupta et al. 2020). The idea of biochar embodies a dual role: it acts as both a stable source of C-rich SOM and a method for restoring depleted soils to fertility. The objective of C-sequestration is pursued through a range of thermochemical conversion techniques, including pyrolysis, which produces biochar as a byproduct (Table 3.3). Resistant C, which includes particles released during the burning of fossil fuels, can constitute a substantial portion of the world's soil C content. Notably, Woolf et al. (2010) reported that augmenting biochar production offers the potential to diminish net anthropogenic greenhouse gas emissions by as much as 12% by substituting pyrolysis oils and gases for fossil fuels. These scientific

Table 3.3 Pyrolysis processes for converting biomass

Process	Temperature (°C)	Heating rate	End Products
Torrefaction	230–300	Very low	Biochar
Slow pyrolysis	380–530	Low	Biochar, bio-oils, gas
Fast pyrolysis	380–530	High	Bio-oils, relatively little biochar
Combustion	700–1400	Very high	Gas
Gasification	>750	Very high	Gas

findings underscore the promising role of biochar in C-sequestration and its implications for CC mitigation.

The future prospects of biochar for enhancing soil fertility are diverse and depend on various factors. Advancements in biochar production technology can enhance its effectiveness in bolstering soil C storage and fertility. Researchers have conducted studies to assess the impact of biochar application on soil fertility, C-sequestration, and the enhancement of crop production. Biochars derived from rice husk at 650 °C and corn stover at 550 °C through a slow pyrolysis process were utilized as soil amendments, alongside unpyrolyzed rice husks and corn stover serving as control treatments. The application of biochars resulted in reduced CO_2 emissions and increased total OC, WHC, and CEC in the soil. Consequently, there was an improvement in soil fertility, and the growth of eggplant crops, including parameters like height, leaf count, and fresh and dry weight. Furthermore, the potential for C mitigation and long-term sequestration was demonstrated through the use of biochar. Teodoro et al. (2020) emphasized the possibility of enhancing the composting process and preserving nutrient availability, diminishing metal mobility in soil, and creating better conditions for plant establishment through the combined application of biochar and compost. Murtaza et al. (2021) noted that incorporating biochar into soil can lead to reductions in soil bulk density, heightened porosity, enhanced infiltration rates, increased microbial activity, elevated pH levels, and improved WHC. Nevertheless, these effects exhibited variations contingent upon soil texture, and in certain instances, the addition of biochar had unfavorable consequences on soil properties and crop productivity. It is essential to recognize that the characteristics and impacts of biochar within agricultural ecosystems are influenced by factors such as biomass type and the conditions of the pyrolysis process. Joseph et al. (2021) reported the addition of biochar resulted in increased P and SOC content, reduced emissions of non-CO_2 greenhouse gases, and an overall improvement of sandy acidic soil by a range of 12–50%. Asadi et al. (2021) found that rice-husk biochar has the potential to reduce nitrate leaching and greenhouse gas emissions (CH_4, CO_2, N_2O) in paddy soil, depending on the pyrolysis heating rate and residence time. Allohverdi et al. (2021) highlighted the challenges in agricultural practices, such as improving water retention, enhancing microbiota in soils, and increasing fertilizer efficacy. Moreover, biochar, recognized as a nutrient-rich substance originating from biomass, has garnered attention as a soil amendment owing to its capacity to enhance soil quality, boost crop yields, and mitigate C emissions. Simiele et al. (2022) noted enhanced root length, surface area, and biomass (including roots, stems, and leaves) with the application of biochar at a 20% rate compared to the control group. Similarly, Hussain et al. (2022) found that utilizing green coconut-modified biochar led to the highest dry biomass of mustard plants in comparison to the control treatment.

3.10 Future Perspectives and Challenges

To guarantee the efficiency and cost-effectiveness of biochar as soil amendments for enhancing fertility and C-sequestration, several factors and viewpoints must be taken into account. These perspectives include optimizing biochar production conditions, predicting biochar decay in the field, understanding plant–microbe–soil–biochar interactions, and determining the sorption capabilities of biochar for nutrient storage. To optimize biochar production, it is crucial to establish suitable implementation rates and standardize production conditions based on factors such as feedstock quality, temperature, and pre- and posttreatment of the biochar. Further research is essential to comprehend the connection between these variables and their performance across diverse soils, crops, and environmental settings. The development of a standardized model for efficient biochar production presents a persistent challenge due to the distinct demands of various applications. Another complex aspect involves predicting the degradation timeline of biochar in the field under various crop rotation patterns. This can be investigated by examining the enduring phase of biochar, which persists in soil for millennia, and contrasting it with the labile phase, which undergoes more rapid decomposition. Field experiments conducted over extended periods are imperative to illustrate the stability of biochar in soil. Understanding the interplay among plant roots, microorganisms, soil, and biochar within the rhizosphere is of paramount importance for elucidating the dynamics of nutrient biogeochemical cycling and release in biochar-amended soil environments. Additionally, it is crucial to ascertain the adsorption and desorption capacities of biochar for soil nutrients to anticipate nutrient accessibility and gradual nutrient delivery to plants within biochar-soil systems. These capacities can vary depending on the specific biochar type employed; hence, diverse biochar types should be assessed in varying soil contexts characterized by distinct properties and crop varieties. In summary, optimizing biochar production, predicting decay in the field, understanding plant–microbe–soil–biochar interactions, and determining nutrient adsorption and desorption capacities are important areas of research to ensure the effective use of biochar in improving soil fertility and C storage.

3.11 Conclusion

The utilization of biochar stands as a vital strategy for sustaining soil health while playing a pivotal role in enhancing, stabilizing, and conserving the essential soil properties required for crop cultivation in a sustainable manner. Nonetheless, there is a pressing need for further research to elucidate the connection between the raw materials employed in biochar production and their influence on enhancing soil nutrients. This becomes particularly pertinent in light of recent advancements in biochar manufacturing methods. Emphasis should be placed on standardizing the production of biochar, forecasting its decomposition, exploring the intricate interactions of biochar within the various elements of the rhizosphere (encompassing roots, microbes, and soil), and probing into biochar's capacity for adsorption and desorption of soil nutrients. These research endeavors aim to establish biochar as an efficient alternative to traditional fertilization approaches and as a potential solution for C-sequestration. Conducting long-term studies is paramount to scrutinize the consequences of activated biochar, in conjunction with cost-effective and environmentally friendly nanomaterials/minerals, on mineralizable N, soil health, and the sustainable production of crops.

References

Agegnehu G, Bass AM, Nelson PN, Muirhead B, Wright G, Bird MI (2015) Biochar and biochar-compost as soil amendments: effects on peanut yield, soil properties and greenhouse gas emissions in tropical North Queensland, Australia. Agric Ecosyst Environ 213:72–85

Agegnehu G, Srivastava AK, Bird MI (2017) The role of biochar and biochar-compost in improving soil quality and crop performance: a review. Appl Soil Ecol 119:156–170. https://doi.org/10.1016/j.apsoil.2017.06.008

Ali F, Amin M, Cobanoglu C (2016) An integrated model of service experience, emotions, satisfaction, and price acceptance: an empirical analysis in the Chinese hospitality industry. J Hosp Market Manag 25(4):449–475. https://doi.org/10.1080/19368623.2015.1019172

Allohverdi T, Mohanty AK, Roy P, Misra M (2021) A review on current status of biochar uses in agriculture. Molecules 26(18):5584. https://doi.org/10.3390/molecules26185584

Al-Rumaihi A, Shahbaz M, Mckay G, Mackey H, Al-Ansari T (2022) A review of pyrolysis technologies and feedstock: a blending approach for plastic and biomass towards optimum biochar yield. Renew Sust Energ Rev 167:112715. https://doi.org/10.1016/j.rser.2022.112715

Al-Wabel MI, Al-Omran A, El-Naggar AH, Nadeem M, Usman AR (2013) Pyrolysis temperature induced changes in characteristics and chemical composition of biochar produced from conocarpus wastes. Bioresour Technol 131:374–379. https://doi.org/10.1016/j.biortech.2012.12.165

Anyanwu IN, Alo MN, Onyekwere AM, Crosse JD, Nworie O, Chamba EB (2018) Influence of biochar aged in acidic soil on ecosystem engineers and two tropical agricultural plants. Ecotoxicol Environ Saf 153:116–126. https://doi.org/10.1016/j.ecoenv.2018.02.005

Arif M, Jan T, Riaz M, Fahad S, Adnan M, Amanullah, Rasul F (2020) Biochar; a remedy for climate change. Environ Clim Plant Veg Growth:151–171. https://doi.org/10.1007/978-3-030-49732-3_8

Asadi H, Ghorbani M, Rezaei-Rashti M, Abrishamkesh S, Amirahmadi E, Chengrong CHEN, Gorji M (2021) Application of rice husk biochar for achieving sustainable agriculture and environment. Rice Sci 28(4):325–343. https://doi.org/10.1016/j.rsci.2021.05.004

Ayaz M, Feizienė D, Tilvikienė V, Akhtar K, Stulpinaitė U, Iqbal R (2021) Biochar role in the sustainability of agriculture and environment. Sustain For 13(3):1330. https://doi.org/10.3390/su13031330

Babu S, Singh R, Kumar S, Rathore SS, Yadav D, Yadav SK, Singh VK (2023) Biochar implications in cleaner agricultural production and environmental sustainability. Environ Sci Adv 2(8):1042–1059. https://doi.org/10.1039/D2VA00324D

Balmuk G, Videgain M, Manyà JJ, Duman G, Yanik J (2023) Effects of pyrolysis temperature and pressure on agronomic properties of biochar. J Anal Appl Pyrolysis:105858. https://doi.org/10.1016/j.jaap.2023.105858

Beesley L, Moreno-Jiménez E, Gomez-Eyles JL, Harris E, Robinson B, Sizmur T (2011) A review of biochars' potential role in the remediation, revegetation and restoration of contaminated soils. Environ Pollut 159(12):3269–3282. https://doi.org/10.1016/j.envpol.2011.07.023

Biederman LA, Harpole WS (2013) Biochar and its effects on plant productivity and nutrient cycling: a meta-analysis. GCB Bioenergy 5(2):202–214. https://doi.org/10.1111/gcbb.12037

Bilias F, Kalderis D, Richardson C, Barbayiannis N, Gasparatos D (2023) Biochar application as a soil potassium management strategy: a review. Sci Total Environ 858:159782. https://doi.org/10.1016/j.scitotenv.2022.159782

Bird MI, Wurster CM, de Paula Silva PH, Bass AM, De Nys R (2011) Algal biochar–production and properties. Bioresour Technol 102(2):1886–1891. https://doi.org/10.1016/j.biortech.2010.07.106

Bogale AA, Melash AA, Percze A (2023) Symbiotic and asymmetric causality of the soil tillage system and biochar application on soil carbon sequestration and crop production. Soil Syst 7(2):48. https://doi.org/10.3390/soilsystems7020048

Bolan N, Hoang SA, Beiyuan J, Gupta S, Hou D, Karakoti A, Van Zwieten L (2022) Multifunctional applications of biochar beyond carbon storage. Int Mater Rev 67(2):150–200. https://doi.org/10.1080/09506608.2021.1922047

Brantley KE, Savin MC, Brye KR, Longer DE (2016) Nutrient availability and corn growth in a poultry litter biochar-amended loam soil in a greenhouse experiment. Soil Use Manag 32(3):279–288. https://doi.org/10.1111/sum.12296

Bridgwater AV (2012) Review of fast pyrolysis of biomass and product upgrading. Biomass Bioenergy 38:68–94. https://doi.org/10.1016/j.biombioe.2011.01.048

Bulmău C, Mărculescu C, Badea A, Apostol T (2010) Pyrolysis parameters influencing the biochar generation from wooden biomass. U Politeh Buch Ser C 72(1):29–38

Burrell LD, Zehetner F, Rampazzo N, Wimmer B, Soja G (2016) Long-term effects of biochar on soil physical properties. Geoderma 282:96–102. https://doi.org/10.1016/j.geoderma.2016.07.019

Cayuela ML, Van Zwieten L, Singh BP, Jeffery S, Roig A, Sánchez-Monedero MA (2014) Biochar's role in mitigating soil nitrous oxide emissions: a review and meta-analysis. Agric Ecosyst Environ 191:5–16. https://doi.org/10.1016/j.agee.2013.10.009

Cayuela ML, Jeffery S, van Zwieten L (2015) The molar H: Corg ratio of biochar is a key factor in mitigating N_2O emissions from soil. Agric Ecosyst Environ 202:135–138. https://doi.org/10.1016/j.agee.2014.12.015

Chatterjee R, Sajjadi B, Chen WY, Mattern DL, Hammer N, Raman V, Dorris A (2020) Effect of pyrolysis temperature on physicochemical properties and acoustic-based amination of biochar for efficient CO_2 adsorption. Front Energy Res 8:85. https://doi.org/10.3389/fenrg.2020.00085

Chen WH, Wang CW, Ong HC, Show PL, Hsieh TH (2019) Torrefaction, pyrolysis and two-stage thermodegradation of hemicellulose, cellulose and lignin. Fuel 258:116168. https://doi.org/10.1016/j.fuel.2019.116168

Chen Q, Zhang Z, Wang Y, Mu G, Wu X, Liu Y, Wen X (2023) Remediation of soil contaminated by heavy metals using biochar: strategies and future prospects. Pol J Environ Stud 32(1). https://doi.org/10.15244/pjoes/153912

Chowdhury ZZ, Karim MZ, Ashraf MA, Khalid K (2016) Influence of carbonization temperature on physicochemical properties of biochar derived from slow pyrolysis of durian wood (*Durio zibethinus*) sawdust. Bioresources 11(2):3356–3372

Crombie K, Mašek O, Sohi SP, Brownsort P, Cross A (2013) The effect of pyrolysis conditions on biochar stability as determined by three methods. GCB Bioenergy 5(2):122–131. https://doi.org/10.1111/gcbb.12030

Dang VM, Van HT, Duong HTM, Nguyen DH, Chao HP, Nguyen LH, Lin CC (2020) Evaluation of fly ash, apatite and rice straw derived-biochar in varying combinations for in situ remediation of soils contaminated with multiple heavy metals. Soil Sci Plant Nutr 66(2):379–388. https://doi.org/10.1080/00380768.2020.1725913

Das SK, Ghosh GK, Avasthe RK, Sinha K (2021) Compositional heterogeneity of different biochar: effect of pyrolysis temperature and feedstocks. J Environ Manag 278:111501. https://doi.org/10.1016/j.jenvman.2020.111501

Denyes MJ, Rutter A, Zeeb BA (2016) Bioavailability assessments following biochar and activated carbon amendment in DDT-contaminated soil. Chemosphere 144:1428–1434. https://doi.org/10.1016/j.chemosphere.2015.10.029

Diatta AA, Fike JH, Battaglia ML, Galbraith JM, Baig MB (2020) Effects of biochar on soil fertility and crop productivity in arid regions: a review. Arab J Geosci 13:1–17. https://doi.org/10.1007/s12517-020-05586-2

Du ZY, Zhang ZH, Xu C, Wang XB, Li WY (2018) Low-temperature steam reforming of toluene and biomass tar over biochar-supported Ni nanoparticles. ACS Sustain Chem Eng 7(3):3111–3119. https://doi.org/10.1021/acssuschemeng.8b04872

Duan W, Oleszczuk P, Pan B, Xing B (2019) Environmental behavior of engineered biochars and their aging processes in soil. Biochar 1:339–351. https://doi.org/10.1007/s42773-019-00030-5

Elbasiouny H, Elbanna BA, Al-Najoli E, Alsherief A, Negm S, Abou El-Nour E, ... Sharabash S (2020) Agricultural waste management for climate change mitigation: some implications to Egypt. In: Waste management in MENA regions, pp 149–169. https://doi.org/10.1007/978-3-030-18350-9_8

Enaime G, Lübken M (2021) Agricultural waste-based biochar for agronomic applications. Appl Sci 11(19):8914. https://doi.org/10.3390/app11198914

Enaime G, Baçaoui A, Yaacoubi A, Lübken M (2020) Biochar for wastewater treatment – conversion technologies and applications. Appl Sci 10(10):3492. https://doi.org/10.3390/app10103492

Eqbalpour M, Andooz A, Kowsari E, Ramakrishna S, Cheshmeh ZA, Chinnappan A (2023) A comprehensive review on how ionic liquids enhance the pyrolysis of cellulose, lignin, and lignocellulose toward a circular economy. Wiley Interdiscip Rev Energy Environ:e473. https://doi.org/10.1002/wene.473

Fan Q, Sun J, Chu L, Cui L, Quan G, Yan J, Iqbal M (2018) Effects of chemical oxidation on surface oxygen-containing functional groups and adsorption behavior of biochar. Chemosphere 207:33–40. https://doi.org/10.1016/j.chemosphere.2018.05.044

Fan Y, Ding R, Kang S, Hao X, Du T, Tong L, Li S (2017) Plastic mulch decreases available energy and evapotranspiration and improves yield and water use efficiency in an irrigated maize cropland. Agric Water Manag 179:122–131

Ferjani AI, Jeguirim M, Jellali S, Limousy L, Courson C, Akrout H, Bennici S (2019) The use of exhausted grape marc to produce biofuels and biofertilizers: effect of pyrolysis temperatures on biochars properties. Renew Sust Energ Rev 107:425–433. https://doi.org/10.1016/j.rser.2019.03.034

Ghorbani M, Amirahmadi E, Neugschwandtner RW, Konvalina P, Kopecký M, Moudrý J, Murindangabo YT (2022) The impact of pyrolysis temperature on biochar properties and its effects on soil hydrological properties. Sustain For 14(22):14722. https://doi.org/10.3390/su142214722

Goldschmidt A, Buffam I (2023) Biochar-amended substrate improves nutrient retention in green roof plots. Nat-Based Solut:100066. https://doi.org/10.1016/j.nbsj.2023.100066

Gonzaga MIS, Mackowiak C, de Almeida AQ, Wisniewski A Jr, de Souza DF, da Silva Lima I, de Jesus AN (2018) Assessing biochar applications and repeated *Brassica juncea* L. production cycles to remediate Cu contaminated soil. Chemosphere 201:278–285. https://doi.org/10.1016/j.chemosphere.2018.03.038

Goswami L, Kushwaha A, Singh A, Saha P, Choi Y, Maharana M, Kim BS (2022) Nano-biochar as a sustainable catalyst for anaerobic digestion: a synergetic closed-loop approach. Catalysts 12(2):186. https://doi.org/10.3390/catal12020186

Griffin GJ, Ward LP, Madapusi S, Shah KV, Parthasarathy R (2022) A study of chemical pretreatment and pyrolysis operating conditions to enhance biochar production from rice straw. J Anal Appl Pyrolysis 163:105455. https://doi.org/10.1016/j.jaap.2022.105455

Gudade BA, Malik GC, Das A, Babu S, Kumar A, Singh R, Bhupenchandra I (2022) Effect of biochar levels and integrated nutrient-management practices on agro-physiological performance and productivity of maize (*Zea mays*). Indian J Agron 67(4):380–385

Gupta DK, Gupta CK, Dubey R, Fagodiya RK, Sharma G, Noor Mohamed MB, Shukla AK (2020) Role of biochar in carbon sequestration and greenhouse gas mitigation. In: Biochar applications in agriculture and environment management, pp 141–165. https://doi.org/10.1007/978-3-030-40997-5_7

Gurmessa B (2021) Soil acidity challenges and the significance of liming and organic amendments in tropical agricultural lands with reference to Ethiopia. Environ Dev Sustain 23:77–99. https://doi.org/10.1007/s10668-020-00615-2

Han L, Sun K, Yang Y, Xia X, Li F, Yang Z, Xing B (2020) Biochar's stability and effect on the content, composition and turnover of soil organic carbon. Geoderma 364:114184. https://doi.org/10.1016/j.geoderma.2020.114184

Hangs RD, Ahmed HP, Schoenau JJ (2016) Influence of willow biochar amendment on soil nitrogen availability and greenhouse gas production in two fertilized temperate prairie soils. Bioenergy Res 9:157–171. https://doi.org/10.1007/s12155-015-9671-5

Haydary J, Jelemenský Ľ, Gašparovič L, Markoš J (2012) Influence of particle size and kinetic parameters on tire pyrolysis. J Anal Appl Pyrolysis 97:73–79. https://doi.org/10.1016/j.jaap.2012.07.003

He M, Xu Z, Hou D, Gao B, Cao X, Ok YS, Tsang DC (2022) Waste-derived biochar for water pollution control and sustainable development. Nat Rev Earth Environ 3(7):444–460. https://doi.org/10.1038/s43017-022-00306-8

Hoang AT, Ong HC, Fattah IR, Chong CT, Cheng CK, Sakthivel R, Ok YS (2021) Progress on the lignocellulosic biomass pyrolysis for biofuel production toward environmental sustainability. Fuel Process Technol 223:106997. https://doi.org/10.1016/j.fuproc.2021.106997

Hol WG, Vestergård M, ten Hooven F, Duyts H, van de Voorde TF, Bezemer TM (2017) Transient negative biochar effects on plant growth are strongest after microbial species loss. Soil Biol Biochem 115:442–451. https://doi.org/10.1016/j.soilbio.2017.09.016

Hou D, Bolan NS, Tsang DC, Kirkham MB, O'Connor D (2020) Sustainable soil use and management: an interdisciplinary and systematic approach. Sci Total Environ 729:138961. https://doi.org/10.1016/j.scitotenv.2020.138961

Hou J, Pugazhendhi A, Sindhu R, Vinayak V, Thanh NC, Brindhadevi K, Yuan D (2022) An assessment of biochar as a potential amendment to enhance plant nutrient uptake. Environ Res 214:113909. https://doi.org/10.1016/j.envres.2022.113909

Huang Z, He W (2023) Impacts of biochar and vermicompost addition on physicochemical characteristics, metal availability, and microbial communities in soil contaminated with potentially toxic elements. Sustain For 15(1):790. https://doi.org/10.3390/su15010790

Hussain S, Khan F, Cao W, Wu L, Geng M (2016) Seed priming alters the production and detoxification of reactive oxygen intermediates in rice seedlings grown under sub-optimal temperature and nutrient supply. Front Plant Sci 7:439. https://doi.org/10.3389/fpls.2016.00439

Hussain T, Ahmed SR, Lahori AH, Mierzwa-Hersztek M, Vambol V, Khan AA, Zengqiang Z (2022) In-situ stabilization of potentially toxic elements in two industrial polluted soils ameliorated with rock phosphate-modified biochars. Environ Pollut 309:119733. https://doi.org/10.1016/j.envpol.2022.119733

Inyang M, Dickenson E (2015) The potential role of biochar in the removal of organic and microbial contaminants from potable and reuse water: a review. Chemosphere 134:232–240. https://doi.org/10.1016/j.chemosphere.2015.03.072

Ippolito JA, Cui L, Kammann C, Wrage-Mönnig N, Estavillo JM, Fuertes-Mendizabal T, Borchard N (2020) Feedstock choice, pyrolysis temperature and type influence biochar characteristics: a comprehensive meta-data analysis review. Biochar 2:421–438. https://doi.org/10.1007/s42773-020-00067-x

Irfan M, Hussain Q, Khan KS, Akmal M, Ijaz SS, Hayat R, Rashid M (2019) Response of soil microbial biomass and enzymatic activity to biochar amendment in the organic carbon deficient arid soil: a 2-year field study. Arab J Geosci 12:1–9. https://doi.org/10.1007/s12517-019-4239-x

Janu R, Mrlik V, Ribitsch D, Hofman J, Sedláček P, Bielská L, Soja G (2021) Biochar surface functional groups as affected by biomass feedstock, biochar composition and pyrolysis temperature. Carbon Resour Convers 4:36–46. https://doi.org/10.1016/j.crcon.2021.01.003

Jeffery S, Abalos D, Prodana M, Bastos AC, Van Groenigen JW, Hungate BA, Verheijen F (2017) Biochar boosts tropical but not temperate crop yields. Environ Res Lett 12(5):053001. https://doi.org/10.1088/1748-9326/aa67bd

Jenkins JR, Viger M, Arnold EC, Harris ZM, Ventura M, Miglietta F, Taylor G (2017) Biochar alters the soil microbiome and soil function: results of next-generation amplicon sequencing across Europe. GCB Bioenergy 9(3):591–612. https://doi.org/10.1111/gcbb.12371

Jones DL, Quilliam RS (2014) Metal contaminated biochar and wood ash negatively affect plant growth and soil quality after land application. J Hazard Mater 276:362–370. https://doi.org/10.1016/j.jhazmat.2014.05.053

Joseph S, Cowie AL, Van Zwieten L, Bolan N, Budai A, Buss W, Lehmann J (2021) How biochar works, and when it doesn't: a review of mechanisms controlling soil and plant responses to biochar. GCB Bioenergy 13(11):1731–1764. https://doi.org/10.1111/gcbb.12885

Kacprzak M, Kupich I, Jasinska A, Fijalkowski K (2022) Bio-based waste'substrates for degraded soil improvement – advantages and challenges in European context. Energies 15(1):385. https://doi.org/10.3390/en15010385

Karhu K, Mattila T, Bergström I, Regina K (2011) Biochar addition to agricultural soil increased CH4 uptake and water holding capacity–results from a short-term pilot field study. Agric Ecosyst Environ 140(1–2):309–313. https://doi.org/10.1016/j.agee.2010.12.005

Karim AA, Kumar M, Mohapatra S, Singh SK (2019) Nutrient rich biomass and effluent sludge wastes co-utilization for production of biochar fertilizer through different thermal treatments. J Clean Prod 228:570–579. https://doi.org/10.1016/j.jclepro.2019.04.330

Karimi A, Moezzi A, Chorom M, Enayatizamir N (2020) Application of biochar changed the status of nutrients and biological activity in a calcareous soil. J Soil Sci Plant Nutr 20:450–459. https://doi.org/10.1007/s42729-019-00129-5

Kätterer T, Roobroeck D, Andrén O, Kimutai G, Karltun E, Kirchmann H, de Nowina KR (2019) Biochar addition persistently increased soil fertility and yields in maize-soybean rotations over 10 years in sub-humid regions of Kenya. Field Crop Res 235:18–26

Kemmou L, Frontistis Z, Vakros J, Manariotis ID, Mantzavinos D (2018) Degradation of antibiotic sulfamethoxazole by biochar-activated persulfate: factors affecting the activation and degradation processes. Catal Today 313:128–133. https://doi.org/10.1016/j.cattod.2017.12.028

Khan S, Irshad S, Mehmood K, Hasnain Z, Nawaz M, Rais A, Ibrar D (2024) Biochar production and characteristics, its impacts on soil health, crop production, and yield enhancement: a review. Plan Theory 13(2):166. https://doi.org/10.3390/plants13020166

Kloss S, Zehetner F, Dellantonio A, Hamid R, Ottner F, Liedtke V, Soja G (2012) Characterization of slow pyrolysis biochars: effects of feedstocks and pyrolysis temperature on biochar properties. J Environ Qual 41(4):990–1000. https://doi.org/10.2134/jeq2011.0070

Kumar SS, Mahale AG, Patil AC (2020) Mitigation of climate change through approached agriculture-soil carbon sequestration (a review). Curr J Appl Sci Technol 39(33):47–64. https://doi.org/10.9734/CJAST/2020/v39i3331017

Kundu B, Kumar R (2024) Enhancing crop resilience to climate change through biochar: a review. Int J Environ Clim Change 14(6):170–184. https://doi.org/10.9734/ijecc/2024/v14i64219

Layek J, Narzari R, Hazarika S, Das A, Rangappa K, Devi S, Mishra VK (2022) Prospects of biochar for sustainable agriculture and carbon sequestration: an overview for Eastern Himalayas. Sustain For 14(11):6684. https://doi.org/10.3390/su14116684

Lehmann J, Kuzyakov Y, Pan G, Ok YS (2015) Biochars and the plant-soil interface. Plant Soil 395:1–5. https://doi.org/10.1007/s11104-015-2658-3

Lei WU, Ruizhi LI, Yan WANG, Zongjun GUO, Jiaheng LI, Hang YANG, Xiaoyi MA (2024) Laboratory and numerical modelling of irrigation infiltration and nitrogen leaching in homogeneous soils. Pedosphere 34(1):146–158. https://doi.org/10.1016/j.pedsph.2023.03.015

Liu W, Chen S, Qin X, Baumann F, Scholten T, Zhou Z, Qin D (2012) Storage, patterns, and control of soil organic carbon and nitrogen in the Northeastern margin of the Qinghai–Tibetan Plateau. Environ Res Lett 7(3):035401. https://doi.org/10.1088/1748-9326/7/3/035401

Liu Z, Dugan B, Masiello CA, Barnes RT, Gallagher ME, Gonnermann H (2016) Impacts of biochar concentration and particle size on hydraulic conductivity and DOC leaching of biochar–sand mixtures. J Hydrol 533:461–472. https://doi.org/10.1016/j.jhydrol.2015.12.007

Lorenz K, Lal R (2014) Biochar application to soil for climate change mitigation by soil organic carbon sequestration. J Plant Nutr Soil Sci 177(5):651–670. https://doi.org/10.1002/jpln.201400058

Luo L, Xu C, Chen Z, Zhang S (2015) Properties of biomass-derived biochars: combined effects of operating conditions and biomass types. Bioresour Technol 192:83–89. https://doi.org/10.1016/j.biortech.2015.05.054

Mandal M, Kamp P, Singh M (2020) Effect of long term manuring on carbon sequestration potential and dynamics of soil organic carbon labile pool under tropical rice-rice agro-ecosystem. Commun Soil Sci Plant Anal 51(4):468–480. https://doi.org/10.1080/00103624.2020.1718690

Masocha BL, Dikinya O (2022) The role of poultry litter and its biochar on soil fertility and *Jatropha curcas* L. growth on sandy-loam soil. Appl Sci 12(23):12294. https://doi.org/10.3390/app122312294

Mayer ZA, Eltom Y, Stennett D, Schröder E, Apfelbacher A, Hornung A (2014) Characterization of engineered biochar for soil management. Environ Prog Sustain Energy 33(2):490–496. https://doi.org/10.1002/ep.11788

McElligott KM, Page-Dumroese DS, Coleman M, McElligott K (2011) Bioenergy production systems and biochar application in forests: potential for renewable energy, soil enhancement, and carbon sequestration. US Department of Agriculture, Forest Service, Rocky Mountain Research Station, Fort Collins. 14 p

Mekuria W, Noble A (2013) The role of biochar in ameliorating disturbed soils and sequestering soil carbon in tropical agricultural production systems. Appl Environ Soil Sci 2013(1):354965. https://doi.org/10.1155/2013/354965

Mishra RK, Kumar DJP, Narula A, Chistie SM, Naik SU (2023) Production and beneficial impact of biochar for environmental application: a review on types of feedstocks, chemical compositions, operating parameters, techno-economic study, and life cycle assessment. Fuel 343:127968. https://doi.org/10.1016/j.fuel.2023.127968

Mohamed BA, Ellis N, Kim CS, Bi X (2017) The role of tailored biochar in increasing plant growth, and reducing bioavailability, phytotoxicity, and uptake of heavy metals in contaminated soil. Environ Pollut 230:329–338. https://doi.org/10.1016/j.envpol.2017.06.075

Mohammadi A (2021) Overview of the benefits and challenges associated with pelletizing biochar. PRO 9(9):1591. https://doi.org/10.3390/pr9091591

Mukherjee A, Zimmerman AR (2013) Organic carbon and nutrient release from a range of laboratory-produced biochars and biochar–soil mixtures. Geoderma 193:122–130. https://doi.org/10.1016/j.geoderma.2012.10.002

Mukherjee A, Zimmerman AR, Harris W (2011) Surface chemistry variations among a series of laboratory-produced biochars. Geoderma 163(3–4):247–255. https://doi.org/10.1016/j.geoderma.2011.04.021

Murtaza G, Ahmed Z, Usman M, Tariq W, Ullah Z, Shareef M, Ditta A (2021) Biochar induced modifications in soil properties and its impacts on crop growth and production. J Plant Nutr 44(11):1677–1691. https://doi.org/10.1080/01904167.2021.1871746

Murtaza G, Ahmed Z, Usman M (2022) Feedstock type, pyrolysis temperature and acid modification effects on physiochemical attributes of biochar and soil quality. Arab J Geosci 15(3):305. https://doi.org/10.1007/s12517-022-09539-9

Murtaza G, Ahmed Z, Eldin SM, Ali I, Usman M, Iqbal R, Tariq A (2023) Biochar as a green sorbent for remediation of polluted soils and associated toxicity risks: a critical review. Separations 10(3):197. https://doi.org/10.3390/separations10030197

Muzyka R, Misztal E, Hrabak J, Banks SW, Sajdak M (2023) Various biomass pyrolysis conditions influence the porosity and pore size distribution of biochar. Energy 263:126128. https://doi.org/10.1016/j.energy.2022.126128

Nelson NO, Agudelo SC, Yuan W, Gan J (2011) Nitrogen and phosphorus availability in biochar-amended soils. Soil Sci 176(5):218–226. https://doi.org/10.1097/SS.0b013e3182171eac

Nguyen TTN, Xu CY, Tahmasbian I, Che R, Xu Z, Zhou X, Bai SH (2017) Effects of biochar on soil available inorganic nitrogen: a review and meta-analysis. Geoderma 288:79–96. https://doi.org/10.1016/j.geoderma.2016.11.004

Nobaharan K, Bagheri Novair S, Asgari Lajayer B, van Hullebusch ED (2021) Phosphorus removal from wastewater: the potential use of biochar and the key controlling factors. Water 13(4):517. https://doi.org/10.3390/w13040517

Novak JM, Ippolito JA, Lentz RD, Spokas KA, Bolster CH, Sistani K, Johnson MG (2016) Soil health, crop productivity, microbial transport, and mine spoil response to biochars. Bioenergy Res 9:454–464. https://doi.org/10.1007/s12155-016-9720-8

Nyambo P, Mupambwa HA, Nciizah AD (2020) Biochar enhances the capacity of climate-smart agriculture to mitigate climate change. In: Handbook of climate change management: research, leadership, transformation, pp 1–18. https://doi.org/10.1007/978-3-030-22759-3_319-1

Pal DB, Singh A, Jha JM, Srivastava N, Hashem A, Alakeel MA, Gupta VK (2021) Low-cost biochar adsorbents prepared from date and delonix regia seeds for heavy metal sorption. Bioresour Technol 339:125606. https://doi.org/10.1016/j.biortech.2021.125606

Panahi HKS, Dehhaghi M, Ok YS, Nizami AS, Khoshnevisan B, Mussatto SI, Lam SS (2020) A comprehensive review of engineered biochar: production, characteristics, and environmental applications. J Clean Prod 270:122462. https://doi.org/10.1016/j.jclepro.2020.122462

Patra BR, Mukherjee A, Nanda S, Dalai AK (2021) Biochar production, activation and adsorptive applications: a review. Environ Chem Lett 19:2237–2259. https://doi.org/10.1007/s10311-020-01165-9

Paz-Ferreiro J, Lu H, Fu S, Mendez A, Gasco G (2014) Use of phytoremediation and biochar to remediate heavy metal polluted soils: a review. Solid Earth 5(1):65–75. https://doi.org/10.5194/se-5-65-2014

Polin JP, Carr HD, Whitmer LE, Smith RG, Brown RC (2019) Conventional and autothermal pyrolysis of corn stover: overcoming the processing challenges of high-ash agricultural residues. J Anal Appl Pyrolysis 143:104679. https://doi.org/10.1016/j.jaap.2019.104679

Pooja D, Prasad S, Singh B, Shrivastava M, Babu S, Vashisth A, Avinash C (2024) Effect of bioaugmented Linz-Donawitz slag and biochar on physiological and yield attributes of wheat (*Triticum aestivum*). Indian J Agric Sci 94(1):021–025. https://doi.org/10.56093/ijas.v94i1.140177

Qadeer S, Anjum M, Khalid A, Waqas M, Batool A, Mahmood T (2017) A dialogue on perspectives of biochar applications and its environmental risks. Water Air Soil Pollut 228:1–26. https://doi.org/10.1007/s11270-017-3428-z

Rahman MH, Holmes AW, Saunders SJ, Islam KR (2011) Biochar impacts on physical and hydrological properties of allophanic soils. In: Proceedings of 2011 Biochar Workshop, Palmerston North, New Zealand, pp 10–11

Raj RE, Kennedy ZR, Pillai BC (2013) Optimization of process parameters in flash pyrolysis of waste tyres to liquid and gaseous fuel in a fluidized bed reactor. Energy Convers Manag 67:145–151. https://doi.org/10.1016/j.enconman.2012.11.012

Rajkovich S, Enders A, Hanley K, Hyland C, Zimmerman AR, Lehmann J (2012) Corn growth and nitrogen nutrition after additions of biochars with varying properties to a temperate soil. Biol Fertil Soils 48:271–284. https://doi.org/10.1007/s00374-011-0624-7

Randolph P, Bansode RR, Hassan OA, Rehrah DJ, Ravella R, Reddy MR, Ahmedna M (2017) Effect of biochars produced from solid organic municipal waste on soil quality parameters. J Environ Manag 192:271–280. https://doi.org/10.1016/j.jenvman.2017.01.061

Rasul M, Cho J, Shin HS, Hur J (2022) Biochar-induced priming effects in soil via modifying the status of soil organic matter and microflora: a review. Sci Total Environ 805:150304. https://doi.org/10.1016/j.scitotenv.2021.150304

Rees F, Sterckeman T, Morel JL (2016) Root development of non-accumulating and hyperaccumulating plants in metal-contaminated soils amended with biochar. Chemosphere 142:48–55. https://doi.org/10.1016/j.chemosphere.2015.03.068

Rogelj J, Shindell D, Jiang K, Fifita S, Forster P, Ginzburg V, … Zickfeld K. (2018) Mitigation pathways compatible with 1.5 C in the context of sustainable development. In: Global warming of 1.5 C. Intergovernmental Panel on Climate Change, pp 93–174

Rogiers SY, Greer DH, Liu Y, Baby T, Xiao Z (2022) Impact of climate change on grape berry ripening: an assessment of adaptation strategies for the Australian vineyard. Front Plant Sci 13. https://doi.org/10.3389/fpls.2022.1094633

Ryu C, Yang YB, Khor A, Yates NE, Sharifi VN, Swithenbank J (2006) Effect of fuel properties on biomass combustion: part I. Experiments – fuel type, equivalence ratio and particle size. Fuel 85(7–8):1039–1046. https://doi.org/10.1016/j.fuel.2005.09.019

Sarma B, Borkotoki B, Narzari R, Kataki R, Gogoi N (2017) Organic amendments: effect on carbon mineralization and crop productivity in acidic soil. J Clean Prod 152:157–166. https://doi.org/10.1016/j.jclepro.2017.03.124

Schmidt J, Schulz E, Michalzik B, Buscot F, Gutknecht JL (2015) Carbon input and crop-related changes in microbial biomarker levels strongly affect the turnover and composition of soil organic carbon. Soil Biol Biochem 85:39–50. https://doi.org/10.1016/j.soilbio.2015.02.024

Shackley S, Hammond J, Gaunt J, Ibarrola R (2011) The feasibility and costs of biochar deployment in the UK. Carbon Manag 2(3):335–356

Sharma HB, Panigrahi S, Dubey BK (2019) Hydrothermal carbonization of yard waste for solid bio-fuel production: study on combustion kinetic, energy properties, grindability and flowability of hydrochar. Waste Manag 91:108–119. https://doi.org/10.1016/j.wasman.2019.04.056

Siedt M, Schäffer A, Smith KE, Nabel M, Roß-Nickoll M, van Dongen JT (2021) Comparing straw, compost, and biochar regarding their suitability as agricultural soil amendments to affect soil structure, nutrient leaching, microbial communities, and the fate of pesticides. Sci Total Environ 751:141607. https://doi.org/10.1016/j.scitotenv.2020.141607

Simiele M, Argentino O, Baronti S, Scippa GS, Chiatante D, Terzaghi M, Montagnoli A (2022) Biochar enhances plant growth, fruit yield, and antioxidant content of cherry tomato (*Solanum lycopersicum* L.) in a soilless substrate. Agriculture 12(8):1135. https://doi.org/10.3390/agriculture12081135

Sohil A, Kichloo MA (2023) Sustainable solutions to animal waste: climate change mitigation and bioproduct harvest. In: Climate changes mitigation and sustainable bioenergy harvest through animal waste: sustainable environmental implications of animal waste. Springer Nature Switzerland, Cham, pp 301–332. https://doi.org/10.1007/978-3-031-26224-1_12

Song G, Qin F, Yu J, Tang L, Pang Y, Zhang C, Deng L (2022) Tailoring biochar for persulfate-based environmental catalysis: impact of biomass feedstocks. J Hazard Mater 424:127663. https://doi.org/10.1016/j.jhazmat.2021.127663

Steiner C, Bayode AO, Ralebitso-Senior TK (2016) Feedstock and production parameters: effects on biochar properties and microbial communities. In: Biochar application. Elsevier, pp 41–54. https://doi.org/10.1016/B978-0-12-803433-0.00002-3

Sujatha MP, Lathika C, Smitha JK (2021) Sustainable and efficient utilization of weed biomass for carbon farming and productivity enhancement: a simple, rapid and ecofriendly approach in the context of climate change scenario. Environ Chall 4:100150. https://doi.org/10.1016/j.envc.2021.100150

Sun Y, Xiong X, He M, Xu Z, Hou D, Zhang W, Tsang DC (2021) Roles of biochar-derived dissolved organic matter in soil amendment and environmental remediation: a critical review. Chem Eng J 424:130387. https://doi.org/10.1016/j.cej.2021.130387

Tammeorg P, Simojoki A, Mäkelä P, Stoddard FL, Alakukku L, Helenius J (2014) Short-term effects of biochar on soil properties and wheat yield formation with meat bone meal and inorganic fertiliser on a boreal loamy sand. Agric Ecosyst Environ 191:108–116. https://doi.org/10.1016/j.agee.2014.01.007

Tauqeer HM, Turan V, Farhad M, Iqbal M (2022) Sustainable agriculture and plant production by virtue of biochar in the era of climate change. In: Managing plant production

under changing environment. Springer Nature Singapore, Singapore, pp 21–42. https://doi.org/10.1007/978-981-16-5059-8_2
Teodoro M, Trakal L, Gallagher BN, Šimek P, Soudek P, Pohořelý M, Mohan D (2020) Application of co-composted biochar significantly improved plant-growth relevant physical/chemical properties of a metal contaminated soil. Chemosphere 242:125255. https://doi.org/10.1016/j.chemosphere.2019.125255
Thengane SK, Kung KS, Gomez-Barea A, Ghoniem AF (2022) Advances in biomass torrefaction: parameters, models, reactors, applications, deployment, and market. Prog Energy Combust Sci 93:101040. https://doi.org/10.1016/j.pecs.2022.101040
Torres-Herrador F, Eschenbacher A, Coheur J, Blondeau J, Magin TE, Van Geem KM (2021) Decomposition of carbon/phenolic composites for aerospace heatshields: detailed speciation of phenolic resin pyrolysis products. Aerosp Sci Technol 119:107079. https://doi.org/10.1016/j.ast.2021.107079
Tripathi M, Sahu JN, Ganesan P (2016) Effect of process parameters on production of biochar from biomass waste through pyrolysis: a review. Renew Sust Energ Rev 55:467–481. https://doi.org/10.1016/j.rser.2015.10.122
Troy SM, Lawlor PG, O'Flynn CJ, Healy MG (2014) The impact of biochar addition on nutrient leaching and soil properties from tillage soil amended with pig manure. Water Air Soil Pollut 225:1–15. https://doi.org/10.1007/s11270-014-1900-6
Tsai CC, Chang YF (2019) Carbon dynamics and fertility in biochar-amended soils with excessive compost application. Agronomy 9(9):511. https://doi.org/10.3390/agronomy9090511
Uras Ü, Carrier M, Hardie AG, Knoetze JH (2012) Physico-chemical characterization of biochars from vacuum pyrolysis of South African agricultural wastes for application as soil amendments. J Anal Appl Pyrolysis 98:207–213. https://doi.org/10.1016/j.jaap.2012.08.007
Videgain-Marco M, Marco-Montori P, Martí-Dalmau C, Jaizme-Vega MDC, Manyà-Cervelló JJ, García-Ramos FJ (2020) Effects of biochar application in a sorghum crop under greenhouse conditions: growth parameters and physicochemical fertility. Agronomy 10(1):104. https://doi.org/10.3390/agronomy10010104
Vijayaraghavan K (2019) Recent advancements in biochar preparation, feedstocks, modification, characterization and future applications. Environ Technol Rev 8(1):47–64. https://doi.org/10.1080/21622515.2019.1631393
Wang M, Zhao Z, Zhang Y (2021) Magnetite-contained biochar derived from Fenton sludge modulated electron transfer of microorganisms in anaerobic digestion. J Hazard Mater 403:123972. https://doi.org/10.1016/j.jhazmat.2020.123972
Wang R, Mattox CM, Phillips CL, Kowalewski AR (2022) Carbon sequestration in turfgrass–soil systems. Plants 11(19):2478. https://doi.org/10.3390/plants11192478
Wani OA, Akhter F, Kumar SS, Babu S, Kanth RH, Mir SA, Rathore SS (2023a) Mitigating soil erosion through biomass-derived biochar: exploring the influence of feedstock types and pyrolysis temperature. Land 12(12):2111. https://doi.org/10.3390/land12122111
Wani OA, Kumar SS, Hussain N, Wani AIA, Subhash BABU, Parvej ALAM, Mansoor S (2023b) Multi-scale processes influencing global carbon storage and land-carbon-climate nexus: a critical review. Pedosphere 33(2):250–267. https://doi.org/10.1016/j.pedsph.2022.07.002
Warnecke R (2000) Gasification of biomass: comparison of fixed bed and fluidized bed gasifier. Biomass Bioenergy 18(6):489–497. https://doi.org/10.1016/S0961-9534(00)00009-X
Woolf D, Amonette JE, Street-Perrott FA, Lehmann J, Joseph S (2010) Sustainable biochar to mitigate global climate change. Nat Commun 1(1):56. https://doi.org/10.1038/ncomms1053
Wu G, Nelson M, Ma S, Meng H, Cui G, Shen PK (2011) Synthesis of nitrogen-doped onion-like carbon and its use in carbon-based CoFe binary non-precious-metal catalysts for oxygen-reduction. Carbon 49(12):3972–3982. https://doi.org/10.1016/j.carbon.2011.05.036
Yaashikaa PR, Kumar PS, Varjani SJ, Saravanan A (2019) Advances in production and application of biochar from lignocellulosic feedstocks for remediation of environmental pollutants. Bioresour Technol 292:122030. https://doi.org/10.1016/j.biortech.2019.122030
Yaashikaa PR, Kumar PS, Varjani S, Saravanan A (2020) A critical review on the biochar production techniques, characterization, stability and applications for circular bioeconomy. Biotechnol Rep 28:e00570. https://doi.org/10.1016/j.btre.2020.e00570

Yang K, Jiang Y, Yang J, Lin D (2018) Correlations and adsorption mechanisms of aromatic compounds on biochars produced from various biomass at 700 C. Environ Pollut 233:64–70. https://doi.org/10.1016/j.envpol.2017.10.035

Yang X, Zhang S, Ju M, Liu L (2019) Preparation and modification of biochar materials and their application in soil remediation. Appl Sci 9(7):1365. https://doi.org/10.3390/app9071365

Yek PNY, Cheng YW, Liew RK, Mahari WAW, Ong HC, Chen WH, Lam SS (2021) Progress in the torrefaction technology for upgrading oil palm wastes to energy-dense biochar: a review. Renew Sust Energ Rev 151:111645. https://doi.org/10.1016/j.rser.2021.111645

Yuan H, Lu T, Wang Y, Chen Y, Lei T (2016) Sewage sludge biochar: nutrient composition and its effect on the leaching of soil nutrients. Geoderma 267:17–23. https://doi.org/10.1016/j.geoderma.2015.12.020

Zhang C, Zeng G, Huang D, Lai C, Chen M, Cheng M, Wang R (2019) Biochar for environmental management: mitigating greenhouse gas emissions, contaminant treatment, and potential negative impacts. Chem Eng J 373:902–922. https://doi.org/10.1016/j.cej.2019.05.139

Zhang Y, He M, Wang L, Yan J, Ma B, Zhu X, Tsang DC (2022) Biochar as construction materials for achieving carbon neutrality. Biochar 4(1):59. https://doi.org/10.1007/s42773-022-00182-x

Zhang B, Biswal BK, Zhang J, Balasubramanian R (2023a) Hydrothermal treatment of biomass feedstocks for sustainable production of chemicals, fuels, and materials: progress and perspectives. Chem Rev. https://doi.org/10.1021/acs.chemrev.2c00673

Zhang X, Shu S, Hou D, Cheng H, Cao W, Mameda N, Liu Q (2023b) Role of the surface characteristics of hyper-crosslinked polymers on the transformation of adsorbed trichlorophenol: implications for understanding the surface reactivity of biochar derived from waste biomass. Sci Total Environ:163864. https://doi.org/10.1016/j.scitotenv.2023.163864

Zhao L, Cao X, Mašek O, Zimmerman A (2013) Heterogeneity of biochar properties as a function of feedstock sources and production temperatures. J Hazard Mater 256:1–9. https://doi.org/10.1016/j.jhazmat.2013.04.015

Zhao J, Ren T, Zhang Q, Du Z, Wang Y (2016) Effects of biochar amendment on soil thermal properties in the North China plain. Soil Sci Soc Am J 80(5):1157–1166. https://doi.org/10.2136/sssaj2016.01.0020

Zhu Q, Wu J, Wang L, Yang G, Zhang X (2015) Effect of biochar on heavy metal speciation of paddy soil. Water Air Soil Pollut 226:1–10. https://doi.org/10.1007/s11270-015-2680-3

Chapter 4
Integrated Farming Systems: A Holistic Approach to Sustainable Agriculture

Anjana J. Atapattu, Tharindu D. Nuwarapaksha, Shashi S. Udumann, and Nuwandhya S. Dissanayaka

Abstract Integrated farming systems aim to achieve synergies by deliberately combining crop, livestock, and other agricultural enterprises within a farm system. Well-managed integration can lead to greater productivity, more efficient resource utilization, and environmental benefits compared to specialized or single-enterprise systems. This chapter provides an overview of key principles, components, major examples, benefits, challenges, and future opportunities associated with integrated farming approaches. The background section defines integrated farming systems as those featuring purposeful complementarities between agricultural, horticultural, and aquatic production components. Several examples are provided, including crop–livestock integration through rotational grazing or stall-feeding of residues, the use of pond water and waste from aquaculture to support crop irrigation and soil nutrient management, and agroforestry combining woody perennials with agricultural crops or livestock. By taking an integrated approach that combines crops, livestock, forestry, aquaculture, and other farm enterprises, integrated farming systems can contribute to multiple sustainable development goals (SDGs). This chapter outlines considerations in enterprise selection along with different resource availability and agroecological contexts. Furthermore, it discusses planning for nutrient balancing, waste and by-product utilization, and coordinated production cycles and infrastructure requirements. While integrated systems can bolster productivity, economic returns, and climate resilience, challenges exist, including heightened management complexity, higher labor and skill requirements, and the need for supporting infrastructure. To realize the full advantages of integrated farming techniques, it is often necessary to expand the scale of farmers' current operations. This expansion requires facilitating farmers' access to knowledge-sharing platforms, financial incentives, new market openings, and transportation infrastructure that can handle multiple agricultural goods. Overall, well-designed and managed integrated farming systems hold promise toward developing multifunctional landscapes yielding environmental gains alongside sustainable livelihoods. The concluding section

A. J. Atapattu (✉) · T. D. Nuwarapaksha · S. S. Udumann · N. S. Dissanayaka
Agronomy Division, Coconut Research Institute, Lunuwila, Sri Lanka

S. Babu et al. (eds.), *Agricultural Diversification for Sustainable Food Production*, Sustainability Sciences in Asia and Africa,
https://doi.org/10.1007/978-981-97-7517-0_4

summarizes the challenges and potential for sustainable intensification of important future directions of food and nutrition security through purposefully managed integrated farming systems.

Keywords Agroforestry · Crop–livestock integration · Nutrient management · Sustainable development goals (SDGs) · Sustainable intensification

4.1 Introduction

Integrated farming systems have emerged as an important approach to sustainable agriculture. These systems refer to agricultural systems that deliberately combine multiple enterprises, mainly crops, livestock, trees, and fish, into a coordinated whole to maximize economic, environmental, and social benefits (Sekaran et al. 2021; Fatima et al. 2023). These integrated systems aim to optimize yields, improve resource efficiency, minimize risks, reduce pollution from agricultural activities, and increase income and nutrition at household and regional levels (Behera and France 2016; Das et al. 2019). The key principles underlying integrated farming systems include system diversification, resource optimization, synergetic relationship, and sustainability (Singh and Kumawat 2023; Shyam et al. 2023). The notion of integrated farming systems has a long history and evolved out of traditional mixed farming practices around the world. In traditional farming systems in Africa and Asia, it was common to integrate crops, trees, and livestock together. This was done as a resilient strategy to deal with risks and uncertainties (Sekaran et al. 2021). However, modern industrialized agriculture favored specialized, large-scale mixed farming systems focused on maximizing total productivity and profits from crops or livestock (Woods 2012).

The definition of an integrated farming system has changed over time depending on the context. There have been differences in understanding how the components of an integrated farming system interact synergistically. Velten et al.'s (2015) definition of an integrated farming system as having at least two components but logically separate crop and livestock operations represents one perspective. Other studies view integrated farming system as a component of broader farming systems aiming to increase yields while minimizing risks. These studies also see integrated farming system as simultaneously raising profits and improving the utilization of organic wastes and crop residues (Singh and Kumawat 2023). The evolving nature of the integrated farming system definition reflects differing views on how its components like crops and livestock interact, as well as its role within wider farming systems in achieving goals like higher yields, lower risks, and better resource utilization (Yadav et al. 2013; Martin et al. 2016). These conceptualizations indicate that integrated farming systems fundamentally focus on resource efficiency, productivity, economic viability, and ecological sustainability at the whole-farm level through the synergistic integration of diverse biological components (Das et al. 2021; Behera and France 2023). Advances in the concepts, principles, and science of integrated farming systems over past decades have firmly established ecological, social, and economic advantages from reintegrating crop and animal production (Garrett et al. 2017). Moreover, significant research is still needed to develop regionally tailored integrated farming models along with supporting policies and incentives to drive faster and broader adoption globally (Fischer and Connor 2018; Babu et al. 2019).

Specific components incorporated in integrated farming systems can vary substantially based on local contexts (Corbeels et al. 2014). But common elements revolve around strategic crop diversification, mixed crop–livestock production, integration of multipurpose trees, adoption of conservation practices, and recycling of organic resources (Rosati et al. 2021). Most existing integrated farming systems research has focused on technical productivity and efficiency parameters. More interdisciplinary studies incorporating economic, social, and environmental trade-offs will help strengthen the foundation for sustainable agricultural systems in diverse contexts across the developed and developing world (Kanter et al. 2018). The social and environmental problems created by industrial agriculture in the mid- to late-twentieth century raised serious doubts about its sustainability, giving impetus to reexamine integrated approaches (Leach et al. 2020). Research also revealed higher productivity, resource efficiency, and environmental benefits from traditional diversified farming compared to conventional agriculture (Kremen and Miles 2012). These factors led to a revival of interest in integrated farming systems as a paradigm for sustainable intensification of agriculture since the 1980s and 1990s (Pretty and Bharucha 2014). The objective of this chapter is to provide a comprehensive view of important concepts, present examples, outline advantages, tackle challenges, and examine potential future prospects associated with integrated farming approaches.

4.2 Principles of Integrated Farming Systems

4.2.1 System Diversification

Diversification is a cornerstone principle of integrated farming systems, emphasizing the cultivation of a diverse range of crops and the integration of multiple agricultural enterprises within a single farming system (Reckling et al. 2023). This principle recognizes the value of biodiversity in promoting resilience, reducing risks, and optimizing resource use. Integrated farms cultivate a mix of crops, which may include cereals, pulses, oilseeds, fruits, and vegetables (Kumar et al. 2022). This diversity helps mitigate the effect of pests and diseases that target specific crops, promoting a more balanced and resilient ecosystem. Integrating livestock, such as poultry, cattle, or goats, into the farming system enhances diversification (Sekaran et al. 2021). Livestock contribute to nutrient cycling by providing manure, which, in turn, improves soil fertility (Schröder et al. 2016). Additionally, livestock offer alternative income streams for farmers (Herrero et al. 2013). Incorporating agroforestry into the farming landscape further diversifies the system. Planting trees alongside crops provides additional benefits such as shade, windbreaks, and potential sources of timber or fruits (Sharma et al. 2016). Agroforestry also contributes to biodiversity conservation and environmental sustainability (Jose 2012). Diversification not only mitigates risks associated with market fluctuations and climate variability but also enhances overall farm productivity by harnessing synergies between different components (Abegunde et al. 2019).

4.2.2 Resource Optimization

Resource optimization is a fundamental principle that underscores the efficient and sustainable use of resources such as water, nutrients, energy, and land (Segovia-Hernández et al. 2023). Integrated farming systems aim to maximize resource use efficiency while minimizing waste, contributing to both economic and environmental sustainability (Abbasi et al. 2024). Integrated farms employ water-efficient practices, including precision irrigation, rainwater harvesting, and the utilization of water-saving technologies. Efficient water management ensures that crops receive adequate moisture while reducing the strain on water resources (Zahoor et al. 2019). Resource optimization extends to nutrient management with a focus on minimizing external inputs and maximizing the use of organic sources. Livestock manure, crop residues, and cover crops are strategically integrated to enhance soil fertility, reducing the need for synthetic fertilizers (Dissanayaka et al. 2022). Integrated farming systems prioritize energy-efficient practices and renewable energy sources. The adoption of solar-powered technologies, efficient machinery, and the incorporation of energy crops all contribute to diminishing the carbon footprint associated with farming operations (Dhonde et al. 2022). Resource optimization not only leads to cost savings for farmers but also minimizes environmental impacts, making farming systems more sustainable and resilient in the long term (Lemaire et al. 2014).

4.2.3 Synergetic Relationship

Synergy is a principle that emphasizes the interconnectedness and mutual benefit derived from integrating different components within a farming system (Pretty and Bharucha 2014). The synergistic relationships between crops, livestock, and other enterprises create a more resilient and productive agricultural ecosystem (Gil et al. 2017). Livestock in integrated farming systems contribute to nutrient cycling by providing manure that serves as an organic fertilizer for crops (Lemaire et al. 2014). This synergy improves soil health, enhances nutrient availability, and reduces the reliance on external inputs. Crop–livestock integration can contribute to natural pest and weed control (Lemaire et al. 2014; Martin et al. 2016). Similarly, certain crops may act as natural pest repellents or attract beneficial insects. The integration of diverse enterprises creates economic synergies (Hatt and Döring 2023). Income from one enterprise can offset losses in another, providing a safety net for farmers during market fluctuations or unexpected challenges (Jerneck and Olsson 2013). The economic resilience plays a vital role in bolstering the overall sustainability of the farming system. In integrated farming systems, synergy fosters a harmonious coexistence among different components, thereby improving overall efficiency and promoting a balanced, resilient agricultural ecosystem (Tittonell et al. 2020).

4.2.4 Sustainability

Sustainability is an overarching principle that encapsulates the long-term viability of integrated farming systems from economic, environmental, and social perspectives (Cosenz et al. 2020). Sustainable agriculture aims to fulfill the current needs without jeopardizing the ability of future generations to meet their own requirements (Smith et al. 2017). Integrated farming systems give precedence to environmentally friendly practices, including agroecology, organic farming, and conservation agriculture (Migliorini and Wezel 2017). Economic sustainability is crucial for the longevity of farming systems (Rydhmer et al. 2014). Diversification, resource optimization, and economic synergies contribute to the economic viability of integrated farming systems, ensuring that farmers can maintain profitable operations over the long term (Yadav et al. 2020; Yadav et al. 2021). Sustainable agriculture recognizes the importance of social well-being. Integrated farming systems promote community engagement, fair labor practices, and social resilience. Socially sustainable farming practices contribute to the well-being of farming communities and foster a sense of shared responsibility (Marchetti et al. 2020; Paramesh et al. 2022). Guided by the principle of sustainability, integrated farming systems ensure not only endurance but also flourishing, offering economic prosperity, environmental stewardship, and social well-being for farmers and their communities (Bacon et al. 2012).

4.3 Key Components of Integrated Farming Systems

This subsection describes the core components underlying integrated agricultural systems. Integrated farming systems succinctly yet comprehensively outline the range of enterprises found on many integrated farms, including diverse crops, livestock, value-added products, ecosystem services, and community-oriented operations (Kakamoukas et al. 2021). However, these systems aim to optimize synergies between different farm components and manage flows of resources to reduce waste, feed all system components, and produce multiple saleable crops and livestock products (Afi and Parsons 2023; Babu et al. 2023). Figure 4.1 shows a graphical illustration of an integrated farming system model. As illustrated in Fig. 4.1, the essential elements of an integrated farming system can be categorized into four

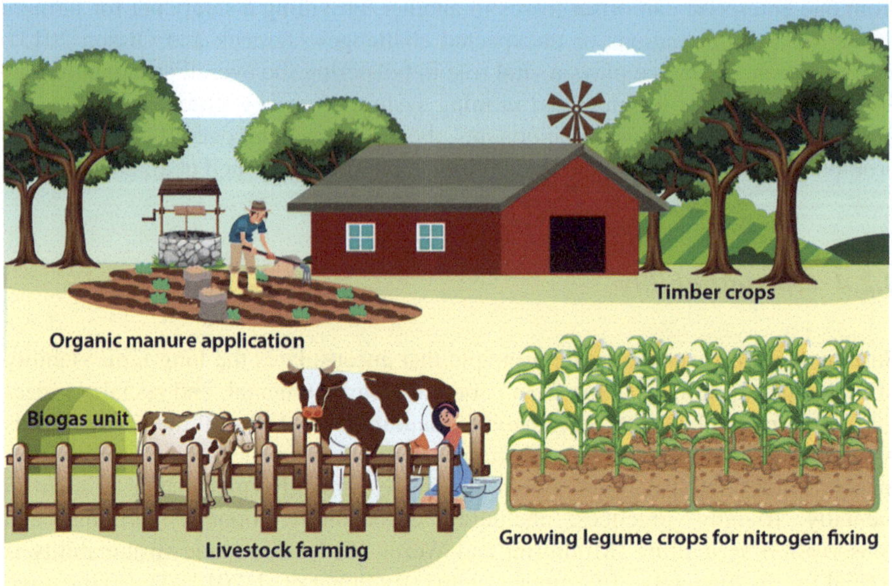

Fig. 4.1 Visual concept of model integrated farming system

main groups: crops and trees husbandry, livestock farming, recycling resources, and environment and community linkages (Bahadur et al. 2016; Kakamoukas et al. 2021).

4.3.1 Crops and Trees Husbandry

Integrating crops and trees into agricultural systems provides many benefits, such as soil fertility enhancement, microclimate regulation for crops, crop pest and disease control, and additional income from tree products (Smith et al. 2013). Common tree integration strategies include agroforestry systems and woodlot systems. Agroforestry systems incorporate trees into the same spaces as crops, either simultaneously or in rotation (Shrestha et al. 2018). The trees are managed to provide benefits such as nitrogen fixation, nutrient cycling, erosion control, windbreaks, shade and microclimate regulation, and tree products (Atapattu et al. 2017; Dissanayaka et al. 2023b; Nuwarapaksha et al. 2023a). For example, pruning branches and foliage from alley cropping or woodlot trees can provide nutrient-rich mulch or green manure for crops (Bhardwaj et al. 2017). Agroforestry systems include alley cropping (trees planted in rows with crops planted in alleys between them), intercropping (alternating rows of trees and crops), and silvopasture (combining trees and livestock grazing) (Udawatta et al. 2019; Udawatta et al. 2021). Woodlot systems establish separate woodland areas on the farm to produce fuelwood, timber, and other tree products (Rathore et al. 2022). Windbreaks protect crops from hot, drying winds and provide sheltered areas with higher humidity and cooler daytime temperatures (Roberts 2017).

Agroforestry systems create cooler, more humid environments than treeless cropland, enhancing crop productivity (Meena et al. 2020). Depending on spacing and arrangement, trees also regulate sunlight for crops. Enough sunlight must pass through the canopy to sustain crops, but filtered light reduces crop water loss while trees transpire groundwater to sustain a humid microclimate (Lawson et al. 2019). In addition to microclimate regulation, diversified systems with higher plant diversity tend to have fewer pest and disease problems (Dissanayaka et al. 2023a). Greater floral diversity provides a habitat for more beneficial insects that prey on crop pests like aphids and beetles (Hassan et al. 2016). Rotating cereal crops with legume trees can similarly break pest and pathogen life cycles (Nadeem et al. 2019). Fast-growing trees and shrubs planted as hedgerows around or within crop fields also provide habitat for bats, birds, and beneficial insects that help control insects and rodents (Montes-Londoño et al. 2022; Rathore et al. 2022). Furthermore, deep tree roots bring up moisture and nutrients from soil depths unavailable to crops with shallow roots, and nutrient-rich leaf litter and nitrogen-fixing trees reduce dependence on synthetic fertilizers (Lebrazi and Fikri-Benbrahim 2022). However, the key characteristic of these components in integrated farming systems lies in the possibility that if one component fails in a given year, the production of other components may continue despite such an occurrence.

4.3.2 Livestock Farming

Integrating livestock and crop production on the same farm can provide numerous synergistic benefits through nutrient cycling while supporting more diverse and resilient rural livelihoods (Sekaran et al. 2021). Mixed farming systems frequently integrate domesticated ruminant animals such as cattle, sheep, and goats, alongside monogastrics like poultry and pigs (Pandey and Upadhyay 2022). When managed appropriately to balance feed sources and nutrient flows, integrating livestock enables more holistic and sustainable food production. Having livestock directly graze annual grain and forage crops in rotation with pasture ecosystems helps control weeds, pests, and diseases across the landscape (Ghanizadeh and Harrington 2019). Animals are moved between paddocks based on precise grass height targets to avoid overgrazing (Sabatier et al. 2015). This rotational system allows pasture grasses to adequately recover while benefiting from natural fertilization through livestock manure deposits. Applying manure appropriately is an essential aspect of livestock integration that enhances soil fertility over seasons (Vanlauwe et al. 2015). This improves yields over time while building soil organic matter, increasing moisture retention, and sequestering carbon. Composting or anaerobic digestion of manures can help stabilize nutrients while killing pathogens (Manyi-Loh et al. 2016).

Incorporating livestock enables using low-value roughages and crop by-products that people cannot consume, like straw, husks, damaged grain, and cotton gin trash.

This converts material with little economic value into high-quality protein foods like meat, milk, eggs, and leather (Costa et al. 2021). However, during certain seasonal pinch periods, special energy or protein-rich supplementary feeds may be required to maintain livestock performance and health (Nechifor et al. 2022). With diverse integrated farming systems, farmers can continue producing nutritional dairy, meat, and eggs even when poor weather or pest damage negatively affects crop yields in a given season (Singh et al. 2023). Livestock provide a production diversification buffer that stabilizes farm incomes across years with fluctuating conditions. However, on very small farms, it may be impractical or inefficient to maintain livestock if crop production cannot satisfy their nutritional requirements year-round (Nuwarapaksha et al. 2023b). Overall, incorporating livestock sustainably on crop farms allows producers to derive multiple gains from the landscape while providing environmental and community benefits (Babu et al. 2023). Well-managed rotational grazing restores grassland ecosystems, while mixed species herding improves pasture quality and parasite control (Forbes and Ellis 2023). Integrated crop–livestock systems make significantly more intensive use of land resources, rainfall, labor, and rural capital investments compared to specialized production (Sekaran et al. 2021; Babu et al. 2023; Fatima et al. 2023; Singh et al. 2023). When implemented adaptively with animal health and nutrient flows appropriately balanced, integrated livestock systems are highly resilient, productive, and regenerative (Teague and Barnes 2017; Yadav et al. 2013).

4.3.3 Recycling Resources

An integrated farming system aims to maximize resource recycling and minimize waste by linking different farm enterprises and closing nutrient, energy, and water loops (Davis et al. 2016). A good example is connecting crop production, livestock rearing, and bioenergy generation into a self-sustaining cycle. The livestock provide manure, which can be converted into biogas in a biodigester to produce renewable energy and nutrient-rich effluent that makes an excellent organic fertilizer for crops (Adegbeye et al. 2020b). The crops in turn yield food, fodder, and residues, which are cycled back as livestock feed, bedding, or substrates for the biodigester, completing the chain (Nwuche et al. 2023). Various recycling flows exist, with nutrient recycling occurring through the systematic collection and composting or biodigestion of livestock manure and human solid waste (Khoshnevisan et al. 2021). The recycling of manure prevents pollution of water bodies by runoff while providing a sustainable internal source of plant nutrients, lowering dependence on costly synthetic fertilizers (Adegbeye et al. 2020a). Crop, tree, and algae biomass harvesting leads to large amounts of residues like straw, husks, bagasse, pruned branches, and processing by-products (Sánchez et al. 2019). Instead of residue burning or landfilling, integrated systems collect and utilize them efficiently.

Integrated farms also emphasize water conservation through rainwater harvesting and the reuse of greywater for irrigation to reduce freshwater use (Radingoana et al. 2019). Greywater from washing, cleaning, and bathing can be diverted to subsurface irrigation channels lined with fruits and vegetable crops (Radingoana et al. 2020). Livestock pens and biodigesters produce large volumes of nutrient-rich wastewater, which when treated in lagoons or constructed wetlands can be profitably reused to water vegetables and crops (Bengtsson-Palme and Heß 2019). Energy recycling occurs when cow dung and leftover farm residues are converted into combustible biogas through anaerobic digestion in biodigesters (Surendra et al. 2014). The methane-rich biogas can replace LP gas fuel for cooking (Black et al. 2021). It can also power internal combustion engines, generators, and pumps to produce electricity and shaft power, displacing diesel and grid electricity. Pyrolysis of dry woody biomass can additionally yield bio-oil for lamps and biochar as a soil conditioner (Dissanayake et al. 2023; Ekanayaka et al. 2023; Lu and El Hanandeh 2019; Shamila et al. 2024). Integrating all these recycling processes allows the farm to approach a circular system that leaves no waste and regenerates resources continually (Ramakrishna et al. 2020). This creates a healthy, sustainable, and economically vibrant farm ecosystem. This builds resilience and autonomy in the farm while enhancing incomes and soil health.

4.3.4 Environmental and Community Linkages

The linkages between the farm environment and the surrounding community are key aspects that need careful consideration when designing integrated systems (Bonaudo et al. 2014). A fundamental environmental linkage is nutrient cycling, and crops take up nutrients from the soil while livestock manure and crop residues can replenish soil fertility (Schröder et al. 2016). When farmers incorporate free-range livestock, cover crops, green manures, buffer strips, and trees on their land, this boosts landscape-level provisioning of these public goods. An integrated farm manages these nutrient flows by balancing feed production, animal stocking rates, and land availability. Applying the appropriate organic matter to match crop needs prevents nutrient overload in soils and eutrophication of nearby water bodies (Ashitha et al. 2021). More diverse agroecosystems demonstrate greater ecological resilience to climate extremes, pest outbreaks, and diseases. Integrated farms additionally promote biodiversity through polycultures, hedge rows, riparian buffer zones, and wildlife habitat preservation (Sangothari et al. 2024). Diversified operations provide regular income and employment opportunities for local families via crop production, animal husbandry, value-added processing, ecotourism, etc. (Rana and Bisht 2023). These community benefits depend on sufficient access to land, capital, technologies, and market systemic constraints that governments should address through appropriate policies, public investments, and extension services.

By stimulating rural development, integrated farming removes pressure for unregulated land conversion while providing an alternative livelihood to urban migration (Ayambire et al. 2019). Local residents additionally rely on ecosystem services from well-managed integrated farms, like water filtration, soil erosion control, carbon sequestration, and agrobiodiversity conservation. In terms of social linkages, integrated farms should be anchored within rural community settings, identity, needs, norms, and knowledge. This necessitates participatory development approaches that engage farmers and their neighbors collaboratively when introducing integrated innovations (Moraine et al. 2017). The appropriate integration of crops, livestock, mechanization, ecological principles, and digital technologies relies on understanding farmer circumstances, constraints, and decision-making (Higgins et al. 2017). The diffusion of integrated practices further depends on peer learning between neighboring farmers. Strong agricultural innovation systems containing government, academic, NGO, and private sector partners can also provide crucial technical assistance, capacity building, and continuing education regarding integrated farming (Kumar et al. 2015). Encouraging farmers through payments for ecosystem services could serve as an incentive for broader adoption of sustainable integrated practices that bring benefits to communities. Ultimately, the adoption and success of integrated systems depend on active community participation, collaborative learning processes, cooperative institutions, and participatory governance structures (Zhang et al. 2023).

4.4 Types of Major Integrated Farming Systems

Several major types of integrated farming systems combine crop cultivation with other agricultural enterprises on the same land (Moraine et al. 2014). Among them identified were mostly common six integrated farming systems. (I) Crop–livestock integration rotates grazing livestock with crop production. (II) Crop–aquaculture systems reuse waste between aquaculture and crops. (III) Crop–poultry integration utilizes poultry manure to fertilize crops, while crops provide feed for poultry. (IV) Agroforestry incorporates trees with crops and livestock, mimicking natural ecosystems. Examples are alley-cropping rows of trees and crops or silvopastoral systems with livestock grazing among trees. (V) Home gardens integrate fruits, vegetables, herbs, and sometimes livestock in household backyards, recycling nutrients onsite. (VI) Integrated mixed farming combines multiple enterprises like crops, livestock, and agroforestry on single farms (Fig. 4.2). In summary, integrated farming systems generate synergies and circular resource flows by merging crop cultivation with other agricultural enterprises such as livestock, aquaculture, agroforestry, and poultry on the same land (Dar et al. 2018). The objective is to enhance productivity, economic returns, and sustainability in contrast to specialized monoculture production. When tailored to local conditions and carefully managed, integrated systems can enhance natural resource efficiency and strengthen farmer livelihoods while sustainably meeting food production goals (Montagnini and Metzel 2017).

Fig. 4.2 Key integrated farming systems: (**a**) crop–livestock integration; (**b**) crop–aquaculture integration; (**c**) crop–poultry integration; (**d**) agroforestry systems; (**e**) home gardens; and (**f**) mixed farming systems

4.4.1 Crop–Livestock Integration

Crop–livestock integration pertains to deliberate agricultural systems wherein both crop and livestock production are intentionally merged within the same area (Duncan et al. 2013). These integrated systems aim to maximize yields and economic returns while improving soil fertility and overall sustainability (Bonaudo et al. 2014). Within an integrated crop–livestock system, there is a favorable interaction between crops and livestock, leading to beneficial outcomes both ecologically and economically (Martin et al. 2016). For example, livestock manure serves as an organic fertilizer for crops, reducing the need for synthetic fertilizers. Meanwhile, crop residues and cover crops provide feed for livestock, minimizing the need to purchase external feeds. Rotating livestock on croplands can also enhance soil structure and break disease and pest cycles (Sanderson et al. 2013). Economically, diversity reduces risk for farmers compared to specialized crop or livestock production. Income is generated from both crop and animal sales, protecting farmers when prices fluctuate. The on-farm benefits also lower production costs over time as fewer external inputs are needed. There are many variations of crop–livestock integration tailored to different agroecosystems. Simple integrations involve rotating cash crops and livestock grazing in the same fields over seasons (Sulc and Franzluebbers 2014). More complex systems interseed cover crops and grain crops while livestock simultaneously graze the cover. In cropland-based integrations, crops remain the primary enterprise and livestock grazing is secondary (Planisich et al. 2021). The cover protects and builds the soil while providing supplemental livestock feed (Dissanayaka et al. 2024).

In rangeland-based integrations, perennial pasture remains the main vegetation, with occasional crop cultivation. For example, one crop may be grown only once every several years (Morales et al. 2016). Alternatively, improved and fertilized pasture may support intermittent crop production. Silvopastoral integrations combine crops, livestock grazing, and forestry (Bussoni et al. 2019). The overstory trees provide fruit, fuelwood, fodder, and timber while offering environmental benefits like carbon storage and improved water cycling. Animals graze the understory herbaceous plants. The benefits of crop–livestock integration depend heavily on how the systems are managed (Martin et al. 2016). Benefits are maximized when production practices match the local environment. Careful management of stocking densities, grazing periods, fertilization, and feed allocation is required (Scott et al. 2013). If overexploited, there are also environmental risks like soil compaction, erosion, and nutrient runoff. Research overwhelmingly shows greater productivity, profitability, and resource efficiency in integrated crop–livestock systems compared to specialized production (Asai et al. 2018). For these reasons, many see crop–livestock integration as a sustainable alternative to specialized production, capable of improving farmer livelihoods while meeting global food demand and conservation goals.

4.4.2 Crop–Aquaculture Integration

Crop–aquaculture integration involves agricultural systems where crop cultivation and aquaculture are combined on the same parcel of land. The objective of these integrated agroecosystems is to optimize productivity and economic returns while enhancing the efficiency of natural resource utilization (Jhariya et al. 2021). In integrated crop–aquaculture systems, the waste from one production component serves as input for another, forming beneficial biological and economic interactions (Hanjra et al. 2012). For example, freshwater fish raised in ponds provide nutrient-rich water and sediment for irrigating and fertilizing cropland (Haque et al. 2016). Meanwhile, crop by-products and processing residues offer farm-made feeds for aquaculture. These synergies create a semiclosed system where waste becomes an asset. This allows farmers to cut costs on synthetic fertilizers and fish feeds while boosting yields. Diversified income from both crops and aquaculture also reduces market risk (Troell et al. 2014). In rice–fish farming, raised beds planted with rice surround fish-stocked trenches in the field. As water fluctuates, fish contribute nutrients and oxygen, regulate pests, and minimize weeds, boosting rice growth (Bao et al. 2022). Vegetable fields can also be fertilized using sediment and wastewater from nearby fish ponds. Alternatively, aquaculture systems like fish cages can be installed within irrigation canals, reservoirs, and water storage ponds.

The freshwater systems enrich the irrigation supply for downstream cropland or hydroponic vegetable production in conservatories. Brackish water ponds for shrimp, shellfish, and seaweed can be integrated with fruit orchards and tree plantations (Hardi et al. 2023). Mangroves or constructed wetlands around coastal fish-ponds filter effluent while providing wood, honey, and fodder. Rice paddies double as shrimp nurseries before salty water is drained for the rice crop. However, crop–aquaculture integration poses sustainability concerns without careful management, especially relating to water use, pollution, and disease. Overstocking aquatic systems risks effluent and methane emissions. Integrated management across components is vital to balance productivity, economics, and environment (Lemaire et al. 2014). Many see crop–aquaculture integration as the future foundation for food security in populous Asian countries with limited land (Liu et al. 2020). The practices are labor-intensive but offer high-value nutrition and livelihoods from small spaces (Gurung et al. 2016). With knowledge transfer and incentivizing policies, crop–aquaculture integration can be further expanded, improved, and intensified.

4.4.3 Crop–Poultry Integration

Crop–poultry integration refers to an intentional, synergistic combination of poultry production with crop farming on individual farms to capture bidirectional benefits between the two enterprises via waste utilization and resource cycling (Garcia et al. 2023). In integrated systems, poultry manure provides organic fertilizer to nourish crops, reducing or eliminating synthetic fertilizer needs, while crops like corn and soy provide nutritious poultry feed, lowering or removing feed purchase requirements (Ashworth et al. 2020; Nkukwana 2018). When strategically designed and managed, integrated crop–poultry systems aim to maximize productivity, economic performance, and environmental sustainability simultaneously by capturing synergies and efficiently utilizing circular resource flows. Core benefits span enhanced soil fertility, increased crop yields, lowered production costs, improved on-farm nutrient capture, reduced pollution, diversified farm income, and optimization of land usage (Balafoutis et al. 2017). Managing integrated systems is complicated by the need to holistically coordinate multiple disparate components fostering interlinked benefits (Liu et al. 2015). Ammonia emissions and odor from poultry proximity to crops must be monitored and controlled while biosecurity measures safeguard against reciprocal disease transmission (Avidov et al. 2021). Nutrient imbalances, contamination, erosion, and compaction threaten soil health if waste flows are mismanaged.

Key tenets for successful crop–poultry systems include thorough site selection, which involves assessing soil, water, and climate conditions (Bhattacharyya et al. 2020). Structural adaptations are crucial for efficient waste cycling, and proactive measures such as pest, odor, and emission controls are necessary. Balanced livestock stocking rates and rotation schedules are essential to distribute manure nutrients evenly across various crop fields based on soil nutrient profiles (Watson et al. 2019). Additionally, careful feed formulation and supplementation are required to maintain the health and productivity of distinct integrated components. When carefully executed under an umbrella of holistic thinking and circular resource flow optimization, integrated crop–poultry farming can unlock immense potential for improving agricultural productivity and sustainability (Poccard-Chapuis et al. 2014). The synergies harnessed between poultry and crop systems via strategic integration hold remarkable promise for simultaneously increasing production, efficiency, and environmental stewardship in agricultural systems needed to feed the world's burgeoning population (Raj et al. 2021). Real-world implementation has confirmed that integrated crop–poultry approaches can exceed results from individual enterprises. As the advantages of agricultural integration gain wider recognition across contexts, integrated crop–poultry farming is likely to assume increasing global importance and warrants expanded research and modeling to further optimize its application (Behera and France 2016). With thoughtful vision and management, the circular linkages in crop–poultry integration can propel more sustainable and self-sufficient agricultural models fulfilling present and future demands.

4.4.4 Agroforestry Systems

Agroforestry systems represent a dynamic and sustainable approach to farming that integrates the cultivation of trees, crops, and livestock within a single agricultural ecosystem (Raj et al. 2019). This intricate and interconnected farming model seeks to optimize the use of available resources, enhance biodiversity, and promote ecological balance. Agroforestry systems are designed to mimic natural ecosystems, where different components synergistically interact, providing multiple benefits to farmers, the environment, and the community at large (Plieninger et al. 2020). This approach offers a diverse range of benefits that extend beyond what conventional monoculture farming systems can achieve (Nuwarapaksha et al. 2022). The existence of trees in agroforestry systems contributes to overall ecosystem resilience by mitigating the impacts of climate change, preventing soil erosion, and providing habitat for beneficial insects and wildlife (Dissanayaka et al. 2023a). While agroforestry systems lack distinct boundaries, they have been categorized based on the extent of integration of their primary components. These categorizations help provide a clearer understanding of the diversity within agroforestry practices, as shown in Table 4.1.

Agroforestry systems are not confined to rural landscapes alone; they can be implemented in peri-urban and urban areas as well (Dagar et al. 2023). Urban agroforestry involves integrating trees and green spaces into city environments, providing multiple benefits such as improved air quality, reduced heat islands, and enhanced aesthetic appeal (Taylor and Lovell 2021). Community gardens and orchards in urban settings exemplify agroforestry principles by combining fruit-bearing trees with vegetable and herb cultivation, creating a harmonious blend of food production and environmental enhancement. The diversity of crops and tree species within these integrated systems provides a buffer against pests and diseases, as the presence of various plants disrupts the buildup of specific pathogens (Ratnadass et al. 2012). This natural pest management reduces the reliance on chemical inputs, making agroforestry an eco-friendly and sustainable alternative to conventional farming methods (Shanka 2020). Furthermore, agroforestry systems contribute to the resilience of farming communities against the impacts of climate

Table 4.1 Classification of major agroforestry systems in the world

Agroforestry	Major components	Examples
Agri-silvicultural systems	Integrating annual and perennial plants and crops which has different growth forms	Shifting cultivation, chena Scattered trees on farms Shelterbelts and windbreaks Plantation-based cropping system Boundary planting and live hedges Woodlots for soil conservation Home gardens
Silvi-pastoral system	Integrating trees, pasture with/without livestock	Plantation crops with pastures Silvi-pastures Horti-pastoral Seasonal forestry grazing Tree on rangelands
Agri-silvi-pastoral systems	Integrating crops, pasture, and trees with livestock	Home gardens Multipurpose woody hedgerows
Apiculture with crops	Integrating crops with bees	

Source: Modified from Ayyam et al. 2019; Dhyani et al. 2021

change. They achieve this by establishing a diversified and robust production system capable of adapting to evolving environmental conditions.

4.4.5 Home Gardens

Home gardens, often referred to as backyard gardens or kitchen gardens, represent a microcosm of integrated farming techniques, bringing together various elements such as fruits, vegetables, herbs, and sometimes small livestock within the confines of a household's premises (Tunc 2017). These intimate and diverse ecosystems serve as a sustainable and resource-efficient model for meeting household food needs while promoting self-sufficiency, biodiversity, and environmental conservation. At the core of home gardens is the integration of different plant species with complementary growth patterns and nutritional profiles (Estrada-Carmona et al. 2020). This meticulous arrangement allows for the optimal use of space and resources. This intercropping approach not only maximizes the yield from limited space but also fosters a dynamic and resilient environment where each component contributes to the overall well-being of the garden (Hallett et al. 2016). One of the key characters of home gardens is the inclusion of fruit trees, providing a perennial source of fresh and nutritious produce (Galhena et al. 2013). Trees such as mango, guava, citrus, and banana are often chosen for their adaptability to small spaces and suitability for home garden environments. The fruits not only contribute to the household's nutrition but also present opportunities for surplus produce to be shared with neighbors or sold in local markets (Galli et al. 2020).

In-home gardens, vegetable cultivation adheres to sustainable and organic practices, placing a strong emphasis on utilizing natural fertilizers and adopting pest control methods that are environmentally friendly (Dorais and Alsanius 2015). Crop rotation and companion planting are commonly employed to manage soil fertility and reduce the risk of pest infestations. Additionally, kitchen waste and compost generated within the household are often recycled to create nutrient-rich organic matter, promoting soil health and reducing the reliance on chemical inputs (O'Connor et al. 2021). Herb gardens within the home garden setting add another layer of functionality, providing not only culinary delights but also medicinal benefits (Hellermann 2017). The integration of medicinal herbs into home gardens aligns with a holistic approach to health and wellness, emphasizing the interconnectedness of food and medicine. Furthermore, home gardens often incorporate small-scale livestock, such as chickens or rabbits, to complement plant-based production (Blay-Palmer et al. 2023). Poultry, for instance, contributes to the system by providing fresh eggs, controlling insect pests, and producing nutrient-rich manure (Alders et al. 2018). The integration of livestock with plant cultivation creates a symbiotic relationship where the waste from animals becomes a valuable resource for fertilizing plants, closing the nutrient loop within the household ecosystem. This combination of plants and animals exemplifies the principles of permaculture, where diverse elements work together to create a self-sustaining and regenerative system (El-Sayed and Cloutier 2022). However, the continual availability of fresh produce from the home garden reduces reliance on external food sources and promotes a more sustainable and resilient food supply.

4.4.6 Integrated Mixed Farming

Integrated mixed farming systems represent a dynamic and sustainable agricultural approach that combines various components of farming, such as crops, livestock, and agroforestry, into a unified and synergistic system (Schiere et al. 2015). This holistic farming model is designed to optimize resource utilization, enhance biodiversity, and promote resilience in the face of varying environmental conditions. Integrated mixed farming systems integrate multiple enterprises on a single farm, offering a more diversified and resilient production system compared to monoculture or single-enterprise farms (Sanderson et al. 2013). At the heart of integrated mixed farming systems is the deliberate integration of different agricultural activities to create a harmonious and mutually beneficial relationship between the components. Crop cultivation, livestock rearing, and agroforestry are strategically combined to capitalize on the complementarities between these enterprises. The diversity in the system not only maximizes resource use efficiency but also mitigates risks associated with market fluctuations, pests, and climate variability (Aguilera et al. 2020). In turn, the availability of nutritious fodder from crop residues enhances the well-being and productivity of the livestock. The diversified system also enables year-round production, addressing food availability issues during lean seasons. Pastures that develop naturally beneath trees comprise native grasses, herbs, and legumes, making them a valuable source of pasture for integrated animal systems (Nuwarapaksha et al. 2023b). Hence, incorporating livestock into agroforestry systems has the potential to further increase productivity and global food security (Nuwarapaksha et al. 2023c).

An essential benefit of integrated mixed farming systems is their capacity to augment the overall farm income and diminish reliance on a solitary source of revenue (Lemaire et al. 2014). Revenue is generated not solely through the sale of crops, but also from livestock products, agroforestry yields, and various other diversified enterprises. This multipronged income approach creates economic stability for farmers and allows them to withstand market fluctuations more effectively (Paus 2012). Integrated mixed farming systems also contribute significantly to environmental sustainability (González-García et al. 2012; Sekaran et al. 2021). By diversifying crops and incorporating agroforestry, these systems promote biodiversity, reduce the need for synthetic inputs, and enhance overall ecosystem health (Rosati et al. 2021). The conservation of natural resources, such as soil and water, is prioritized, leading to more sustainable farming practices. As agriculture faces difficulties related to climate change, resource scarcity, and food security, integrated mixed farming systems emerge as a viable and forward-thinking solution (Benedek et al. 2023). These systems draw on the principles of agroecology, emphasizing the importance of integrating various components to create resilient, sustainable, and economically viable farming landscapes (Tittonell 2020). The adaptability of integrated mixed farming systems to diverse agroecological zones and their potential to address multiple challenges make them a promising avenue for the future of agriculture.

4.5 Integrated Farming Systems to Achieving Sustainable Development Goals

Integrated farming systems play a pivotal role in advancing sustainable development and contributing to the attainment of various United Nations Sustainable Development Goals (SDGs) (Atapattu et al. 2024; Montagnini and Metzel 2017). Integrated farming systems can promote the several SDGs mainly (as shown in Fig. 4.3). These systems take a holistic approach that combines multiple agricultural practices to enhance productivity and sustainability. Diversified integrated farming systems improve farmers' economic resilience to price and weather variability, supporting **SDG 1** to **no poverty** (Kiani et al. 2021). For instance, integrating crops and livestock enables more efficient nutrient cycling, as animal manure can fertilize crops while crop residues can feed animals. This boosts productivity while decreasing the need for synthetic fertilizers, advancing **SDG 2** to **zero hunger** through sustainable agriculture (Ladha et al. 2020). Moreover, these systems contribute to food security by diversifying agricultural production (Sekaran et al. 2021). The inclusion of multiple components ensures resilience against unforeseen challenges, such as crop failure or disease outbreaks. The diversity of production from integrated farming can provide a reliable and varied food supply, decreasing susceptibility to food shortages. Methods like intercropping, crop rotations, and increased biodiversity help manage pests while reducing pesticide use. This progresses **SDG 3** on **good health and well-being** by promoting sustainable and healthy diets (He et al. 2019).

Integrated farming systems support **SDG 6** on **clean water and sanitation** by enhancing groundwater recharge, reducing soil erosion and water contamination, increasing water use efficiency, and improving water access through integrated system design (Pavlidis and Tsihrintzis 2018). Planting trees and maintaining permanent soil cover through pasture or mulch helps regulate water cycles and mitigates droughts and floods (Asbjornsen et al. 2014). Integrated crop–livestock systems enable the use of livestock manure as organic fertilizer, decreasing dependence on synthetic fertilizers made with fossil fuels. This reduces greenhouse gas emissions, aligning with **SDG 13** on **climate action** (Bell and Moore 2012). Additionally, several practices common in integrated farming systems can help adapt to and build resilience against climate change effects. This further supports SDG 13 (Altieri et al. 2015). Incorporating trees, shrubs, and pasturelands into croplands can prevent soil erosion and enhance soil health by building up organic matter. This supports **SDG 15** to protect and restore **life on land** (Fahad et al. 2022). Integrated systems

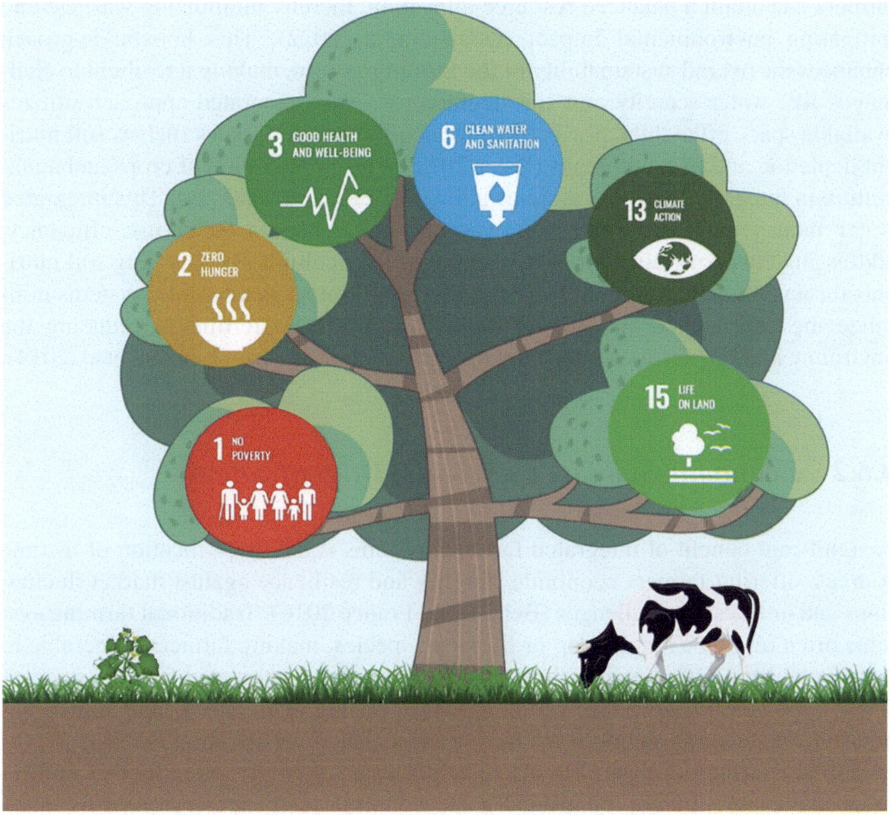

Fig. 4.3 Visual concept: SDGs directly achievable through integrated farming systems

are knowledge-intensive and require context-specific design based on local conditions, but they represent a very promising path towards sustainable agriculture worldwide. If thoughtfully implemented and adapted globally, integrated farming systems can make significant contributions toward achieving broad sustainable development objectives.

4.6 Benefits of Integrated Farming Systems

4.6.1 Resource Use Efficiency and Sustainability

Integrated farming systems are designed to optimize the utilization of resources, ensuring a sustainable and efficient agricultural model (Shyam et al. 2023). One of the primary benefits lies in the judicious use of land, water, and nutrients. Through the integration of diverse components like crops, livestock, and aquaculture, farmers can attain a balanced resource allocation, thereby minimizing wastage and mitigating environmental impact (Obaisi et al. 2022). This holistic approach enhances the overall sustainability of the farming system, making it resilient to challenges like water scarcity and soil degradation. This integrated approach utilizes available space efficiently, preventing monoculture-related issues such as soil nutrient depletion and pest outbreaks (Solér 2012). The combination of crops and aquaculture in integrated farming systems allows for efficient water use. This integrated water management reduces water wastage and enhances water use efficiency, addressing concerns related to water scarcity in agriculture. By recycling soil nutrients through the integration of livestock and crops, integrated farming systems minimize the dependence on external inputs like synthetic fertilizers, reducing the environmental impact associated with their production and use (Lemaire et al. 2014).

4.6.2 Economic Stability by Product Diversification

A significant benefit of integrated farming systems is the diversification of income sources, offering farmers economic stability and resilience against market fluctuations and unforeseen challenges (Behera and France 2016). Traditional farming systems often rely on a single crop or livestock species, making farmers vulnerable to price volatility (Altieri and Nicholls 2017). In contrast, integrated farming systems involve the simultaneous cultivation of crops, raising of livestock, and sometimes aquaculture, creating multiple income streams. The diversification not only serves to stabilize farmers' income but also establishes a safety net in the face of unforeseen challenges, thereby contributing to economic resilience (Petersen-Rockney et al. 2021). Diversification in integrated farming systems acts as a risk mitigation technique (Szymczak et al. 2020). If one component of the farming system faces challenges such as pest outbreaks, diseases, or adverse weather conditions, other components may continue to thrive, providing a buffer against complete crop failure or financial loss. This risk-spreading mechanism enhances the overall resilience of the farming enterprise.

4.6.3 Soil Health Improvement and Regeneration

Integrated farming systems undeniably and significantly contribute to the enhancement of soil health and quality through the synergistic relationships and connections between crops, livestock animals, and the recycling of organic matter (Sarkar et al. 2021). The inclusion and integration of organic matter significantly improve various aspects of soil structure and composition. This includes the capacity to retain moisture, an increase in water-holding ability, and the enhancement of the availability and supply of essential macro and micronutrients that are crucial for the growth and thriving of plants (Bashir et al. 2021). Biological nitrogen fixation by leguminous

Table 4.2 The nitrogen-fixing potential of legume and actinorhizal trees and shrubs in integrated farming systems

Tree species	Nitrogen fixation (kg/ha/year)
Legume trees	
Gliricidia sepium	13–50
Sesbania sesban	260
Leucaena leucocephala	100–500
Erythrina sp.	57–66
Acacia mearnsii	200
Faidherbia albida	20–50
Grevellea robusta	100–300
Albizia lebbeck	260
Inga edulis	35–40
Actinorhizal trees	
Allocasuarina littoralis	218
Coriaria arborea	192
Alnus glutinosa	40–53
Alnus nepalensis	30–120
Alnus rubra	85–320
Alnus acuminata	279
Casuarina equisetifolia	60–110

Source: Ram et al. (2017)

and actinorhizas tree species is an important agroforestry function (Table 4.2). The deliberate incorporation of commonly grown leguminous crops like *Gliricidia sepium, Sesbania sesban*, *Leucaena leucocephala*, and *Crotalaria juncea.* L with nitrogen-fixing bacteria species into integrated diverse farming systems markedly contributes to natural biological nitrogen fixation processes (Jangir et al. 2022; Udumann et al. 2023). Instead of relying on synthetic fertilizers, providing a coconut palm with 30 kg of fresh leaves from either *Gliricidia sepium* or *Leucaena leucocephala* can fulfill 100% of its annual nitrogen requirement and contribute to 20% of its annual phosphorus and potassium needs. Furthermore, it plays a role in fixing atmospheric nitrogen, leading to a rise in soil nitrogen levels (Dissanayaka et al. 2022).

4.6.4 Climate Change Mitigation and Adaptation

Integrated farming systems assume a critical role in addressing the barriers posed by climate change, providing both mitigation and adaptation strategies (Altieri et al. 2015). The diverse nature of integrated farming systems equips farmers with increased adaptability to cope with changing climatic conditions (Kremen et al. 2012). Farmers, within integrated farming systems, can modify crop choices and management practices to respond to variations in temperature, precipitation, and other climate-related factors. This adaptability serves to diminish the vulnerability

Table 4.3 Soil carbon stocks in major agroforestry systems under an ecological-based approach

Agroforestry system	Soil carbon stock (Mg/C/ha)
Humid lowlands	10–25
Alley cropping systems	61–75
Woodlots	123–149
Improved fallow lands	108–119
Home gardens	27–78
Tree intercropping	96–173
Silvopasture	21–235
Shaded perennial cropping systems	
Tropical highlands	
Silvopastoral systems	132–173
Shaded perennial systems	21–97
Arid and semiarid lowlands	
Live fences	24
Grazing lands	27–33
Fodder–tree dominated	25–80
Crop dominated	20–70
Forage banks	33
Fuelwood dominated	30–90

Source: Nair et al. (2009)

of the farming system to extreme weather events, ensuring more stable and secure food production (Altieri et al. 2015). The incorporation of trees, agroforestry practices, and perennial crops in integrated farming systems contributes significantly to carbon sequestration (De Stefano and Jacobson 2018). Table 4.3 presents a spectrum of documented soil carbon stock values within prominent agroforestry systems across three ecological regions: tropical highlands, humid lowlands, and arid and semiarid lowlands. It is important to highlight that some major systems lack reported values or have close approximations. An apparent incongruity in Table 4.3 arises from the focus on soil carbon stock values specifically within the 0–50 cm soil depth. This selection aligns with the prevailing practice in the literature, where soil carbon stock values are predominantly documented for the upper soil layers, typically reaching a depth of less than 50 cm (Nair et al. 2009). Additionally, through the process of photosynthesis, trees and crops capture and store carbon in their biomass and the soil. This carbon sequestration helps alleviate the impacts of greenhouse gas emissions from agriculture, rendering integrated farming systems a valuable component in climate change mitigation efforts (Pathak 2015).

4.7 Challenges to Integrated Farming Systems

4.7.1 Economic Viability and Financial Constraints

Despite the potential for increased economic stability through diversification, there are notable challenges related to the initial investment, market access, and financial sustainability of integrated farming systems (Sánchez et al. 2022). The establishment of integrated systems may require significant initial investments in infrastructure, equipment, and specialized training. Farmers, especially those with limited financial resources, may find it challenging to make these upfront investments. This financial barrier can impede the widespread adoption of integrated farming systems, particularly in regions where access to capital is limited. Integrated farming systems often produce a diverse range of products, including crops, livestock, and aquaculture products (Nuwarapaksha et al. 2023c). Establishing efficient market linkages and value chains for this diverse output can be challenging. Farmers may face difficulties in accessing markets that appreciate the integrated nature of their products or encounter limitations in processing facilities that can handle the diversity of goods produced (Lee et al. 2012).

4.7.2 Technological and Infrastructure Challenges

The successful implementation of integrated farming systems relies heavily on appropriate technologies and supportive infrastructure. Challenges in this domain include the availability of suitable technology, access to necessary resources, and the integration of modern techniques into traditional farming practices (Sharma et al. 2022). Embracing technology, including precision farming, sensor-based monitoring, and automated systems, is crucial for optimizing the efficiency of integrated farming systems (Adeyemi et al. 2017). However, farmers may encounter challenges in embracing and adjusting to these technologies, attributed to factors such as high initial costs, limited technical expertise, and concerns about the reliability of technology in diverse agroecological settings (Altieri et al. 2015). The successful implementation of integrated farming systems may necessitate specific infrastructure, including agroprocessing units, storage facilities, and water management systems. In many cases, existing agricultural infrastructure may not be suitable for the diverse needs of integrated farming systems. The lack of supportive infrastructure can hinder the efficient functioning of integrated systems, limiting their scalability and impact (Berti and Mulligan 2016).

4.7.3 Environmental and Ecological Challenges

While integrated farming systems aim to enhance sustainability, there are challenges related to environmental impacts, particularly in terms of water usage, biodiversity conservation, and potential unintended consequences. The integration of aquaculture into integrated farming systems introduces water management challenges (Edwards 2015). Maintaining water quality for both crops and aquatic organisms is crucial, and improper water management can lead to pollution and the spread of diseases (Akhtar et al. 2021). Balancing the water requirements of different components in integrated farming systems poses a delicate challenge, especially in regions where water scarcity is a prevalent concern. The introduction of new components or practices in an integrated system may have unintended ecological consequences (Pretty 2018). For example, the use of certain crops or the introduction of specific livestock breeds may disrupt local ecosystems, leading to changes in pest populations, soil composition, or biodiversity (Emmerson et al. 2016). Careful consideration and monitoring are necessary to prevent or mitigate these unintended consequences.

4.7.4 Policy and Institutional Barriers

The regulatory and policy environment plays a crucial role in the successful adoption and scaling of integrated farming systems (Šūmane et al. 2018). Nevertheless, several challenges, such as policy gaps, insufficient institutional support, and inadequate incentives, impede the widespread adoption of integrated farming systems. Existing agricultural policies may not be conducive to the integration of diverse farming components. Policymakers may need to revisit and adapt regulations to accommodate the unique characteristics of integrated farming systems. This includes addressing issues related to land use planning, water rights, and zoning regulations that may impact the feasibility of integrated systems (Delphin et al. 2022). Achieving successful implementation of integrated farming systems necessitates support from agricultural extension services, research institutions, and other pertinent agencies (Singh and Burman 2019). However, these institutions may not be adequately equipped to provide the necessary guidance and support for farmers transitioning to integrated systems. Strengthening institutional capacities and creating dedicated support mechanisms are essential for overcoming this challenge.

4.8 Future Prospects for Integrated Farming Systems

4.8.1 Technological Advancements in Precision Agriculture

One of the most promising prospects for the future of integrated farming systems lies in the ongoing advancements of precision agriculture technologies. Precision farming encompasses the utilization of sensors, drones, GPS technology, and data analytics to optimize resource management and enhance decision-making on the farm (Karunathilake et al. 2023). As technology continues to advance, integrated farming systems can leverage these tools to monitor crop health, soil conditions, and water usage in real time (Bayih et al. 2022). The progress in precision agriculture will enable more precise and efficient application of inputs, such as fertilizers and pesticides, thereby reducing waste and minimizing environmental impact

(Balafoutis et al. 2017). Autonomous machinery and robotics may play a significant role in tasks like planting, harvesting, and weeding, further enhancing operational efficiency on integrated farms (Fountas et al. 2020). The integration of artificial intelligence (AI) and machine learning can provide predictive analytics, helping farmers make informed decisions for optimal yield and resource utilization (Redhu et al. 2022).

4.8.2 Agroecological Approaches for Sustainable Agriculture

Anticipated future integrated farming systems are likely to place increased emphasis on agroecological principles, aiming to promote sustainable and resilient agricultural practices. Agroecology involves designing farming systems that work in harmony with natural ecosystems, fostering biodiversity, and minimizing the use of external inputs (Bezner Kerr et al. 2022). Implementing agroecological practices like agroforestry, cover cropping, and organic farming can contribute significantly to soil health, water conservation, and pest control in agricultural systems (Wezel et al. 2014). Incorporating trees and shrubs into farming landscapes not only enhances biodiversity but also provides additional ecosystem services, such as carbon sequestration and improved water quality. Advocating for agroecology aligns with the growing recognition of the need for regenerative agricultural practices that aim to restore and enhance ecosystem functions (Hathaway 2016).

4.8.3 Climate-Resilient Farming Practices

As climate change poses challenges to global agriculture, the future of integrated farming systems will likely focus on developing and implementing climate-resilient farming practices (Singh et al. 2021). This involves choosing and breeding crop varieties that are well-suited to evolving climatic conditions, alongside the adoption of water-efficient irrigation techniques and enhanced soil management practices. Integrated farming systems play a pivotal role in bolstering resilience through the diversification of crops, the integration of agroforestry, and the implementation of water harvesting strategies. These practices contribute to a more robust and adaptable agricultural system in the face of various challenges (Altieri and Nicholls 2017). Incorporating livestock into farming systems can also enhance resilience, as diversified systems tend to be more adaptable to changing environmental conditions (Altieri et al. 2015). Nonetheless, the adoption of climate-smart agriculture practices will be crucial to guarantee food security and sustainable agricultural production amid the challenges posed by a changing climate (Scherr et al. 2012).

4.8.4 Digital Agriculture and Smart Farming Solutions

The incorporation of digital technologies and intelligent farming solutions represents a promising avenue for the evolution of integrated farming systems. Digital agriculture entails leveraging information and communication technologies to enhance farm management, decision-making processes, and overall operational efficiency (Demestichas and Daskalakis 2020). This encompasses the utilization of farm management software, sensor networks, and Internet of Things (IoT) devices (Sanjeevi et al. 2020). Farmers engaged in integrated systems can reap advantages from real-time data on weather conditions, crop health, and market trends. This information empowers them to make informed and timely decisions, enhancing their overall operational efficiency and productivity (Cheema and Khan 2019). The widespread acceptance of integrated farming systems hinges on the creation of user-friendly and cost-effective digital tools, coupled with the extension of connectivity in rural areas (Gómez-Carmona et al. 2023). Smart farming solutions can enhance precision in resource use, reduce waste, and improve the overall economic viability of integrated farming systems.

4.9 Conclusion

Integrated farming systems embrace a comprehensive and interconnected approach to agriculture, incorporating various components such as crops, livestock, trees, aquaculture, and the generation of renewable energy. The core principles underlying integrated systems are diversification, resource optimization, fostering synergy between components, and enhancing sustainability across economic, environmental, and social dimensions. These multifunctional systems aim to maximize productivity and efficient resource utilization while minimizing external inputs and waste. There are numerous well-documented benefits of thoughtfully designed integrated agricultural systems compared to specialized monoculture operations. These include enhanced resource use efficiency, economic stability via enterprise diversification, soil regeneration, and resilience against climate disruptions. Integrated farming has higher knowledge requirements and labor needs compared to conventional specialized farms. Several priority areas exist to further optimize integrated systems. Continued research should focus on developing regionally tailored models based on local agroecological and socioeconomic conditions. This requires the active participation of farming communities to match innovations with local needs and constraints. Policy reform and institutional capacity building for technical training, financial services, and market linkages must accompany research to enable broader technology adoption. However, technology deployment must be accompanied by infrastructure development and capacity building suited for local contexts. Overall, integrated farming systems hold immense potential to make agriculture more sustainable, resilient, and multifunctional. However, realizing this potential requires greater public and private investments into scientific research, institutional capacity building, financial incentives, market linkages, and participatory farmer engagement to develop locally tailored systems. Policy reforms to value ecosystem services and incentivize regenerative practices are also needed. Integrated approaches are knowledge- and management-intensive, demanding a whole-system perspective radically different from specialized commodity production models. This paradigm shift requires committed leadership and patient change management across every level of the food system. However, the holistic perspective and synergistic integration of diverse biological components in integrated farming offer a very promising pathway to transform agriculture globally. Through strategic and location-specific implementation, along with coordinated advancements in technology, ecology, policies, and social aspects, integrated systems have the potential to significantly contribute to achieving ecological sustainability, economic viability, and social justice in agriculture. The potential returns on investing in integrated farming transitions are immense in terms of productivity, environmental quality, food security, and rural livelihoods. Unlocking these potential demands recognizing and rewarding the multiple ecological, economic, and social benefits generated by integrated farming systems.

References

Abbasi IA, Shamim A, Shad MK, Ashari H, Yusuf I (2024) Circular economy-based integrated farming system for indigenous chicken: fostering food security and sustainability. J Clean Prod:140368. https://doi.org/10.1016/j.jclepro.2023.140368

Abegunde VO, Sibanda M, Obi A (2019) The dynamics of climate change adaptation in Sub-Saharan Africa: a review of climate-smart agriculture among small-scale farmers. Climate 7(11):132. https://doi.org/10.3390/cli7110132

Adegbeye MJ, Ravi Kanth Reddy P, Obaisi AI, Elghandour MMMY, Oyebamiji KJ, Salem AZM, Morakinyo-Fasipe OT, Cipriano-Salazar M, Camacho-Díaz LM (2020a) Sustainable agriculture options for production, greenhouse gasses and pollution alleviation, and nutrient recycling in emerging and transitional nations – an overview. J Clean Prod 242:118319. https://doi.org/10.1016/j.jclepro.2019.118319

Adegbeye MJ, Salem AZM, Reddy PRK, Elghandour MMM, Oyebamiji KJ (2020b) Waste recycling for the eco-friendly input use efficiency in agriculture and livestock feeding. In: Kumar S, Meena RS, Jhariya MK (eds) Resources use efficiency in agriculture. Springer, Singapore, pp 1–45. https://doi.org/10.1007/978-981-15-6953-1_1

Adeyemi O, Grove I, Peets S, Norton T (2017) Advanced monitoring and management systems for improving sustainability in precision irrigation. Sustain For 9(3):353. https://doi.org/10.3390/su9030353

Afi M, Parsons J (2023) Integrated vs. specialized farming systems for sustainable food production: comparative analysis of systems' technical efficiency in Nebraska. Sustain For 15(6):5413. https://doi.org/10.3390/su15065413

Aguilera E, Díaz-Gaona C, García-Laureano R, Reyes-Palomo C, Guzmán GI, Ortolani L, Sánchez-Rodríguez M, Rodríguez-Estévez V (2020) Agroecology for adaptation to climate change and resource depletion in the Mediterranean region. A review. Agric Syst 181:102809. https://doi.org/10.1016/j.agsy.2020.102809

Akhtar N, Syakir Ishak MI, Bhawani SA, Umar K (2021) Various natural and anthropogenic factors responsible for water quality degradation: a review. Water 13(19):2660. https://doi.org/10.3390/w13192660

Alders RG, Dumas SE, Rukambile E, Magoke G, Maulaga W, Jong J, Costa R (2018) Family poultry: multiple roles, systems, challenges, and options for sustainable contributions to household nutrition security through a planetary health lens. Matern Child Nutr 14:12668. https://doi.org/10.1111/mcn.12668

Altieri MA, Nicholls CI (2017) The adaptation and mitigation potential of traditional agriculture in a changing climate. Clim Chang 140(1):33–45. https://doi.org/10.1007/s10584-013-0909-y

Altieri MA, Nicholls CI, Henao A, Lana MA (2015) Agroecology and the design of climate change-resilient farming systems. Agron Sustain Dev 35(3):869–890. https://doi.org/10.1007/s13593-015-0285-2

Asai M, Moraine M, Ryschawy J, de Wit J, Hoshide AK, Martin G (2018) Critical factors for crop-livestock integration beyond the farm level: a cross-analysis of worldwide case studies. Land Use Policy 73:184–194. https://doi.org/10.1016/j.landusepol.2017.12.010

Asbjornsen H, Hernandez-Santana V, Liebman M, Bayala J, Chen J, Helmers M, Ong CK, Schulte LA (2014) Targeting perennial vegetation in agricultural landscapes for enhancing ecosystem services. Renewable Agric Food Syst 29(2):101–125. https://doi.org/10.1017/S1742170512000385

Ashitha A, Rakhimol KR, Mathew J (2021) Fate of the conventional fertilizers in environment. In: Controlled release fertilizers for sustainable agriculture. Academic, London, pp 25–39. https://doi.org/10.1016/B978-0-12-819555-0.00002-9

Ashworth AJ, Chastain JP, Moore PA Jr (2020) Nutrient characteristics of poultry manure and litter. In: Waldrip HM, Pagliari PH, He Z (eds) Animal manure: production, characteristics, environmental concerns, and management, vol 67. ASA and SSSA, Madison, pp 63–87. https://doi.org/10.2134/asaspecpub67.c5

Atapattu AAAJ, Senarathne SHS, Raveendra SAST, Egodawatte WCP, Mensah S (2017) Effect of short term agroforestry systems on soil quality in marginal coconut lands in Sri Lanka. Agri Res J 54(3):324. https://doi.org/10.5958/2395-146X.2017.00060.6

Atapattu AJ, Ranasinghe CS, Nuwarapaksha TD, Udumann SS, Dissanayaka NS (2024) Sustainable agriculture and sustainable development goals (SDGs). In: Garwi J, Masengu R, Chiwaridzo O (eds) Emerging technologies and marketing strategies for sustainable agriculture. IGI Global, Hershey, pp 1–27. https://doi.org/10.4018/979-8-3693-4864-2.ch001

Avidov R, Sudharsan Varma V, Saadi I, Hanan A, Yoselevich I, Lublin A, Chen Y, Laor Y (2021) Physical and chemical indicators of transformations of poultry carcass parts and broiler litter during short term thermophilic composting. Waste Manag 119:202–214. https://doi.org/10.1016/j.wasman.2020.09.040

Ayambire RA, Amponsah O, Peprah C, Takyi SA (2019) A review of practices for sustaining urban and peri-urban agriculture: implications for land use planning in rapidly urbanising Ghanaian cities. Land Use Policy 84:260–277. https://doi.org/10.1016/j.landusepol.2019.03.004

Ayyam V, Palanivel S, Chandrakasan S (2019) Agroforestry for livelihood and biodiversity conservation. In: Ayyam V, Palanivel S, Chandrakasan S (eds) Coastal ecosystems of the tropics-adaptive management. Springer Singapore, Singapore, pp 363–389. https://doi.org/10.1007/978-981-138926-9_16

Babu S, Das A, Mohapatra KP, Yadav GS, Singh R, Tahashildar M, Devi MT, Das S, Panwar AS, Prakash N (2019) Pond dyke utilization: an innovative means for enhancing productivity and income under integrated farming system in North East Hill Region of India. Indian J Agric Sci 89(1):117–122. https://doi.org/10.56093/ijas.v89i1.86190

Babu S, Das A, Singh R, Mohapatra KP, Kumar S, Rathore SS, Yadav SK, Yadav P, Ansari MA, Panwar AS, Wani OA (2023) Designing an energy efficient, economically feasible, and environmentally robust integrated farming system model for sustainable food production in the Indian Himalayas. Sustainable Food Technol 1(1):126–142. https://doi.org/10.1039/D2FB00016D

Bacon CM, Getz C, Kraus S, Montenegro M, Holland K (2012) The social dimensions of sustainability and change in diversified farming systems. Ecol Soc 17(4). https://doi.org/10.5751/ES-05226-170441

Bahadur V, Chauhan S, Pati K (2016) Low cost protected cultivation of vegetable crops for sustainable farm income. https://www.researchgate.net/publication/333198368

Balafoutis A, Beck B, Fountas S, Vangeyte J, Wal T, Soto I, Gómez-Barbero M, Barnes A, Eory V (2017) Precision agriculture technologies positively contributing to GHG emissions mitigation, farm productivity and economics. Sustain For 9(8):1339. https://doi.org/10.3390/su9081339

Bao J, Jiang H, Li X (2022) Thirty years of rice-crab coculture in China – research progress and prospects. Rev Aquac 14(3):1597–1612. https://doi.org/10.1111/raq.12664

Bashir O, Ali T, Baba ZA, Rather GH, Bangroo SA, Mukhtar SD, Naik N, Mohiuddin R, Bharati V, Bhat RA (2021) Soil organic matter and its impact on soil properties and nutrient status. In: Dar GH, Bhat RA, Mehmood MA, Hakeem KR (eds) Microbiota and biofertilizers, 2. Springer, Cham, pp 129–159. https://doi.org/10.1007/978-3-030-61010-4_7

Bayih AZ, Morales J, Assabie Y, de By RA (2022) Utilization of internet of things and wireless sensor networks for sustainable smallholder agriculture. Sensors 22(9):3273. https://doi.org/10.3390/s22093273

Behera UK, France J (2016) Integrated farming systems and the livelihood security of small and marginal farmers in india and other developing countries. Adv Agron 138:235–282. https://doi.org/10.1016/bs.agron.2016.04.001

Behera UK, France J (2023) Farming systems research: concepts, design and methodology. Adv Agron 177:1–49. https://doi.org/10.1016/bs.agron.2022.08.001

Bell LW, Moore AD (2012) Integrated crop–livestock systems in Australian agriculture: trends, drivers and implications. Agric Syst 111:1–12. https://doi.org/10.1016/j.agsy.2012.04.003

Benedek A, Rokicki T, Szeberényi A (2023) Bibliometric evaluation of energy efficiency in agriculture. Energies 16(16):5942. https://doi.org/10.3390/en16165942

Bengtsson-Palme J, Heß S (2019) Strategies to reduce or eliminate resistant pathogens in the environment. Antibiot Drug Resist:637–673. https://doi.org/10.1002/9781119282549.ch24

Berti G, Mulligan C (2016) Competitiveness of small farms and innovative food supply chains: the role of food hubs in creating sustainable regional and local food systems. Sustain For 8(7):616. https://doi.org/10.3390/su8070616

Bezner Kerr R, Liebert J, Kansanga M, Kpienbaareh D (2022) Human and social values in agroecology: a review. Elem Sci Anth 10(1):00090. https://doi.org/10.1525/elementa.2021.00090

Bhardwaj DR, Navale MR, Sharma S (2017) Agroforestry practices in temperate regions of the world. In: Dagar J, Tewari V (eds) Agroforestry. Springer, Singapore, pp 163–187. https://doi.org/10.1007/978-981-10-7650-3_6

Bhattacharyya P, Pathak H, Pal S (2020) Crop management for climate-smart agriculture. In: Climate smart agriculture. Springer, Singapore, pp 85–111. https://doi.org/10.1007/978-981-15-9132-7_7

Black MJ, Roy A, Twinomunuji E, Kemausuor F, Oduro R, Leach M, Sadhukhan J, Murphy R (2021) Bottled biogas – an opportunity for clean cooking in Ghana and Uganda. Energies 14(13):3856. https://doi.org/10.3390/en14133856

Blay-Palmer A, van Veenhuizen R, Santini G, Hoffmann D, Wodni P, Stierand P, Stahlbrand L (2023) Urban–rural food system convergence as a vector for sustainable transformation. Urban Reg Agric:75–92. https://doi.org/10.1016/B978-0-12-820286-9.00015-7

Bonaudo T, Bendahan AB, Sabatier R, Ryschawy J, Bellon S, Leger F, Magda D, Tichit M (2014) Agroecological principles for the redesign of integrated crop–livestock systems. Eur J Agron 57:43–51. https://doi.org/10.1016/j.eja.2013.09.010

Bussoni A, Alvarez J, Cubbage F, Ferreira G, Picasso V (2019) Diverse strategies for integration of forestry and livestock production. Agrofor Syst 93(1):333–344. https://doi.org/10.1007/s10457-017-0092-7

Cheema MJM, Khan MA (2019) Information technology for sustainable agriculture. In: Farooq M, Pisante M (eds) Innovations in sustainable agriculture. Springer, Cham, pp 585–597. https://doi.org/10.1007/978-3-030-23169-9_19

Corbeels M, de Graaff J, Ndah TH, Penot E, Baudron F, Naudin K, Andrieu N, Chirat G, Schuler J, Nyagumbo I, Rusinamhodzi L, Traore K, Mzoba HD, Adolwa IS (2014) Understanding the impact and adoption of conservation agriculture in Africa: a multi-scale analysis. Agric Ecosyst Environ 187:155–170. https://doi.org/10.1016/j.agee.2013.10.011

Cosenz F, Rodrigues VP, Rosati F (2020) Dynamic business modeling for sustainability: exploring a system dynamics perspective to develop sustainable business models. Bus Strateg Environ 29(2):651–664. https://doi.org/10.1002/bse.2395

Costa C, Azoia NG, Coelho L, Freixo R, Batista P, Pintado M (2021) Proteins derived from the dairy losses and by-products as raw materials for non-food applications. Food Secur 10(1):135. https://doi.org/10.3390/foods10010135

Dagar JC, Gupta SR, Sileshi GW (2023) Urban and peri-urban agroforestry to sustain livelihood and food security in the face of global environmental change and epidemic threats. In: Agroforestry for sustainable intensification of agriculture in Asia and Africa. Springer Nature Singapore, Singapore, pp 89–118. https://doi.org/10.1007/978-981-19-4602-8_4

Dar NA, Lone BA, Alaie BA, Dar ZA, Gulzafar, Bahar FA, Haque SA, Singh KN (2018) Integrated farming systems for sustainable agriculture. In: Sengar R, Singh A (eds) Eco-friendly agrobiological techniques for enhancing crop productivity. Springer, Singapore, pp 111–127. https://doi.org/10.1007/978-981-10-6934-5_6

Das A, Layek J, Ramkrushna G, Subhash B, Devi MT, Dey U, Suting D, Yadav GS, Lyngdoh DBD, Prakash N (2019) Integrated organic farming system: an innovative approach for enhancing productivity and income of farmers in North Eastern Hill Region of India. Indian J Agric Sci 89(8):1267–1272. https://doi.org/10.56093/ijas.v89i8.92843

Das A, Datta D, Samajdar T, Idapuganti RG, Islam M, Choudhury BU, Mohapatra KP, Layek J, Subhash B, Yadav GS (2021) Livelihood security of small holder farmers in eastern Himalayas, India: pond based integrated farming system a sustainable approach. Curr Res Environ Sustain. https://doi.org/10.1016/j.crsust.2021.100076

Davis SC, Kauneckis D, Kruse NA, Miller KE, Zimmer M, Dabelko GD (2016) Closing the loop: integrative systems management of waste in food, energy, and water systems. J Environ Stud Sci 6(1):11–24. https://doi.org/10.1007/s13412-016-0370-0

De Stefano A, Jacobson MG (2018) Soil carbon sequestration in agroforestry systems: a meta-analysis. Agrofor Syst 92:285–299. https://doi.org/10.1007/s10457-017-0147-9

Delphin S, Snyder KA, Tanner S, Musálem K, Marsh SE, Soto JR (2022) Obstacles to the development of integrated land-use planning in developing countries: the case of Paraguay. Land 11(8):1339. https://doi.org/10.3390/land11081339

Demestichas K, Daskalakis E (2020) Data lifecycle management in precision agriculture supported by information and communication technology. Agronomy 10(11):1648. https://doi.org/10.3390/agronomy10111648

Dhonde M, Sahu K, Murty VVS (2022) The application of solar-driven technologies for the sustainable development of agriculture farming: a comprehensive review. Rev Environ Sci Biotechnol 21(1):139–167. https://doi.org/10.1007/s11157-022-09611-6

Dhyani S, Murthy IK, Kadaverugu R, Dasgupta R, Kumar M, Gadpayle KA (2021) Agroforestry to achieve global climate adaptation and mitigation targets: are south Asian countries sufficiently prepared? Forests 12(3):303. https://doi.org/10.3390/f12030303

Dissanayaka DMNS, Nuwarapaksha TD, Udumann SS, Dissanayake DKRPL, Atapattu AJ (2022) A sustainable way of increasing productivity of coconut cultivation using cover crops: a review. Circ Agric Syst 2(1):1–9. https://doi.org/10.48130/CAS-2022-0007

Dissanayaka DMNS, Dissanayake DKRPL, Udumann SS, Nuwarapaksha TD, Atapattu AJ (2023a) Agroforestry a key tool in the climate-smart agriculture context: a review on coconut cultivation in Sri Lanka. Front Agron 5:1162750. https://doi.org/10.3389/fagro.2023.1162750

Dissanayaka NS, Udumann SS, Nuwarapaksha TD, Atapattu AJ (2023b) Agroforestry: an avenue for resilient and productive farming through integrated crops and livestock production. In: Kiba DI (ed) Transitioning to zero hunger. MDPI Books, Basel. https://www.mdpi.com/books/edition/1311-transitioning-to-zero-hunger

Dissanayaka DMNS, Udumann SS, Nuwarapaksha TD, Atapattu AJ (2024) Harnessing the potential of Mucuna cover cropping: a comprehensive review of its agronomic and environmental benefits. Circ Agric Syst. https://doi.org/10.48130/cas-0024-0001

Dissanayake DKRPL, Dissanayaka DMNS, Udumann SS, Nuwarapaksha TD, Atapattu AJ (2023) Is biochar a promising soil amendment to enhance perennial crop yield and soil quality in the tropics? Technol Agron 3:4. https://doi.org/10.48130/TIA-2023-0004

Dorais M, Alsanius B (2015) Advances and trends in organic fruit and vegetable farming research. Hortic Rev 43:185–268. https://doi.org/10.1002/9781119107781.ch04

Duncan AJ, Tarawali SA, Thorne PJ, Valbuena D, Descheemaeker K, Homann-Kee Tui S (2013) Integrated crop-livestock systems – a key to sustainable intensification in Africa. Trop Grasslands 1(2):202. https://doi.org/10.17138/TGFT(1)202-206

Edwards P (2015) Aquaculture environment interactions: past, present and likely future trends. Aquaculture 447:2–14. https://doi.org/10.1016/j.aquaculture.2015.02.001

Ekanayaka EMGN, Dissanayake DKRPL, Udumann SS, Dissanayaka DMNS, Nuwarapaksha TD, Herath HMSK, Atapattu AJ (2023) Sustainable utilization of king coconut husk as a feedstock in biochar production with the highest conversion efficiency and desirable properties. IOP Conf Ser Earth Environ Sci 1235(10):012009. https://doi.org/10.1088/1755-1315/1235/1/012009

El-Sayed S, Cloutier S (2022) Weaving disciplines to conceptualize a regenerative food system. J Agric Food Sys Community Dev:1–29. https://doi.org/10.5304/jafscd.2022.112.003

Emmerson M, Morales MB, Oñate JJ, Batáry P, Berendse F, Liira J, Aavik T, Guerrero I, Bommarco R, Eggers S, Pärt T, Tscharntke T, Weisser W, Clement L, Bengtsson J (2016) How agricultural intensification affects biodiversity and ecosystem services. Adv Ecol Res 55:43–97. https://doi.org/10.1016/bs.aecr.2016.08.005

Estrada-Carmona N, Raneri JE, Alvarez S, Timler C, Chatterjee SA, Ditzler L, Kennedy G, Remans R, Brouwer I, den Berg KB, Talsma EF, Groot JCJ (2020) A model-based exploration of farm-household livelihood and nutrition indicators to guide nutrition-sensitive agriculture interventions. Food Secur 12(1):59–81. https://doi.org/10.1007/s12571-019-00985-0

Fahad S, Chavan SB, Chichaghare AR, Uthappa AR, Kumar M, Kakade V, Pradhan A, Jinger D, Rawale G, Yadav DK, Kumar V, Farooq TH, Ali B, Sawant AV, Saud S, Chen S, Poczai P (2022) Agroforestry systems for soil health improvement and maintenance. Sustainability 14(22):14877. https://doi.org/10.3390/su142214877

Fatima A, Singh VK, Subhash B, Singh RK, Upadhyay PK, Rathore SS, Kumar B, Hasanain M, Parween H (2023) Food production potential and environmental sustainability of different farming system models in northwest India. Front Sustainable Food Syst. https://doi.org/10.3389/fsufs.2023.959464

Fischer RA, Connor DJ (2018) Issues for cropping and agricultural science in the next 20 years. Field Crop Res 222:121–142. https://doi.org/10.1016/j.fcr.2018.03.008

Forbes A, Ellis K (2023) Parasite control in regenerative livestock farming. Livestock 28(3):112–120. https://doi.org/10.12968/live.2023.28.3.112

Fountas S, Mylonas N, Malounas I, Rodias E, Hellmann Santos C, Pekkeriet E (2020) Agricultural robotics for field operations. Sensors 20(9):2672. https://doi.org/10.3390/s20092672

Galhena DH, Freed R, Maredia KM (2013) Home gardens: a promising approach to enhance household food security and wellbeing. Agric Food Secur 2(1):8. https://doi.org/10.1186/2048-7010-2-8

Galli F, Grando S, Adamsone-Fiskovica A, Bjørkhaug H, Czekaj M, Duckett DG, Almaas H, Karanikolas P, Moreno-Pérez OM, Ortiz-Miranda D, Pinto-Correia T, Prosperi P, Redman M, Rivera M, Toma I, Sánchez-Zamora P, Šūmane S, Żmija K, Żmija D, Brunori G (2020) How do small farms contribute to food and nutrition security? Linking European small farms, strategies and outcomes in territorial food systems. Glob Food Sec 26:100427. https://doi.org/10.1016/j.gfs.2020.100427

Garcia K, Halmos V, Thongjued K, Dupuis JR, Gonthier DJ (2023) Net effects of pasture-raised poultry on arthropod communities driven by top-down and bottom-up forces in a mixed-cover crop system. Front Sustainable Food Syst 7:1162753. https://doi.org/10.3389/fsufs.2023

Garrett RD, Niles MT, Gil JDB, Gaudin A, Chaplin-Kramer R, Assmann A, Assmann TS, Brewer K, de Faccio Carvalho PC, Cortner O, Dynes R, Garbach K, Kebreab E, Mueller N, Peterson C, Reis JC, Snow V, Valentim J (2017) Social and ecological analysis of commercial integrated crop livestock systems: current knowledge and remaining uncertainty. Agric Syst 155:136–146. https://doi.org/10.1016/j.agsy.2017.05.003

Ghanizadeh H, Harrington KC (2019) Weed management in New Zealand pastures. Agronomy 9(8):448. https://doi.org/10.3390/agronomy9080448

Gil JDB, Cohn AS, Duncan J, Newton P, Vermeulen S (2017) The resilience of integrated agricultural systems to climate change. WIREs Clim Change 8(4). https://doi.org/10.1002/wcc.461

Gómez-Carmona O, Buján-Carballal D, Casado-Mansilla D, López-de-Ipiña D, Cano-Benito J, Cimmino A, Poveda-Villalón M, García-Castro R, Almela-Miralles J, Apostolidis D, Drosou A, Tzovaras D, Wagner M, Guadalupe-Rodriguez M, Salinas D, Esteller D, Riera-Rovira M, González A, Clavijo-Ágreda J, … Bujalkova N (2023) Mind the gap: the AURORAL ecosystem for the digital transformation of smart communities and rural areas. Technol Soc 74:102304. https://doi.org/10.1016/j.techsoc.2023.102304

González-García E, Gourdine JL, Alexandre G, Archimède H, Vaarst M (2012) The complex nature of mixed farming systems requires multidimensional actions supported by integrative research and development efforts. Animal 6(5):763–777. https://doi.org/10.1017/S1751731111001923

Gurung K, Bhandari H, Paris T (2016) Transformation from rice farming to commercial aquaculture in Bangladesh: implications for gender, food security, and livelihood. Gend Technol Dev 20(1):49–80. https://doi.org/10.1177/0971852415618747

Hallett S, Hoagland L, Toner E (2016) Urban agriculture: environmental, economic, and social perspectives. Hortic Rev:65–120. https://doi.org/10.1002/9781119281269.ch2

Hanjra MA, Blackwell J, Carr G, Zhang F, Jackson TM (2012) Wastewater irrigation and environmental health: implications for water governance and public policy. Int J Hyg Environ Health 215(3):255–269. https://doi.org/10.1016/j.ijheh.2011.10.003

Haque MM, Belton B, Alam MM, Ahmed AG, Alam MR (2016) Reuse of fish pond sediments as fertilizer for fodder grass production in Bangladesh: potential for sustainable intensification and improved nutrition. Agric Ecosyst Environ 216:226–236. https://doi.org/10.1016/j.agee.2015.10.004

Hardi EH, Susmiyati HR, Diana R, Palupi NP, Agriandini M, Saptiani G, Asikin AN, Agustina (2023) Traditional polyculture as a mangrove restoration solution in Delta Mahakam, East Kalimantan Indonesia. IOP Conf Ser Earth Environ Sci 1208(1):012056. https://doi.org/10.1088/1755-1315/1208/1/012056

Hassan K, Pervin M, Mondal F, Mala M (2016) Habitat management: a key option to enhance natural enemies of crop pest. Univers J Plant Sci 4(4):50–57. https://doi.org/10.13189/ujps.2016.040402

Hathaway MD (2016) Agroecology and permaculture: addressing key ecological problems by rethinking and redesigning agricultural systems. J Environ Stud Sci 6(2):239–250. https://doi.org/10.1007/s13412-015-0254-8

Hatt S, Döring TF (2023) Designing pest suppressive agroecosystems: principles for an integrative diversification science. J Clean Prod 432:139701. https://doi.org/10.1016/j.jclepro.2023.139701

He H, Liu L, Munir S, Bashir NH, Wang Y, Yang J, Li C (2019) Crop diversity and pest management in sustainable agriculture. J Integr Agric 18(9):1945–1952. https://doi.org/10.1016/S2095-3119(19)62689-4

Hellermann M (2017) Gardening on the fringes: a case study at city tech. Food, Cult Soc 20(4):651–669. https://doi.org/10.1080/15528014.2017.1357952

Herrero M, Grace D, Njuki J, Johnson N, Enahoro D, Silvestri S, Rufino MC (2013) The roles of livestock in developing countries. Animal 7:3–18. https://doi.org/10.1017/S1751731112001954

Higgins V, Bryant M, Howell A, Battersby J (2017) Ordering adoption: materiality, knowledge and farmer engagement with precision agriculture technologies. J Rural Stud 55:193–202. https://doi.org/10.1016/j.jrurstud.2017.08.011

Jangir CK, Thakur A, Bijani H, Thakur P, Kumar S, Meena RS, Bedwal S, Rani K, Shukla UN, Meena AK, Dev P (2022) Residual nitrogen for succeeding crops in legume-based crop-

ping system. In: Advances in legumes for sustainable intensification. Academic, London, pp 113–132. https://doi.org/10.1016/B978-0-323-85797-0.00024-0
Jerneck A, Olsson L (2013) More than trees! Understanding the agroforestry adoption gap in subsistence agriculture: insights from narrative walks in Kenya. J Rural Stud 32:114–125. https://doi.org/10.1016/j.jrurstud.2013.04.004
Jhariya MK, Meena RS, Banerjee A (2021) Ecological intensification of natural resources towards sustainable productive system. In: Jhariya MK, Meena RS, Banerjee A (eds) Ecological intensification of natural resources for sustainable agriculture. Springer, Singapore, pp 1–28. https://doi.org/10.1007/978-981-33-4203-3_1
Jose S (2012) Agroforestry for conserving and enhancing biodiversity. Agrofor Syst 85:1–8. https://doi.org/10.1007/s10457-012-9517-5
Kakamoukas G, Sarigiannidis P, Maropoulos A, Lagkas T, Zaralis K, Karaiskou C (2021) Towards climate smart farming – a reference architecture for integrated farming systems. Telecom 2(1):52–74. https://doi.org/10.3390/telecom2010005
Kanter DR, Musumba M, Wood SLR, Palm C, Antle J, Balvanera P, Dale VH, Havlik P, Kline KL, Scholes RJ, Thornton P, Tittonell P, Andelman S (2018) Evaluating agricultural tradeoffs in the age of sustainable development. Agric Syst 163:73–88. https://doi.org/10.1016/j.agsy.2016.09.010
Karunathilake EMBM, Le AT, Heo S, Chung YS, Mansoor S (2023) The path to smart farming: innovations and opportunities in precision agriculture. Agriculture 13(8):1593. https://doi.org/10.3390/agriculture13081593
Khoshnevisan B, Duan N, Tsapekos P, Awasthi MK, Liu Z, Mohammadi A, Angelidaki I, Tsang DCW, Zhang Z, Pan J, Ma L, Aghbashlo M, Tabatabaei M, Liu H (2021) A critical review on livestock manure biorefinery technologies: sustainability, challenges, and future perspectives. Renew Sust Energ Rev 135:110033. https://doi.org/10.1016/j.rser.2020.110033
Kiani AK, Sardar A, Khan WU, He Y, Bilgic A, Kuslu Y, Raja MAZ (2021) Role of agricultural diversification in improving resilience to climate change: an empirical analysis with Gaussian paradigm. Sustain For 13(17):9539. https://doi.org/10.3390/su13179539
Kremen C, Miles A (2012) Ecosystem services in biologically diversified versus conventional farming systems: benefits, externalities, and trade-offs. Ecol Soc 17:4. https://doi.org/10.5751/ES-05035-170440
Kremen C, Iles A, Bacon C (2012) Diversified farming systems: an agroecological, systems-based alternative to modern industrial agriculture. Ecol Soc 17(4). https://doi.org/10.5751/ES-05103-170444
Kumar V, Wankhede KG, Gena HC (2015) Role of cooperatives in improving livelihood of farmers on sustainable basis. Am J Educ Res 3(10):1258–1266. https://doi.org/10.12691/education-3-10-8
Kumar R, Krishna B, Sundaram PK, Kumawat N, Jeet P, Singh AK (2022) Crop diversification. In: Sustainable agriculture systems and technologies. Wiley, Hoboken, pp 63–80. https://doi.org/10.1002/9781119808565.ch5
Ladha JK, Jat ML, Stirling CM, Chakraborty D, Pradhan P, Krupnik TJ, Sapkota TB, Pathak H, Rana DS, Tesfaye K, Gerard B (2020) Achieving the sustainable development goals in agriculture: the crucial role of nitrogen in cereal-based systems. Adv Agron 163:39–116. https://doi.org/10.1016/bs.agron.2020.05.006
Lawson G, Dupraz C, Watté J (2019) Can silvoarable systems maintain yield, resilience, and diversity in the face of changing environments? In: Agroecosystem diversity. Academic, London, pp 145–168. https://doi.org/10.1016/B978-0-12-811050-8.00009-1
Leach M, Nisbett N, Cabral L, Harris J, Hossain N, Thompson J (2020) Food politics and development. World Dev 134:105024. https://doi.org/10.1016/j.worlddev.2020.105024
Lebrazi S, Fikri-Benbrahim K (2022) Potential of tree legumes in agroforestry systems and soil conservation. In: Advances in legumes for sustainable intensification. Academic, London, pp 461–482. https://doi.org/10.1016/B978-0-323-85797-0.00004-5
Lee J, Gereffi G, Beauvais J (2012) Global value chains and agrifood standards: challenges and possibilities for smallholders in developing countries. Proc Natl Acad Sci 109(31):12326–12331. https://doi.org/10.1073/pnas.0913714108

Lemaire G, Franzluebbers A, Carvalho PCDF, Dedieu B (2014) Integrated crop–livestock systems: strategies to achieve synergy between agricultural production and environmental quality. Agric Ecosyst Environ 190:4–8. https://doi.org/10.1016/j.agee.2013.08.009

Liu J, Mooney H, Hull V, Davis SJ, Gaskell J, Hertel T, Lubchenco J, Seto KC, Gleick P, Kremen C, Li S (2015) Systems integration for global sustainability. Science 347(6225). https://doi.org/10.1126/science.1258832

Liu S, Li X, Chen D, Duan Y, Ji H, Zhang L, Chai Q, Hu X (2020) Understanding land use/land cover dynamics and impacts of human activities in the Mekong Delta over the last 40 years. Global Ecol Conserv 22:e00991. https://doi.org/10.1016/j.gecco.2020.e00991

Lu HR, El Hanandeh A (2019) Life cycle perspective of bio-oil and biochar production from hardwood biomass; what is the optimum mix and what to do with it? J Clean Prod 212:173–189. https://doi.org/10.1016/j.jclepro.2018.12.025

Manyi-Loh C, Mamphweli S, Meyer E, Makaka G, Simon M, Okoh A (2016) An overview of the control of bacterial pathogens in cattle manure. Int J Environ Res Public Health 13(9):843. https://doi.org/10.3390/ijerph13090843

Marchetti L, Cattivelli V, Cocozza C, Salbitano F, Marchetti M (2020) Beyond sustainability in food systems: perspectives from agroecology and social innovation. Sustain For 12(18):7524. https://doi.org/10.3390/su12187524

Martin G, Moraine M, Ryschawy J, Magne M-A, Asai M, Sarthou J-P, Duru M, Therond O (2016) Crop–livestock integration beyond the farm level: a review. Agron Sustain Dev 36(3):53. https://doi.org/10.1007/s13593-016-0390-x

Meena RS, Kumar S, Yadav GS (2020) Soil carbon sequestration in crop production. In: Meena R (ed) Nutrient dynamics for sustainable crop production. Springer Singapore, Cham, pp 1–39. https://doi.org/10.1007/978-981-13-8660-2_1

Migliorini P, Wezel A (2017) Converging and diverging principles and practices of organic agriculture regulations and agroecology. A review. Agron Sustain Dev 37(6):63. https://doi.org/10.1007/s13593-017-0472-4

Montagnini F, Metzel R (2017) The contribution of agroforestry to sustainable development goal 2: end hunger, achieve food security and improved nutrition, and promote sustainable agriculture. In: Montagnini F (ed) Integrating landscapes: agroforestry for biodiversity conservation and food sovereignty. Advances in agroforestry, 12. Springer, Cham, pp 11–45. https://doi.org/10.1007/978-3-319-69371-2_2

Montes-Londoño I, Calle A, Montes O, Montes A (2022) Hacienda Pinzacuá: an example of regenerative agriculture amidst a transformed landscape in the Colombian Andes. In: Biodiversity islands: strategies for conservation in human-dominated environments. Springer, Cham, pp 305–335. https://doi.org/10.1007/978-3-030-92234-4_12

Moraine M, Duru M, Nicholas P, Leterme P, Therond O (2014) Farming system design for innovative crop-livestock integration in Europe. Animal 8(8):1204–1217. https://doi.org/10.1017/S1751731114001189

Moraine M, Melac P, Ryschawy J, Duru M, Therond O (2017) A participatory method for the design and integrated assessment of crop-livestock systems in farmers' groups. Ecol Indic 72:340–351. https://doi.org/10.1016/j.ecolind.2016.08.012

Morales H, Coronato F, Carvalho S, Saravia A, Schweitzer A, Burlamaqui A, Tourrand JF (2016) Building new human–natural systems for sustainable pasture management in South America. In: Dong S, Kassam KA, Tourrand J, Boone R (eds) Building resilience of human-natural systems of pastoralism in the developing world. Springer, Cham, pp 177–208. https://doi.org/10.1007/978-3-319-30732-9_5

Nadeem F, Nawaz A, Farooq M (2019) Crop rotations, fallowing, and associated environmental benefits. In: Oxford research encyclopedia of environmental science. Oxford University Press, New York. https://doi.org/10.1093/acrefore/9780199389414.013.197

Nair PKR, Nair VD, Kumar BM, Haile SG (2009) Soil carbon sequestration in tropical agroforestry systems: a feasibility appraisal. Environ Sci Pol 12(8):1099–1111. https://doi.org/10.1016/j.envsci.2009.01.010

Nechifor I, Florea MA, Radu-Rusu R-M, Pascal C (2022) Influence of supplemental feeding on body condition score and reproductive performance dynamics in Botosani karakul sheep. Agriculture 12(12):2006. https://doi.org/10.3390/agriculture12122006

Nkukwana TT (2018) Global poultry production: current impact and future outlook on the South African poultry industry. S Afr J Anim Sci 48(5):869–884. https://doi.org/10.4314/sajas.v48i5.7

Nuwarapaksha TD, Udumann SS, Dissanayaka DMNS, Dissanayake DKRPL, Atapattu AJ (2022) Coconut based multiple cropping systems: an analytical review in Sri Lankan coconut cultivations. Circ Agric Syst 2(1):1–7. https://doi.org/10.48130/CAS-2022-0008

Nuwarapaksha TD, Dissanayaka NS, Udumann SS, Atapattu AJ (2023a) Gliricidia as a beneficial crop in resource-limiting agroforestry systems in Sri Lanka. Indian J Agrofor 25:12–18

Nuwarapaksha TD, Rajapaksha UN, Ekanayake J, Weerasooriya SA, Atapattu AJ (2023b) Exploring the economic viability of integrating Jamnapari goat into underutilized pastures under coconut cultivations in Coconut Research Institute, Sri Lanka. Biol Life Sci Forum 27:27. https://doi.org/10.3390/IECAG2023-14990

Nuwarapaksha TD, Udumann SS, Dissanayaka NS, Atapattu AJ (2023c) Coconut-based livestock farming: a sustainable approach to enhancing food security in Sri Lanka. In: Kiba DI (ed) Transitioning to zero hunger. MDPI Books, Basel. https://www.mdpi.com/books/edition/1311-transitioning-to-zero-hunger

Nwuche CO, Gupta S, Akor J, Nweze JE, Nweze JA, Unah VU (2023) Biogas from manure: the future of renewable natural gas and its implications. In: Arshad M (ed) Climate changes mitigation and sustainable bioenergy harvest through animal waste. Springer Nature Switzerland, Cham, pp 171–214. https://doi.org/10.1007/978-3-031-26224-1_7

O'Connor J, Hoang SA, Bradney L, Dutta S, Xiong X, Tsang DCW, Ramadass K, Vinu A, Kirkham MB, Bolan NS (2021) A review on the valorisation of food waste as a nutrient source and soil amendment. Environ Pollut 272:115985. https://doi.org/10.1016/j.envpol.2020.115985

Obaisi AI, Adegbeye MJ, Elghandour MMMY, Barbabosa-Pliego A, Salem AZM (2022) Natural resource management and sustainable agriculture. In: Lackner M, Sajjadi B, Chen WY (eds) Handbook of climate change mitigation and adaptation. Springer, Cham, pp 2577–2613. https://doi.org/10.1007/978-3-030-72579-2_133

Pandey HO, Upadhyay D (2022) Global livestock production systems: classification, status, and future trends. In: Emerging issues in climate smart livestock production. Academic, London, pp 47–70. https://doi.org/10.1016/B978-0-12-822265-2.00017-X

Paramesh V, Kumar P, Shamim M, Ravisankar N, Arunachalam V, Nath AJ, Mayekar T, Singh R, Prusty AK, Rajkumar RS, Panwar AS, Reddy V, Malay P, Das A, Kannappa MK, Subhash B, Kashyap P (2022) Integrated farming systems as an adaptation strategy to climate change: case studies from diverse agro-climatic zones of India. Sustain For 14:11629. https://doi.org/10.3390/su141811629

Pathak H (2015) Greenhouse gas emission from Indian agriculture: trends, drivers and mitigation strategies. Proc Indian Natl Sci Acad 81(5):1133–1149. https://doi.org/10.16943/ptinsa/2015/v81i5/48333

Paus E (2012) Confronting the middle-income trap: insights from small latecomers. Stud Comp Int Dev 47(2):115–138. https://doi.org/10.1007/s12116-012-9110-y

Pavlidis G, Tsihrintzis VA (2018) Environmental benefits and control of pollution to surface water and groundwater by agroforestry systems: a review. Water Resour Manag 32(1):1–29. https://doi.org/10.1007/s11269-017-1805-4

Petersen-Rockney M, Baur P, Guzman A, Bender SF, Calo A, Castillo F, De Master K, Dumont A, Esquivel K, Kremen C, LaChance J, Mooshammer M, Ory J, Price MJ, Socolar Y, Stanley P, Iles A, Bowles T (2021) Narrow and brittle or broad and nimble? Comparing adaptive capacity in simplifying and diversifying farming systems. Front Sustainable Food Syst 5:564900. https://doi.org/10.3389/fsufs.2021.564900

Planisich A, Utsumi SA, Larripa M, Galli JR (2021) Grazing of cover crops in integrated crop-livestock systems. Animal 15(1):100054. https://doi.org/10.1016/j.animal.2020.100054

Plieninger T, Muñoz-Rojas J, Buck LE, Scherr SJ (2020) Agroforestry for sustainable landscape management. Sustain Sci 15(5):1255–1266. https://doi.org/10.1007/s11625-020-00836-4

Poccard-Chapuis R, Navegantes Alves L, Grise MM, Bâ A, Coulibaly D, Ferreira LA, Lecomte P (2014) Landscape characterization of integrated crop–livestock systems in three case studies of the tropics. Renewable Agric Food Syst 29(3):218–229. https://doi.org/10.1017/S174217051400009X

Pretty J (2018) Intensification for redesigned and sustainable agricultural systems. Science 362(6417):eaav0294. https://doi.org/10.1126/science.aav0294

Pretty J, Bharucha ZP (2014) Sustainable intensification in agricultural systems. Ann Bot 114(8):1571–1596. https://doi.org/10.1093/aob/mcu205

Radingoana MP, Dube T, Mollel MHN, Letsoalo JM (2019) Perceptions on greywater reuse for home gardening activities in two rural villages of Fetakgomo Local Municipality, South Africa. Phys Chem Earth Parts A/B/C 112:21–27. https://doi.org/10.1016/j.pce.2019.02.009

Radingoana MP, Dube T, Mazvimavi D (2020) Progress in greywater reuse for home gardening: opportunities, perceptions and challenges. Phys Chem Earth Parts A/B/C 116:102853. https://doi.org/10.1016/j.pce.2020.102853

Raj A, Jhariya MK, Yadav DK, Banerjee A, Meena RS (2019) Agroforestry: a holistic approach for agricultural sustainability. In: Jhariya M, Banerjee A, Meena R, Yadav D (eds) Sustainable agriculture, forest and environmental management. Springer Singapore, Singapore, pp 101–131. https://doi.org/10.1007/978-981-13-6830-1_4

Raj A, Jhariya MK, Khan N, Banerjee A, Poonam, Meena RS, Rani K (2021) Resource conservation for sustainable development. In: Jhariya MK, Banerjee A, Meena RS, Kumar S, Raj A (eds) Sustainable intensification for agroecosystem services and management. Springer Singapore, Singapore, pp 457–492. https://doi.org/10.1007/978-981-16-3207-5_14

Ram A, Dev I, Uthappa AR, Kumar D, Kumar N, Chaturvedi OP, Dotaniya ML, Meena BP (2017) Reactive nitrogen in agroforestry systems of India. In: The Indian nitrogen assessment. Elsevier, Duxford, pp 207–218. https://doi.org/10.1016/B978-0-12-811836-8.00014-8

Ramakrishna S, Ngowi A, De Jager H, Awuzie BO (2020) Emerging industrial revolution: symbiosis of industry 4.0 and circular economy: the role of universities. Sci Technol Soc 25(3):505–525. https://doi.org/10.1177/0971721820912918

Rana JC, Bisht IS (2023) Reviving smallholder hill farming by involving rural youth in food system transformation and promoting community-based agri-ecotourism: a case of Uttarakhand State in North-Western India. Sustain For 15(11):8816. https://doi.org/10.3390/su15118816

Rathore SS, Subhash B, El-Sappah AH, Shekhawat K, Singh VK, Singh RK, Upadhyay PK, Singh R (2022) Integrated agroforestry systems improve soil carbon storage, water productivity, and economic returns in the marginal land of the semi-arid region. Saudi J Biol Sci. https://doi.org/10.1016/j.sjbs.2022.103427

Ratnadass A, Fernandes P, Avelino J, Habib R (2012) Plant species diversity for sustainable management of crop pests and diseases in agroecosystems: a review. Agron Sustain Dev 32(1):273–303. https://doi.org/10.1007/s13593-011-0022-4

Reckling M, Watson CA, Whitbread A, Helming K (2023) Diversification for sustainable and resilient agricultural landscape systems. Agron Sustain Dev 43(4):44. https://doi.org/10.1007/s13593-023-00898-5

Redhu NS, Thakur Z, Yashveer S, Mor P (2022) Artificial intelligence: a way forward for agricultural sciences. Bioinform Agric:641–668. https://doi.org/10.1016/B978-0-323-89778-5.00007-6

Roberts E (2017) Agroforestry for the Northeastern United States: research, practice, and possibilities. In: Montagnini F (ed) Integrating landscapes: agroforestry for biodiversity conservation and food sovereignty. Advances in agroforestry, 12. Springer, Cham, pp 79–126. https://doi.org/10.1007/978-3-319-69371-2_4

Rosati A, Borek R, Canali S (2021) Agroforestry and organic agriculture. Agrofor Syst 95(5):805–821. https://doi.org/10.1007/s10457-020-00559-6

Rydhmer L, Gourdine JL, de Greef K, Bonneau M (2014) Evaluation of the sustainability of contrasted pig farming systems: breeding programmes. Animal 8(12):2016–2026. https://doi.org/10.1017/S175173111400216X

Sabatier R, Oates LG, Brink GE, Bleier J, Jackson RD (2015) Grazing in an uncertain environment: modeling the trade-off between production and robustness. Agron J 107(1):257–264. https://doi.org/10.2134/agronj14.0357

Sánchez J, Curt MD, Robert N, Fernández J (2019) Biomass resources. In: The role of bioenergy in the bioeconomy. Academic, London, pp 25–111. https://doi.org/10.1016/B978-0-12-813056-8.00002-9

Sánchez AC, Kamau HN, Grazioli F, Jones SK (2022) Financial profitability of diversified farming systems: a global meta-analysis. Ecol Econ 201:107595. https://doi.org/10.1016/j.ecolecon.2022.107595

Sanderson MA, Archer D, Hendrickson J, Kronberg S, Liebig M, Nichols K, Schmer M, Tanaka D, Aguilar J (2013) Diversification and ecosystem services for conservation agriculture: outcomes from pastures and integrated crop–livestock systems. Renewable Agric Food Syst 28(2):129–144. https://doi.org/10.1017/S1742170512000312

Sangothari A, Archana HA, Vasuki A, Surya R, Keerthana T (2024) Biodiversity conservation in agricultural landscapes: the role of integrated farming systems. Int J Environ Clim Chang 14(2):577–583. https://doi.org/10.9734/ijecc/2024/v14i23972

Sanjeevi P, Prasanna S, Siva Kumar B, Gunasekaran G, Alagiri I, Vijay Anand R (2020) Precision agriculture and farming using Internet of Things based on wireless sensor network. Trans Emerg Telecommun Technol 31(12). https://doi.org/10.1002/ett.3978

Sarkar D, Dubey PK, Chaurasiya R, Sankar A, Shikha, Chatterjee N, Ganguly S, Meena VS, Meena SK, Parewa HP, Rakshit A (2021) Organic interventions conferring stress tolerance and crop quality in agroecosystems during the United Nations Decade on Ecosystem Restoration. Land Degrad Dev 32(17):4797–4816. https://doi.org/10.1002/ldr.4094

Scherr SJ, Shames S, Friedman R (2012) From climate-smart agriculture to climate-smart landscapes. Agric Food Secur 1(1):12. https://doi.org/10.1186/2048-7010-1-12

Schiere HJB, Louis Baumhardt R, Van Keulen H, Whitbread AM, Bruinsma AS, Goodchild TAV, Gregorini P, Slingerland MMA, Hartwell B (2015) Mixed crop-livestock systems in semiarid regions. In: Dryland agriculture. ASA, CSSA, and SSSA, Madison, pp 227–291. https://doi.org/10.2134/agronmonogr23.2ed.c8

Schröder JJ, Schulte RPO, Creamer RE, Delgado A, van Leeuwen J, Lehtinen T, Rutgers M, Spiegel H, Staes J, Tóth G, Wall DP (2016) The elusive role of soil quality in nutrient cycling: R review. Soil Use Manag 32(4):476–486. https://doi.org/10.1111/sum.12288

Scott JM, Gaden CA, Edwards C, Paull DR, Marchant R, Hoad J, Sutherland H, Coventry T, Dutton P (2013) Selection of experimental treatments, methods used and evolution of management guidelines for comparing and measuring three grazed farmlet systems. Anim Prod Sci 53(8):628. https://doi.org/10.1071/AN12265

Segovia-Hernández JG, Contreras-Zarazúa G, Ramírez-Márquez C (2023) Sustainable design of water–energy–food nexus: a literature review. RSC Sustainability 1(6):1332–1353. https://doi.org/10.1039/D3SU00110E

Sekaran U, Lai L, Ussiri DAN, Kumar S, Clay S (2021) Role of integrated crop-livestock systems in improving agriculture production and addressing food security – a review. J Agric Food Res 5:100190. https://doi.org/10.1016/j.jafr.2021.100190

Shamila SK, Udumann SS, Dissanayaka NS, Rajaratnam K, Atapattu AJ (2024) Assessing the impact of king coconut husk ash and biochar, combined with chemical fertilizer application, on enhancing soil fertility in coconut plantations. Crops 4(2):227–241. https://doi.org/10.3390/crops4020017

Shanka D (2020) Roles of eco-friendly low input technologies in crop production in sub-Saharan Africa. Cogent Food Agric 6(1):1843882. https://doi.org/10.1080/23311932.2020.1843882

Sharma N, Bohra B, Pragya N, Ciannella R, Dobie P, Lehmann S (2016) Bioenergy from agroforestry can lead to improved food security, climate change, soil quality, and rural development. Food Energy Secur 5(3):165–183. https://doi.org/10.1002/fes3.87

Sharma V, Tripathi AK, Mittal H (2022) Technological revolutions in smart farming: current trends, challenges & future directions. Comput Electron Agric 201:107217. https://doi.org/10.1016/j.compag.2022.107217

Shrestha B, Chang S, Bork E, Carlyle C (2018) Enrichment planting and soil amendments enhance carbon sequestration and reduce greenhouse gas emissions in agroforestry systems: a review. Forests 9(6):369. https://doi.org/10.3390/f9060369

Shyam CS, Shekhawat K, Rathore SS, Subhash B, Singh RK, Upadhyay PK, Dass A, Fatima A, Kumar S, Sanketh GD, Singh VK (2023) Development of integrated farming system model – a step towards achieving biodiverse, resilient and productive green economy in agriculture for small holdings in India. Agronomy 13:955. https://doi.org/10.3390/agronomy13040955

Singh AK, Burman RR (2019) Agricultural extension reforms and institutional innovations for inclusive outreach in India. In: Agricultural extension reforms in South Asia. Academic, London, pp 289–315. https://doi.org/10.1016/B978-0-12-818752-4.00016-3

Singh S, Kumawat KC (2023) Integrated farming system for improving sustainability in agriculture. Indian Farming 73(9):40–43

Singh R, Machanuru R, Singh B, Shrivastava M (2021) Climate-resilient agriculture. In: Global climate change. Elsevier, Amsterdam, pp 45–61. https://doi.org/10.1016/B978-0-12-822928-6.00016-2

Singh M, Patton RN, Mollier RT, Pongener N, Yadav R, Singh V, Katiyar R, Singh GD, Deori S, Doley S, Chaudhary JK, Subhash B, Kalita H, Mishra VK (2023) Indigenous chicken production system in different agro-ecology of Indian Himalayan Region: implication on food and economic security. Front Nutr 10:1244413. https://doi.org/10.3389/fnut.2023.1244413

Smith J, Pearce BD, Wolfe MS (2013) Reconciling productivity with protection of the environment: is temperate agroforestry the answer? Renewable Agric Food Syst 28(1):80–92. https://doi.org/10.1017/S1742170511000585

Smith G, Nandwani D, Kankarla V (2017) Facilitating resilient rural-to-urban sustainable agriculture and rural communities. Int J Sustain Dev World Ecol 24(6):485–501. https://doi.org/10.1080/13504509.2016.1240723

Solér C (2012) Conceptualizing sustainably produced food for promotional purposes: a sustainable marketing approach. Sustain For 4(3):294–340. https://doi.org/10.3390/su4030294

Sulc RM, Franzluebbers AJ (2014) Exploring integrated crop–livestock systems in different ecoregions of the United States. Eur J Agron 57:21–30. https://doi.org/10.1016/j.eja.2013.10.007

Šūmane S, Kunda I, Knickel K, Strauss A, Tisenkopfs T, des Ios Rios I, Rivera M, Chebach T, Ashkenazy A (2018) Local and farmers' knowledge matters! How integrating informal and formal knowledge enhances sustainable and resilient agriculture. J Rural Stud 59:232–241. https://doi.org/10.1016/j.jrurstud.2017.01.020

Surendra KC, Takara D, Hashimoto AG, Khanal SK (2014) Biogas as a sustainable energy source for developing countries: opportunities and challenges. Renew Sust Energ Rev 31:846–859. https://doi.org/10.1016/j.rser.2013.12.015

Szymczak LS, Carvalho PCDF, Lurette A, de Moraes A, Nunes PADA, Martins AP, Moulin CH (2020) System diversification and grazing management as resilience-enhancing agricultural practices: the case of crop-livestock integration. Agric Syst 184:102904. https://doi.org/10.1016/j.agsy.2020.102904

Taylor JR, Lovell ST (2021) Designing multifunctional urban agroforestry with people in mind. Urban Agric Reg Food Syst 6(1). https://doi.org/10.1002/uar2.20016

Teague R, Barnes M (2017) Grazing management that regenerates ecosystem function and grazingland livelihoods. Afr J Range Forage Sci 34(2):77–86. https://doi.org/10.2989/10220119.2017.1334706

Tittonell P (2020) Assessing resilience and adaptability in agroecological transitions. Agric Syst 184:102862. https://doi.org/10.1016/j.agsy.2020.102862

Tittonell P, Piñeiro G, Garibaldi LA, Dogliotti S, Olff H, Jobbagy EG (2020) Agroecology in large scale farming – a research agenda. Front Sustainable Food Syst 4. https://doi.org/10.3389/fsufs.2020.584605

Troell M, Naylor RL, Metian M, Beveridge M, Tyedmers PH, Folke C, Arrow KJ, Barrett S, Crépin A-S, Ehrlich PR, Gren Å, Kautsky N, Levin SA, Nyborg K, Österblom H, Polasky S, Scheffer M, Walker BH, Xepapadeas T, de Zeeuw A (2014) Does aquaculture add resilience

to the global food system? Proc Natl Acad Sci 111(37):13257–13263. https://doi.org/10.1073/pnas.1404067111
Tunc TE (2017) Martha Turnbull's garden diary: unearthing the domestic sphere at rosedown plantation. Home Cult 14(2):167–191. https://doi.org/10.1080/17406315.2017.1396752
Udawatta R, Rankoth L, Jose S (2019) Agroforestry and biodiversity. Sustain For 11(10):2879. https://doi.org/10.3390/su11102879
Udawatta RP, Rankoth LM, Jose S (2021) Agroforestry for biodiversity conservation. In: Udawatta RP, Jose S (eds) Agroforestry and ecosystem services. Springer International Publishing, Cham, pp 245–274. https://doi.org/10.1007/978-3-030-80060-4_10
Udumann SS, Dissanayaka NS, Nuwarapaksha TD, Thelwadana EP, Atapattu AJ (2023) Assessing the growth potential of Sunn hemp (*Crotalaria juncea* L.) as a cover crop for major coconut-growing soils. Trends Horticult 6(2):3579. https://doi.org/10.24294/th.v6i2.3579
Vanlauwe B, Descheemaeker K, Giller KE, Huising J, Merckx R, Nziguheba G, Wendt J, Zingore S (2015) Integrated soil fertility management in sub-Saharan Africa: unravelling local adaptation. Soil 1(1):491–508. https://doi.org/10.5194/soil-1-491-2015
Velten S, Leventon J, Jager N, Newig J (2015) What is sustainable agriculture? A systematic review. Sustain For 7(6):7833–7865. https://doi.org/10.3390/su7067833
Watson CA, Topp CFE, Ryschawy J (2019) Linking arable cropping and livestock production for efficient recycling of N and P. In: Agroecosystem diversity. Academic, London, pp 169–188. https://doi.org/10.1016/B978-0-12-811050-8.00010-8
Wezel A, Casagrande M, Celette F, Vian JF, Ferrer A, Peigné J (2014) Agroecological practices for sustainable agriculture. A review. Agron Sustain Dev 34(1):1–20. https://doi.org/10.1007/s13593-013-0180-7
Woods A (2012) Rethinking the history of modern agriculture: British pig production, c.1910-65. Twent Century Br Hist 23(2):165–191. https://doi.org/10.1093/tcbh/hwr010
Yadav GS, Chandan D, Datta M, Ngachan SV, Yadav JS, Subhash B (2013) Comparative evaluation of traditional and improved farming practices in Tripura. Indian J Agric Sci 83(3):310–314
Yadav GS, Subhash B, Das A, Mohapatra KP, Singh R, Avasthe RK, Roy S (2020) No till and mulching enhance energy use efficiency and reduce carbon footprint of a direct-seeded upland rice production system. J Clean Prod. https://doi.org/10.1016/j.jclepro.122700
Yadav GS, Das A, Kandpal BK, Subhash B, Rattan L, Datta M, Das B, Singh R, Singh VK, Mohapatra KP, Chakraborty M (2021) The food-energy-water-carbon nexus in a maize-maize-mustard cropping sequence of the Indian Himalayas: an impact of tillage-cum-live mulching. Renew Sust Energ Rev. https://doi.org/10.1016/j.rser.2021.111602
Zahoor SA, Ahmad S, Ahmad A, Wajid A, Khaliq T, Mubeen M, Hussain S, Din MSU, Amin A, Awais M, Nasim W (2019) Improving water use efficiency in agronomic crop production. In: Hasanuzzaman M (ed) Agronomic crops. Springer, Singapore, pp 13–29. https://doi.org/10.1007/978-981-32-9783-8_2
Zhang Z, Yu J, Tian J (2023) Community participation, social capital cultivation and sustainable community renewal: a case study from Xi'an's Southern Suburbs, China. J Knowl Econ:1–34. https://doi.org/10.1007/s13132-023-01536-x

Chapter 5
Genetic Gain Enhancement in Maize Breeding for Sustained Food and Nutritional Security

Ganapati Mukri, Jayant S. Bhat, Shilpa Kumari, Gowtham K. V. Subhash Babu, and Chandu Singh

Abstract Maize is one of the important cereal crops of the world and has diverse uses as a source of energy, viz., food, fodder, feed, and fuel. In the past few decades, the genetic gain loss has slowly increased, which ultimately declined the yield growth in maize and also other cultivated crops. There are various approaches to enhancing genetic potential associated with different parameters like unlocking and creating genetic variation, improving heritability, increasing selection intensity, and shortening breeding cycle time. To enhance genetic gain, there were many approaches like modern breeding techniques and platforms, implemented with improved agronomic practices, field-based phenotyping, a better understanding of the genetic architecture of traits, and enhanced public–private association to ensure that the improved crop products effectively reach the farmers. Recent years have witnessed a significant shift towards genomics-driven approaches in maize, both qualitative and quantitative trait improvement like precision phenotyping, genomic selection, next-generation sequencing, genotyping-by-sequencing (GBS), marker-assisted breeding employing trait-associated traits, speed breeding, and doubled haploid techniques are in trends these days to enhance genetic gain of maize yield and nutritional quality. These advanced technologies not only improve genetic gain but also enhance nutritional security for fulfilling the demand for food and nutritional requirements of the growing billions of the population across the world.

Keywords Genomic selection · Next-generation sequencing · Genetic variance · Marker-assisted selection

G. Mukri · S. Kumari · G. K. V. · S. Babu · C. Singh (✉)
ICAR-Indian Agricultural Research Institute, New Delhi, India

J. S. Bhat
ICAR-Indian Agricultural Research Institute, Dharwad, India

S. Babu et al. (eds.), *Agricultural Diversification for Sustainable Food Production*, Sustainability Sciences in Asia and Africa,
https://doi.org/10.1007/978-981-97-7517-0_5

5.1 Introduction

Maize (*Zea mays* L.) is the most important cereal crop worldwide after wheat and rice. Maize is a highly versatile crop that has the potential to adapt to diverse agro-climatic conditions. Maize is utilized as a food commodity by humans, livestock, and feed industry and is also used as a raw material for industrial and processed food products (Gupta et al. 2015). Maize provides over 20% of total calories alone in human diets in more than 21 countries and over 30% in 12 countries, which are home to more than 310 million people. The major corn-producing countries are given in Table 5.1.

India is one of the top sixth maize producers in the world, contributing 2–3% of the total maize production worldwide, and among the top five maize exporters of the world, India contributes almost 14% of the total maize exported to different countries around the globe. South-East Asia is one of the largest markets for Indian corn, i.e., Indonesia, Malaysia, and Vietnam. India produces 38.09 million tons (MT) of maize from the crop cultivated in an area of 10.74 million hectares, and productivity is approximately 3.54 metric tons/ha (*Foreign Agricultural Service/USDA-2022-23*). The production trends of Indian corn are increasing in India from 2010 to 2022 (Fig. 5.1).

Maize also plays a key role in the sustained development of the poultry industry, which contributes around 47% of the share in poultry feed, a direct consumption share of 20%, cattle feed (14%), starch (14%), and food and beverage industry shares with 7%. By looking at its diverse utility, it is important to enhance its genetic potential thereby increasing production and productivity in the limited availability of land.

Table 5.1 The top corn-producing countries

Country	Area (million hectare)	Productivity (metric tons per hectare)	Production (million metric tons)	Rank	Percent of world production
United States	31.85	10.89	346.74	1	32
China	43.07	6.44	277.2	2	23
Brazil	22.40	6.12	137.0	3	11
European Union	8.84	5.93	52.40	4	5
Argentina	6.7	5.22	35.0	5	4
India	10.74	3.54	38.09	6	3
Mexico	6.89	4.07	28.08	7	2
Ukraine	4.05	6.67	27.0	8	2
South Africa	2.95	5.81	17.10	9	1
Russia	2.64	6.0	15.83	10	1

Source: Foreign Agricultural Service/USDA-2022-23

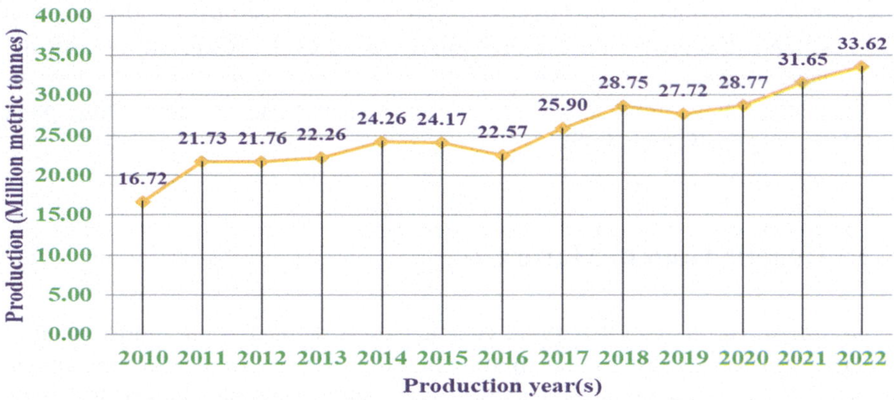

Fig. 5.1 Maize production scenario of India

In India, maize is cultivated in *kharif*, *rabi*, and *spring*. Among the three growing seasons of maize, the *kharif* season is the most important season, which covers around 80% of the total area of maize in India and contributes approximately 70% of maize production, followed by the *rabi* season, which covers around 15% of the total area and rest of the area occupied under spring season.

There are various biotic and abiotic and socioeconomic factors that influence the productivity of maize, viz., climate change, drought, crop protection, small land holding, and availability of seeds. Natural disasters are the major reasons, which fall in the category of weather-induced shocks leading to agricultural losses (Smith and Katz 2013). During 2003, the European heat wave resulted in a yield penalty of up to 13%, and the total direct economic losses in the agricultural sector were more than US$12 billion (COPA-COGECA 2003). In 2012, drought conditions caused a reduction in maize yields of >22% in the USA (Glotter and Elliott 2016; Elliott et al. 2018), which resulted in direct economic damages through crop losses alone of almost US$30 billion (Elliott et al. 2018). During the year 2019, a new biotic stress emerged in India, i.e., fall armyworm, and this pest was spread as far as Mizoram in the northeast, Uttar Pradesh in the north, Gujarat in the west, Chhattisgarh in central India, and also different states in the south India. The biggest victims were the farmers of northeastern states of India, where a total of 10,772 ha of maize crops were affected by fall armyworm (Sekar 2020).

To enhance the genetic gain, it has to explore the pre-breeding approach, modern tools and techniques, i.e., the genetic and genomic approaches, base editing approach, and selection approach to enhance the genetic gain, which ultimately enhances the productivity of maize. Various genetics and genomics approaches including precision phenotyping, genomic selection, next-generation sequencing, genotyping-by-sequencing (GBS), SNP genotyping, marker-assisted breeding employing trait-associated traits, speed breeding, and doubled haploidy techniques are in trend these days to enhance the genetic gain of maize yield.

Despite the complexity of the yield trait, it has been researched that it can be increased by the modification of some of the related genes, alleles, etc. Positive results have been obtained by targeting different yield constituents, indicating that there is plenty of scope for further yield improvement by genetic means. Among the yield-related genes, the successful targets are photosynthesis and starch biosynthetic genes, plant architecture, and transcriptional networks controlling plant development (Van Camp 2005).

5.2 Genetic Gain in Maize Crop

According to Hazel and Lush 1942, genetic gain is defined as an improvement in average genetic value in a population or the improvement in average phenotypic value due to selection within a population over cycles of breeding known as genetic gain. Genetic gain is not a new term; it was described earlier by scientists during crop domestication. As crop domestication means bringing the crop under cultivation, then onward the genetic gain of crop breeding emerged. The basic/fundamental principles of inheritance were established by Gregor Johann Mendel, and the same principles can explain the inheritance of traits (monogenic or oligogenic traits also known as qualitative traits), which fall into discrete categories.

The idea of the application of quantitative genetics for plant breeding is around 80 years old. Numerous researchers have established diverse plant breeding programs; however, the potential of this knowledge has not yet been fully utilized to enhance the rates of genetic improvement. This may be due to poor knowledge of genes or genomic regions that affect traits but now it is very easy for precisely stacking favorable genes. Bernardo (2008) reported genetic gain approach may be useful for monogenic or even oligogenic traits, it is not a solution for the improvement of quantitative traits only, these traits are known to be influenced by a large number of genes. For quantitative traits such as grain yield, genetic improvement over cycles of selection is necessary so that favorable alleles can be brought together gradually; hence, it is very much required to achieve genetic gain in the breeding programs (Fisher 1930). The enhancement of genetic gain can be achieved for any trait, but the trait must be heritable and also conferred by multiple loci.

5.2.1 *Status of Genetic Gain in Maize*

The approaches to enhance genetic potential are associated with different parameters: (i) unlocking and creating genetic variation, (ii) improving heritability, (iii) increased selection intensity, and (iv) shortening breeding cycle time, whereas the approaches to gap-closing are associated with different parameters are breeding practices, crop and field management, and socioeconomic policies (Xu et al. 2017). The positive linear increases in average yield from 1930 to 2012 have been achieved for three major crops in the USA such as maize, wheat, and soybean. Among these crops, the highest annual gain for yield is in maize crop (0.11 tons) followed by wheat crop (0.028 tons) and soybean crop (0.023 tons) (USDA, N 2013). However, the pace of crop yield has been slowing down recently. Yield growth for maize, rice, wheat, soybean, sugarcane, and vegetables has reduced from 2.20, 2.19, 2.95, 1.79, 0.70, and 1.55 for the period 1960–1990 to 1.74, 1.07, 0.79, 1.49, 0.69, and 1.10 for the period 1990–2010, respectively, and it is expected (projected genetic gain loss) that the yield growth for the period 2010–2050 will be reduced to 1.33, 0.62, 0.63, 0.91, 0.73, and 0.71, respectively (Pardey et al. 2014).

Further, improvement in genetic gain is of utmost priority to meet our future demand for agricultural products; hence, there is a need to be the deployment of modern breeding techniques and platforms, implemented with improved agronomic practices, field-based phenotyping, a better understanding of the genetic architecture of traits, and enhanced public-private association to ensure that the improved crop products effectively reach the farmers (Prasanna 2016). In the present scenario, the unfolding of various concepts in the areas of genomics, bioinformatics, and phenomics is generating innovative tools, resources, and technologies for the transformation of crop breeding programs. It is envisioned that these strategies and tools will be incorporated into maize breeding programs, thereby advancing these programs and addressing current and future challenges, and are expected to enhance genetic gains and impact on smallholder farmers.

5.2.2 Breeding Cycle of Genetic Gain

The genetic gain can only be realized by performing at least one cycle or more than one cycle of breeding required to enhance the genetic gain in the population. The breeding cycle required to realize genetic gain is given in Fig. 5.2.

The cycle of genetic gain encompasses four parts: generation, evaluation, selection, and recombination (Jessica and Rutkoski 2018), which are required to achieve

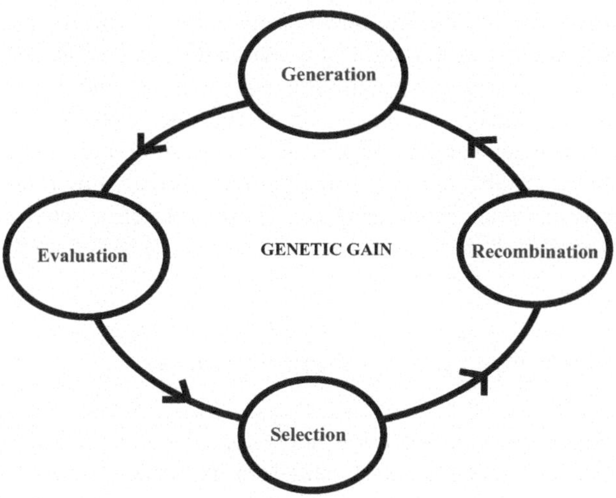

Fig. 5.2 The breeding cycle for targeted genetic gain

genetic gain over time, this entire process is known as population improvement (recurrent selection).

5.2.3 *The Genetic Gain Over Cycles of Breeding (e.g., Northern Corn Leaf Blight)*

Figure 5.3 shows distributions of phenotypic values for the base and selected populations. A vertical dotted line shows the mean distribution between both populations. The realized genetic gain was explained in an outcrossing population by using a large randomly mating population of maize (*Zea mays*), and this population is resistant to northern corn leaf blight (NLB) caused by *Exserohilum turcicum*. Poland reported that the resistance to NLB is a quantitative trait and more plants are chosen to isolate resistant ones.

Neither the selection process initiated nor pedigree relationships exist among the individuals, such population is known as the base population. This base population could serve to start a new breeding program for NLB resistance. Approximately 500 base population was selected and grown (generation), then phenotypically evaluated for resistance to NLB (evaluation). Based on phenotypic observation, a total of 50 individuals showed high resistance and were selected (selection). The selected individuals were allowed to open pollination in such a way that the male parents were randomly selected from the population (recombination). The progeny of these crosses forms a new population; hence, this new population is known as a selected population from the first cycle of selection (Fig. 5.1). The same cycle is repeated up to five times, till we achieve the separation of the selected population from base population and form the independent mean of the selected population. The genetic gain in NLB resistance was realized by the example maize breeding program after each cycle of selection can be observed by comparing the average level of NLB resistance within the selected populations with the average level of NLB resistance of the base population as shown in (Fig. 5.2). The mean value of the base population and the selected population was observed, and the mean values for both the population equal to genetic gain in the one cycle of selection. In the advanced cycle, the mean value of NLB resistance slightly increases means that the resistance factor is heritable in nature and does not influence environmental factors. As more cycles (up to five cycles) of selection, the difference between the selected population mean and the base population mean increases (Jessica and Rutkoski 2018).

5.2.4 *Breeder's Equation for Genetic Gains*

The genetic gain depends on selection intensity, selection accuracy, genetic variance, and number of years per cycle (Lush 1937). The description of these four core factors is as follows (Fig. 5.4).

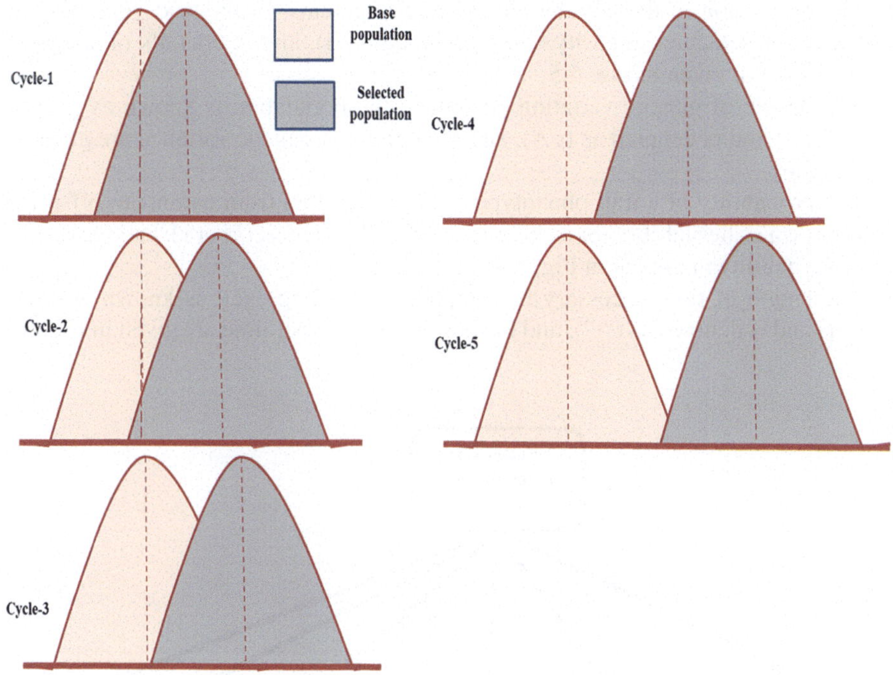

Fig. 5.3 The realized genetic gain over five cycles of selection

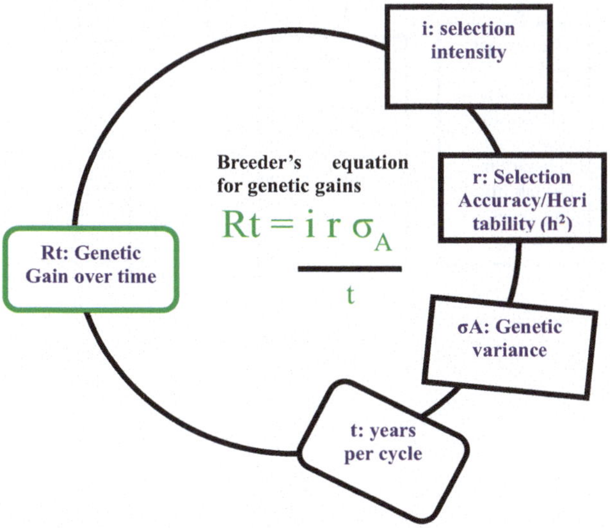

Fig. 5.4 Breeder's equation for genetic gains

1. The proportion of the population selected as parents for the next generation is known as selection intensity and is denoted as (i), and the details of selection intensity are given in Fig. 5.5
2. The degree of genetic variation present in the population is known as genetic variance and is denoted as (σA), and the details of genetic variance are given in Fig. 5.6
3. The probability of a trait phenotype to be transmitted from parents to offspring is known as heritability (narrow sense) and is denoted as (h^2 or r), and the details of heritability are given in Fig. 5.7
4. The length of time necessary to complete a breeding cycle is known as cycle time and is denoted as ("t"), and the details of cycle per time are given in Fig. 5.8

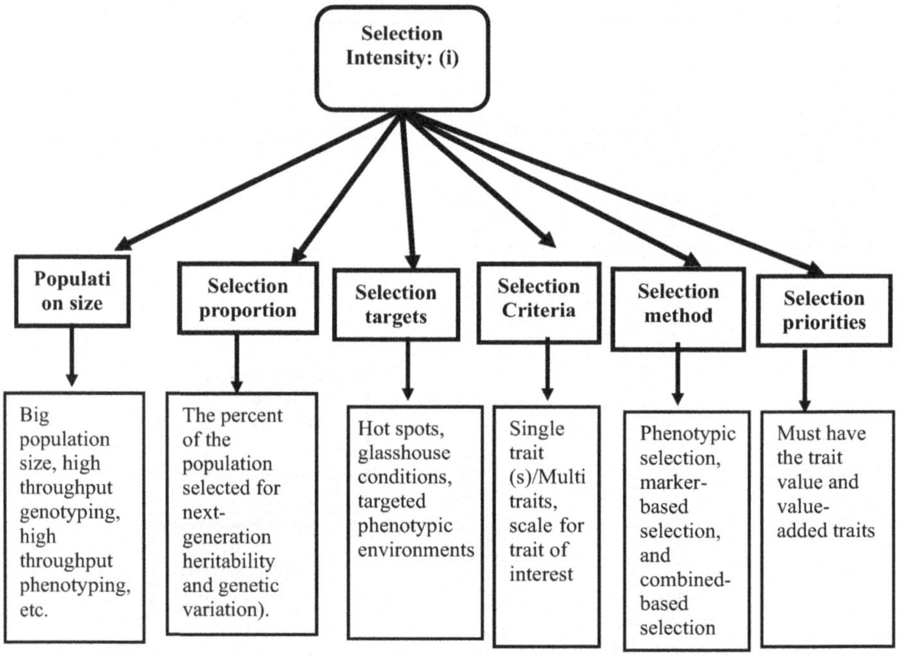

Fig. 5.5 Selection intensity (i)

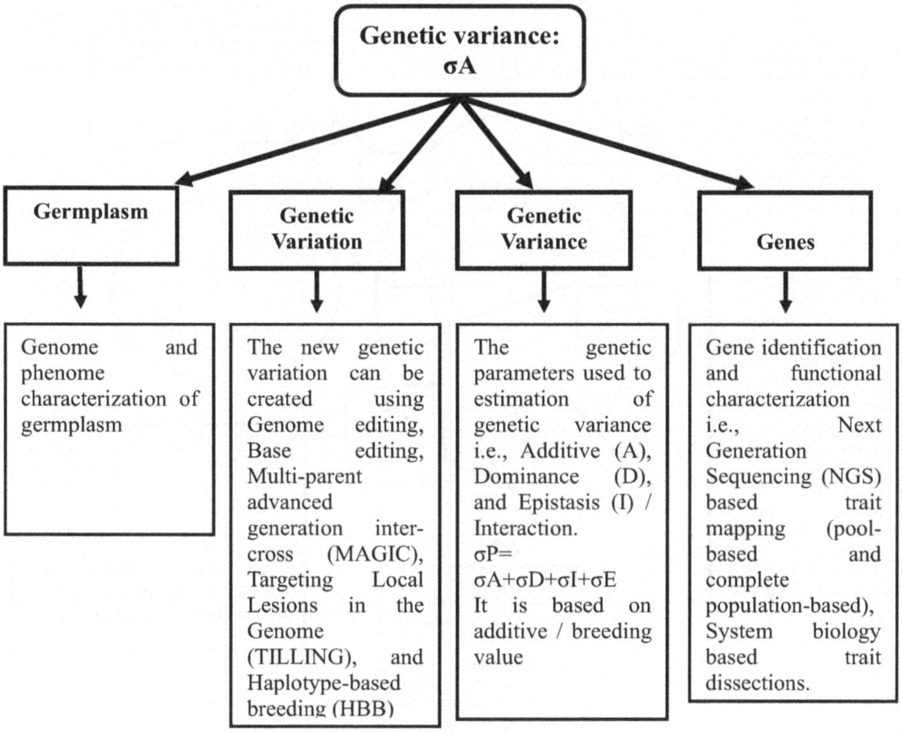

Fig. 5.6 Genetic variance (σA)

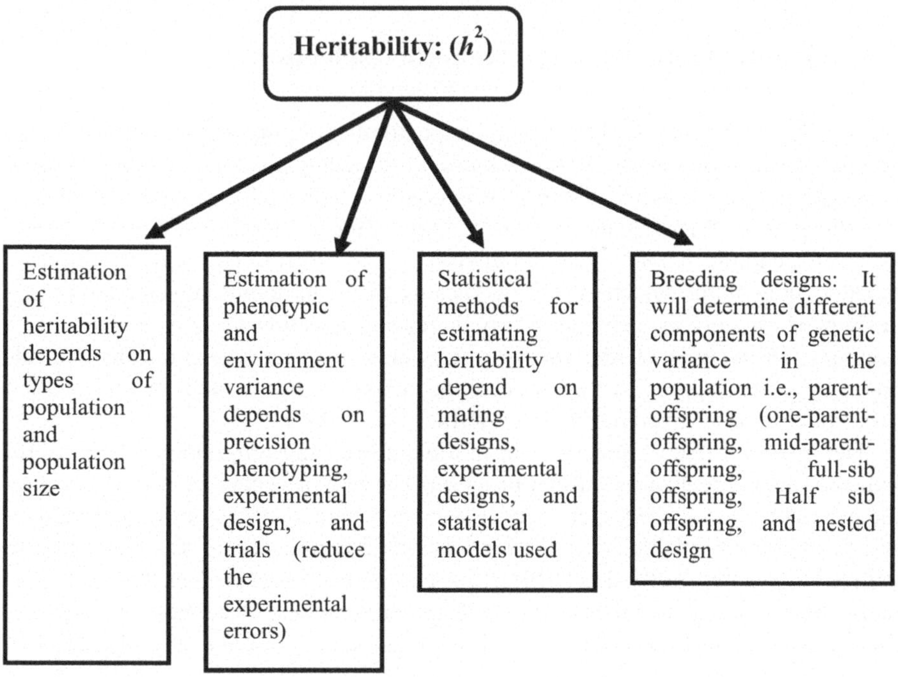

Fig. 5.7 Heritability (narrow sense—h^2 or r)

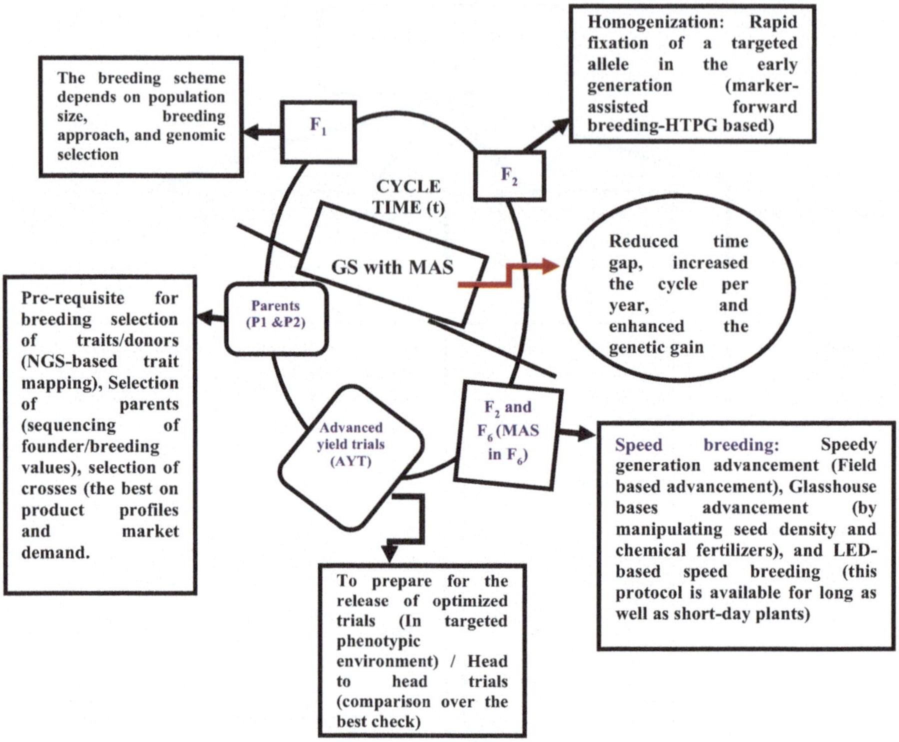

Fig. 5.8 Breeding cycle per time (GS: genomic selection, MAS: marker-assisted selection)—('t')

5.3 Genetic Gain Through Genomic Selection

Previously, the breeder used to screen a large number of potential individuals based on phenotypic observations. To better know, the breeding potential of those screened genotypic and phenotypic selection was coupled with pedigree-based prediction (Piepho et al. 2008) and genomic tools to harness the availability of novel and cost-effective genomic technologies (Meuwissen et al. 2001). The wider interest of breeders in genomic selection (GS) is because of the hastening of selection procedures by shortening the generation gap, increasing selection strength, and also high accuracy (Heslot et al. 2015). The short- as well as long-term genetic gains are balanced by various approaches while selecting crosses in the GS method. The detailed procedure of GS is given in Fig. 5.9 (Heffner et al. 2009).

The observed rate of genetic gain in a plant breeding program can be a useful indicator for the realization of that program. The rate of genetic gain is affected by the genetic nature of the target traits as well as genetic drift and also other factors, which are uncontrollable. The five fundamental components for assessing genetic gain were given by Rutkoski (2019); these components are (1) clear objectives, (2) meticulous data management, and (3) effective operations. Once these are in place, then it is possible to attain the next level of essential components, (4) effective

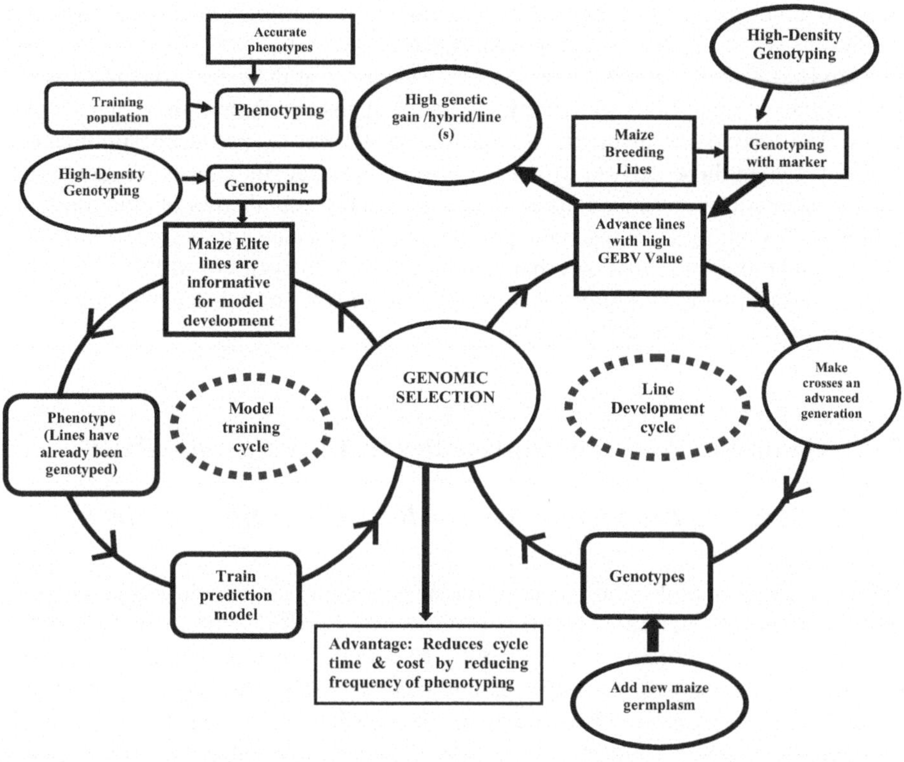

Fig. 5.9 Genetic gain through genomic selection

phenotyping, and (5) selection based on best linear unbiased prediction (BLUP). If a breeding program follows a four-step cycle process, i.e., generation, evaluation, selection, and recombination, it can be expected to realize genetic gain as long as a set of key components is in place.

The International Institute of Tropical Agriculture (IITA) under the maize improvement program improved a biparental maize population through marker-assisted recurrent selection (MARS) by employing single nucleotide polymorphism (SNPs) markers to boost grain yield and other yield attributes under abiotic (drought stress) condition (Bankole et al. 2017). Eathington (2005) and Crosbie et al. (2006) recommended that the genetic gain in maize crops can be attained using MARS by two-fold that of phenotypic selection in some reference populations. The comparison between the usefulness of one- and two-stage phenotypic selection over MARS in test crosses of double haploid lines was made by Mayor and Bernardo (2009), and they also observed that multiple cycles of MARS resulted in the same level of genetic gain as two-stage phenotypic selection but at a much lower cost. Beyene et al. (2015) compared the effectiveness of MARS with conventional pedigree selection using hybrids derived from 10 biparental maize populations and found higher gain for MARS than that for pedigree selection under both drought stress and optimum growing conditions. In the study carried out by Bankole et al. (2017), the average genetic gain in grain yield (GY) for the test crosses under well-watered conditions was 43 kg per cycle representing a relative gain of 1% per cycle. A separate study was undertaken to assess these test crosses at three locations for 2 years under rain-fed conditions to determine genetic gains in GY and other agronomic traits. The testcrosses involving S_1 lines derived from C_2 produced significantly higher GY than those derived from C_0 with a relative genetic gain of 3% per cycle. This investigation scrutinized the genetic gain in a biparental cross of elite drought-tolerant lines with good agronomic parameters for resistance to maize streak virus disease and other foliar diseases that had undergone two cycles of MARS. The population did respond to MARS for improvement in GY and other agronomic attributes in maize (Bankole et al. 2017).

5.4 Genomics-Assisted Approaches to Target Productivity

5.4.1 Targeted Production System-Based Breeding for Yield

With the advent of molecular breeding and new breeding tools such as genome/gene editing, promising signs are emerging that this could provide a new way forward to enhance the genetic gain in maize crops. Molecular breeding has been and will continue to be a staple of the improvement process. Both genome/gene editing and molecular breeding require knowledge of the associated gene sequences and, even better, the function and importance of the gene(s) in determining yield. This requires advances in the fundamental science of gene and process function. There is also a need to consider how processes are integrated, like source-sink balance within the whole plant in the agricultural environment. Molecular breeding for enhancing yield can rely upon many factors or various yield-related target genes. It would pave the way for augmenting the breeding by providing more dividends since the system decides the type of traits to be phenotyped and maximizes the genetic gain of the crop. As has already been discussed, yield is a multifactorial trait that has several genes involved in different biosynthetic pathways, and they play a significant role in enhancing grain yield.

Molecular breeding-based targeted traits combine objectivism with reductionism, resulting in a clear picture of how plants grow (van Bueren et al., 2018) and how to improve them (Yin et al. 2004; Munns and Richards 2007). Some yield-trait-associated genes that have been the area of research to target yield in different cereal crops since the last decade are as follows.

5.4.2 Increasing Source Strength by Targeting Photosynthesis Pathway Genes

Photosynthesis has been a preferred target in recent years for yield improvement concerns. There are several components of photosynthetic leaf gas exchange that have been targeted by Simkin et al. (2019) in line with photosynthetic models of where limitations lie (Zhu et al., 2010). Photosynthesis is the prime source of carbon assimilation in crop plants. The relative strength of various source and sink organs determines the movement of carbohydrates and other nutrients within the plant and therefore delimitate the growth to a larger extent. If we take the example of crop canopy, the association of photosynthesis with crop yield is not much understood (Paul et al. 2017). Keeping this point in mind, Wu et al. (2019) developed a model for diurnal canopy photosynthesis stomatal conductance, which connected leaf photosynthesis to crop yield in a crop canopy typical of an Australian cropping region. This investigation was carried out in both spring as well as summer seasons using wheat and sorghum crops for considering water-limited and water-unlimited conditions. The model is comprised of three main targets that have been proposed for enhancing C_3 and C_4 photosynthesis ability to enhance the crop yields: (i) maximum carboxylation rate of Rubisco; (ii) electron transport capacity; and (iii) mesophyll conductance for CO_2. It was found that the model suggested that increasing all three components together lifted yield improvements to 9.2% and 12.2% for sorghum and wheat, respectively, under full irrigation, but negligible improvement on increasing these components individually by 20%. Maximal improvements in crop yield were 7–8% for a 20% increase in the rate of electron transport in wheat as well as sorghum under full irrigation. Various researchers have altered the biochemical pathways of sink organs, and the enzymes involved were targeted for the conversion of sucrose into starch.

Downregulation of the plastidial isoform of adenylate kinase resulted in a 60% increase in starch in potato and also enhanced the tuber yield by 39% (Regierer et al. 2002). The manipulation of starch metabolism is important for improving plant yield for various reasons (Lloyd and Kossmann 2019). As starch is a main form of calories within plants, increasing starch concentrations in plant tissues can mean that less food or fodder has to be spent to supply the same energetic value (Ruckle et al. 2017). It is also a major sink within storage organs, so enhancing starch content, which leads to increased plant yield, as it simply shows more accumulation. Starch plays a significant role in overcoming abiotic stress (Thalmann and Santelia 2017) and transcriptional upregulation of stress-related genes (Stitt et al., 2007). In the end, starch has been shown to affect the developmental processes of plants (Matsoukas et al., 2013), and therefore, its manipulation could increase yield through altering the plant or seed development stage. ADP-glucose pyrophosphorylase (AGP) produces the substrate for starch synthesis and is thought to catalyze a rate-limiting step in the starch biosynthetic pathway.

The introduction of a bacterial or mutant form of the large subunit of AGP, which could be less sensitive to allosteric inhibition, can help to increase starch biosynthesis and enhance yield in maize (Giroux et al., 1996), potato (Stark et al., 1992), wheat (Smidansky et al., 2002), and rice (Smidansky et al., 2003). In all the cases, seed yield was enhanced by more than 20%, primarily as the result of an increase in the total number of seeds.

Further, alteration of starch branching enzyme (SBE) activity in Arabidopsis leaves has shown that it is a potential method for improving growth. The replacement of endogenous Arabidopsis activities with two from maize led to plants with an increased starch in their leaves, which could improve seed yield, most likely also due to increased nighttime sugar levels of the plant leaves (Liu et al. 2016).

Recently, it was also demonstrated that repression of the starch-phosphorylating Glucan Water-Dikinase (GWD) enzyme in the wheat endosperm improves both growth and seed production. This case study was demonstrated convincingly in glasshouse trials (Ral et al. 2012; Bowerman et al. 2016), but the same lines were examined in the open field condition, and yield reduction was observed (Whan et al. 2017). In this case, repressing genes that are involved in this process could improve forage and silage crops by improving their calorific value for animal feed (Weise et al. 2012; Ruckle et al. 2018).

5.4.3 Genetic Gain Through Marker-Assisted Recurrent Selection (MARS) and Marker-Assisted Selection (MAS)

Across the Globe, maize is cultivated over 160 Mha under rain-fed conditions, and the annual yield loss due to abiotic stresses (mainly drought) was about 25% (Edmeades 2008). The yield losses may be expected more in tropical countries, as these countries rely on unpredictable and erratic rainfall (Mhike et al. 2010). The gap between potential yield and yield under drought stress (DS) is often large; this gap can be reduced up to 20–25% through genetic improvement in drought tolerance, and further, the gap is also reduced (20–25%) through water-conserving agronomic practices (Edmeades 2013). The remaining gap of up to 50–60% can only be met by irrigation, when available and affordable (Edmeades et al. 2006). Edmeades et al. (1999) analyzed that the average yield loss due to drought was 17–60%, while *Striga* spp. caused about 10–100% yield losses (Lagoke et al., 1991). Fall Armyworm (*Spodoptera* spp.) causes yield loss up to 34% (Cruz et al. 1999; Lima et al. 2010); however, marker-assisted recurrent selection (MARS) and marker-assisted selection (MAS) in amalgamation with high-throughput technologies, precise phenotyping, and year-round nurseries (Fig. 5.10 and Table 5.2) could help to develop a

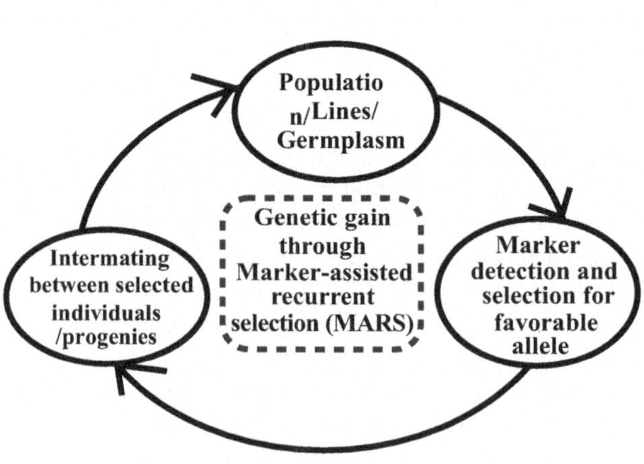

Fig. 5.10 Genetic gain through marker-assisted recurrent selection

Table 5.2 Applications of MAS in maize for yield and yield attributes traits

Marker	Target alleles/genes	Traits	References
RFLP	Introgression of favorable QTL alleles	Yield, plant height, and ear number	Stuber (1995) and Stuber et al. (1999)
RFLP	Early generation test of inbreeding	Yield and stalk lodging percentage	Eathington et al. (1997)
RFLP	Foreground selection of three loci	Yield and earliness	Bouchez et al. (2002)
RFLP	Introgression of three QTL alleles	Seedling emergency	Yousef and Juvik (2002)
SSR	Introgression of fertility restorer gene	Rf3 gene selection	Xia and Zheng (2002)
SSR	Introgression of four QTL alleles	Yield	Ho et al. (2002)

Table 5.3 Applications of MAS in maize for biotic and abiotic stress traits

Marker	Target alleles/genes	Traits	References
RFLP	Tracing of five loci during backcross	Southwestern corn borer resistance	Willcox et al. (2002)
RFLP	Introgression of beneficial QTL alleles	Stalk strength and European corn borer resistance	Flint-Garcia et al. (2003)
RFLP	Introgression of beneficial QTL alleles	Drought tolerance	Ribaut and Ragot (2007)
RFLP, SSR	Introgression of five beneficial QTL alleles	Drought tolerance	Ribaut et al. (2002)
SSR	Introgression of qMrdd8 gene	Rough dwarf disease resistance	Xu et al. (2020)
SSR	Introgression of msv1 gene	Maize streak virus (MSV) disease resistance	Mafu (2013)
SSR	Introgression of Ht1 gene	Northern corn leaf blight (NCLB) resistance	Puttarach et al. (2016)
SSR	Introgression of SDM resistance QTLs	Sorghum downy mildew (SDM) resistance	Sumathi et al. (2020)
SSR	Introgression of qHSR1 gene	Head smut resistance	Zhao et al. (2012, b)

climate-resilient maize germplasm, and this germplasm can be utilized to improve biotic and abiotic tolerance (drought) in maize (Xu et al. 2012; Prasanna et al. 2013).

The improvement of biparental maize populations for tolerance to biotic stress and abiotic stress through MAR and MARS (Table 5.3) is an advanced breeding program, as it harnesses several QTLs, which were carrying the most desirable combinations of favorable alleles using only significant and tightly linked markers to envisage performance of the population (Meuwissen et al. 2001; Bohra 2013). MARS is effective and superior to phenotypic selection (PS); through this, it can be accumulate favorable alleles for multiple trait improvement in maize and other crops (Johnson 2001, 2004; Yousef and Juvik 2001; Eathington et al., 2007; Crossa et al. 2013; Massman et al. 2013).

Frascaroli and Landi (1998) stated that the changes in the frequency of favorable alleles within a biparental population improved through MARS and would provide information on specific genomic regions that have responded to selection. Furthermore, the assessment of genetic gains in breeding programs provides an opportunity to critically analyze efficiency and plan new actions and strategies. The MARS is an effective tool to breed for complex traits because it enables the harnessing of those gene (s)/QTL (s), which even exhibit minor effects on the phenotype (Xu et al. 2012).

Edwards and Johnson investigated the changes in the frequency of favorable alleles in the F_2 MARS population of sweet corn, and they reported an increase in the frequency of the favorable alleles ranging from 0.50 to ≥0.80 at 11 out of 35 markers used, and one marker locus is fixed for the favorable allele in the studied population of sweet corn. Similarly, an increase in the frequency of favorable alleles for grain yield ranged from C0 (0.50) to C2 (0.65) in a double haploid population improved through MARS by Bernardo and Mayor. Semagn et al. (2015) and Beyene et al. (2016a, b) explored MARS techniques for enhancing the genetic gains in grain yield of tropical maize populations under drought stress, and they reported an average gain of 184 kg/ha/cycle under well-watered (WW) and 45 kg/ha under drought stress conditions, whereas an average gain of 105 kg/ha year^{-1} under WW and 51 kg/ha/year under drought stress using biparental maize populations, respectively.

5.5 Genetic Enhancement for Nutritional Traits in Maize

The maize kernels hold significant economic and nutritional value due to having high starch content, making up 75% of the matured seed weight. The maize seed contains protein, which constitutes 10% of the matured seed weight, and also oil content, which makes up 4.6%. Interestingly, oil is the most valuable byproduct resulting from the industrial processing of maize grain, offering a high-quality source of oil for human consumption (Hartings et al. 2012).

5.5.1 Genetics and Molecular Approach for Enhancement of Oil Content Traits in Maize

The maize crop oil is of superior quality and stands out as a preferable option compared to other edible oils due to its unique composition of fatty acids and its capacity to remain stable during both storage and cooking. In the United States, corn oil contributed 4.3% of total vegetable oil production and approximately 7.2% of domestic oil consumption. However, in India, the utilization of corn oil is less common. This is largely due to the wider availability of various other edible oils but, with the higher cost and lack of public awareness about corn oil, factors that limit its popularity in the country. Although some high-oil corn varieties/hybrids reported ranging from 3% to 7%, it is worth noting that these high-oil lines often come with reduced overall yield (White 2007).

In practical applications, high-oil corn has shown positive effects on livestock. Cattle fed with high-oil maize have revealed an increased body weight; while laying hens tend to have improved egg production and pigmentation. Additionally, lactating dairy cattle have shown enhanced feed intake. High-oil hybrids also tend to possess higher levels of yellow pigments like carotenoids and xanthophylls, which have positive implications for poultry feed and overall biological functions.

Germ oil is one of the highly valued products of the maize processing industry. Genetic enhancement of oil quantity and quality plays an important role in meeting the growing demand. Although the molecular basis for oil QTLs remains largely unknown, some key insights have been gained. Such as, a QTL known as qHO6 affecting maize seed oil and oleic acid content has been identified, and it encodes an enzyme responsible for the final step of oil synthesis. Subsequently, with the development of sequencing technology (Chia et al., 2012; Bukowski et al., 2018), an increasing number of molecular markers have been applied to QTL mapping, which helps to identify a large number of chromosomal regions and QTLs affecting oil concentration and fatty acid composition using different populations in maize (Song et al., 2004, Clark et al., 2006), and these studies indicated that the oil concentration and fatty acid composition were controlled by a few major genes and many minor genes with mainly additive effects. In addition, epistatic interactions also contribute to variations in oil content in specific populations (Yang et al. 2010; Fang et al., 2020).

Together with the use of molecular and biotechnology techniques, various breeding approaches have been employed to enhance oil content in maize kernels. Genetic diversity and recurrent selection have led to significant oil content improvements. For example, Alexander started a breeding program that achieved an oil concentration of 21.2% after 27 selection cycles, while another program increased oil content from 4.71% to 15.55% through recurrent selection and the main efforts to enhance the maize kernel oil content through breeding, which had yielded success, but in case of high-oil lines often suffer from reduced yield as compared to normal corn yield due to the energy-intensive process of oil synthesis, hence the yield reduced up to 5–10% compared to normal hybrids (Song et al. 2004). Singh et al. (2014) reported that the yield gap can be narrowed down through continuous selection for improved agronomic performance and yield in high-oil hybrids.

In 1896, at the University of Illinois, Hopkins began a divergent selection experiment, the main focus on oil and protein concentrations within the open-pollinated cultivar 'Burr's White,' which had white endosperm in corn kernels. Over the course of 70 generations, a mass selection was applied to choose ears with the highest oil content in maize kernels. Strangely, this process resulted in a substantially enhanced oil content, i.e., 3.6-fold in a population known as "Illinois High Oil Line." This groundbreaking experiment provided compelling evidence that oil content is an inheritable trait in maize. This crucial study entitled "The structure of the corn kernel and the composition of its different parts" laid the foundation for subsequent divergent mass selection research conducted by Hopkins and his colleagues in 1903.

Maize crop plays a key role as raw material for agroindustry, especially for the production of starch, and several mutants have been effectively employed for the production of modified starches. The most significant of this crop is perhaps waxy maize, which was discovered around the beginning of the twentieth century. The dry weight of maize kernels encompasses up to 70% starch, and it is a complex carbohydrate that stands as the primary energy reserve in the grain (Burrell 2003; Hannah and Boehlein 2017). Therefore, breeding the high-yielding maize cultivars offers a lot of potential to enhance starch content in maize kernels. A multibillion-dollar industry involved in the processing of starch and used in biofuels, paper production, high fructose corn syrup, and pharmaceuticals, in addition to its vital role as a source of calories for humans and animals (Smith 2008; Ranum et al. 2014; Egharevba 2019). Starch is a highly valued commodity, hence in order to improve grain yields as well as starch quality in maize, which supports efforts to feed a growing human population, it is crucial to enhance the efficiency of starch biosynthesis pathways.

Starch/carbohydrate serves as an immediate energy source for humans and also livestock, which contributes significantly to caloric intake. Moreover, starch has found widespread industrial applications. The food industry relies on maize starch for its thickening, gelling, and binding properties, making it a crucial ingredient in products ranging from soups to confectionery (Xu and Chang 2009). Beyond food, starch plays a key role in biofuel production, offering a renewable and sustainable source of energy. Additionally, innovations in biodegradable plastics leverage maize starch to create environmentally friendly alternatives, further emphasizing its industrial importance (Fasahat et al. 2019). The foundation of starch improvement breeding in maize lies in traditional breeding methods that began centuries ago. Early maize cultivars selectively propagated plants with desirable traits, such as larger kernels and higher starch content. These efforts laid the groundwork for the development of improved maize lines, even though the underlying genetic mechanism was not completely understood at the time.

5.5.2 *Molecular and Biotechnological Approach for Enhancement of Starch Content Traits in Maize*

The advent of modern genetics and molecular biology brought a deeper understanding of the genetic basis of starch synthesis in maize. The scientific community identified key genes and quantitative trait loci (QTLs) associated with starch content. For instance, the wx gene (waxy gene) was identified as a main player in determining the amylose content of maize starch, and it is an important quality trait (BeMiller & Whistler 2009).

Marker-assisted selection (MAS) revolutionized maize starch improvement by enabling breeders to select plants with desired genetic traits more accurately and efficiently. By using molecular markers linked to specific starch-related genes or QTLs, breeders could identify and select promising individuals for further breeding programs (Holland et al. 2003). Several QTLs were mapped and investigated the genetic architecture of starch content in maize kernels. For identifying different QTLs related to starch, they explored different QTL mapping methods and various biparent populations. The studies by Clark et al. (2006), Cook et al. (2012), Karn et al. (2017), Li et al. (2009), Lin et al. (2019), and Yang et al. (2013) have contributed to the identification of QTLs associated with starch content in maize kernels.

The genetic basis of the variation in kernel starch content was shown to be influenced by a combination of large effect as well as small effect additive quantitative trait loci (QTLs) and a restricted number of pairs of epistatic QTLs. The study based on pathways revealed that ZmTPS9, a gene responsible for encoding trehalose-6-phosphate synthase in the trehalose pathway, was identified as the causative gene for the observed quantitative trait locus (QTL) qSTA4-2. The knockout of ZmTPS9 resulted in an elevation in kernel starch content and subsequently, an increase in kernel weight in maize. These findings indicate the potential use of ZmTPS9 in enhancing maize starch production and overall yield. The present findings contribute to the existing body of research about the genetic underpinnings of starch content in maize kernels, hence offering significant insights for enhancing the amount and quality of starch in maize using genetic improvement strategies. Recent years have witnessed a significant shift towards genomics-driven approaches in maize starch improvement. High-throughput sequencing technologies, coupled with advanced computational tools, have enabled researchers to comprehensively study the maize genome and identify key regulatory elements and genes involved in starch synthesis. This knowledge has facilitated the development of maize lines with improved starch content and quality (Zhao et al. 2020). Biotechnology, including genetic modification (GM), has played a crucial role in maize starch improvement. Researchers have engineered maize lines with enhanced starch biosynthesis pathways, resulting in higher starch content and altered starch quality as well. These innovations have expanded the potential industrial applications of maize starch (Li et al. 2017).

The starch improvement breeding has increasingly been integrated with other traits, such as yield and yield attributing traits, biotic and abiotic resistance, and agronomic traits. This holistic approach ensures that improved starch content does not compromise the overall performance of maize lines, making them more appealing to farmers and also end-users (Duvick 2005).

The narrow genetic basis of starch content within tropical maize endosperm is the major bottleneck for genetic improvement. The screening involved high cost and short turnaround time of high-performance liquid chromatography-based quantification, which is laborious. However, with the help of predictive/associated markers, the identification of suitable genotypes for enhancing the genetic gain in maize's starch content becomes possible.

Genetic enhancement, through selective breeding, base editing technology, and biotechnology-driven approaches, has become an essential strategy to surge maize's nutritional value. As the global population increases, so does the demand for nutritious food sources, making it imperative to enhance the nutritional content of major staple crops. Genetic gain, the incremental improvement of desired traits in crop varieties over successive generations, has emerged as a central theme in modern maize breeding programs (Crossa et al. 2017).

5.5.3 Molecular and Biotechnological Approach for Enhancement Protein Content Traits in Maize

Another important component in maize is protein, which is indispensable for human nutrition, providing essential amino acids for growth, repair, and overall health (Shiferaw et al. 2011). Maize protein, while often considered of lower quality compared to certain other grains due to its limiting amino acid profile, remains indispensable for human nutrition. Protein is vital for the synthesis of enzymes, hormones, and structural components in the human body. While maize protein is not a complete protein on its own, it can complement other dietary protein sources to provide a well-rounded amino acid profile (Edmeades et al. 1992). Moreover, maize protein is a significant component of animal feed and contributes to the production of meat and dairy products that form essential elements of the human diet.

Several researchers studied and analyzed the genetic foundation of protein content in maize kernels (Dudley et al. 2004; Zhang et al. 2007; Wassom et al. 2008). Zhang et al. (2007) explored the QTL Cartographer and mapped six proteins that were shown to be linked. The range of phenotypic variation accounted for by individual QTL, and it was found to be between 5.19% and 6.66% of protein. These QTLs associated with protein concentrations exhibited a recurring pattern of detection within equivalent intervals, and their effects were consistently aligned with the sign of their phenotypic correlation. It was suggested that the inclusion of a wider range of germplasm is imperative to effectively identify more quantitative trait loci (QTL) and uncover more advantageous alleles about the chemical composition of maize grain (Li et al. 2004). The details of protein and oil content enhancement through MAS are given in Table 5.4.

Table 5.4 Applications of MAS in maize protein and oil content traits

Molecular	Target alleles/genes	Traits	References
SSR	Introgression of opaque2 gene	QPM	Pukalenthy et al. (2019), Kaur et al. (2020)
SSR	Introgression of crtRB1 gene	QPM	Goswami et al. (2019);Natesan et al. (2020)
SSR	Introgression of opaque2 and crtRB1 genes	QPM	Mehta et al. (2020)
SSR	Introgression of crtRB1, lcyE, and opaque2 genes	QPM	Zunjare et al. (2018)
SSR	Introgression of lpa2-2 allele	QPM	Sureshkumar et al. (2014a), Tamilkumar et al. (2014)
SSR	Introgression of lpa2 allele	QPM	Sureshkumar et al. (2014b)
SSR	Introgression of qHSR1 gene	QPM	Zhao et al. (2012)
SSR	Introgression of the ZmVTE4 gene	QPM	Feng et al. (2015)
SSR	Introgression of the lcyE gene	ProA	Yang et al. (2018)
SSR	Pyramiding of opaque2 and opaque16 genes	QPM	Sarika et al. (2018)
SSR	Mapping and validation of sugary1 and shrunken2 genes	QPM	Hossain et al. (2015)
SSR	Pyramiding of crtRB1 and lcyE genes into opaque2	QPM	Singh et al. (2021)
SSR	Introgression of qHO6 locus	Oil content	Hao et al. (2014)

QPM quality protein maize, *ProA* provitamin A

5.6 Conclusion

The maize crops not only provide food and nutritional security but also have tremendous industrial applications. In industrial applications especially, a lot of importance is given to starch extraction, protein, and oil content of the dry-weight maize kernels. In the modern era, genomic-driven tools and techniques (genomic selection, next-generation sequence technology, speed breeding, and marker-assisted recurrent selection) can be explored easily to enhance the number of cycles per year in a crop, which could help to improve the genetic gain of the crop. The ongoing genetic improvement of maize for enhanced starch and protein content represents a promising avenue to address the nutritional needs of a growing global population, promote sustainable agriculture, and support a range of industries that rely on this versatile cereal crop, and it is one of the complex challenges for crop breeders.

References

Bankole F, Menkir A, Olaoye G, Crossa J, Hearne S, Unachukwu N, Gedil M (2017) Genetic gains in yield & yield-related traits under drought stress & favorable environments in a maize population improved using marker-assisted recurrent selection. Front Plant Sci 8:808

BeMiller JN, Whistler RL (2009) Starch: chemistry & technology, Food Science & Technology International series, 3rd edn. Academic Press, Cambridge

Bernardo R (2008) Molecular markers & selection for complex traits in 768 plants: learning from the last 20 years. Crop Sci 48:1649–1664

Beyene Y, Semagn K, Mugo S, Tarekegne A, Babu R, Meisel B, Gakunga J (2015) Genetic gains in grain yield through genomic selection in eight bi-parental maize populations under drought stress. Crop Sci 55(1):154–163

Beyene Y, Semagn K, Crossa J, Mugo S, Gary N, Tarekegne A (2016a) Improving maize grain yield under drought stress & non-stress environments in sub-Saharan Africa using marker-assisted recurrent selection. Crop Sci 56:344–353. https://doi.org/10.2135/cropsci2015.02.0135

Beyene Y, Semagn K, Mugo S, Prasanna BM, Tarekegne A, Gakunga J (2016b) Performance & grain yield stability of maize populations developed using marker-assisted recurrent selection & pedigree selection procedures. Euphytica 208:285–297. https://doi.org/10.1007/s10681-015-1590-1

Bohra A (2013) Emerging paradigms in genomics-based crop improvement. Sci World J:585467. https://doi.org/10.1155/2013/585467

Bouchez A, Hospital F, Causse M, Gallais A, Charcosset A (2002) Marker-assisted introgression of favorable alleles at quantitative trait loci between maize elite lines. Genetics 162(4):1945–1959

Bowerman AF, Newberry M, Dielen AS, Whan A, Larroque O, Pritchard J, Ral JP (2016) Suppression of glucan, and water dikinase in the endosperm alters wheat grain properties, germination & coleoptile growth. Plant Biotechnol J 14(1):398–408

Bukowski R, Guo X, Lu Y, Zou C, He B, Rong Z, Wang B, Xu D, Yang B, Xie C, Fan L (2018) Construction of the third-generation Zea mays haplotype map. Gigascience 7(4):gix134

Burrell MM (2003) Starch: the need for improved quality or quantity – an overview. J Exp Bot 54(382):451–456

Chia JM, Song C, Bradbury PJ, Costich D, De Leon N, Doebley J, Elshire RJ, Gaut B, Geller L, Glaubitz JC, Gore M (2012) Maize HapMap2 identifies extant variation from a genome in flux. Nat Genet 44(7):803–807

Clark RM, Wagler TN, Quijada P, Doebley J (2006) A distant upstream enhancer at the maize domestication gene tb1 has pleiotropic effects on plant & inflorescent architecture. Nat Genet 38(5):594–597

Cook JP, McMullen MD, Holl JB, Tian F, Bradbury P, Ross-Ibarra J et al (2012) Genetic architecture of maize kernel composition in the nested association mapping & inbred association panels. Plant Physiol 158:824–834. https://doi.org/10.1104/pp.111.185033

COPA-COGECA (2003) Assessment of the impact of the heat wave & drought of the summer on agriculture & forestry. Tech. Rep., COPA-COGECA

Crosbie TM, Eathington SR, Johnson GR Sr, Edwards M, Reiter R, Stark S, Dobert R (2006) Plant breeding: past, present, & future. In: Plant breeding: the Arnel R. Hallauer International Symposium, Ames, Iowa, USA, pp 3–50

Crossa J, Beyene Y, Kassa S, Pérez P, Hickey JM, Chen C (2013) Genomic prediction in maize breeding populations with genotyping-by- sequencing. G3 3:1903–1926. https://doi.org/10.1534/g3.113.008227

Crossa J, Perez-Rodriguez P, Cuevas J, Montesinos-Lopez O, Jarquin D, de Los Campos G, Burgueno J (2017) Genomic selection in plant breeding: methods, models, & perspectives. Trends Plant Sci 22(11):961–975

Cruz I, Figueiredo MLC, Oliveira AC, Vasconcelos CA (1999) Damage of *Spodoptera frugiperda* (Smith) in different maize genotypes cultivated in soil under three levels of aluminum saturation. Inter J Pest Manag 45:293–296. https://doi.org/10.1080/096708799227707

Dudley J, Dijkhuizen A, Paul C, Coates S, Rocheford T (2004) Effects of random mating on marker–QTL associations in the cross of the Illinois high protein × Illinois low protein maize strains. Crop Sci 44(4):1419–1428

Duvick DN (2005) The contribution of breeding to yield advances in maize (*Zea mays* L.). Adv Agron 86:83–145

Eathington SR (2005) Practical applications of molecular technology in the development of commercial maize hybrids. In: Proceedings of the 60th Annual Corn & Sorghum Seed Research Conferences. American Seed Trade Association, Washington, DC
Eathington SR, Dudley JW, Rufener GK II (1997) Usefulness of marker-QTL associations in early generation selection. Crop Sci 37(6):1686–1693
Eathington SR, Crosbie TM, Edwards MD, Reiter RS, Bull JK (2007) Molecular markers in a commercial breeding program. Crop Sci 47:154–163. https://doi.org/10.2135/cropsci2007.04.0015IPBS
Edmeades GO (2008) Drought tolerance in maize: an emerging reality. In: James C (ed) Global status of commercialized biotechnology/GM crops. ISAAA, Ithaca, pp 1–12
Edmeades GO (2013) Progress in achieving & delivering drought tolerance in maize-an update. International Service for the Acquisition of Agri-Biotech Applications, Ithaca
Edmeades GO, Bolanos J, Lafitte HR (1992) Progress of breeding for drought tolerance in maize. In: Wilkinson D (ed) Proceedings of the 47th Annual Corn & Sorghum Industry Research Conference. ASTA, Washington, DC, pp 93–111
Edmeades GO, Bolanos J, Chapman SC, Lafitte HR, Bänziger M (1999) Selection improves drought tolerance in tropical maize populations: I. Gains in biomass, grain yield, & harvest index. Crop Sci 39:1306–1315. https://doi.org/10.2135/cropsci1999.3951306x
Edmeades GO, Bänziger M, Campos H, Schussler J (2006) Improving tolerance to abiotic stresses in staple crops: a random or planned process? In: Lamkey KR, Lee M (eds) Proceedings of the Plant Breeding: the Arnel R. Hallauer International Symposium. Blackwell Publishing, Ames, pp 293–309. https://doi.org/10.1002/9780470752708.ch21
Egharevba HO (2019) Chemical properties of starch & its application in the food industry. In: Chemical properties of starch, vol 9. https://doi.org/10.5772/intechopen.87777
Elliott J, Glotter M, Ruane AC, Boote KJ, Hatfield JL, Jones JW, ... Foster I (2018) Characterizing agricultural impacts of recent large-scale US droughts & changing technology & management. Agric Syst 159:275–281
Fang H, Fu X, Wang Y, Xu J, Feng H, Li W, Xu J, Jittham O, Zhang X, Zhang L, Yang N (2020) Genetic basis of kernel nutritional traits during maize domestication & improvement. Plant J 101(2):278–292
Fasahat P, Tafreshi ZK, Aghebati N (2019) Maize starch-based biodegradable materials: a state-of-the-art review. Int J Biol Macromol 134:408–419
Feng F, Wang Q, Liang C (2015) Enhancement of tocopherols in sweet corn by marker assisted backcrossing of ZmVTE4. Euphytica 206:513–521
Fisher RA (1930) The genetical theory of natural selection. Clarendon Press. https://doi.org/10.5962/bhl.title.27468
Flint-Garcia SA, Darrah LL, McMullen MD, Hibbard BE (2003) Phenotypic versus marker assisted selection for stalk strength & second-generation European corn borer resistance in maize. Theor Appl Genet 107(7):1331–1336
Frascaroli E, Landi P (1998) Allelic frequency changes of P1 gene in a maize population after recurrent selection for grain yield. Crop Sci 38:1391–1394. https://doi.org/10.2135/cropsci1998.0011183X003800050039x
Giroux MJ, Shaw J, Barry G, Cobb BG, Greene T, Okita T, Hannah LC (1996) A single mutation that increases maize seed weight. Proc Natl Acad Sci 93(12):5824–5829
Glotter M, Elliott J (2016) Simulating US agriculture in a modern Dust Bowl drought. Nat Plants 3(1):1–6
Goswami R, Rajkumar Uttamrao Z, Suphiya K, Aanchal B, Vignesh M, Firoz H, Thomas L (2019) Marker-assisted introgression of rare allele of β-carotene hydroxylase (crtRB1) gene into elite quality protein maize inbred for combining high lysine, tryptophan & provitamin A in maize. Plant Breed 138(2):174–183
Gupta HS, Hossain F, Muthusamy V (2015) Biofortification of maize: an Indian perspective. Indian J Genet 75(1):1–22

Hannah LC, Boehlein S (2017) Starch biosynthesis in maize endosperm. In: Maize kernel development. CABI, Wallingford, pp 149–159

Hao X, Li X, Yang X (2014) Transferring a major QTL for oil content using marker-assisted backcrossing into an elite hybrid to increase the oil content in maize. Mol Breed 34:739–748

Hartings H, Fracassetti M, Motto M (2012) Genetic enhancement of grain quality-related traits in maize. In: Transgenic plants–advances & limitations: monograph. InTechOpen, Rijeka, pp 191–218

Hazel LN, Lush JL (1942) The efficiency of three methods of selection. J Hered:393–399

Heffner EL, Sorrells ME, Jannink JL (2009) Genomic selection for crop improvement. Crop Sci 49:1–12

Heslot N, Jannink JL, Sorrells ME (2015) Perspectives for genomic selection applications & research in plants. Crop Sci 55(1):1–12

Ho JC, McCouch SR, Smith ME (2002) Improvement of hybrid yield by advanced backcross QTL analysis in elite maize. Theor Appl Genet 105(2–3):440–448

Holland JB, Nyquist WE, Cervantes-Martinez CT (2003) Estimating & interpreting heritability for plant breeding: an update. In: Proceedings of the Plant & Animal Genome Conference, vol 11, pp 1–17

Hossain F, Nepolean T, Vishwakarma AK (2015) Mapping & validation of microsatellite markers linked to sugary1 & shrunken2 genes in maize (*Zea mays* L.). J Plant Biochem Biotechnol 24:135–142

Jessica E, Rutkoski (2018) A practical guide to genetic gain. International Rice Research Institute, Rice Breeding Platform, Manila

Johnson L (2001) Marker assisted sweet corn breeding: a model for specialty crops. In: Proceedings of 56th Annual Corn & Sorghum Industry Research Conference, Chicago, IL, vol 56, pp 25–30

Johnson R (2004) Marker-assisted selection. Plant Breed Rev 24:293–310

Karn A, Gillman JD, Flint-Garcia SA (2017) Genetic analysis of teosinte alleles for kernel composition traits in maize. G3 Genes Genomes Genetics 7:1157–1164

Kaur R, Kaur G, Vikal Y (2020) Genetic enhancement of essential amino acids for nutritional enrichment of maize protein quality through marker assisted selection. Physiol Mol Biol Plants 26:2243–2254

Lagoke STO, Parkinson V, Agunbiade RM (1991) Parasitic weed control methods in Africa. In: Kim SK (ed) Proceedings in International Workshop (IITA, ICRISAT & IDRC): Combating Striga in Africa, Ibadan, Nigeria 22–24 August 1998. IITA, Ibadan, pp 3–14

Li JM, Thomson M, McCouch SR (2004) Fine Mapping of a grain-weight quantitative trait locus in the peri-centromeric region of rice chromosome 3. Genetics 168:2187–2195

Li Y, Wang Y, Wei M, Li X, Fu J (2009) QTL identification of grain protein concentration & its genetic correlation with starch concentration & grain weight using two populations in maize (*Zea mays* L.). J Genet 88:61–67

Li Y, Wang W, Du J, Jin C, Song Q (2017) Genetic modification of starch quality in crops. Biotechnol Lett 39(3):421–431

Lima MS, Silva PSL, Oliveira OF, Silva KMB, Freitas FCL (2010) Corn yield response to weed & fall armyworm controls. Planta Daninha 28:103–111. https://doi.org/10.1590/S0100-83582010000100013

Lin F, Zhou L, He B, Zhang X, Dai H, Qian Y, Ruan L (2019) QTL mapping for maize starch content & candidate gene prediction combined with co-expression network analysis. Theor Appl Genet 132:1931–1941

Liu F, Zhao Q, Mano N, Ahmed Z, Nitschke F, Cai Y, Emes MJ (2016) Modification of starch metabolism in transgenic Arabidopsis thaliana increases plant biomass & triples oilseed production. Plant Biotechnol J 14(3):976–985

Lloyd JR, Kossmann J (2019) Starch trek: the search for yield. Front Plant Sci 9:1930

Lush JL (1937) Animal breeding plans, 3rd edn. Iowa State College Press, Ames

Mafu NF (2013) Marker-assisted selection for maize streak virus resistance & concomitant conventional selection for downy mildew resistance in a maize population. 177 pp. MSc Thesis, University of KwaZulu-Natal

Massman JM, Jung HJG, Bernardo R (2013) Genomewide selection versus marker-assisted recurrent selection to improve grain yield & stover-quality traits for cellulosic ethanol in maize. Crop Sci 53:58–66

Matsoukas IG, Massiah AJ, Thomas B (2013) Starch metabolism & antiflorigenic signals modulate the juvenile-to-adult phase transition in Arabidopsis. Plant Cell Environ 36(10):1802–1811

Mayor PJ, Bernardo R (2009) Doubled haploids in commercial maize breeding: one-stage & two-stage phenotypic selection versus marker-assisted recurrent selection. Maydica 54(4):439–448

Mehta BK, Vignesh M, Zunjare RU, Aanchal B, Chauhan HS, Chhabra R, Singh AK, Hossain F (2020) Biofortifcation of sweet corn hybrids for provitamin-A, lysine & tryptophan using molecular breeding. J Cereal Sci 96:103093

Meuwissen TH, Hayes BJ, Goddard M (2001) Prediction of total genetic value using genome-wide dense marker maps. Genetics 157(4):1819–1829

Mhike X, Okori P, Magorokosho C (2010) Genetic inheritance studies & early generation Marker selection in maize breeding for drought tolerance. In: Proceeding of the Second Ruforum Biennial Meeting 20–24th September, Entebbe, pp 389–393

Munns R, Richards RA (2007) Recent advances in breeding wheat for drought & salt stresses. In: Jenks MA, Hasegawa PM, Jain SM (eds) Advances in molecular breeding toward drought & salt tolerant crops, pp 565–586

Natesan S, Duraisamy T, Pukalenthy B, Chran S, Nallathambi J, Adhimoolam K, Manickam D, Sampathrajan V, Muniyi SJ, Meitei LJ, Thirunavukkarasu N, Kalipatty Nalliappan G, Rajasekaran R (2020) Enhancing β-Carotene concentration in parental lines of CO6 maize hybrid through marker-assisted backcross breeding (MABB). Front Nutr 7:134

Pardey PG, Beddow JM, Hurley TM, Beatty TK, Eidman VR (2014) A bounds analysis of world food futures: global agriculture through to 2050. Aust J Agric Resour Econ 58(4):571–589

Paul MJ, Oszvald M, Jesus C, Rajulu C, Griffiths CA (2017) Increasing crop yield & resilience with trehalose 6-phosphate: targeting a feast–famine mechanism in cereals for better source-sink optimization. J Exp Bot 68(16):4455–4462

Piepho HP, Richter C, Williams E (2008) Nearest neighbour adjustment & linear variance models in plant breeding trials. Biom J 50(2):164–189

Prasanna BM (2016) Developing & deploying abiotic stress-tolerant maize varieties in the tropics: challenges & opportunities. In: Molecular breeding for sustainable crop improvement. Springer, Cham, pp 61–77

Prasanna BM, Cairns J, Xu Y (2013) Genomic tools & strategies for breeding climate resilient cereals. In: Kole C (ed) Genomics & breeding for climate resilient crops. concepts & strategies. Springer-Verlang, Berlin, pp 213–239. https://doi.org/10.1007/978-3-642-37045-8_5

Pukalenthy B, Manickam D, Chran S, Adhimoolam K, Sampathrajan V, Rajasekaran R, Arunachalam K, Ganapathyswamy H, Chocklingam V, Muthusamy V, Hossain F, Natesan S (2019) Incorporation of opaque-2 into 'UMI 1200', an elite maize inbred line, through marker-assisted backcross breeding. Biotechnol Equip 33(1):144–153

Puttarach J, Puddhanon P, Siripin S, Sangtong V, Songchantuek S (2016) Marker assisted selection for resistance to northern corn leaf blight in sweet corn. Sabrao J Breed Genet 48(1):72–79

Ral JP, Bowerman AF, Li Z, Sirault X, Furbank R, Pritchard JR, Morell MK (2012) Down-regulation of glucan and water-dikinase activity in wheat endosperm increases vegetative biomass & yield. Plant Biotechnol J 10(7):871–882

Ranum P, Peña-Rosas JP, Garcia-Casal MN (2014) Global maize production, utilization, & consumption. Ann N Y Acad Sci 1312(1):105–112

Regierer B, Fernie AR, Springer F, Perez-Melis A, Leisse A, Koehl K, Kossmann J (2002) Starch content & yield increase as a result of altering adenylate pools in transgenic plants. Nat Biotechnol 20(12):1256–1260

Ribaut JM, Ragot M (2007) Marker-assisted selection to improve drought adaptation in maize: the backcross approach, perspectives, limitations, & alternatives. J Exp Bot 58(2):351–360

Ribaut JM, Anziger MB, Betran J (2002) Use of molecular markers in plant breeding: drought tolerance improvement in tropical maize. In: Kang MS (ed) Quantitative genetics, genomics, & plant breeding. CABI Publishing, Wallingford, pp 85–99

Ruckle ME, Meier MA, Frey L, Eicke S, Kolliker R, Zeeman SC, Studer B (2017) Diurnal leaf starch content: an orphan trait in forage legumes. Agronomy 7(1):16

Ruckle ME, Bernasconi L, Kölliker R, Zeeman SC, Studer B (2018) Genetic diversity of diurnal carbohydrate accumulation in white clover (*Trifolium repens* L.). Agronomy 8(4):47

Rutkoski J (2019) A practical guide to genetic gain. Adv Agron 157:217–249

Sarika K, Hossain F, Muthusamy V, Zunjare RU, Baveja A, Goswami R, Bhat JS, Saha S, Gupta HS (2018) Marker-assisted pyramiding of opaque2 & novel opaque16 genes for further enrichment of lysine & tryptophan in sub-tropical maize. Plant Sci 272:142–152

Sekar S (2020). https://india.mongabay.com/2019/09/fall-armyworm-destroyer-of-maize-farms-causes-concern-in-india/

Semagn K, Beyene Y, Babu R, Nair S, Gowda M, Das B (2015) Quantitative trait loci mapping & molecular breeding for developing stress resilient maize for sub-Saharan Africa. Crop Sci 55:1. https://doi.org/10.2135/cropsci2014.09.0646

Shiferaw B, Prasanna BM, Hellin J, Bänziger M (2011) Crops that feed the world 6. Past successes & future challenges to the role played by maize in global food security. Food Secur 3(3):307–327

Simkin AJ, López-Calcagno PE, Raines CA (2019) Feeding the world: improving photosynthetic efficiency for sustainable crop production. J Exp Bot 70(4):1119–1140

Singh N, Vasudev S, Yadava DK, Chaudhary DP, Prabhu KV (2014) Oil improvement in maize: potential & prospects. In: Maize: nutrition dynamics & novel uses. Springer, New York, pp 77–82

Singh J, Sharma S, Kaur A (2021) Marker-assisted pyramiding of lycopene-ε-cyclase, β-carotene hydroxylase1 & opaque2 genes for development of bio-fortified maize hybrids. Sci Rep 11:12642

Smidansky ED, Clancy M, Meyer FD, Lanning SP, Blake NK, Talbert LE, Giroux MJ (2002) Enhanced ADP-glucose pyrophosphorylase activity in wheat endosperm increases seed yield. Proc Natl Acad Sci 99(3):1724–1729

Smidansky ED, Martin JM, Hannah CL, Fischer AM, Giroux MJ (2003) Seed yield & plant biomass increases in rice are conferred by deregulation of endosperm ADP-glucose pyrophosphorylase. Planta 216(4):656–664

Smith AM (2008) Prospects for increasing starch & sucrose yields for bioethanol production. Plant J 54(4):546–558

Smith AB, Katz RW (2013) US billion-dollar weather & climate disasters: data sources, trends, accuracy & biases. Nat Hazards 67(2):387–410

Song XF, Song TM, Dai JR, Rocheford T, Li JS (2004) QTL mapping of kernel oil concentration with high-oil maize by SSR markers. Maydica 49(1):41–48

Stark DM, Timmerman KP, Barry GF, Preiss J, Kishore GM (1992) Regulation of the amount of starch in plant tissues by ADP glucose pyrophosphorylase. Science 258(5080):287–292

Stitt M, Gibon Y, Lunn JE, Piques M (2007) Multilevel genomics analysis of carbon signalling during low carbon availability: coordinating the supply & utilization of carbon in a fluctuating environment. Funct Plant Biol 34(6):526–549

Stuber CW (1995) Mapping & manipulating quantitative traits in maize. Trends Genet 11(12):477–481

Stuber CW, Polacco M, Senior ML (1999) Synergy of empirical breeding, marker-assisted selection, & genomics to increase crop yield potential. Crop Sci 39(6):1571–1583

Sumathi K, Ganesan KN, Aarthi P (2020) Introgression of QTLs determining sorghum downy mildew (SDM) resistance into elite maize line UMI 79 through marker-assisted backcross breeding (MABC). Australas Plant Pathol 49:159–165

Sureshkumar S, Tamilkumar P, Senthil N, Nagarajan P, Thangavelu AU, Raveendran M, Vellaikumar S, Ganesan KN, Balagopal R, Vijayalakshmi G, Shobana V (2014a) Marker assisted selection of low phytic acid trait in maize (*Zea mays* L.). Hereditas 151(1):20–27

Sureshkumar S, Paramasivam T, Thangavelu AU, Senthil N, Nagarajan P, Vellaikumar S, Ganesan KN, Balagopal R, Raveendran M, Lubberstedt T (2014b) Marker-assisted introgression of lpa2

locus responsible for low-phytic acid trait into an elite tropical maize inbred (*Zea mays* L.). Plant Breed 133(5):566–578
Tamilkumar P, Senthil N, Sureshkumar S, Thangavelu AU, Nagarajan P, Vellaikumar S, Ganesan KN, Natarajan N, Balagopal R, Nepolean T, Raveendran M (2014) Introgression of low phytic acid locus (lpa2-2) into an elite Maize (*Zea mays* L.) inbred through marker assisted backcross breeding. Aust J Crop Sci 8(8):1224–1231
Thalmann M, Santelia D (2017) Starch as a determinant of plant fitness under abiotic stress. New Phytol 214(3):943–951
USDA, N (2013) USDA National Agricultural Statistics Service quick stats 2.0. Retrieved March, 6, 2013
Van Bueren ETL, Struik PC, van Eekeren N, Nuijten E (2018) Towards resilience through systems-based plant breeding. A review. Agron Sustain Dev 38(5):42
Van Camp W (2005) Yield enhancement genes: seeds for growth. Curr Opin Biotechnol 16(2):147–153
Wassom JJ, Wong JC, Martinez E, King JJ, DeBaene J, Hotchkiss JR, Mikkilineni V, Bohn MO, Rocheford TR (2008) QTL associated with maize kernel oil, protein, & starch concentrations; kernel mass; & grain yield in Illinois high oil × B73 backcross-derived lines. Crop Sci 48(1):243–252
Weise SE, Aung K, Jarou ZJ, Mehrshahi P, Li Z, Hardy AC, Sharkey TD (2012) Engineering starch accumulation by manipulation of phosphate metabolism of starch. Plant Biotechnol J 10(5):545–554
Whan AP, Verbyla AP, Mieog JC, Howitt CA, Ral JP (2017) Transferring a biomass enhancement biotechnology from glasshouse to field: a case study on wheat GWD RNAi. Agronomy 7(4):82
White PJ (2007) Fatty acids in oilseeds (vegetable oils). In: Fatty acids in foods & their health implications. CRC Press, pp 241–276
Willcox MC, Khairallah MM, Bergvinson D, Crossa J, Deutsch J, Edmeades G, González-deLeón D, Jiang C, Jewell D, Mihm J, Williams W, Hoisington D (2002) Selection for resistance to southwestern corn borer using marker assisted & conventional backcrossing. Crop Sci 42(5):1516–1528
Wu J, Lawit SJ, Weers B, Sun J, Mongar N, Van Hemert J, Collet KH (2019) Overexpression of zmm28 increases maize grain yield in the field. Proc Natl Acad Sci 116(47):23850–23858
Xia JH, Zheng YL (2002) Molecular marker-assisted backcross breeding of maize Rf3 NIL & its efficient analysis. Acta Agron Sin 28:339–344
Xu B, Chang SK (2009) Phytochemical profiles & health-promoting effects of cool-season food legumes as influenced by thermal processing. J Agric Food Chem 57(22):10718–10731
Xu Y, Lu Y, Xie C, Gao S, Wan J, Prasanna BM (2012) Whole-genome strategies for marker-assisted plant breeding. Mol Breed 29:833–854. https://doi.org/10.1007/s11032-012-9699-6
Xu Y, Li P, Zou C, Lu Y, Xie C, Zhang X, Olsen MS (2017) Enhancing genetic gain in the era of molecular breeding. J Exp Bot 68(11):2641–2666
Xu Z, Hua J, Wang F, Cheng Z, Meng Q, Chen Y, Han X, Tie S, Liu C, Li X, Wang Z, Weng J (2020) Marker-assisted selection of qMrdd8 to improve maize resistance to rough dwarf disease. Breed Sci 70(2):183–192
Yang X, Guo Y, Yan J, Zhang J, Song T, Rocheford T, Li JS (2010) Major & minor QTL & epistasis contribute to fatty acid compositions & oil concentration in high-oil maize. Theor Appl Genet 120:665–678
Yang G, Dong Y, Li Y, Wang Q, Shi Q, Zhou Q (2013) Verification of QTL for grain starch content & its genetic correlation with oil content using two connected RIL populations in high-oil maize. PLoS One 8:e53770
Yang R, Yan Z, Wang Q, Li X, Feng F (2018) Marker-assisted backcrossing of lcyE for enhancement of proA in sweet corn. Euphytica 214:130
Yin X, Struik PC, Kropff MJ (2004) Role of crop physiology in predicting gene-to-phenotype relationships. Trends Plant Sci 9(9):426–432
Yousef GG, Juvik JA (2001) Comparison of phenotypic & marker-assisted selection for quantitative traits in sweet corn. Crop Sci 41:645–655. https://doi.org/10.2135/cropsci2001.413645x

Yousef GG, Juvik JA (2002) Enhancement of seedling emergence in sweet corn by marker-assisted backcrossing of beneficial QTL. Crop Sci 42(1):96–104

Zhang J, Lu XQ, Song XF, Yan JB, Song TM, Dai JR, Rocheford T, Li JS (2007) Mapping quantitative trait loci for oil, starch, & protein concentrations in grain with high-oil maize by SSR markers. Euphytica 162(3):335–344

Zhao X, Tan G, Xing Y, Wei L, Chao Q, Zuo W, Lübberstedt T, Xu M (2012) Marker-assisted introgression of qHSR1 to improve maize resistance to head smut. Mol Breed 30(2):1077–1088

Zhao Q, Zhu D, Iglesias A, Sun L, Hung YS, Shewry PR, Li Z (2020) Improved starch accumulation in rice stems by overexpression of the maize starch branching enzyme IIb gene. Int J Mol Sci 21(7):2403

Zhu XG, Long SP, Ort DR (2010) Improving photosynthetic efficiency for greater yield. Annu Rev Plant Biol 61:235–261

Zunjare RU, Hossain F, Muthusamy V, Baveja A, Chauhan HS, Bhat JS, Thirunavukkarasu N, Saha S, Gupta HS (2018) Development of biofortified maize hybrids through marker-assisted stacking of β-Carotene Hydroxylase, Lycopene-ε-Cyclase & Opaque2 Genes. Front Plant Sci 9:178

Chapter 6
Diversification of Sugarcane-Based Cropping System with Medicinal and Aromatic Plants

S. K. Yadav, S. K. Shukla, A. P. Dwivedi, V. P. Yadav, Pooja Yadav, Sanjay Singh Rathore, Sanjeev Kumar, Roopa Patel, and Subhash Babu

Abstract Sugarcane is one of the most important cash crops in the world. The crop has the potential to perform comparatively better under adverse climatic conditions. Employment opportunities, livelihood security, and environmental concerns are the main focus of interest amidst changing climatic scenarios with increasing population pressure and deteriorating natural resources. Hence, it becomes essential to achieve the target of sustainable development goals (SDGs). Meeting food energy demand and supply gaps without interrupting environmental sustainability is the key challenge to scientists, policymakers, and environmental planners worldwide. The farming community is highly sensitive to climatic abnormality that is highly impacted by resource scarcity such as depletion of water resources and land degradation. This situation may further worsen for sustainability in the future. In recent years, consequences of both intensive agricultural practices and anthropogenic activities have deteriorated the natural agroecosystem substantially. At present, modern agriculture has targeted the adoption of advanced technological interventions that can serve as more efficient for resource utilization besides time-saving production-enhancing processes and practices. Multiple cropping has been practiced in many parts of the world as a way to maximize land productivity and resource use efficiency. Sugarcane is a long-duration crop and takes about 90–120 days for canopy development, which allows for growing intercrops during

S. K. Yadav (✉) · S. K. Shukla · A. P. Dwivedi · V. P. Yadav · P. Yadav
ICAR-Indian Institute of Sugarcane Research, Lucknow, Uttar Pradesh, India

S. S. Rathore · S. Babu
Division of Agronomy, ICAR-Indian Agricultural Research Institute, New Delhi, India

S. Kumar
Department of Soil Science & Agricultural Chemistry, Sardar Vallabhbhai Patel University of Agriculture and Technology, Meerut, Uttar Pradesh, India

R. Patel
Department of Soil Science & Agricultural Chemistry, Banaras Hindu University, Varanasi, Uttar Pradesh, India

S. Babu et al. (eds.), *Agricultural Diversification for Sustainable Food Production*, Sustainability Sciences in Asia and Africa,
https://doi.org/10.1007/978-981-97-7517-0_6

the early stage, especially in autumn planted crops. The main purpose of sugarcane cultivation is to generate more farm income to meet the basic financial requirements of a growing community. High-value crops, especially medicinal and aromatic plants, are known as industrial crops having tremendous scope to generate farm income. The length of the growing period of selected high-value crops is not more than 6 months, and few of them like *Tulsi* and marigold have crop duration of just 60–70 days, which may be easily adjusted in a sugarcane-based cropping system. There is a great need to develop a system approach by diversifying the proportion of existing cropping systems of sugarcane through commercial crops in system mode to provide employment opportunities throughout the year and generate higher income at each level of the farming community. Hence, this book chapter will include the concept, merit, and demit of crop diversification, agronomy of various component crops under diversification, and identification of the best cropping system for higher sugarcane yield, energy efficiency, environmental friendliness, and more profitable through diversification of existing cropping system with high-value crops.

Keywords Cropping systems · Economic returns · Medicinal plants · Short-duration crops

6.1 Introduction

Sugarcane is a prominent cash crop used to produce sugar, jaggery, as well as bioenergy (heat, electricity, and ethanol) and by-products such as press mud and bagasse ash (Shukla et al. 2021). It is a perennial bunchgrass and is mostly grown in tropical and subtropical countries. It is critically important in many developing nations, with over 75% of worldwide production concentrated in Brazil, India, China, Thailand, and Pakistan (FAO 2019). Sugarcane is native to India and is referenced in the Atharva Veda (Hinduism's) most holy text. Sugarcane is one of India's most important foreign currency crops, accounting for 51.14 lakh hectares of land (Shukla et al. 2019). Brazil is the world's largest producer of sugarcane and produces more than 40% of the global crop. Sugarcane is the finest example of a renewable natural agricultural resource since it turns solar energy into chemical energy while also producing sugar, ethanol, fiber, fertilizer, and other by-products. Nowadays, it has become one of the important energy crops widely grown globally. It is a high biomass energy crop, with the sugar stored in its stalk and the lingocellulosic residue remaining after the extraction of sugar used to produce biofuel or other bioproducts. Biofuels are increasingly seen as an alternative to petroleum derivates to reduce carbon emissions to the atmosphere and recent escalating petroleum prices. Recently, the Government of India has formulated a policy to encourage sugar mills to produce ethanol directly from sugarcane juice, which could avoid surplus sugar and also improve sugar mills viability besides fulfilling multiple needs. Bagasse, produced during the sugarcane crushing process, generates heat and electricity through the cogeneration process. The current demand for sugarcane and its many valuable by-products remains projected to rise globally.

Integrating high-value crops, such as medicinal and aromatic plants, into sugarcane-based cropping systems can significantly enhance agricultural revenue and support the financial needs of growing societies (Taghouti et al. 2022). High-value crops, particularly medicinal and aromatic plants, are known as industrial crops and have enormous potential to create farm revenue. High-value crops like *Tulsi* (Holy Basil) and marigold can generate substantial income due to their medicinal, aromatic, and ornamental uses. These crops often have higher market prices compared to traditional crops, leading to better profitability for farmers. The growth period of chosen high-value crops is less than 6 months, with some, such as *Tulsi* and marigold, having crop durations of only 60–70 days, which may be readily changed in a sugarcane-based cropping system. Diversification reduces dependency on a single crop, mitigating risks associated with market fluctuations, pests, and diseases. It promotes sustainable farming practices by improving soil health and reducing the need for chemical inputs. Diversifying crop systems can create year-round employment opportunities as different stages of crop cultivation, such as planting, maintenance, harvesting, and processing, require labor, thus generating jobs at various levels. This chapter will include the concept, merit, and demerit of crop diversification, agronomy of various component crops under diversification, and identification of the best cropping system for higher sugarcane yield, energy efficiency, environmental friendliness, and profitability through diversification.

6.2 Challenges of Prevalent Cropping System

The rice–wheat system is followed under a sugarcane-based cropping system as a succeeding crop in the subtropical region of India at a large scale. While this method has been favored for its simplicity and efficiency in crop management, it presents numerous challenges that can significantly impact agricultural sustainability, economic stability, and environmental health. Various challenges are associated with the rice–wheat system, particularly focusing on the context of sugarcane cultivation in India.

6.2.1 Soil Degradation

One of the most significant challenges of the rice–wheat system is soil degradation. Continuous cultivation of the same crop depletes specific nutrients from the soil, leading to reduced soil fertility over time. In the case of sugarcane:

***Nutrient depletion**:* Sugarcane has high nutrient requirements, particularly nitrogen, phosphorus, and potassium. Rice–wheat system exhausts these nutrients, necessitating increased use of chemical fertilizers.

Soil structure: The repeated planting of sugarcane after the rice–wheat system disrupts the soil structure, reducing its ability to retain water and support diverse microbial life.

Organic matter reduction: Lack of crop rotation reduces the return of organic matter to the soil, which is vital for maintaining soil health.

6.2.2 Pest and Disease Build-Up

Rice–wheat system systems are particularly vulnerable to pests and diseases. The absence of crop diversity creates an ideal environment for pests and pathogens to thrive and spread:

Pest proliferation: Pests that are specialized to the rice–wheat–sugarcane system crop find an abundant and continuous food source, leading to population explosions.

Disease cycles: Pathogens can build up in the soil and plant residue, leading to recurrent disease outbreaks that become harder to control over time.

Increased pesticide use: Farmers often resort to higher quantities of chemical pesticides to manage pest and disease pressures, which can lead to pesticide resistance and environmental contamination.

6.2.3 Water Resource Depletion

Rice is a water-intensive crop, and its rice–wheat–sugarcane system can lead to significant water resource depletion:

High water demand: Continuous rice–wheat–sugarcane system cultivation requires substantial irrigation, leading to the over-extraction of groundwater.

Water table decline: Excessive groundwater use lowers the water table, making it difficult for farmers to access water for irrigation.

Competition for water: Rice and sugarcane out of the rice–wheat–sugarcane system of water-intensive crops can create competition for water resources, affecting the availability of water for other crops and domestic use.

6.2.4 Reduced Biodiversity

Rice–wheat–sugarcane system significantly reduces biodiversity, both above and below ground:

Loss of plant diversity: Growing the same crop eliminates the variety of plants that could otherwise provide habitat and food for a range of organisms.

Impact on soil microbial life: The lack of plant diversity affects soil microbial communities, which are essential for nutrient cycling and soil health.

Ecological imbalance: Reduced biodiversity can disrupt the ecological balance, making the agricultural ecosystem more vulnerable to environmental stresses.

6.2.5 *Environmental Pollution*

Rice–wheat–sugarcane system practices often lead to increased use of chemical inputs, contributing to environmental pollution:

Fertilizer runoff: Excessive use of chemical fertilizers can lead to runoff into water bodies, causing eutrophication and harming aquatic life.

Pesticide contamination: Pesticides can contaminate soil, water, and air, posing risks to human health and nontarget organisms.

Soil salinization: Overreliance on irrigation, especially in poorly drained areas, can lead to soil salinization, reducing soil productivity.

6.2.6 *Economic Risks*

Relying on the same crop can pose significant economic risks for farmers:

Market fluctuations: Farmers are vulnerable to price fluctuations in the market. A drop in the price of sugarcane can have devastating economic consequences.

Income stability: Lack of crop diversity means farmers do not have alternative sources of income, increasing their financial vulnerability.

Investment in Inputs: The rice–wheat–sugarcane system often requires heavy investment in fertilizers, pesticides, and irrigation infrastructure, which can strain farmers' financial resources.

6.3 Need for Diversification of Sugarcane-Based Cropping System

Sugarcane is a pivotal crop in the Indian agricultural landscape, significantly contributing to the economy and livelihoods of millions of farmers. However, the traditional monoculture of sugarcane presents several challenges that necessitate a shift toward diversification. This chapter explores the critical need for diversifying sugarcane-based cropping systems, highlighting the economic, environmental, and agronomic benefits that diversification can offer.

6.3.1 Economic Necessities

Income stability: Rice–wheat system exposes farmers to the risks associated with price volatility. The global sugar market is highly susceptible to fluctuations due to factors like international trade policies, production levels in other major sugar-producing countries, and changing demand. Diversifying crops can provide a buffer against these price shocks, ensuring a more stable income.

Risk mitigation: Agricultural diversification spreads the risk across multiple crops. If one crop fails due to pests, diseases, or unfavorable weather conditions, other crops can still provide income. This risk mitigation is crucial for small and marginal farmers who are heavily dependent on agriculture for their livelihoods.

Market opportunities: The rising demand for medicinal and aromatic plants (MAPs) presents lucrative market opportunities (Booker et al. 2012). MAPs are sought after in the pharmaceutical, cosmetic, and food industries, both domestically and internationally. By integrating these high-value crops, farmers can tap into new markets and enhance their profitability.

6.3.2 Environmental Imperatives

Soil health improvement: Continuous cultivation of sugarcane depletes soil nutrients and organic matter, leading to soil degradation. Diversification with MAPs can improve soil health through varied root structures and nutrient requirements. For example, leguminous plants can fix nitrogen, enhancing soil fertility.

Pest and disease management: Monoculture systems are more vulnerable to pests and diseases. Diverse cropping systems disrupt the lifecycle of pests and reduce the incidence of diseases. Many MAPs possess natural pest-repellent properties, which can decrease the reliance on chemical pesticides.

Water conservation: Sugarcane is a water-intensive crop. Kumar et al. (2010) reported that the depletion rate of the water table was much faster as below 9 m was increased from 90% in 2004 and almost 100% in 2010.

Integrating drought-resistant MAPs like Aloe vera and lemongrass can reduce water usage and pressure on local water resources. These plants can thrive with minimal irrigation, promoting sustainable water management.

6.3.3 Social Benefits

Employment generation: Diversified farming systems can create additional employment opportunities. The cultivation, processing, and marketing of MAPs require labor, thereby generating more jobs and reducing rural unemployment (Shahidulla and Emdad Haque 2010).

Community health: Access to medicinal plants can improve community health by providing natural remedies for various ailments. This is particularly beneficial in rural areas with limited access to healthcare facilities. Medicinal plants play vital roles in disease prevention in rural communities that have limited access to modern medicine (Smith-Hall et al. 2012; Sofowora et al. 2013).

Preservation of traditional knowledge: Diversifying crops encourages the use of traditional agricultural practices and knowledge associated with MAPs. This helps preserve cultural heritage and promotes sustainable farming practices (Kerr et al. 2018).

6.3.4 Agronomic Advantages

Enhanced biodiversity: Crop diversification enhances biodiversity, which is crucial for ecological stability (Beillouin et al. 2021). A variety of plants support different types of insects, birds, and other wildlife, creating a balanced ecosystem that is more resilient to environmental changes.

Soil erosion control: Crops like vetiver and lemongrass have strong root systems that help in controlling soil erosion. Integrating these plants in sugarcane fields can prevent soil loss and improve soil structure.

Improved crop yields: Crop rotation and intercropping can lead to improved yields (Raseduzzaman and Jensen 2017). For instance, rotating sugarcane with nitrogen-fixing plants can enhance soil fertility and result in better sugarcane yields in subsequent seasons (Singh et al. 2020).

6.4 Sugarcane and Its Diversification Scenario

Sugarcane is a long-duration crop that takes 12–16 months to mature, cultivated in wider row spacing. The intercropping of short-duration high-value crops with befitting advantages may enhance systems productivity and profit. Most of the region's preceding crops of sugarcane are rice, maize, and wheat. Conventional short-duration crops like pulses, oilseeds, and cereals are not as productive as compared to high-value crops such as medicinal and aromatic plants. Crop diversification with medicinal and aromatic plants can be a better option to solve the problems of low profitability per unit area per unit time and enhance crop profitability of the sugarcane-based production system. Intercropping or sequential cropping in sugarcane-based production systems of medicinal and aromatic plants having short-duration crops concerning system yield enhancement and economic return, particularly in north-central India. Crop diversification in sugarcane with medicinal and aromatic plants is a potential option for meeting the world's growing need for medicine and aroma for different uses. Comparatively, farmers get more returns from their investment within the shortest possible time by replacing conventional systems with medicinal and aromatic plants in sugarcane-based cropping systems.

Diversification of the intensive cropping systems through the inclusion of pulses and oilseeds has become inevitable as their soil ameliorative properties and minimum requirement of external input, especially nitrogen (Singh et al. 2021; Yadav et al. 2021; Babu et al. 2023a). For mitigating the ill effects of sugarcane monoculture, maintaining soil health, checking declining factor productivity, and arresting plateauing yield levels, the inclusion of short-duration high-value intercrops in sugarcane-based cropping systems holds great promise for mid-season income generation and household nutrition, as well as the economic security of small and marginal farmers (Shukla et al. 2018). Given the unique characteristics of the deep root system, higher leaf fall, biological nitrogen fixation, and release of root exudates, grain legumes offer the potential option for improving soil health and ecosystem restoration (Babu et al. 2020; Babu et al. 2023b). This cropping pattern needs intensification to increase system productivity and increase the income of the Uttar Pradesh farmers. Therefore, intensive and diversified use of the lands will help to increase production and ease market pressure on commodities, their availability, farmer's income generation, employment opportunities, and livelihood improvement. Sugarcane is one of the most popular crops in India, holding more than 60% of India's land area under cultivation. About 50 million farmers' families are dependent on sugarcane cultivation, and our country produces approximately 35.5 million tons of sugar annually, which makes it one of the largest producers in the world. Sugarcane, an important agroindustrial crop in India, plays a pivotal role in the national economy.

6.4.1 A Viable Option for Diversification of Sugarcane-Based Cropping System

Reduction in efficiency of fertilizers and agrochemicals along with excess water in the paddy–wheat cropping system and delay in sowing of sugarcane. Groundwater is the major source of irrigation in India for sugarcane-based cropping systems in which paddy crops occupy significantly more acreage. Both sugarcane and paddy deplete more groundwater as compared to major crops and become major concerns to water conservation and ground recharge (Rodell et al. 2009). The effect of wild animals, insects, and diseases on medicinal and aromatic plants is much less as compared to rice and wheat. In the subtropical part of India, sugarcane is planted in April–May after harvesting the wheat. In the rice–wheat–sugarcane system, the planting of sugarcane is delayed due to the harvesting of wheat in April. Delayed planting of sugarcane resulted in poor tillering, which adversely affected the number of millable canes per unit area. Individual cane weight and cane thickness are also reduced due to the shortened growth duration, consequently a reduction in sugarcane yield (Shukla et al. 2022). In such a situation, diversification of sugarcane-based cropping systems, either sequential or intercropping of high-value crops such as pulses, vegetables, oilseeds, and other remunerative crops like medicinal and aromatic plants in place of prominent rice and wheat cropping systems, could provide a better option for improving the productivity of these component crops at a greater level. Medicinal and aromatic plants that are cultivated for a short duration can be easily accommodated in place of paddy and wheat in sugarcane-based crop rotation. In the situation of decreasing employment opportunities, increasing air pollution, degradation in soil fertility, and low yield, maximum profit can be earned by diversifying sugarcane-based cropping systems with medicinal plants. Intercropping high-value crops such as gladiolus, marigolds, vegetables, and pulse crops are also being adopted in various sugarcane-growing regions (Shukla et al. 2017). In subtropical India, several short-duration high-value crops have been attempted as component crops under a sugarcane-based cropping system. At the IISR, Lucknow, sequential cropping of medicinal and aromatic plants such as wild marigold recorded the highest wheat equivalent yield (95.5 t/ha) under *Tulsi*–wild marigold (two cuts)–sugarcane (Spring)–sugarcane ratoon–mint cropping sequence. All the cropping sequences of medicinal and aromatic plants (except two cuttings of wild marigold and Stevia) recorded significantly higher cane yield than rice–wheat–sugarcane (Spring)–sugarcane ratoon–wheat cropping sequence.

6.4.1.1 Substitution of Rice and Wheat Through Short-Duration Crops

A field experiment was concluded at the Indian Institute of Sugarcane Research, Lucknow, to diversify sugarcane-based cropping systems with medicinal and aromatic plants. Rice, *Tulsi*, Marigold, and Kalmegh (*Andrographis paniculata*) have been grown during the Kharif season. Wheat, wild marigold, Stevia, and sugarcane (autumn with wild marigold) were planted during *Rabi* season and harvested during the spring. The result highlighted that the highest rice equivalent yield (6.80 Mg/ha) of *Tulsi* (Ocimum) was recorded when *Tulsi* was grown in place of rice during the Kharif season. Similarly, the maximum wheat equivalent yield (9.55/ha) of wild marigolds was recorded when—wheat was replaced by wild marigold (two cuts) during the Rabi season before planting of sugarcane (Table 6.1).

Data presented in Table 6.2 show that the significantly highest cane yield (90.0 Mg/ha) of plant sugarcane crop was recorded under the *Tulsi*–Stevia–sugarcane cropping sequence as compared to cane yield (54.0 Mg/ha) in rice–wheat–sugarcane cropping sequence. Significantly the highest cane yield (85.5 Mg/ha) of sugarcane ratoon crop was recorded under the *Tulsi*–Stevia–sugarcane cropping sequence as compared to ratoon cane yield (68.8 Mg/ha) in rice–wheat–sugarcane cropping sequence. All the cropping sequences having a crop of medicinal and aromatic plants recorded higher cane yields than the rice–wheat–sugarcane cropping sequence. Higher crop yields of about 67% and 39% of sugarcane were recorded in plant crops and ratoon crops due to diversifying the rice–wheat system through *Tulsi* and Stevia under a sugarcane-based cropping system. Growing stevia after harvesting of *Tulsi* ready to harvest in the month of February–March and allowing sugarcane to be planted timely in the same field while in the rice–wheat system,

Table 6.1 Yield (Mg/ha) of various component crops under sugarcane-based cropping system

Kharif season	Yield (Mg/ha)	Rice equivalent yield (Mg/ha)	Rabi season	Yield (Mg/ha)	Wheat equivalent yield (Mg/ha)
Rice	4.53	4.50	Wheat	3.26	3.26
Marigold	6.30	5.15	Stevia	0.72	5.61
Tulsi	15.5	6.76	Wild marigold	40.4.0	7.35
Tulsi	15.60	6.8.0	Stevia	0.64	4.99
Kalmegh	2.20	4.8.0	Wild marigold	50.60	9.55

Table 6.2 Effect of crop diversification through MAPs on sugarcane yield

	Sugarcane yield (Mg/ha)		
Cropping system	Plant	Ratoon	Mean
Rice–wheat–sugarcane	54.0	68.8	61.4
Marigold-–stevia–sugarcane	88.0	78.4	83.3
Tulsi–wild marigold–sugarcane	84.0	77.2	80.6
Tulsi–stevia–sugarcane	90.0	85.5	87.8
Kalmegh–wild marigold–sugarcane	84.6	79.6	82.1

Table 6.3 Fertilizers dose of (NPK) to component crops of sugarcane-based cropping systems

Kharif crop	Fertilizers (kg/ha)	*Rabi* crop	Fertilizers (kg/ha)	Cropping sequence	Fertilizers (kg/ha)
Rice	150:60.60	Wheat	150:60:40	Rice–wheat	300:120:100
Marigold	120:60:40	Stevia	120:60:60	Marigold–stevia	240:120:100
Tulsi	80:40:40	Wild marigold	100:60:40	*Tulsi*–wild marigold	180:100:80
Tulsi	80:40:40	Stevia	120:60:60	*Tulsi*–stevia	200:100:100
Kalmegh	80:40:40	Wild marigold	100:60:40	Kalmegh–wild marigold	180:100:80

harvesting of wheat was done in April and planting of sugarcane was done in May resulted in poor tillering and number of millable canes.

Fertilizer consumption in the rice–wheat system is also challenging for the nutrient imbalance due to the mining of major nutrients (N, P, K, and S) from the soil, leading to the deterioration of soil fertility and soil quality. Data presented in Table 6.3 represent that generally, fertilizer requirements of medicinal and aromatic plants are quite low as compared to the rice–wheat system. The highest fertilizers of about 300 kg of nitrogen, 120 kg of phosphorus, and about 100 kg of potassium were applied in the rice–wheat cropping system, whereas the *Tulsi*–wild marigold system had the least quantity of fertilizers applied for producing a higher level of yield in the term of monetary returns as mentioned in Table 6.1. Therefore, inclusion of MAPs such as *Tulsi* and wild marigold in place of rice–wheat system may be very much helpful to reduce the quantity of fertilizer consumption in the sugarcane-based cropping system.

6.5 Disadvantages of Traditional Sugarcane-Based Cropping System

Sugarcane-based paddy–wheat cultivation requires more irrigation, due to which a continuous decline in groundwater level is being seen (Kumar et al. 2010). Therefore, the risk of future water crises is increasing. Due to the adoption of paddy and wheat crop rotation, the efficiency of main nutrients is decreasing. Apart from the main nutrients, there also seems to be a deficiency in some micronutrients and soil organic matter (Bhatt et al. 2016). Further chance of increasing the disease susceptibility of the soil due to continuous cultivation of the same type of crop (Kwak and Weller 2013), the efficiency of various productive resources like fertilizers and pesticides is decreasing, due to which the cost is increasing due to increase in per unit usage. Apart from the damage caused by wild animals day by day, the infestation of insects is also increasing in paddy and wheat crops. Burning of crop residues in some areas of Punjab, Haryana, and Uttar Pradesh is causing air pollution as well as loss of beneficial microorganisms present in the soil. Due to excessive water logging in canal areas, soil salinity is increasing, due to which the quality and productivity of the soil are decreasing in rice-based prominent cropping system of sugarcane cultivation (Singh 2015).

6.6 The Usefulness of Sugarcane-Based Crop Diversification Through MAPs

Sugarcane ranks among the world's cash crops and is used primarily for sugar, jaggery, fuel, and animal feed. The government's attention is now moving towards making liquid fuel (ethanol) from sugarcane, due to which the income and employment possibilities of the farmers are also increasing. To prevent the outbreak of coronavirus, a sanitizer prepared from sugarcane is being used. Medicinal plants and other products made from them are also being used in abundance; as a result, the demand for medicinal plants is increasing day by day in the national and international markets. Commercial cultivation of sugarcane and medicinal plants provides employment opportunities to a large number of people.

6.7 Diversification Method of Sugarcane-Based Crop Rotation

Mainly in North India, traditionally paddy and wheat cultivation is done in the months of *Kharif* and *Rabi*, respectively. First of all, short-term varieties of medicinal and aromatic plants should be selected whose duration should be a maximum of 150 days. Recommended varieties of medicinal and aromatic crops suited for

Table 6.4 Recommended varieties of medicinal and aromatic plants

Crop	Varieties
Tulsi	CIM Saumay, CIM Jyoti, CIM Sharda, CIM Surabhi
Kalmegh	CIM Megha
Stevia	CIMAP Mithi, CIMAP Madhu, Him Stevia
Wild marigold	Vanphool
Mentha	CIM Kranti, CIM Saryu, CIM Koshi, CIM Kranti

intercropping with sugarcane are given in Table 6.4. Instead of paddy in *Kharif*, *Tulsi* or Kalmegh can be successfully planted mainly as medicinal or aromatic plants in the land with proper drainage. After harvesting *Tulsi* or Kalmegh in Rabi season, Stevia or wild marigold can be planted successfully. Sugarcane crops can be planted immediately after harvesting stevia or wild marigolds in February or March. The quantity of seeds, fertilizers, plant-to-plant distance, and other recommended agricultural practices should be done in the crop according to that area.

6.8 Benefits of Crop Diversification

In the cultivation of medicinal and aromatic plants, 70–80% of water is saved compared to paddy and wheat. The data given in Table 6.5 show the water used pattern in the sugarcane-based cropping system and comparative observation to be recorded as applied irrigation water in each crop during its growth and development, excluding the rainfall contributing to each crop. Maximum irrigation water was applied to rice, while the least irrigation water was to Kalmegh followed by *Tulsi* during the Kharif season. The water requirement for *Tulsi* and Kalmegh was very much essential at the time of transplanting. The rest water requirement at Lucknow Center has been fully filled by the rainfall. Proper drainage during the Kharif season has been given top priority for higher yield for medicinal and aromatic plants such as *Tulsi* and Kalmegh. Water requirement through irrigation water was observed to be the highest for wheat crops over other crops during the Rabi season. All the medicinal crops of the Rabi season have higher water requirements as compared to Kharif season crops due to less rainfall. The total water requirement of the rice and wheat system is about 2000 mm of irrigation was applied. Rice is the most water-demanding crop, which is responsible for groundwater depletion. However, the highest irrigation was applied to the *Tulsi*–Stevia cropping system which was just about 25% of the rice–wheat cropping system. Hence, diversification of rice and wheat through these medicinal sugarcane-based cropping systems will be very crucial to water conservation and groundwater recharge by reducing the pumping of irrigation water from the natural resources.

The economics of selected medicinal and aromatic plants in the sugarcane-based cropping system have been given in Table 6.6. Cultivation of medicinal and aromatic plants gives about 40–50% more income as compared to paddy and wheat.

Table 6.5 Irrigation water requirement of various component crops of sugarcane-based cropping system

Kharif season	Water applied (ha mm)	Rabi season	Water applied (ha mm)	Cropping sequence	Total water requirement (ha mm)
Rice	1500	Wheat	500	Rice–wheat	2000
Marigold	400	Stevia	350	Marigold–stevia	750
Tulsi	150	Wild marigold	300	*Tulsi*–wild marigold	450
Tulsi	150	Stevia	350	*Tulsi*–stevia	500
Kalmegh	100	Wild marigold	300	Kalmegh–wild marigold	400

Table 6.6 Economics of crop diversification through MAPS

Cropping sequence	Cost of cultivation (Rs × 10^3/ha)	Net returns (Rs × 10^3/ha)	B:C ratio
Rice–wheat	105.0	90.0	1.86
Marigold–Stevia	115.0	95.0	1.83
Tulsi–wild marigold	65.0	195.0	4.00
Tulsi–Stevia	110.0	130.0	2.18
Kalmegh–wild marigold	60.0	170.0	3.83

The cost of cultivation of medicinal and aromatic plants such as *Tulsi*/Kalmegh–wild marigolds was lower than another cropping system. However, the net profit and benefit–cost ratio were found the highest in the *Tulsi*–wild marigold sequence. The net benefit–cost ratio of *Tulsi* and wild marigold or Kalmegh–wild marigold cropping sequence was found more than two times in the rice–wheat cropping system.

In comparison to paddy and wheat, the incidence of wild animals, insects, and diseases is very low. Medicinal and aromatic plants require very little water in comparison to paddy and wheat, fertilizers, and agricultural chemicals, due to which the pollution is also reduced. The decline in the water table is also less due to low irrigation requirements in medicinal and aromatic plants as compared to paddy and wheat. By adopting proper crop rotation, there is a satisfactory improvement in soil health and an increase in soil fertility, which also leads to a satisfactory increase in crop yield. There is also an opportunity to create more employment opportunities due to participation in agricultural activities as well as product processing and marketing. The selling price of medicinal and aromatic plants is more than the minimum support price of paddy and wheat. Therefore, there is no problem related to minimum support price in the cultivation of these plants.

6.9 Initiative Toward Self-Reliant India

First of all, a group of progressive farmers should be formed, and a small unit for product processing and packaging should be established. By forming a farmer group, it becomes easier to get support for the farming-related projects run by the

Government of India. Under Bharat Nirman Abhiyan, a provision has also been made to provide grants for starting agriculture-related businesses. Cultivation of medicinal and aromatic plants can be easily started in large areas in sugarcane-based cropping systems due to the existence of farmer groups. There is ease in plant management due to a large number of products, easy availability of big machines, and contact with big industries for marketing. Hence, cultivation of medicinal and aromatic plants on a large scale, one can easily register in the Aroma Mission run by the Government of India, due to which facilities from seeds to distillation are easily available for making more profitability and sustainability of sugarcane-based high-value cropping system including short-duration medicinal and aromatic crops under suitable agroclimatic condition. In the above process, people can get more opportunities for the generation of employment at the local level, which can also reduce the migration of farmers and rural laborers toward cities and can also help in improving their economic condition besides improving their living standards.

6.9.1 Precautions in the Cultivation of Medicinal and Aromatic Plants

Before cultivating medicinal plants, the farmer group should establish contact with any buyer/industry for purchasing its products. For information related to the availability and varieties of seeds of medicinal and aromatic plants, before starting farming, contact can be made with Central Institute of Medicinal and Aromatic Plants, Lucknow, or another similar reputed institution. Under no circumstances should a situation of waterlogging arise for a long time in the standing crops of plants. That is, for successful farming, it is very important to have proper arrangements for proper drainage in the field. For detailed information on suitable varieties of sugarcane, availability of seeds, and any topic related to crop management, ICAR-Indian Sugarcane Research Institute, Lucknow, Uttar Pradesh, can be contacted.

6.10 Future Scope

There is tremendous scope for increasing area under medicinal and aromatic crops in sugarcane-based cropping systems under Indian subtropics. The inclusion of medicinal and aromatic plants in place of the traditional rice–wheat system also mitigates the many challenging issues of water scarcity especially depletion ground water table and environmental sustainability besides enhancing higher system productivity, profitability, and sustainability of sugarcane-based cropping system.

References

Babu S, Singh R, Avasthe RK, Yadav GS, Mohapatra KP, Selvan T, Das A, Singh VK, Valente D, Petrosillo I (2020) Soil carbon dynamics in Indian Himalayan intensified organic rice-based cropping sequences. Ecol Indic. https://doi.org/10.1016/j.ecolind.2020.106292

Babu S, Singh R, Avasthe R, Rathore SS, Kumar S, Das A, Layek J, Sharma V, Wani OA, Singh VK (2023a) Conservation tillage and diversified cropping enhance system productivity and eco-efficiency and reduce greenhouse gas intensity in organic farming. Front Sustain Food Syst 7:1114617. https://doi.org/10.3389/fsufs.2023.1114617

Babu S, Singh R, Avasthe RK, Kumar S, Rathore SS, Singh VK, Ansari MA, Valente D, Petrosillo I (2023b) Soil carbon dynamics under organic farming: impact of tillage and cropping diversity. Ecol Indic. https://doi.org/10.1016/j.ecolind.2023.109940

Beillouin D, Ben-Ari T, Malezieux E, Seufert V, Makowski D (2021) Positive but variable effects of crop diversification on biodiversity and ecosystem services. Glob Change Biol. https://doi.org/10.1111/gcb.15747

Bhatt R, Kukal S, Mutiu S, Busari A, Sanjay A, Mathura Y (2016) Sustainability issues on rice–wheat cropping system. Int Soil Water Conserv Res 4(1):64–74

Booker A, Johnston D, Heinrich M (2012) Value chains of herbal medicines-research needs and key challenges in the context of ethnopharmacology. J Ethnopharmacol 140(3):624–633. https://doi.org/10.1016/j.jep.2012.01.039

FAO (2019) Food and Agriculture Organization of the United Nations FAO. http://www.fao.org/3/CA6030EN/CA6030EN.pdf

Kerr RB, Nyantakyi-Frimpong H, Dakishoni L, Lupafya E, Shumba L, Luginaah I, Snapp SS (2018) Knowledge politics in participatory climate change adaptation research on agroecology in Malawi. Renew Agr Food Syst 33(3):238–251

Kumar R, Gopal R, Jat ML, Gupta RK (2010) Conservation agriculture-based strategies for sustainable weed management in maize (*Zea mays*) training manual. Directorate of Maize Research (DMR), New Delhi

Kwak YS, Weller DM (2013) Take-all of wheat and natural disease suppression: a review. Plant Pathol J 29:125–135

Raseduzzaman M, Jensen ES (2017) Does intercropping enhance yield stability in arable crop production? A meta-analysis. Eur J Agron 91:25–33. https://doi.org/10.1016/j.eja.2017.09.009

Rodell M, Velicogna I, Famiglietti JS (2009) Satellite-based estimates of groundwater depletion in India. Nature 466:999–1002

Shahidulla AKM, Emdad Haque C (2010) Linking medicinal plant production with livelihood enhancement in Bangladesh: implications of a vertically integrated value chain. J Transdiscipl Stud 9(2)

Shukla SK, Singh KK, Pathak AD, Jaiswal VP, Solomon S (2017) Crop diversification options involving pulses and sugarcane for improving crop productivity, nutritional security and sustainability in India. Sugar Tech 19(1):1–10

Shukla SK, Yadav SK, Menhi Lal, Pathak AD (2018) Alternate/skip furrow irrigation. In low cost technologies in sugarcane agriculture. Technical Bulletin, IISR Lucknow, p 44

Shukla SK, Solomon S, Sharma L, Jaiswal VP, Pathak AD, Priyanka S (2019) Green technologies for improving cane sugar productivity and sustaining soil fertility in sugarcane based cropping system. Sugar Tech 21(2):186–196

Shukla SK, Sharma L, Jaiswal VP, Awasthi SK, Pathak AD, Adil Z, Yadav SK (2021) Identification of appropriate agri-technologies minimizing yield gaps in different sugarcane growing states of India. Sugar Tech 23(3):580–595

Shukla SK, Sharma L, Jaiswal VP et al (2022) Diversification options in sugarcane-based cropping systems for doubling farmers' income in subtropical India. Sugar Tech 24:1212–1229. https://doi.org/10.1007/s12355-022-01127-1

Singh A (2015) Soil salinization and waterlogging: a threat to environment and agricultural sustainability. Ecol Indic 57:128–130

Singh RK, Singh P, Li HB et al (2020) Diversity of nitrogen-fixing rhizobacteria associated with sugarcane: a comprehensive study of plant-microbe interactions for growth enhancement in *Saccharum* spp. BMC Plant Biol 20:220. https://doi.org/10.1186/s12870-020-02400-9
Singh R, Subhash B, Avasthe RK, Meena RS, Yadav GS, Das A, Mohapatra KP, Rathore SS, Kumar A, Singh C (2021) Conservation tillage and organic nutrients management improve soil properties, productivity, and economics of the maize-vegetable pea system in the Eastern Himalayas. Land Degrad Dev. https://doi.org/10.1002/ldr.4066
Smith-Hall C, Larsen HO, Pouliot M (2012) People, plants and health: a conceptual framework for assessing changes in medicinal plant consumption. J Ethnobiol Ethnomed 8(1):1–11
Sofowora A, Ogunbodede E, Onayade A (2013) The role and place of medicinal plants in the strategies for disease prevention. Afr J Tradit Complement Altern Med 10(5):210–229
Taghouti I, Cristobal R, Brenko A, Stara K, Markos N, Chapelet B, Hamrouni L, Burši'c D, Bonet JAA (2022) The market evolution of medicinal and aromatic plants: a global supply chain analysis and an application of the Delphi method in the Mediterranean area. Forests 13:808
Yadav GS, Das A, Kandpal BK, Subhash B, Rattan L, Datta M, Das B, Singh R, Singh VK, Mohapatra KP, Chakraborty M (2021) The food-energy-water-carbon nexus in a maize-maize-mustard cropping sequence of the Indian Himalayas: an impact of tillage-cum-live mulching. Renew Sust Energ Rev. https://doi.org/10.1016/j.rser.2021.111602

Chapter 7
Diversification of Rainfed Maize-Based Systems in the Indian Himalayas

Amit Kumar, Saurav Saha, Devideen Yadav, Subhash Babu, Raghavendra Singh, T. L. Bhutia, E. L. Devi, Ramgopal Devadas, and B. A. Gudade

Abstract Rainfed maize cultivation plays a pivotal role in ensuring both nutritional and economic security for the local communities in the Indian Himalayan Region (IHR). However, the existing mono-cropping practices, coupled with the challenges posed by climate change and resource constraints, necessitate a shift toward a more diversified and resilient agricultural system. Crop diversification approach not only mitigates the impact of climate uncertainties but also contributes to improved soil health through crop rotation. Furthermore, the inclusion of leguminous crops enhances nitrogen fixation, thereby reducing the dependency on external fertilizers, and making the system more sustainable and cost-effective for farmers. Nutritional security is a critical aspect addressed by diversification, as it ensures a more balanced and diverse diet for local communities. Economic security is another key dimension addressed by the proposed diversification strategy. Mixed cultivation of local cultivars, farmers can tap into different market demands, reducing the vulnerability associated with reliance on a single commodity. This approach aligns with broader goals of promoting food security, preserving agro-biodiversity, and fostering economic sustainability in the unique context of the IHR.

A. Kumar (✉) · S. Saha · T. L. Bhutia · E. L. Devi · R. Devadas
ICAR Research Complex for NEH Region, Sikkim Centre, Tadong, Gangtok, Sikkim, India

ICAR-Indian Institute of Soil and Water Conservation, Dehradun, Uttarakhand, India

D. Yadav

Division of Agronomy, ICAR – Indian Agricultural Research Institute, New Delhi, India

ICAR-Indian Institute of Soil and Water Conservation, Dehradun, Uttarakhand, India

S. Babu
Division of Agronomy, ICAR – Indian Agricultural Research Institute, New Delhi, India

R. Singh
ICAR – Indian Institute of Farming Systems Research, Meerut, India

B. A. Gudade
Spices Park Chhindwara, Spices Board of India, Chhindwara, Madhya Pradesh, India

S. Babu et al. (eds.), *Agricultural Diversification for Sustainable Food Production*, Sustainability Sciences in Asia and Africa,
https://doi.org/10.1007/978-981-97-7517-0_7

Keywords Economic security · Nutritional security · Productivity · Rainfed maize systems

7.1 Introduction

Cereal grains, including wheat (*Triticum aestivum* L.), rice (*Oryza sativa* L.), and maize (*Zea mays* L.), are universally recognized as staple foods, providing approximately 50–60% of the daily energy needs for people worldwide (Babu et al. 2020a; Kumar et al. 2023). Globally, maize is ranked as the third most significant cereal and earns its distinction through both its abundant yield and nutritional richness, earning it the title of the premier cereal crop. Maize stands out as a staple cereal, widely favored for its substantial nutritional value in hill agriculture. It is rich in essential macronutrients such as starch, fiber, protein, and fat. Additionally, maize boasts a wealth of micronutrients, including vitamin B complex, β-carotene, and vital minerals such as magnesium, zinc, phosphorus, and copper, among others. The diverse array of nutrients in maize contributes to its popularity and its role in providing a well-rounded and nourishing dietary source (Babu et al. 2016). Furthermore, maize includes a natural antioxidant booster, offering protection against a range of degenerative diseases. This antioxidant component adds another layer of health benefits to maize consumption, contributing to its potential in supporting overall well-being and disease prevention (Kennett et al. 2020).

Maize is a more versatile crop when it comes to serving multiple purposes, compared to wheat and rice. Its main use in industrialized nations is as a crop for cattle feed, but it also has a variety of uses as an industrial and energy crop. In developing countries, corn finds its largest application in the poultry industry, accounting for approximately 47% of its usage. Direct human consumption follows at 20%, with cattle feed and starch extraction each contributing 14%. The food and beverage industry utilizes about 7% of maize in these regions (Grote et al. 2021). Maize plays a dynamic and diverse role in global agri-food systems, significantly contributing to food and nutrition security worldwide. Its versatility in applications, ranging from livestock feed and industrial uses to direct human consumption, underscores its importance in ensuring a resilient and secure global food supply (Poole et al. 2021). The agri-food system has garnered increasing attention in the last 10 years (Jones and Yosef 2015; Fanzo et al. 2021). The pivotal role of agri-food systems is evident in their contribution to the 2030 Agenda for Sustainable Development and its 17 Sustainable Development Goals (Fanzo et al. 2021). Global maize cultivation spans 197 million hectares, resulting in a production of 1137 million tons worldwide (FAO Stat 2021). The cultivation of maize is widespread, encompassing both emerging economies and developed nations. It is practiced in 165 countries across the continents of the Americas, Asia, Europe, and Africa (FAO Stat 2021). Maize exhibits significant variations in yields across different regions. The Americas play a substantial role, contributing half of the global maize production, followed by Asia at 32%. The remaining share is mainly distributed between Europe (11%) and Africa

(7.4%). For India, maize cultivation covers an area of approximately 9.47 million hectares, yielding around 28.6 million tons and achieving a productivity rate of 3190 kg/ha (DACNET 2022). In India, Madhya Pradesh and Karnataka lead in maize cultivation, each contributing 15% of the total maize-growing area, followed by Maharashtra (10%), Rajasthan (9%), Uttar Pradesh (8%), and other states. The productivity of maize in India stands at approximately 3.19 Mg per hectare, slightly exceeding half of the global average of 5.6 Mg per hectare (DACNET 2022). In the IHR, maize holds the position of the second most crucial food grain after rice. Maize holds distinctive importance in the region as a key ingredient in concentrate feeds for poultry, fish, pigs, and cattle. Furthermore, there is a growing demand for green cob and popcorn among the local population in the IHR (Singh et al. 2018). This underscores the versatile uses of maize, not only in animal feeds but also as a popular food item for direct consumption (Das et al. 2019). Rainfed maize cultivation in the IHR is a dynamic and challenging agricultural practice deeply rooted in the diverse topography and climatic conditions of this ecologically sensitive area (Singh et al. 2021a). The region's farmers must navigate the challenges posed by unpredictable weather patterns, including the risk of droughts or excessive rainfall that can adversely affect maize yields (Das and Mehar 2019). Altitude variation plays a significant role in shaping rainfed maize cultivation practices. The diverse soil types in the region, ranging from loamy to sandy, add another layer of complexity to rainfed maize cultivation. Maize varieties cultivated in the IHR are often traditional and well-adapted to local conditions. The selection of maize varieties is influenced by local factors such as altitude, temperature, and moisture availability. Cropping practices in rainfed maize cultivation often involve intercropping with legumes or other cereals (Das et al. 2018; Babu et al. 2020c). This diversification not only helps in maximizing land use but also contributes to soil health through nitrogen fixation by leguminous crops (Singh et al. 2021c; Babu et al. 2022). Crop rotation is another essential aspect of rainfed agriculture in the Himalayan region, with farmers alternating maize cultivation with other crops like wheat, barley, or pulses to maintain soil health status and reduce the risk of biotic and abiotic stress. Monocultures are susceptible to the rapid spread of pests and diseases. Crop diversification disrupts the continuous presence of a specific host, making it more challenging for pests and diseases to establish themselves and spread (Walia et al. 2022). This reduces the reliance on chemical inputs and promotes more sustainable pest management practices (Kumar et al. 2022a; Yadav et al. 2023). Diversification opens up opportunities for farmers to explore new markets and value chains. Different crops may have distinct market demands, and diversification enables farmers to respond to changing consumer preferences and capture niche markets. Diversifying the crops grown alongside rainfed maize introduces a variety of nutrient profiles into local diets (Sharma et al. 2022). Different crops contribute different vitamins, minerals, and macronutrients, addressing potential nutritional gaps that might exist in a diet primarily reliant on maize. Intercropping with legumes, such as beans or peas, can complement the amino acid profile, providing a more balanced source of protein in the diet. This combination enhances the nutritional quality of the overall food intake. Diversification allows for the cultivation of micronutrient-rich crops such as leafy greens, root vegetables, and fruits. These crops contribute essential vitamins and minerals crucial for overall health, addressing specific micronutrient deficiencies that may be prevalent in IHR (Singh et al. 2018).

7.2 Agroecological Context of Indian Himalayan Region

Approximately 22% of the Earth's land surface comprises mountain ecosystems, where millions of people depend primarily on agriculture to secure their livelihoods. The Himalayas are the tallest and youngest mountain range on the globe. It spans eight South Asian regions, encompassing Afghanistan, Pakistan, China, India, Nepal, Bhutan, Bangladesh, and Myanmar, with coordinates ranging from 15.95° to 39.31° N and 60.85° to 105.04° E (Das and Meher 2019). This extensive mountain range is over 3000 km in length and 250–300 km in width, with a height difference of 300–>8000 m above mean sea level. Its total area is close to 7,50,000 km^2. The IHR contains the majority of the Himalayan range. The IHR spans approximately 2400 km along India's northern border, extending from the eastern border of Pakistan in the west to the frontiers of Myanmar in the east. These geologically dynamic and intricate young mountains of the IHR hold a unique and significant position within the world's mountain ecosystems. From a geographical standpoint, the IHR situated between the Indus and Brahmaputra river systems, encompasses an area of approximately 533,604 km^2. This accounts for about 16.2% of the total geographical area of the country. In this region, approximately 40 million people reside, constituting around 3.5% of the total population of the country (Babu et al. 2023a).

The IHR has been categorized into three distinct vertical divisions based on altitude, physiography, and geological history. These divisions are (a) The Outer or Sub-Himalaya (the Shiwaliks), the Lesser Himalaya (Himachal), and the Greater or Higher Himalaya (the Himadri). To the north of the Himalayan region in Tibet lies the Trans-Himalaya. Medlicott's historic classification of the northwest Himalaya is an early system that divides the Indian Himalaya into two main groups: the Sub-Himalayan category, encompassing Upper, Middle, and Lower Shiwalik subdivisions, and the Himalayan category, which further distinguishes materials as either metamorphic or non-metamorphic (Yadav et al. 2018; Kumar et al. 2022). The IHR encompasses either partially or in full 12 states of the country. These states include Jammu & Kashmir, Himachal Pradesh, Uttarakhand, Sikkim, Arunachal Pradesh, Nagaland, Manipur, Mizoram, Tripura, Meghalaya, as well as certain hill regions in Assam (Karbi Anglong and Dima Hasao) and West Bengal (Darjeeling and Kalimpong). Agro-climatic zone of the IHR is given in Table 7.1. According to a recent FAO report (IMI FAO Report 2019), nearly 70% of the population in the IHR is rural, while approximately 30% resides in urban areas. Notably, Himachal Pradesh has the highest rural population at 90%, followed by Uttarakhand at 70%, and Mizoram at 50%. The IHR holds significant importance within India, as it supplies water to a substantial portion of the Indian subcontinent, offers a potential

Table 7.1 Agro-climatic zone of the Indian Himalayan Region

Agro-climatic zone	Climate	Rainfall (mm)	State/regions
Zone I	High-altitude temperate (humid to cold arid)	<1200	Jammu & Kashmir
Zone I	Hill temperate (per humid to subhumid)	1200–1800	Himachal Pradesh and Uttaranchal
Zone II	Per humid to humid	1800–2200	Nagaland, Mizoram, Manipur, and Tripura
Zone II	Per humid to humid	2200–2800	Sub-Himalayan West Bengal, Sikkim, Assam, and Meghalaya
Zone II	Per humid to humid	>2800	Arunachal Pradesh

Source: Agro-climatic Regional Planning, Planning Commission (1989)

reservoir of medicinal plants, and boasts rich genetic diversity across various forms of flora and fauna (Wani et al. 2023). Each year, the rivers in the Indian Himalayas discharge approximately 1,600,000 million cubic meters of water, serving various purposes such as drinking, irrigation, and hydropower generation. Additionally, a substantial proportion of crude medicinal substances available in the Indian market originate from this Himalayan region (Panwar et al. 2019). Despite the abundance of natural resources, a significant portion of its population remains marginalized and continues to subsist at a basic level of existence. The land resources in IHR may be grouped into two categories: (i) *the croplands*, which remain under cultivation of annual field crops and/or perennial horticultural crops, and (ii) *the support lands*, which support farming-based livelihoods. The agricultural lands in different ecological regions of the IHR exhibit varying limitations and opportunities. Meanwhile, the support lands can play a significant role in ensuring the livelihood security of the local population, either partially or entirely. For example, in the northwest Himalayas (Ladakh, Kashmir, and Sikkim), the people completely depend on pastoral and agropastoral (support) lands. In the northeastern Himalayan region, shifting cultivation stands out as a significant livelihood system based on support lands. Additionally, across much of the Indian Himalayan Region (IHR), forest and range lands have been utilized as support lands, contributing to the livelihoods of local communities.

The National Land Use Policy of 2016 has recognized the rights of local communities, marginalized farmers, and ethnic minorities to access available land resources. However, it is worth noting that the average size of agricultural land holdings has notably decreased over the years. Presently, the average land holding per family is less than 1.0 hectares (IMI FAO Report 2019). Land-use patterns and livelihood strategies in the Indian Himalayan Region are often profoundly influenced not only by environmental factors but also by the ethnic, sociocultural affiliations, and territorial rights of local communities. To gain a more comprehensive understanding of the diversity, complexity, and evolving nature of regional land-use systems, it is essential to identify and evaluate the historical dimension of the interactions between humans and their environment.

The IHR is blessed with a rich variety of natural vegetation, which can be broadly categorized into four groups based on climatic conditions: tropical, subtropical, temperate, and alpine. The local terrain and exposure to sunlight play a significant role in shaping microclimates and the types of vegetation found in different areas. In the humid foothills of the eastern and central Himalayas, tropical evergreen rainforests, featuring species like ceylon ironwood, bamboo, and alder trees are very common. In contrast, the western Himalayas predominantly host tropical deciduous forests, with trees like *sal* being prominent. Moving further westward, you encounter forest steppe characterized by grasslands interspersed with trees. At the lowest altitudes of the mountains, the climate is tropical, but at the highest points, it is permanently covered in snow and ice. This region's environment is ideal for a variety of agricultural pursuits, such as flower gardening, horticulture, and the production of different crops (Das et al. 2017). The negative consequences of climate change, characterized by fluctuations in temperature and rainfall patterns, are already becoming evident in both the biophysical aspects such as mountainous terrain, rivers, landslides, forests, and land degradation, as well as in the socioeconomic systems, including hill communities, agriculture, and animal husbandry within the region (IPCC 2014). Himalayan communities rely heavily on climate-sensitive rainfed agriculture, with limited alternative livelihood opportunities. The IHR exhibits a diverse range of soil types, with varying sediment thicknesses found on both ridges and valleys. These soil variations are a consequence of the region's complex geological, physiographic, and climatic features. The present soil resources have developed over extended periods through erosion and deposition processes linked to the relatively recent mountain-building activities in the Himalayas. The soil in the IHR is fertile, but it is also delicate and, as a result, susceptible to erosion (Chandra et al. 2011). The composition and quality of soil in the IHR are influenced by a combination of intrinsic factors, such as soil-forming elements and the parent rock, as well as extrinsic factors, including agricultural management practices (Choudhary et al. 2013). The steep inclines and ongoing tectonic activity in the IHR heighten its susceptibility to landslides, rockfalls, and soil erosion. Unfortunately, there is a scarcity of data concerning the region's soil resources. In the high-altitude areas, soils predominantly originate from morainic deposits.

7.3 Agriculture of Indian Himalayan Region

Agriculture in the Himalayas has a strong foundation in traditional and indigenous farming practices. Terrace farming is a common technique used on the steep slopes of the Himalayan hills to optimize arable land and prevent soil erosion (Layek et al. 2018). Traditionally, the agricultural systems in the Himalayan region have been characterized as subsistent family farming systems (Babu et al. 2023a; Yadav et al. 2024). These farming practices are deeply rooted in the integration of crops, livestock, and forest products, forming a holistic and interconnected approach to agriculture. Moreover, these systems heavily rely on locally available natural resources to meet the agricultural needs of the community. Agriculture remains the primary source of livelihood for the population in the Himalayan region, playing a crucial role in ensuring food security for local communities. While traditional mixed farming systems are still practiced to some extent, there has been a noticeable shift in crop choices toward cash crops (Layek et al. 2015). This transition in livelihood strategies has had significant consequences for agro-biodiversity and the nutritional security of households in the region. The practice of rice/maize–fallow monoculture stands out as a notable cropping system in the region. However, this system exhibits low agronomic productivity and is susceptible to the impact of dry spells. There is a potential for increasing cropping intensity and enhancing soil quality by introducing short-duration crops during the winter season (Yadav et al. 2023, 2024). Traditional agriculture in the IHR faces challenges as it is not economically rewarding, posing a threat to the livelihood security of hill farmers (Kumar et al. 2022b). Therefore, there is a pressing need for the development and implementation of strategies aimed at enhancing both agricultural productivity and the economic well-being of farmers in the region (Das et al. 2022). This could involve adopting more sustainable and resilient farming practices, promoting diversified cropping systems, and providing support for alternative income-generating activities to ensure the overall prosperity of the farming communities in the IHR (Singh et al. 2021b). Hence, there is an urgent requirement to enhance the current maize/rice–fallow system by incorporating diverse and alternative cropping systems. This initiative aims to boost both farm productivity and profitability while also preserving the health of the soil (Saha et al. 2012; Yadav et al. 2013). However, many of the intensification strategies primarily concentrate on introducing winter crops into the maize/rice–fallow system. Therefore, utilizing leftover fallows after maize or rice cultivation for planting a suitable winter crop in sequence has the potential to generate additional biomass. This approach not only enhances carbon sequestration capabilities but also contributes to achieving food security (Babu et al. 2020c).

7.4 Crop Diversification

Traditional farming often relies on monoculture, where a single crop dominates large expanses of land (Singh et al. 2023b). While this approach has shown short-term gains in terms of yield, it often leads to environmental degradation, soil depletion, and increased vulnerability to pests and diseases. In this situation, crop diversification is the paramount importance to prevent these emerging problems (Panwar et al. 2003, 2019). Crop diversification refers to the practice of growing a variety of crops on the same piece of land over time or in each cropping season. It involves planting different types of crops in a sequence or in a mixed cropping system instead of monoculture, which is the practice of planting only one type of crop repeatedly in the same field (Fig. 7.1). In this strategy, the introduction of new crops or cropping systems to the farm's agricultural production considers the diverse returns derived from value-added crops that offer complementary marketing

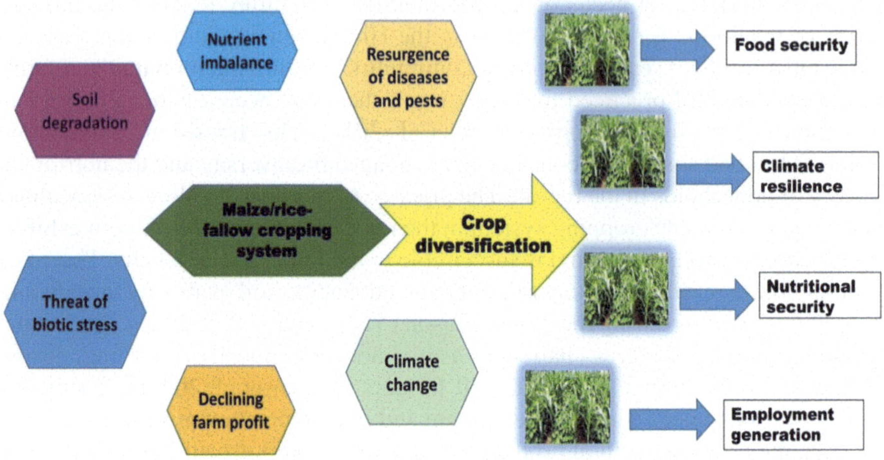

Fig. 7.1 Benefits of crop diversification

opportunities. The objective of crop diversification is to broaden the range of crops in a portfolio, reducing farmers' reliance on a single crop for income generation (Ngangom et al. 2020). Relying on a single crop exposes farmers to high risks in the face of unforeseen climate events, such as the emergence of pests or sudden occurrences of frost or drought, which can significantly impact agricultural production (Babu et al. 2022). Plant productivity, quality, health, and nutritional value can all be enhanced through technical means, such as the introduction of newly farmed species and improved crop varieties. Additionally, it builds resilience in crops against diseases, pests, and environmental stresses. This strategy minimizes the risk of total crop failure and creates additional revenue streams, as different crops respond uniquely to diverse climate conditions (Prachi et al. 2017). For example, while one crop may suffer from cold weather, another might experience increased production. In the Indian Himalayas, crop diversification is generally seen as a transition from traditionally grown, less profitable crops to more lucrative ones. The primary benefits of this approach encompass enhanced income for small-scale farms, reduced exposure to price fluctuations and climate uncertainties, equilibrium in food supply and demand, amplified production of high-quality livestock fodder, positive impacts on natural resource conservation, decreased environmental pollution, lowered reliance on external inputs, and heightened community food security (Ramakrushna et al. 2022; Rathore et al. 2022a).

Research institutions should contribute to the development of novel technological advancements aimed at expanding production capabilities, optimizing input efficiency, and implementing precision farming practices to significantly enhance farmers' income and overall productivity (Tong et al. 2017). In the IHR, increasing the diversity and intensity of cereal-centric production systems by adding short-duration crops and using efficient soil and crop management techniques has resulted in more efficient use of land, water, and nutrients (Fig. 7.2) (Singh et al. 2021a). In

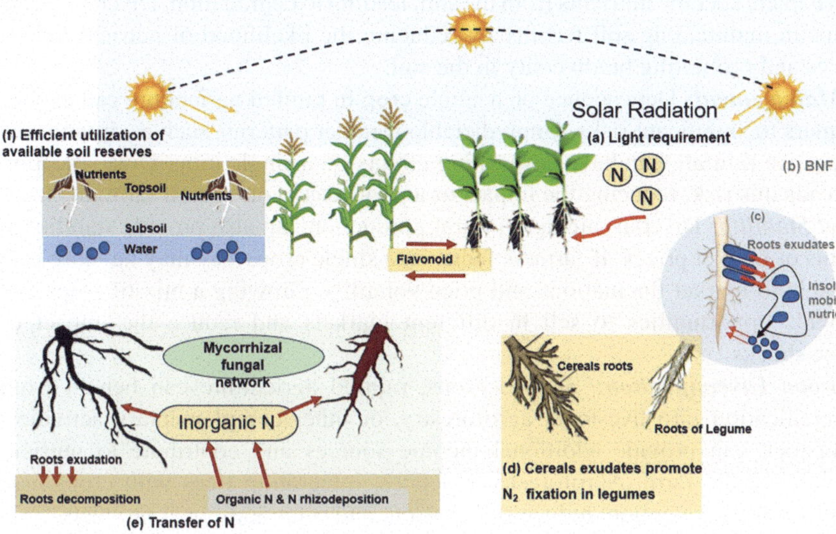

Fig. 7.2 Impact of intensified cropping systems on resource use efficiency. (Modified from Yang et al. 2021)

this context, the goal of crop diversification is not only to maximize yields but also to ensure long-term food security, protect natural resources, and promote rural livelihoods. This introduction sets the stage for a comprehensive exploration of the benefits, challenges, and strategies associated with agricultural diversification for sustainable food production (Yadav et al. 2019).

7.4.1 Concept of Diversification and Its Relevance in Rainfed Agriculture

Diversification in agriculture refers to the practice of growing a variety of crops or raising multiple types of livestock on a farm. This strategy is employed to spread risk and increase resilience against various environmental, market, and economic uncertainties. In the context of rainfed agriculture, where crop yields are heavily dependent on rainfall, diversification becomes particularly relevant for several reasons:

Climate Variability: Rainfall is inherently variable, both in terms of timing and quantity. Diversifying crops allows farmers to hedge against the risk of crop failure due to insufficient or untimely rainfall. Different crops have varying water requirements and tolerances, so having a mix can help ensure that at least some crops will perform well under prevailing conditions.

Soil Health and Nutrient Management: Different crops have different nutrient requirements and interactions with the soil. Monoculture (growing a single crop) can deplete specific nutrients from the soil, leading to degradation. Diversification helps in maintaining soil fertility by reducing the likelihood of nutrient imbalances and promoting biodiversity in the soil.

Risk Management: Dependence on a single crop in rainfed agriculture can expose farmers to significant risks. Unpredictable weather patterns, such as droughts or excessive rainfall, can have devastating effects on a single crop. Diversification spreads this risk, reducing the impact of a poor season on overall farm income.

Market Stability: Diversifying agricultural production can also provide stability in terms of market prices. If farmers focus on a single crop, they may be more vulnerable to market fluctuations and price volatility. Growing a mix of crops can provide opportunities to sell in different markets and reduce the impact of price shocks.

Livelihood Diversification: Beyond crops, rainfed agriculture can benefit from diversification into livestock, agroforestry, or other complementary activities. Livestock can provide additional income sources and contribute to nutrient cycling on the farm. Agroforestry practices, integrating trees with crops, can improve water retention, reduce soil erosion, and offer additional products.

Water-Use Efficiency: Different crops have varying water-use efficiencies. Diversifying with crops that have different water requirements can help maximize water utilization, especially in regions where water availability is a limiting factor.

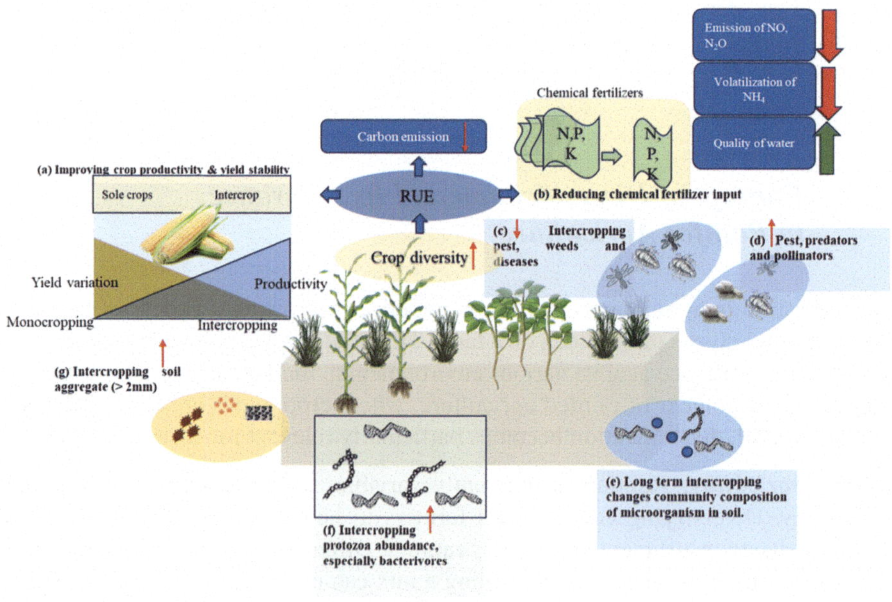

Fig. 7.3 Impact of crop diversifications on agroecosystem functions. (Modified from Yang et al. 2021)

Ecosystem Services: Diversification can enhance ecosystem services such as pollination, pest control, and soil conservation. A diverse agricultural landscape supports a variety of beneficial organisms, contributing to a more resilient and sustainable farming system (Fig. 7.3).

7.5 Opportunity of Rainfed Maize-Based Cropping System in IHR

In India, around 50% of the net cultivated land is under *rainfed* crops, contributing 40% of the country's food production and accounting for 61% of the farmers' population (Department of Agriculture and Farmers Welfare, Government of India; https://agricoop.nic.in/en/divisiontype/rainfed-farming-system). In contrast, about 80% cultivated area in IHR is *rainfed*, >80% of farmers are small and marginal farmers contributing 30% of the gross domestic product in the region (Yadav et al. 2020). The sustainability of *rainfed* agricultural systems is at risk due to inefficient resource usage, which also drives up cultivation costs and reduces farm profitability. IHR has a very erratic rainfall pattern with uneven distributions. About 75% of the yearly rainfall falls between May and September, with very little falling in the post- and pre-rainy seasons. In IHR regions, maize-based cropping systems have a lot of potential to support hill farming sustainability. Rainfed conditions offer considerable potential for maize during the rainy season, followed by pulses and oilseeds during the winter (Kumar et al. 2022a). In IHR, several cropping options are traditionally followed by the farmers in rainfed condition, that is, maize–millet, maize–urd/moong bean/soybean, maize–cowpea/rice bean, sorghum+ arhar, sorghum–French bean/toria/buckwheat, maize–buckwheat/oat/barley, maize–sunflower/til/toria. In order to intensify the maize–fallow system, cowpea + maize–garden pea with field residue management is a sustainable crop production practice in *rainfed* sub-temperate regions of Himalaya (Singh et al. 2021a). Prominent and potential maize-based cropping systems of IHR and invention are provided in Table 7.2.

7.5.1 Impact of Rainfed Maize-Based Cropping Systems on System Productivity

Implementing sustainable intensification in cereal-based systems, coupled with tailored soil and crop management practices, enhances the productivity of land, crops, and water across diverse agroecosystems. Introducing legumes as intercrops and incorporating short-duration winter crops, along with in situ moisture conservation measures, has notably increased maize productivity compared to the traditional maize–fallow system in hilly regions. Beyond the additional economic gains,

Table 7.2 Efficient maize cropping systems for IHR

State	Existing cropping systems	Potential maize-based cropping systems	
Sikkim	Maize–fallow Rice–fallow	Maize-*Pahenlo* dal (Green seeded urd bean)–buckwheat	Kumar et al. (2022)
		Maize–vegetable pea/French bean	
		Maize + cowpea–rapeseed maize + cowpea–buckwheat maize + cowpea–barley maize + cowpea–garden pea maize + cowpea–rajmash maize–rapeseed/buckwheat	Singh et al. (2021b) Babu et al. (2022)
Manipur	Rice–fallow	Maize–vegetable pea/rapeseed and mustard maize–vegetables maize + groundnut/soybean maize + urd bean/green gram/ rice bean	Ansari et al. (2022)
	Vegetable–fallow		
	Rice–rice		
Meghalaya	Rice–fallow	Maize–French bean/black gram/rapeseed	Das et al. (2017)
	Ginger/turmeric–fallow		
Tripura	Rice–fallow	Maize–rice–pea/lentil	Yadav et al. (2021a)
	Vegetable–fallow		
Mizoram	*Jhum* maize–fallow	Sweet corn–soybean/vegetable mustard/French bean (Zorin bean)	Sati and Vangchhia (2017)
	Jhum rice– Fallow	Rice+maize (sweetcorn/ popcorn)–broccoli/chili/brinjal/ cabbage/French bean (Zorinbean)	
	Lowland rice–fallow	Cowpea/okra–rice–pea/ vegetable mustard	
Himachal Pradesh	Potato–fallow Maize–fallow	Maize–gram Maize–mustard Maize –barely Maize–buckwheat–gram/barley Maize–vegetables	Choudhary et al. (2021)
Jammu and Kashmir	Rice–fallow	Maize–oilseed Maize–Oat Maize–potato Maize + bean	Bhat et al. (2019)
		Maize + lentil (*rabi*) Maize + pea (*rabi*)	Sharma et al. (2022)
Uttarakhand	Rice–wheat	Maize–vegetable pea/lentil/ barley/toria Maize + soybean/cowpea	Prachi et al. (2017)

including legumes as an intercrop in maize, provides various advantages such as weed suppression, prevention of nutrient loss from intense rainfall, atmospheric nitrogen fixation, and an overall enhancement of soil health (Yadav et al. 2018). According to the findings of Singh et al. (2021a), co-cultivating cowpea with maize and introducing a winter crop resulted in increased maize yield ranging from 6.20% to 23.5% compared to the traditional maize–fallow system. Among these systems, the maize + cowpea–green pea cropping system exhibited the highest maize grain yield, surpassing the maize–fallow system by 23.5%, and outperforming other combinations like maize + cowpea–rapeseed, maize + cowpea–buckwheat, maize + cowpea–barley, and maize + cowpea–rajmash systems by 15.5%, 10.4%, 2.6%, and 2.6%, respectively. Intercropping cowpea with maize not only increased maize grain yield but also provided additional pod yield ranging from 1.45 to 1.60 Mg per hectare. The maize + cowpea–green pea system exhibited the highest pod yield at 1.60 Mg per hectare, followed closely by maize + cowpea–rajmash at 1.59 Mg per hectare, surpassing other cropping systems. Numerous studies have shown that adding legumes to intensive systems increases agricultural output (Ghosh et al. 2020). According to Ngangom et al. (2020), the average maize equivalent yields (MEYs) in the maize–French bean (pole type) system, with in situ maize stalk mulching + white hoary pea, and maize stalk mulching + ragweed mulches, were 218% and 202% higher, respectively, compared to the maize–rapeseed cropping sequence without mulch. Similarly, the MEYs for the maize–French bean (bunch type), maize–black gram, and maize–rapeseed systems, with maize stalk mulching + white hoary pea, were 150%, 136%, and 52.8% higher, respectively, than those recorded with maize–fallow without mulch. As per the findings of Yadav et al. (2015), implementing no-tillage (NT) methods for crop establishment improved the productivity and profitability of the maize–maize–field pea cropping system. The average system productivity was 6.1% greater under NT systems compared to conventional tillage (CT) systems. Yadav et al. (2020) mentioned that the adoption of the maize–maize–field pea cropping system resulted in a 28% increase in system productivity, along with a 16% improvement in carbon sequestration and overall enhancement of soil properties in the eastern Himalayas and comparable eco-regions. All of the intensive cropping systems were generally more productive than the local farmers' techniques (maize–fallow). When comparing the total productivity of intensified cropping systems in terms of maize grain equivalent yield, the more intense cropping system maize (green cobs)–urd bean–buckwheat yielded a significantly higher yield (8.89 Mg/ha), followed by the maize–rajmash and maize–toria systems. Conversely, the field that was left fallow in the years following maize harvesting produced the lowest yield (3.3 Mg/ha) (Babu et al. 2016). Ramkrushna et al. (2022) noted that all the maize–legume cropping systems significantly increased maize equivalent yield (MEY) compared to sole maize. Intercropping maize with groundnut had a positive and significant impact on French bean yield in both bush and pole types in both years of the study, except for pole type in the first year. The observed yield improvement was in the range of 0.30–0.50 Mg per hectare for pod yield and 0.42–0.68 Mg per hectare for MEY in bush types. For pole types, the enhancement was 0.29–0.30 Mg per hectare for pod yield and 0.32–0.41 Mg per hectare for

MEY. System productivity, measured in terms of MEY, ranged from 2.54 to 10.36 Mg per hectare, with the highest and lowest values observed in the maize + groundnut–French bean (bush type) and sole maize systems, respectively. In Tripura, the inclusion of cowpea live mulch under no-till and reduced tillage in the summer maize–rainy season maize–mustard sequence enhanced agronomic productivity, while reducing greenhouse gas emissions (Yadav et al. 2021b). This finding is very useful to develop a sustainable high-productivity policy through efficient resource use in the Eastern Himalayan Ecosystems that have the potential to become the corn belts of India.

7.5.2 *Economic Security Through Rainfed Maize-Based Cropping System*

Economic security and income diversification in the Indian Himalayan region can be achieved through the promotion of maize-based rainfed cropping systems (Sati and Vangchhia 2017). Maize, being a versatile and hardy crop, can serve as a key component in sustainable agricultural practices. Promote the cultivation of climate-resilient maize varieties that are adapted to the agro-climatic conditions of the Himalayan region. This ensures a stable maize yield even under variable weather patterns (Singh et al. 2023a). Economic security is a crucial aspect of sustainable development, and income diversification plays a pivotal role in achieving this goal. This analysis explores the potential for income diversification through crop diversification, with a focus on regional contexts. Here are strategies to enhance economic security and income diversification through maize-based rainfed cropping systems such as encouraging integrated farming systems that include maize as a primary crop. Integrate livestock farming, agroforestry, and horticulture to diversify income sources and create a more resilient agricultural system. Promote agritourism and farm stays around maize farms. This diversifies income by attracting tourists interested in experiencing rural life, participating in farming activities, and purchasing local produce. Organic maize fetches premium prices in the market, providing an opportunity for income diversification while promoting sustainable farming practices (Babu et al. 2020b). Yadav et al. (2015) reported that maize–maize–field pea cropping system recorded average system profitability was 18.1% higher under NT systems than CT systems. In comparison to the maize–fallow cropping system, the system gross return, net return, and B:C ratio were best in the maize + groundnut–French bean (bush type) cropping system, at 23%, 27%, and 35%, respectively (Ramakrushna et al. 2022). As per Babu et al. (2016), the maize–urd bean–buckwheat system demonstrated superior resource efficiency within intensive cropping sequences. In contrast to alternative cropping systems, it exhibited notably elevated figures for net return (303.4×10^3 Rs per hectare), benefit–cost ratio (2.59), system employment generation (285 man-days per hectare), relative system production efficiency (24.36 kg per hectare per day), and system productivity (8.89 Mg per hectare). The maize–rajmash–buckwheat system demonstrated the highest gross returns within the system, reaching a total of 4643 US dollars per hectare. Following closely was the maize–black gram–buckwheat system, generating 4014 US dollars per hectare (Babu et al. 2023b). The maize–vegetable pea cropping system grown under no-tillage (NT) system resulted in system production efficiency being 9.86% and 1.96% higher compared to conventional tillage (CT) and reduced tillage (RT), respectively. Economically, under NT, the system gross returns, system net returns, and system benefit–cost (B:C) ratio were 9.36%, 23.2%, and 23.3% higher than under CT, respectively (Singh et al. 2021b).

7.5.3 Nutritional Security Through Rainfed Maize-Based Cropping System

The Indian Himalayas are adopting a multifaceted strategy to promote nutritional security through the diversification of rainfed maize-based agricultural systems. This strategy includes expanding crop diversity, improving farming techniques, and raising knowledge of the nutritional benefits of various crops (Babu et al. 2016; Ansari et al. 2022). Nutritional security through maize diversification is a multifaceted approach aimed at enhancing food security and addressing nutritional deficiencies by promoting the cultivation and consumption of a diverse range of maize varieties in maize-based cropping systems. Maize, also known as corn, is a staple crop with a rich nutritional profile that significantly contributes to local diets around the world. Its versatility, affordability, and adaptability make it a crucial source of nutrients for various communities. Maize is a powerhouse of essential nutrients, providing a balanced mix of carbohydrates, proteins, and fibers. The majority of its caloric content comes from carbohydrates, mainly in the form of starch. This makes maize an excellent energy source, crucial for meeting the daily energy requirements of individuals, especially in regions where it serves as a dietary staple (Babu et al. 2023b). One of the key nutritional components of maize is dietary fiber. While the outer layer of the kernel, called the bran, contains insoluble fiber, the inner part, called the endosperm, is rich in soluble fiber. Both types of fiber play essential roles in promoting digestive health. Insoluble fiber adds bulk to stool, aiding in the prevention of constipation, while soluble fiber helps regulate blood sugar levels and lower cholesterol. Maize also contains a moderate amount of protein, albeit with lower levels of certain essential amino acids compared to animal proteins (Yadav et al. 2015). However, when paired with complementary proteins from other sources, such as legumes, the overall protein quality of maize-based diets can be enhanced. This is particularly important in regions where access to animal protein is limited. Beyond macronutrients, maize is a good source of various micronutrients, including vitamins and minerals. It contains significant amounts of vitamin A, crucial for maintaining healthy vision, skin, and the immune system. In areas where vitamin A deficiency is a prevalent health issue, the incorporation of maize into the diet can contribute to addressing this nutritional challenge (Jones and Yosef 2015). Additionally, maize is a source of B vitamins, including thiamine (B1), riboflavin (B2), niacin (B3), and folate (B9). These vitamins play essential roles in energy metabolism, the formation of red blood cells, and the development of the nervous system. The presence of these B vitamins in maize makes it a valuable component of a well-rounded diet. Minerals such as iron, magnesium, and phosphorus are also present in maize. Iron is vital for preventing anemia, while magnesium and phosphorus contribute to bone health and various metabolic processes within the body. Including maize in the local diet helps ensure a diverse intake of essential minerals necessary for overall well-being (Goglio et al. 2014).

Diversifying maize involves encouraging the cultivation of various maize varieties with different nutrient profiles (Yadav et al. 2021b). This approach addresses the specific dietary needs of populations and contributes to a more resilient and sustainable agricultural system. Maize, a staple food for millions globally, is a critical component of diets in various regions, particularly in Africa, Latin America, and parts of Asia. However, an overreliance on a limited number of maize varieties can lead to nutritional imbalances, vulnerability to pests and diseases, and hinder overall food security. First, understanding the nutrient content of different maize varieties is crucial. Maize comes in various colors, including white, yellow, and red, each indicating distinct nutritional properties (Chandra et al. 2011). By promoting the cultivation of a mix of these varieties, it becomes possible to increase the availability of essential nutrients such as vitamins, minerals, and antioxidants in the diet. Biofortification is another critical aspect of maize diversification. Through breeding programs, maize can be enhanced to contain higher levels of key nutrients, such as vitamin A, iron, and zinc. This biofortification strategy directly addresses specific micronutrient deficiencies in populations heavily reliant on maize as a primary food source, contributing to improved overall health. Several key strategies contribute to the success of nutritional security through maize diversification. Introduce and promote the cultivation of nutrient-rich crops alongside maize. Include crops such as pulses, millets, and locally adapted vegetables to diversify the nutritional profile of the farming systems. Integrating maize into diverse cropping systems, including crop rotation and intercropping, is a holistic approach that enhances soil fertility, reduces the risk of pests and diseases, and contributes to higher maize yields (Kumar et al. 2022a). This not only improves the overall productivity of agricultural systems but also enhances food security by diversifying the sources of nutrition. Implement intercropping systems that combine maize with legumes, oilseeds, and vegetables. This practice not only enhances soil fertility but also provides a diverse range of nutrients in the harvested produce. Integrate agroforestry practices that incorporate fruit-bearing trees and shrubs. This adds diversity to the cropping systems, providing fruits, nuts, and other tree-based products that contribute to nutritional security (Negi et al. 2016; Rathore et al. 2022b).

7.5.4 *Impact of Maize-Based Cropping Systems on Soil Health*

It is important to note that the success of a rainfed maize-based cropping system in improving soil health depends on various factors, including local climate, soil types, crop management practices, and the adoption of sustainable agricultural techniques. Additionally, farmers should consider holistic approaches that integrate multiple strategies to enhance soil health and promote long-term sustainability (Das et al. 2018). Cropping systems, which affect soil health and quality from several geographical and temporal dimensions, include crop diversification, crop rotation, intercropping, and associated agronomic methods used in agriculture (Kumar et al. 2022b). Cropping systems were originally created to maximize agro-system productivity, but in contemporary agriculture, concerns regarding cropping systems' environmental sustainability have grown (Babu et al. 2016). Under the five essential function evaluation standards: nutrient cycling, water relations, biodiversity and habitat, filtering and buffering, and physical stability and support, the aim of maintaining soil health is to ensure long-term stable high productivity and environmental sustainability of cropping systems (Singh et al. 2021b). Maize plants produce a significant amount of crop residue. When this residue is left on the field after harvest, it acts as a natural mulch. Mulch helps to reduce soil erosion, retain soil moisture, and regulate soil temperature. The crop residue from maize, when incorporated into the soil, adds organic matter. Organic matter improves soil structure, water retention, and nutrient-holding capacity (Ansari et al. 2022). For soil microbes, it also serves as a source of energy, encouraging their diversity and activity. Some varieties of maize, especially legume intercrops or cover crops, have the capability to fix atmospheric N. This BNF enhances soil fertility by supplying N to the soil, benefiting subsequent crops in the rotation. Implementing a crop rotation system with maize and other crops can break pest and disease cycles, reducing the risk of soil-borne pathogens. Diversifying crops in the rotation also helps maintain a balanced nutrient profile in the soil. Cultivation of maize + cowpea–winter crops significantly improved the available N (3.2–12.9%), P (3.6–12.7%), K (1.9–26.3%), organic carbon (9.2–16.8%), microbial biomass carbon–MBC (15.2–43.9%), and dehydrogenases–DHA (17.2–42.3%) in soil at 0–15 cm depth as compared to maize–fallow cropping system (Singh et al. 2021a). In various cropping systems, incorporating French bean as a sequential crop and adopting groundnut intercropping demonstrated positive effects on sustaining soil organic carbon (SOC). This highlights the importance of intercropping and conservation measures in preserving SOC status, thereby contributing to extending the productive capacity duration of land after forest clearing for cultivation. Simultaneously, the examination of different SOC fractions (labile, non-labile, and recalcitrant) is crucial for understanding the turnover rate of SOC over the shifting cultivation duration (Ramkrushna et al. 2022). The highest levels of SMBC and dehydrogenase at 22.1 μg TPF g^{-1} soil h^{-1} and fluorescein diacetate at 67.1 μg FDA g^{-1} soil h^{-1} enzymatic activities were observed when farmyard manure (6 t ha^{-1}) was combined with vermicompost

(2 t ha^{-1}). Specifically, the maize–French bean (PT) and maize–French bean (BT) systems showed 11.1% and 9.4% higher soil organic carbon (SOC) stock compared to the maize–fallow system at a depth of 0–15 cm after two cropping cycles. Additionally, the concentrations of dissolved organic carbon (DHA) under maize–French bean (PT) and maize–French bean (BT) were 57% and 40.8% higher, respectively, than those in the soil under the maize–fallow system (Ngangom et al. 2020).

7.6 Future Perspectives

The diversification of rainfed maize systems in the IHR holds immense promise for ensuring both nutritional and economic security. As we look to the future, the integration of diverse crops into these systems presents a transformative pathway that not only safeguards against food and income vulnerabilities but also aligns with sustainable agricultural practices essential for the unique Himalayan conditions. In the coming years, a significant prospective development is the expansion of the crop portfolio in rainfed maize systems. The introduction and promotion of nutrient-rich crops such as pulses, millets, and indigenous vegetables alongside maize can create a more resilient agricultural landscape. This diversification not only enhances the nutritional diversity of diets but also acts as a natural adaptation strategy against the uncertainties posed by climate change. The growth of value-added processing industries is another promising avenue. As farmers cultivate a variety of crops, opportunities arise for the establishment of food processing units. These units can transform the diverse produce into fortified flours, snacks, and beverages, contributing not only to economic growth but also to the availability of convenient and nutritious food products. The prospect of tourism and agritourism stands out as a potential economic boon for the region. The aesthetically pleasing landscapes resulting from crop diversification can attract tourists interested in experiencing rural life. Agritourism initiatives, including farm stays, guided tours, and the sale of local produce, can provide additional income streams for farmers and contribute to the overall economic vibrancy of the region. Improved livelihoods and rural development are expected outcomes of diversifying rainfed maize systems. By integrating various crops and agricultural activities, communities can build resilient and sustainable economies. This holistic approach not only enhances the economic well-being of farmers but also creates ripple effects, positively impacting the entire rural landscape. Market linkages and improved access to diverse, nutrient-rich foods are also on the horizon. Diversification efforts can stimulate demand for a variety of crops in both local and regional markets. Farmers stand to benefit from increased market opportunities and improved incomes as consumers become more conscious of the nutritional value of their food. The future of rainfed maize systems in the Himalayan region also holds the promise of improved nutritional health. The introduction of diverse crops contributes to more balanced and nutrient-dense diets, addressing nutritional deficiencies prevalent in many communities. This shift toward diversified and locally sourced foods is fundamental for enhancing the overall health and well-being of the population. As we look ahead, climate-resilient agriculture emerges as a key focus. To maximize the impact of diversification projects, cooperation between government agencies, nongovernmental organizations (NGOs), research institutes, and the corporate sector is necessary. Pooling resources, expertise, and knowledge through partnerships can address challenges more effectively and create a holistic development framework for the region. Capacity building and education programs will be crucial for the successful implementation of crop diversification. Training initiatives that focus on sustainable practices, value addition, and marketing can empower farmers with the skills and knowledge needed for successful diversification, ensuring the long-term success of these initiatives. In the era of digital agriculture, the integration of technology and improved information access will play a pivotal role. Farmers can benefit from real-time data, weather forecasts, and market trends, facilitating informed decision-making and enhancing overall productivity. As these initiatives unfold, they have the potential to not only transform agricultural landscapes but also improve the overall well-being of communities in the IHR.

References

Ansari MA, Choudhury BU, Mandal S, Jat SL, Meitei C, Bungbungcha. (2022) Converting primary forests to cultivated lands: long-term effects on the vertical distribution of soil carbon and biological activity in the foothills of eastern Himalaya. J Environ Manag 301:113886. https://doi.org/10.1016/j.jenvman.2021.113886

Babu S, Singh R, Avasthe RK, Yadav GS, Rajkhowa DJ (2016) Intensification of maize (*Zea mays*)-based cropping sequence in rainfed ecosystem of Sikkim Himalayas for improving system productivity, profitability, employment generation and energy-use efficiency under organic management condition. Indian J Agric Sci 86(6):778–784

Babu S, Singh R, Avasthe RK, Yadav GS, Mohapatra KP, Selvan T, Das A, Singh VK, Valente D, Petrosillo I (2020a) Soil carbon dynamics in Indian Himalayan intensified organic rice-based cropping sequences. Ecol Indic 114:106292. https://doi.org/10.1016/j.ecolind.2020.106292

Babu S, Singh R, Avasthe RK, Yadav GS, Das A, Singh VK, Mohapatra KP, Rathore SS, Chandra P, Kumar A (2020b) Impact of land configuration and organic nutrient management on productivity, quality and soil properties under baby corn in eastern Himalayas. Sci Rep 10:16129. https://doi.org/10.1038/s41598-020-73072-6

Babu S, Mohapatra KP, Das A, Yadav GS, Tahasildar M, Singh R, Panwar AS, Yadav V, Chandra P (2020c) Designing energy-efficient, economically sustainable and environmentally safe cropping system for the rainfed maize-fallow land of the eastern Himalayas. Sci Total Environ 722:137874. https://doi.org/10.1016/j.scitotenv.2020.137874

Babu S, Das A, Singh R, Mohapatra KP, Kumar S, Rathore SS, Yadav SK, Yadav P, Ansari MA, Panwar AS, Wani OA, Singh M, Ravishankar N, Layek J, Chandra P, Singh VK (2022) Designing an energy efficient, economically feasible, and environmentally robust integrated farming system model for sustainable food production in the Indian Himalayas. Sustain Food Technol 1:126. https://doi.org/10.1039/d2fb00016d

Babu S, Singh R, Avasthe RK, Kumar S, Singh VK RSS, Ansari MA, Valente D, Petrosillo I (2023a) Soil carbon dynamics under organic farming: impact of tillage and cropping diversity. Ecol Indic 147:109940. https://doi.org/10.1016/j.ecolind.2023.109940

Babu S, Singh R, Avasthe R, Rathore SS, Kumar S, Das A, Layek J, Sharma V, Wani OA, Singh VK (2023b) Conservation tillage and diversified cropping enhance system productivity and eco-efficiency and reduce greenhouse gas intensity in organic farming. Front Sustain Food Syst 7:1114617. https://doi.org/10.3389/fsufs.2023.1114617

Chandra AP, Saradhi P, Maikhuri RK, Saxena KG, Rao KS (2011) Traditional agrodiversity management: a case study of central Himalayan village ecosystem. J Mt Sci 8(1):62–74

Choudhary AK, Thakur SK, Suri VK (2013) Technology transfer model on integrated nutrient management technology for sustainable crop production in high value cash crops and vegetables in north-western Himalayas. Commun Soil Sci Plant Anal 44:1684–1699

Choudhary AK, Yadav DS, Sood P, Dua VK, Singh A, Rahi S (2021) Influence of integrated crop management technology on potato productivity, profitability, energy dynamics and carbon footprints in north-western Himalayas. Potato J 48(2):148–160

DACNET (2022) Directorate of Economics and Statistics, Ministry of Agriculture and Farmers Welfare Government of India. https://eands.dacnet.nic.in/PDF/Agricultural%20Statistics%20at%20a%20Glance%20%202021%20(Hindi%20version).pdf

Das L, Meher JK (2019) Drivers of climate over the Western Himalayan region of India: a review. Earth Sci Rev 198:102935

Das A, Ramakrishna GI, Layek J, Babu S, Ngangom B, Suiting D (2017) Package of practices of Meghalaya. In: Ravisankar N, Panwar AS, Kamta P, Vipin K, Bhaskar S (eds) Organic farming crop production guide, network project on organic farming. ICAR Indian Institute of Farming Systems Research, Modipuram, Meerut, pp 1–592

Das A, Thoithoi DM, Babu S, Ansari M, Layek J, Bhowmick SN, Yadav GS, Singh R (2018) Cereal-legume cropping system in Indian Himalayan region for food and environmental sustainability. In: Meena R, Das A, Yadav G, Lal R (eds) Legumes for soil-health and sustainable management. Springer, Singapore. https://doi.org/10.1007/978-981-13-0253-4_2

Das A, Layek J, Babu S, Ramkrushna GI, Baiswar P, Krishnappa R, Thoithoi DM, Kumar M, Prakash N (2019) Package of practices for organic production of important crops in NEH region of India. Meghalaya, ICAR Research Complex for NEH Region, Umiam, p 228

Das A, Babu S, Singh R, Kumar S, Rathore SS, Singh VK, Datta M, Yadav SK, Wani OA, Yadav D (2022) Impact of live mulch-based conservation tillage on soil properties and productivity of summer maize in Indian Himalayas. Sustainability 14:12078. https://doi.org/10.3390/su141912078

Fanzo J, Haddad L, Schneider KR, Béné C, Covic NM, Guarin A, Herforth AW, Herrero M, Sumaila UR, Aburto NJ, Amuyunzu-Nyamongo M, Barquera S, Battersby J, Beal T, Bizzotto MP, Brusset E, Cafero C, Campeau C, Caron P, Rosero MJ (2021) Viewpoint: rigorous monitoring is necessary to guide food system transformation in the countdown to the (2030) global goals. Food Policy 104:102163. https://doi.org/10.1016/j.foodpol.(2021).102163

FAO Stat (2021) FAO Stat. FAO, Rome. http://www.fao.org/faosta

Ghosh PK, Hazra KK, Venkatesh MV, Praharaj CS, Kumar N, Nath CP, Singh U, Singh SS (2020) Grain legume inclusion in cereal-cereal rotation increased base crop productivity in the long run. Exp Agric 56(1):142–158

Goglio P, Grant BB, Smith WN, Desjardins RL, Worth DE, Zentner R, Malhi SS (2014) Impact of management strategies on the global warming potential at the cropping system level. Sci Total Environ 490:921–933

GOI (1989) Agro-climatic regional planning: an overview. Planning Commission, New Delhi

Grote U, Fasse A, Nguyen TT, Erenstein O (2021) Food security and the dynamics of wheat and maize value chains in Africa and Asia. Front Sustain Food Syst 4:617009. https://doi.org/10.3389/fsufs.2020.617009

IMI FAO Report (2019) State of the Himalayan farmers and farming. A part of the project "Strengthening Institutional Capacities for Sustainable Mountain Development in the Indian Himalayan Region" awarded by FAO under the TCP/IND/3601/C1 to IMI

IPCC (2014) Climate change (2014): impacts, adaptation, and vulnerability. Part a: global and sectoral aspects. In: summary for policymakers. In: Field CB, Barros VR, Dokken DJ, Mach KJ,

Mastrandrea MD, Bilir TE, Chatterjee M, Ebi KL, Estrada YO, Genova RC, Girma B, Kissel ES, Levy AN, Mac Cracken S, Mastrandrea PR, White LL (eds) Contribution of working group II to the fifth assessment report of the intergovernmental panel on climate change. Cambridge University Press, Cambridge, UK/New York, pp 1–32

Jones AD, Yosef S (2015) The implications of a changing climate on global nutrition security. In: Sahn DE (ed) The fight against hunger and malnutrition. Oxford University Press, pp 432–466

Kennett DJ, Prufer KM, Culleton BJ, George RJ, Robinson M, Trask WR, Buckley GM, Moes E, Kate EJ, Harper TK, O'Donnell L, Ray EE, Hill EC, Alsgaard A, Merriman C, Meredith C, Edgar HJH, Awe JJ, Gutierrez SM (2020) Early isotopic evidence for maize as a staple grain in the Americas. Sci Adv 6:eaba3245. https://doi.org/10.1126/sciadv.aba3245

Kumar SS, Mir SA, Wani OA, Babu S, Yeasin M, Bhat MA, Hussain N, Wani AA, Kumar R, Yadav D, Dar SR (2022) Land-use systems regulate carbon geochemistry in the temperate Himalayas, India. J Environ Manag 320:115811. https://doi.org/10.1016/j.jenvman.2022.115811

Kumar A, Babu S, Singh R, Avasthe RK, Laha R, Mishra VK, Bhatt M, Saha S, Devi EL, Chandra P, Gudade BA, Verma G (2022a) Intercropping: an innovative agronomic approach for food, nutrition and livelihood security of marginal and small farmers in terraces land of Sikkim Himalaya. Agricolation 2(3):01–08

Kumar A, Saha S, Das A, Babu S, Layek J, Mishra VK, Singh R, Chowdhury S, Yadav DD, Bhutia TL, Verma G, Das S, Pandey P (2022b) Resource management for enhancing nutrient use efficiency in crops and cropping systems of rainfed hill ecosystems of the north-eastern region of India. Indian J Fertilisers 18(11):1090–1111

Kumar A, Avasthe RK, Singh R, Babu S, Das SK, Ravisankar N, Panwar AS, Qureshi AA, Laha RG, Mishra VK, Bhupenchandra I, Prasad SK (2023) Productivity and profitability of maize (*Zea mays*) cultivars under organic management condition in mid hills of Sikkim. Indian Journal of Agronomy 68(1):21–25

Layek J, Shivakumar BG, Rana DS, Munda S, Lakshman K, Das A, Ramkrushna GI (2015) Soybean-cereal intercropping systems as influenced by nitrogen nutrition. Agron J 106(7):1–14

Layek J, Das A, Mitran T, Nath CP, Meena RS, Yadav GS, Lal R (2018) Cereal + legume intercropping: an option for improving productivity and sustaining soil health. In: Meena RS et al (eds) Legumes for soil health and sustainable management. Springer Nature Singapore Pte Ltd, pp 347–386

Bhat SF, Saad AA, Masood A, Nazir A, Rasool R (2019) Yield and yield attributes of maize and bean as influenced by intercropping under various tillage and fertilizer management practices in Kashmir valley. Res J Agric Sci 10(3):552–557

Negi VS, Maikhuri RK, Pharswan D, Thakur S, Dhyani PP (2016) Climate change impact in the western Himalaya: people's perception and adaptive strategies. J Mt Sci 14(2):403–416

Ngangom B, Das A, Lal R, Idapuganti RG, Layek J, Basavaraj S, Babu S, Yadav GS, Ghosh PK (2020) Double mulching improves soil properties and productivity of maize-based cropping system in eastern Indian Himalayas. Int Soil Water Cons Res 8:308. https://doi.org/10.1016/j.iswcr.2020.07.001

Panwar AS, Singh NP, Saxena DC, Munda GC (2003) Agricultural status and cropping systems in NEH region. In: Bhatt BP, Bujarbaruah KM, Sharma YP, Patiram (eds) Proceedings of approaches for increasing agricultural productivity in hill and mountain ecosystem. ICAR Research Complex for NEH Region, Umiam, pp 191–195

Panwar AS, Babu S, Noopur K, Tahashildar M, Kumar S, Singh A (2019) Vertical cropping to enhance productivity and profitability of dry terraces in north eastern Indian Himalayas. Indian J Agric Sci 89(12):2020–2024

Poole N, Donovan J, Erenstein O (2021) Agri-nutrition research: revisiting the contribution of maize and wheat to human nutrition and health. Food Policy 100:101976. https://doi.org/10.1016/j.foodpol.2020.101976

Prachi N, Priya P, Pooja P, Kusum A, Arunachalam A (2017) Cropping systems and agromanagement practices in relation to soil carbon in the temperate regime of Uttarakhand. Int J Environ Rehabil Cons 1:60–72

Ramakrushna GI, Das A, Layek J, Verma BC, Babu S, Mohapatra KP, Shahan AA (2022) Assessing maize based cropping systems for higher productivity and income under shifting cultivation in eastern Indian Himalayas. Indian J Soil Cons 50(2):101–106

Rathore SS, Babu S, Shekhawat K, Singh R, Yadav SK, Singh VK, Singh C (2022a) Designing energy cum carbon-efficient environmentally clean production system for achieving green economy in agriculture. Sustain Energy Technol Assess 52:102190. https://doi.org/10.1016/j.seta.2022.102190

Rathore SS, Babu S, El-Sappah AH, Shekhawat K, Singh VK, Singh RK, Upadhyay PK, Singh R (2022b) Integrated agroforestry systems improve soil carbon storage, water productivity, and economic returns in the marginal land of the semi-arid region. Saudi J Biol Sci 29:103427. https://doi.org/10.1016/j.sjbs.2022.103427

Saha R, Chaudhary RS, Somasundaram J (2012) Soil-health management under hill agroecosystem of north East India. Appl Environ Soil Sci 2012:1. https://doi.org/10.1155/2012/696174

Sati VP, Vangchhia L (2017) Food security status in rural areas of Mizoram, Northeast India. J Mt Sci 14:795–805. https://doi.org/10.1007/s11629-016-4092-2

Sharma PK, Bali AS, Sharma BC (2022) Effect of intercropping systems and fertility levels of winter maize (*Zea mays* L.) on production potential and nutrient uptake by succeeding maize crop. J Pharm Innov 11(10):1860–1866

Singh R, Babu S, Avasthe RK, Singh M, Yadav GS, Chettri TK (2018) Residual effect of organic amendments on growth, productivity, economics and agri-energetics of local popcorn (*Zea mays* everta) in toria (*Brassica campestris*) – popcorn cropping system under mid hills of Sikkim Himalayas. Indian J Agric Sci 88(12):1887–1892

Singh R, Babu S, Avasthe RK, Yadav GS, Das A, Mohapatra KP, Kumar A, Singh VK, Chandra P (2021a) Crop productivity, soil-health, and energy dynamics of Indian Himalayan intensified organic maize-based systems. Int Soil Water Cons Res 9(2):260–270

Singh R, Babu S, Avasthe RK, Meena RS, Yadav GS, Das A, Mohapatra KP, Rathore SS, Kumar A, Singh C (2021b) Conservation tillage and organic nutrients management improve soil properties, productivity, and economics of a maize-vegetable pea system in the eastern Himalayas. Land Degrad Dev 1–18:4637. https://doi.org/10.1002/ldr.4066

Singh R, Babu S, Avasthe RK, Das A, Praharaj CS, Layek J, Kumar A, Rathore SS, Kancheti M, Kumar S, Yadav SK, Pashte V (2021c) Organic farming in North-East India: status and strategies. Indian J Agron 66(5th IAC Special issue):S163–S179

Singh D, Mishra AK, Patra S, Dwivedi AK, Ojha CSP, Singh VP, Mariappan S, Babu S, Singh N, Yadav D, Ojasvi PR, Kumar G, Madhu M, Sena DR, Chand L, Kumar S (2023a) Effect of long-term tillage practices on runoff and soil erosion in sloping croplands of Himalaya, India. Sustainability 15:8285. https://doi.org/10.3390/su15108285

Singh R, Kumar A, Babu S, Avasthe R, Das A, Rathore SS, Kumar S, Singh C, Sharma V, Bhupenchandra I (2023b) Development of organic nutrients management system for profitable and soil-supportive French bean (*Phaseolus vulgaris* L.) farming in northeastern Himalayas, India. Front Sustain Food Syst 7:1115521. https://doi.org/10.3389/fsufs.(2023).1115521

Tong Y, Liu J, Li X, Sun J, Herzberger A, Wei D, Zhang W, Dou Z, Zhang F (2017) Cropping system conversion led to organic carbon change in China's Mollisols regions. Sci Rep 7:18064. https://doi.org/10.1038/s41598-017-18270-5

Walia SS, Babu S, Gill RS, Kaur T, Kohima N, Panwar AS, Yadav DK, Ansari MA, Ravishankar N, Kumar S, Kaur K, Ansari MH (2022) Designing resource-efficient and environmentally safe cropping systems for sustainable energy use and economic returns in indo-Gangetic Plains, India. Sustainability 14:14636. https://doi.org/10.3390/su142114636

Wani OA, Sharma V, Kumar SS, Babu S, Sharma KR, Rathore SS, Marwaha S, Ganai NA, Dar SR, Yeasin M, Singh R, Tomar J (2023) Climate plays a dominant role over land management in governing soil carbon dynamics in North Western Himalayas. J Environ Manag 338:117740. https://doi.org/10.1016/j.jenvman.2023.117740

Yadav SK, Singh Y, Yadav MK, Babu S, Singh K (2013) Effect of organic nitrogen sources on yield, nutrient uptake and soil health under rice (*Oryza sativa*) based cropping sequence. Indian J Agric Sci 83(2):170–175

Yadav GS, Datta M, Babu S, Das A, Bhowmik SN, Ranebennur H, Debnath C, Saha P (2015) Effect of tillage and crop-establishment techniques on productivity, profitability and soil health under maize (*Zea mays*)-maize-field pea (*Pisum sativum*) cropping system. Indian J Agron 60(3):360–364

Yadav GS, Das A, Lal R, Babu S, Meena RS, Saha P, Singh R, Datta M (2018) Energy budget and carbon footprint in a no-till and mulch-based rice-mustard cropping system. J Clean Prod 191:144–157

Yadav GS, Das A, Lal R, Babu S, Datta M, Swaroop MR, Patil SB, Singh R (2019) Impact of no-till and mulching on soil carbon sequestration under rice (*Oryza sativa* L.)-rapeseed (*Brassica campestris* L. var. rapeseed) cropping system in hilly agro-ecosystem of the eastern Himalayas, India. Agric Ecosyst Environ 275:81–92

Yadav GS, Babu S, Das A, Mohapatra KP, Singh R, Avasthe RK, Roy S (2020) No-till and mulching enhances energy use efficiency and reduce carbon footprint of a direct-seeded upland rice production system. J Clean Prod 271:122700. https://doi.org/10.1016/j.jclepro.(2020).122700

Yadav GS, Babu S, Das A, Datta M, Mohapatra KP, Singh R, Singh VK, Rathore SS, Chakraborty M (2021a) Productivity, soil-health, and carbon management index of Indian Himalayan intensified maize-based cropping systems under live mulch-based conservation tillage practices. Field Crop Res 264:108080

Yadav GS, Das A, Kandpal BK, Babu S, Lal R, Datta M, Das B, Singh R, Singh VK, Mohapatra KP, Chakraborty M (2021b) The food-energy-water-carbon nexus in a maize-maize-mustard cropping sequence of the Indian Himalayas: an impact of tillage-cum-live mulching. Renew Sust Energ Rev 151:111602. https://doi.org/10.1016/j.rser.2021.111602

Yadav D, Singh D, Babu S, Madegowda M, Singh DV, Mandal D, Rathore AC, Sharma VK, Singhal V, Kumawat A, Yadav DK, Yadav RK, Kumar S (2023) Intensified cropping reduces soil erosion and improves rainfall partitioning and soil properties in the marginal land of the Indian Himalayas. Int Soil Water Cons Res 12:521. https://doi.org/10.1016/j.iswcr.2023.10.002

Yadav D, Singh D, Madhu M, Singh DV, Islam S, Kumar G, Gupta AK, Mandal D, Babu S, Barh A, Kumawat A, Dubey R, Pati K, Bhargavi B, Yadav DK (2024) Interplanted crops and rainfall variability regulate soil erosion in rainfed maize of the Indian Himalayas. Land Degrad Dev 35:4271. https://doi.org/10.1002/ldr.5221

Yang H, Zhang W, Long LI (2021) Intercropping: feed more people and build more sustainable agroecosystem. Front Agric Sci Eng 8(3):373–386

Chapter 8
Enhancing Nutrient Use Eficiency in Pulses Under Diversified Farming through Agronomic Manipulations

Narendra Kumar, Mrunalini Kancheti, K. K. Hazra, C. P. Nath, Mohammad Hashim, and Man Mohan Deo

Abstract Pulse crops, renowned for their nitrogen-fixing abilities, play a pivotal role in replenishing soil nutrients, offering versatile solutions for crop diversification and intensification across various agroecosystems. Strategies such as intercropping and crop rotation prove effective in optimizing nutrient utilization in pulses, while agronomic interventions like precision water management, crop rotation, and microbial inoculation contribute significantly to nutrient use efficiency (NUE). The chapter underscores the importance of precision nutrient management through practices like soil testing and balanced fertilization tailored to pulse crop requirements. Noteworthy environmentally friendly approaches include microbial inoculation with rhizobia bacteria, enhancing nitrogen fixation, and nutrient solubilization. The critical impact of planting techniques, crop residue management, and sowing windows on NUE is explored, highlighting their quantifiable contributions to soil organic matter enhancement, nutrient recycling, and erosion control.

Keywords Pulses · Planting methods · Paira cropping · Precision · Nutritional security

8.1 Introduction

Pulses, encompassing a wide range of leguminous crops have long been recognized for their nutritional significance and their contributions to sustainable agriculture. These crops are endowed with the unique ability to fix atmospheric nitrogen through symbiotic relationships with rhizobia bacteria, thus enriching the soil with essential nutrients (Ali et al. 2012; Kumar et al. 2022a). However, as the global population continues to expand, the demand for nutrient-rich pulses intensifies, emphasizing

N. Kumar (✉) · M. Kancheti · K. K. Hazra · C. P. Nath · M. Hashim · M. M. Deo
ICAR-Indian Institute of Pulses Research, Kanpur, Uttar Pradesh, India

S. Babu et al. (eds.), *Agricultural Diversification for Sustainable Food Production*, Sustainability Sciences in Asia and Africa,
https://doi.org/10.1007/978-981-97-7517-0_8

the pivotal role of pulse crops in ensuring food and nutritional security. In this context, enhancing nutrient use efficiency (NUE) in pulse crops has become a paramount objective. Nutrient use efficiency, refers to the capacity of a crop to maximize its growth and yield while minimizing the input of fertilizers and the potential for nutrient losses to the environment, is critical for the economic and environmental sustainability of agricultural systems. Achieving high nutrient use efficiency in pulses not only increases crop productivity but also mitigates the negative environmental impacts associated with excessive fertilizer use, such as nutrient runoff and greenhouse gas emissions (Pandey et al. 2012; Kumar et al. 2022b).

Diversified farming which integrate a variety of crops within the same agricultural landscape, present a promising avenue for optimizing nutrient use efficiency in pulse crops. Intercropping with legumes can enhance nutrient resource utilization and boost the growth of both intercrops, particularly in low-input farming systems (Tsubo et al. 2005; Kumar et al. 2012). The rhizobium-legume association provides benefits to nonleguminous crops in mixed intercropping and crop rotations through direct nitrogen transfer (Hayat et al. 2010; Singh et al. 2021). A recent meta-analysis conducted by Tang et al. (2021) focusing on phosphorus (P) efficiency in cereal/legume intercropping revealed a notable increase in P absorption, with an average equivalent ratio-soil for P uptake of 1.24. In situations where vulnerability is a concern, this intercropping approach offers several advantages, including the enhancement of soil fertility through increased nutrient availability and improved soil quality (Betencourt et al. 2012). Through intercropping, crop rotation, and the promotion of agroecological diversity, diversified farming practices hold the potential to enhance the sustainability of pulse cropping, ensuring that these crops continue to play a crucial role in providing nutrient-rich food while simultaneously preserving the health of the ecosystem. The overview of NUE is depicted in Fig. 8.1. This chapter explores the multifaceted realm of enhancing nutrient use efficiency in pulse crops under diversified farming systems, elucidating the pivotal role of these practices in redefining the future of pulse agriculture.

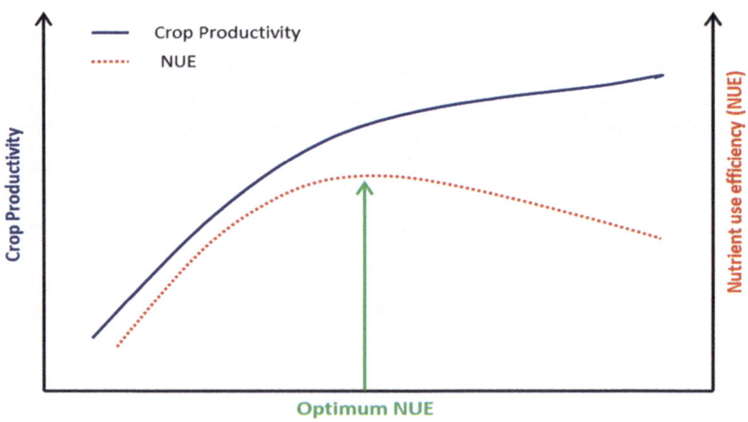

Fig. 8.1 An overview of nutrient use efficiency (NUE)

8.2 Pulses in Diversified Cropping Systems

Pulses are characterized by their exceptional ability to efficiently utilize limited soil moisture and nutrients while positively influencing the physical and chemical properties of the soil, making them an essential component of sustainable crop production systems (Kumar et al. 2016). Notable agricultural strategies for incorporating legumes into cropping systems encompass crop rotation, concurrent intercropping, improved fallow practices, green manure utilization, and alley cropping (Kumar et al. 2012; Kebede 2021). Leveraging the relatively short cropping window, approximately 2 months, between major *Rabi* and *kharif* crops offers a practical means for farmers to generate additional income through the inclusion of spring/summer mungbean or urdbean. The versatility of pulses allows them to seamlessly integrate with various crop types within prominent cropping systems, including intercropping, cropping sequences, and paira/utera cultivation (Singh et al. 2017). Key intercropping methods involving pulses entail mixed intercropping (in homesteads), row intercropping (e.g., maize + green gram/Horsegram), and relay intercropping (sequentially planting green gram, maize, potato, and wheat) (Kumar et al. 2010a, b). Pulse inclusion in sequential cropping systems encompasses double and triple cropping. Paira/utera cultivation optimizes the use of residual moisture from the preceding crop. It is worth noting that cropping systems that include pulses, such as rice-rice-green gram, rice-rice-sesame, rice-rice-onion, maize-rice-sesame, and groundnut-rice-black gram, exhibit higher organic carbon and available soil nitrogen content, reducing the risks of soil erosion and depletion (Porpavai et al. 2011). Furthermore, pulses, when used as cover crops, substantially contribute to increased soil carbon accumulation rates, outperforming cereals and grasses. Crop rotations that incorporate pulses demonstrate a higher potential for soil carbon sequestration compared to monocrop systems (Kumar et al. 2019a).

Grain legume-cereal intercrops exemplify the efficient utilization of resources, particularly nitrogen sources. Leguminous species harness atmospheric N_2, while nonleguminous counterparts rely on soil inorganic nitrogen. The interaction between different crops in intercropping systems revolves around shared resources like soil, residual moisture, light, and nutrients. Crops efficiently harness these resources by adapting their growth and rooting patterns. Intercropping legumes with cereals, especially pigeon peas with maize, enhances nutrient availability and uptake, particularly phosphorus. The physical root interaction in the rhizosphere and the biochemical activities positively influence soil structure and nutrient storage in P-rich soils (Garland et al. 2017). When maize and green gram are intercropped, their resource requirements peak at different times, minimizing competition. As a result, the system as a whole offers benefits such as effective weed suppression, reduced disease and pest incidence, enhanced yield, increased test weight, and higher available nitrogen content post-harvest, all while reducing weed growth. Rajeshkumar et al. (2017) and Kumar et al. (2013) also reported the weed-suppressing effectiveness of pulses in intercropping systems. Additionally, they were found to lower pest incidence (Uzokwe et al. 2016) and disease occurrence (Schoeny et al. 2010). Popular intercropping patterns of India are given in Table 8.1.

Table 8.1 Popular intercropping patterns of pulses in different states of India

State	Pulse crop	Intercropped crop	Intercropping pattern
Punjab	Chickpea	Wheat, mustard	Wheat/mustard + chickpea
Haryana	Lentil	Mustard, wheat	Wheat/mustard + lentils
Gujarat	Red gram	Groundnut, cotton	Groundnut/cotton + red gram
Maharashtra	Red gram	Soybean, groundnut	Red gram + soybean/ groundnut
Andhra Pradesh, Telangana	Red gram	Groundnut, sorghum, cotton	Groundnut/sorghum/ cotton + red gram
Tamil Nadu	Green gram	Pearl millet, sorghum	Pearl millet/sorghum + green gram
Karnataka	Chickpea	Pearl millet, sorghum	Pearl millet/ sorghum + chickpea
	Red gram	Pearl millet	Red gram + pearl millet
Uttar Pradesh	Lentil	Mustard, wheat	Wheat/mustard + lentils
	Chickpea	Wheat, mustard	Wheat/mustard + chickpea
	Green gram/ black gram	Sugarcane	Sugarcane + green gram/black gram
Uttarakhand	Black gram	Maize	Black gram + maize
Rajasthan	Green gram	Pearl millet, sorghum	Pearl millet/sorghum + green gram
	Moth bean	Pearl millet	Moth bean + pearl millet
	Cluster bean	Pearl millet	Cluster bean + pearl millet
	Black gram	Maize	Maize + black gram
Bihar, Jharkhand	Red gram	Black gram	Red gram + black gram
Odisha	Red gram	Finger millet	Red gram + finger millet
West Bengal	Green gram/ black gram	Baby corn	Baby corn + green gram/black gram
	Chickpea	Wheat	Wheat + chickpea

Source: Praharaj and Blaise (2016); Kumar et al. (2023)

8.3 Agronomic Interventions to Enhance NUE in Pulses

8.3.1 Crop Rotation

Crop rotation is a fundamental practice in modern agriculture that involves the sequential planting of different crops in the same field over time. Among the myriad benefits of crop rotation, its role in enhancing nutrient use efficiency in pulse farming is particularly noteworthy. Nutrient management is a critical aspect of sustainable agriculture. Different crops have varying nutrient requirements, and over time, continuous cultivation of the same crop can lead to nutrient imbalances in the soil. Pulses, with their unique ability to fix atmospheric nitrogen through symbiotic relationships with rhizobia bacteria, have the potential to improve soil fertility. However, they also require other essential nutrients, such as phosphorus and potassium. Crop rotation helps balance nutrient demands.

For example, when pulse crops are incorporated into crop rotations with cereals, the residual nitrogen left behind by the previous pulse crop benefits the subsequent cereals, reducing the reliance on synthetic nitrogen fertilizers (Babu et al. 2020). In return, cereals contribute organic matter and additional nutrients to the soil, creating a more favorable environment for the next pulse crop. This practice of integrating pulses like chickpeas, lentils, and peas into cereal and oilseed cropping systems replaces fallow periods, promoting sustainability in agriculture (Singh et al. 2017). Research conducted in Saskatchewan, Canada, demonstrated that this approach enhanced nitrogen use efficiency by 33% and increased grain yields by 35.5% (Gan et al. 2015). The inclusion of pulses in crop rotations significantly boosted wheat grain yield, nitrogen removal, nitrogen fertilizer use efficiency (NFUE), and water use efficiency (WUE) by 16%, 31%, 21%, and 14%, respectively, compared to systems without pulses (Lasisi and Liu 2023). In a long-term study, pulse-based systems (such as pigeon pea–wheat and maize–wheat–mungbean) exhibited higher levels of available nitrogen (8–11%), phosphorus (42–73%), and potassium (8–12%) compared to the maize–wheat system (Nath et al. 2023). The incorporation of pulses into crop rotations also had a positive impact on soil microbial diversity due to their unique rhizobia associations and nitrogen-fixing abilities, ultimately enhancing overall soil fertility and structure. A 17-year study in Saskatchewan, Canada, substituting summer fallow with lentils in wheat cropping systems, resulted in a significant 14–18% increase in soil organic carbon (SOC), as reported by Campbell et al. (2000). Inclusion of intensified leguminous live mulch systems has improved system productivity, carbon management index (CMI), and soil health in Tripura (Yadav et al. 2021). Crop rotation remains a proven and essential strategy in modern, nutrient-efficient agriculture.

8.3.2 *Tillage Practices*

Conservation tillage methods have garnered considerable attention in sustainable agriculture for their potential to enhance soil health, boost crop productivity, and improve nutrient utilization efficiency (Kumar et al. 2019a). This approach minimizes soil disturbance and retains crop residues on the soil surface, creating favorable conditions for the growth of pulse crops and their effective utilization of nitrogen. Conservation tillage also plays a vital role in conserving soil moisture by reducing evaporation and runoff, which is critical for ensuring a steady supply of nitrogen to pulse crops, thus enhancing their nutrient use efficiency (NUE) by minimizing the risk of nitrogen leaching. The retention of nitrogen-rich crop residues on the soil surface contributes to increased organic matter content, promoting improved soil structure, microbial activity, and nutrient retention, particularly nitrogen. Pulse crops thrive in such organic-rich soils and utilize nitrogen more effectively. Long-term conservation tillage systems, spanning 10–15 years, lead to higher levels of soil organic matter and an increased concentration of mineralizable organic nutrients at the soil surface, thereby enhancing the soil's capacity to supply nutrients

(Sirivedhin and Gray 2006). Additionally, the practice of retaining decaying roots through no-tillage preserves legume rhizobia populations and diversity, resulting in higher crop yields and improved NUE compared to conventional tillage systems (Ferreira et al. 2000; Stahl et al. 2019; Zhou et al. 2020). The cumulative benefits of reduced soil erosion, reduced nutrient removal by weeds, improved soil physical health, optimized water management, and the restoration of soil moisture contribute to increased yield attributes and overall yield in conservation tillage plots (Mukherjee et al. 2023).

8.3.3 Sowing Window

The sowing window, or the timing of when pulse crops are planted, significantly affects nutrient use efficiency. Optimal sowing times ensure suitable soil temperature and moisture for nutrient uptake, promote nitrogen fixation with rhizobia bacteria, and enhance phosphorus availability, improving nutrient utilization. Timely planting reduces nutrient losses due to leaching or runoff, minimizes competition with weeds, and avoids exposure to pests and diseases. It also contributes to crop maturity and harvest timing alignment with favorable environmental conditions for nutrient uptake, ultimately leading to higher yields and improved nutrient use efficiency in pulse crops. Earliest sowing dates encountered more optimum environmental conditions and allowed the plant to accumulate more dry matter accumulation in chickpeas. Vegetative growth continued into the reproductive stage for longer under normal than late sowing and observed that the number of days to flowering decreased with delay in sowing (Thombre et al. 2019).

8.3.4 Planting Methods

The choice of planting methods, including Furrow Irrigation with Raised Beds (FIRB), broad bed and furrow (BBF), and Ridge and Furrow, plays a pivotal role in nutrient use efficiency in pulse crops. FIRB systems help retain nutrients and reduce soil erosion, enhancing NUE by preserving essential nutrients and minimizing leaching. BBF methods encourage root development and reduce soil compaction, improving NUE through efficient nutrient access and utilization. Kumar et al. (2015) confirmed that raised bed planting reduces seed rate requirements and provides a favorable environment for pulse crop growth and reported considerable savings in seeds and fertilizer, about 25%, through raised bed planting. The research conducted at IIPR demonstrated that in a rice-lentil cropping system, the use of raised beds resulted in a 16.5% increase in lentil yield compared to flatbed sowing (Kumar et al. 2012). Connor et al. (2003) observed that raised bed planting significantly increased grain yields, with substantial improvements in urdbean (33.6%), mungbean (21.8%), green peas (14.5%), pigeon pea (46.7%), and chickpea (37.0%) compared to flat planting. Enhanced nodulation, root and shoot growth, and yield attributes under the FIRB raised bed system (Bhooshan and Singh 2014). Ridge and furrow systems concentrate nutrients on ridges, potentially boosting NUE, and also assist in water management and weed control. The highest increase in available nitrogen, phosphorus, potassium, and sulfur in the topsoil (0–15 cm) was observed in the broad bed and furrow (BBF) system for pigeon pea cultivation (Singh et al. 2018b). Set-furrow cultivation in pigeon pea + green gram intercropping system exhibited significantly greater nutrient uptake compared to the flatbed method. This could be attributed to the favorable physical conditions created by set-furrow cultivation, which enhanced the mineralization and mobility of nutrients. This improvement was observed in increased availability of nitrogen, phosphorus, and potassium, ultimately resulting in higher nutrient uptake (Sharma and Guled 2012).

8.3.5 Varietal Selection

The selection of specific pulse crop varieties can significantly influence nutrient use efficiency. Varietal choice impacts nutrient uptake, crop resilience to nutrient deficiencies, and overall agricultural sustainability. Pulse varieties with efficient root systems and the ability to thrive under suboptimal nutrient conditions enhance NUE. Furthermore, selecting high-NUE varieties can reduce reliance on synthetic fertilizers, leading to cost savings and decreased environmental impacts. For instance, efficient crop cultivars demonstrate the capacity to effectively utilize soil micronutrients even under deficient conditions and do not show a significant response to the application of micronutrient fertilizers. In contrast, inefficient cultivars respond positively to external micronutrient fertilizer application but exhibit limited ability to access native soil micronutrients. Genetically, micronutrient-inefficient cultivars can play a crucial role in increasing micronutrient content in seeds.

Research indicates that modern cultivars of crop plants, specifically those designed for high-input and intensive agricultural systems, may have lost the ability to acquire certain soil nutrients under conditions of low availability (Wissuwa et al. 2009). This phenomenon likely stems from changes in efficiency mechanisms, which have evolved as soil nutrient availability increased over time. This allows efficient cultivars to be utilized by breeders for identifying quantitative trait loci (QTLs) responsible for nutrient efficiency and developing high-yielding, micronutrient-enriched cultivars through genetic biofortification. Inefficient cultivars can be employed for agronomic biofortification, where they are used to enrich the grains of highly responsive cultivars with micronutrients (Patra et al. 2017). Overall, varietal selection in pulses is a pivotal factor in optimizing nutrient management, improving crop performance, and fostering environmentally friendly agricultural practices.

8.3.6 Microbial Inoculation

Incorporating rhizobia bacteria through inoculation is a crucial practice in pulse crop cultivation, as these crops can establish symbiotic relationships with specific soil-dwelling rhizobia. This nitrogen fixation process not only helps pulse crops thrive in nitrogen-deficient soils but also offers an eco-friendly alternative to chemical fertilizers, reducing their environmental impact. In South Africa, cowpeas effectively formed symbiotic relationships with native *Bradyrhizobium* strains, allowing them to obtain over 60% of their nitrogen from the atmosphere, and showcasing their adaptability to nutrient-poor environments (Makaure et al. 2023). Selecting the right rhizobia strains for specific pulse crop species (Table 8.2), following best practices, and using high-quality inoculants are essential for unlocking the full potential of this symbiotic relationship, resulting in healthier soils, increased yields, and a more sustainable agricultural future. Additionally, the inoculation of legumes with effective rhizobia strains can influence nutrient uptake in intercropping systems by affecting root spread, which, in turn, impacts water and nutrient absorption (Gao et al. 2010).

In the root region, a diverse microbial community plays a pivotal role in promoting plant growth and enhancing nutrient uptake, as well as conferring tolerance to various stresses. These microorganisms produce plant growth regulators, including auxins, cytokinins, and gibberellins, which stimulate plant growth. Additionally, they improve nutrient availability in the soil by solubilizing essential elements, secreting organic acids, siderophores, exopolysaccharides, and hydrolytic enzymes like proteases, pectinases, lipases, and chitinases. This, in turn, results in enhanced nutrient absorption, root development, and overall plant well-being, contributing significantly to sustainable agriculture. For instance, a single or co-inoculation with halotolerant *Bacillus subtilis* and *Bacillus thurigiensis* from halophytic plants in Egypt mitigated salt stress effects by promoting shoot and root growth, along with increased mineral (N, P, K, Ca, and Mg) absorption (Mahgoub et al. 2021). Stenotrophomonas spp. strains have been shown to stimulate the expression of sulfate transport and sox genes in the root and rhizosphere of pigeon peas, enhancing sulfur absorption and translocation from roots to the aerial parts. This leads to increased uptake of nitrogen, phosphorus, and potassium compared to uninoculated plants (Malviya et al. 2022). Rhizobacteria in pulse-based cropping systems exhibit various plant growth-promoting activities, including P solubilization, nitrogen fixation, and production of beneficial compounds like indole 3-acetic acid and siderophores. These activities, typical of plant growth-promoting rhizobacteria, help

Table 8.2 List of rhizobium species for corresponding pulse crops

Crop	Rhizobia species
Chickpea	*Rhizobium ciceri*
Lentil, pea, faba bean	*Rhizobium leguminosarum*
Common bean	*Rhizobium phaseoli*
Cowpea, mungbean	*Bradyrhizobium* spp.

plants cope with common stresses such as nutrient deficiency, salinity, and drought (Kumar et al. 2019b).

8.3.7 *Nutrient Management*

Balanced nutrient management in crops refers to the practice of providing essential nutrients in a carefully calibrated and harmonious manner to optimize crop growth and yield while minimizing nutrient imbalances or excesses (Kumar et al. 2011). This approach involves assessing the specific nutrient requirements of pulse crops, considering factors such as soil conditions, crop type, and growth stages, and then applying the right types and quantities of nutrients to meet those requirements. The goal is to ensure that pulse crops receive an appropriate balance of macronutrients (e.g., nitrogen, phosphorus, and potassium) and micronutrients (e.g., zinc, iron, boron, and molybdenum) throughout their growth cycle. Nutrient uptake of different pulse crops is given in Fig. 8.2. Ali and Gupta (2012) have highlighted the widespread occurrence of micronutrient deficiencies, particularly sulfur and zinc, in regions where pulse crops are cultivated. Their study revealed that among 137 districts known for pulse cultivation, a substantial 87 districts, roughly representing 20–60% of the total, experience S deficiency. Furthermore, approximately half of the pulse-growing districts in the study exhibited deficiencies in zinc. The application of a multi-micronutrient blend to the soil led to a substantial increase in the grain yield of black gram while also maintaining the availability of essential micronutrients, as demonstrated in the study conducted by Poongothai and Chitdeshwari (2003).

The Soil Test Crop Response (STCR) approach is a scientifically backed method for optimizing nutrient management in pulse crops. Soil testing is the foundational step in this approach, allowing for precise nutrient recommendations tailored to the specific needs of pulse crops. Research evidence supports the effectiveness of the STCR approach. For instance, studies by Singh et al. (2016) and Gudadhe et al. (2015) found that by using soil testing to guide nutrient management, farmers could significantly improve nutrient use efficiency by maintaining optimum soil available

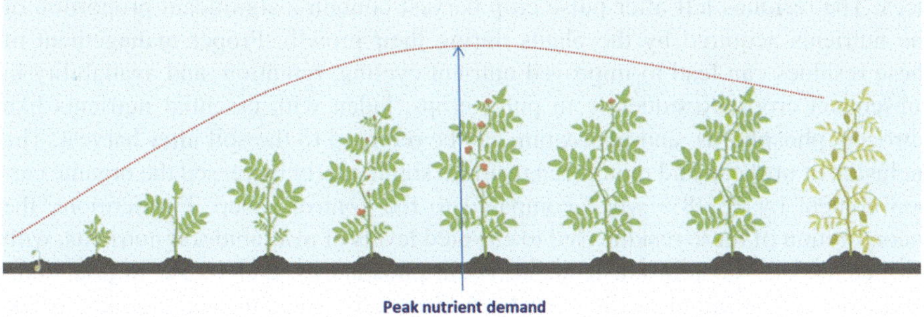

Fig. 8.2 Variations in nutrient requirement during different crop growth stages

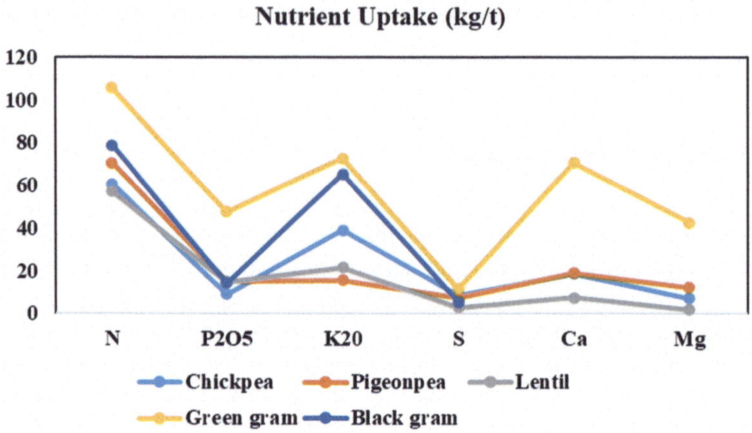

Fig. 8.3 Total nutrient uptake in pulses (FAI statistics [2012–2013])

plant nutrients in chickpea cultivation, resulting in higher yields. Another research conducted by Singh et al. (2018a) demonstrated that the STCR approach led to a substantial reduction in excess fertilizer application in lentil crops, resulting in cost savings and reduced environmental impacts. The utilization of N, P, and K, guided by the STCR equation, in combination with vermicompost and zinc sulfate, resulted in improved nutrient metabolism, enhanced biological activity, and favorable growth parameters. This approach stimulated the development of vegetative branches, leading to increased values in all yield-related attributes and ultimately contributing to higher seed yields in chickpeas (Kokate et al. 2022). Nutrient uptake kg/ton by major pulses is presented in Fig. 8.3.

8.3.8 Crop Residue Management

Crop residue management is an underexplored avenue in this pursuit, with the potential to enhance nutrient recycling and promote sustainable agricultural practices. The residues left after pulse crop harvest contain a significant proportion of the nutrients acquired by the plants during their growth. Proper management of these residues can lead to improved nutrient cycling, retention, and availability in subsequent crops. Residues from pulse crops, laden with essential nutrients like nitrogen, phosphorus, and potassium, can be returned to the soil after harvest. The inclusion of urdbean and mungbean residues significantly increased the organic carbon content by 35.48% when compared to the control group. Furthermore, the incorporation of these residues led to elevated levels of available soil nutrients, with an increase of 24.6% in nitrogen, 11.5% in phosphorus, and 18.5% in potassium compared to the initial fertility levels. Additionally, soil physical characteristics, including bulk density, particle density, percentage of pore space, and water-holding capacity, exhibited improvements in plots where crop residues were incorporated, outperforming plots where residues were removed (Kumar et al. 2012). Over time, these residues release their nutrient content, making them available for the subsequent crop. This recycling process reduces the need for external nutrient inputs, effectively improving nutrient use efficiency. Incorporating residues from urdbean and mungbean into the soil boosted the yield of the subsequent mustard crop by 6–7%. In a rice–chickpea sequence, chickpea yield was most influenced by the incorporation of rice residues, with the highest seed yield observed when rice residues were chopped and incorporated with irrigation (Venkatesh et al. 2014). Additionally, the incorporation of crop residues into the soil increases soil organic matter, contributing to improved nutrient retention and availability. Soil organic matter acts as a nutrient reservoir, preventing leaching and runoff, which can lead to nutrient losses. Moreover, crop residues stimulate microbial activity in the soil, fostering nutrient mineralization and cycling, ultimately enhancing nutrient availability for plants. These practices also play a crucial role in erosion prevention, as residue cover on the soil surface acts as a protective layer, minimizing nutrient-rich topsoil loss due to erosion. Furthermore, the moisture-retaining capacity of crop residue cover ensures adequate soil moisture for nutrient uptake by crops. By conserving moisture, crop residues indirectly support nutrient absorption by plants. Ultimately, effective crop residue management can reduce the dependency on synthetic fertilizers, promoting more sustainable agricultural practices while improving nutrient use efficiency and overall soil health in pulse-based cropping systems.

8.3.9 Nanotechnology

Conventional synthetic fertilizers often incur significant losses, leading to low nutrient use efficiency, typically below 30%. Nano-fertilizers have emerged as a cutting-edge solution to address the challenges of nutrient use efficiency in pulse crops. Nano-fertilizers represent a synthesized form derived from traditional fertilizers, extracted from various plant parts using diverse chemical, physical, mechanical, or biological methods facilitated by nanotechnology. By leveraging nanotechnology, these fertilizers enable precise and targeted delivery of essential nutrients to pulse crops, minimizing losses and optimizing nutrient utilization. The nanoscale particles facilitate improved nutrient absorption at the cellular level, promoting more efficient utilization of fertilizers by pulse plants (Faraz et al. 2019; Srinivasarao 2021). Nano-fertilizers exhibit superior nutrient transport and delivery capabilities through plasmodesmata, resulting in heightened NUE and substantially reduced nutrient losses. The cumulative advantages of nano-fertilizers ultimately manifest in a noteworthy increase in crop productivity, ranging from 6% to 17% (Iqbal 2019). The application of Nano-N, specifically, is poised to effectively cut down urea consumption by 50% through the application of just two sprays in lentils and field peas (Kumar et al. 2020). The foliar application of 0.5% ZnO nanoparticles (NPs) + 0.5% Fe_2O_3 NPs at the pre-flowering stage demonstrated superior efficacy, significantly increasing grain yield, Zn and Fe content, and nutrient uptake in chickpeas compared to mineral and chelated form (Dhaliwal et al. 2021). This innovative approach not only holds the promise of increasing agricultural yields but also contributes to sustainable farming practices by reducing the environmental impact associated with excessive fertilizer application.

8.3.10 Precision Water Management

Precision irrigation like drip fertigation and sprinkler irrigation enhances nutrient use efficiency in pulse crops by delivering water and nutrients directly to the root zone, reducing nutrient leaching, optimizing water-nutrient uptake, and enabling nutrient timing aligned with the crop's growth stages (Hagin and Lowengart 1996; Gärdenäs et al. 2005). It reduces nutrient competition from neighboring weeds, resulting in efficient nutrient absorption by pulse plants. Furthermore, precision irrigation systems offer data-driven decision-making tools that enable growers to adjust irrigation and fertilization based on real-time soil moisture, nutrient levels, and crop performance. Mini sprinkler irrigation resulted in the highest grain and haulm yield for green gram at 1.10 and 3.42 t/ha, respectively, while border irrigation led to lower yields (Dabhi et al. 1998). Fertigation with micro-sprinklers reduced water usage by 60.9% and increased pea green pod yield by 10.0% compared to surface irrigation with recommended fertilizer doses, resulting in a 3.12-fold increase in water productivity (Saroch et al. 2015). In the rainy season, a yield of 0.8 t/ha was achieved, while the highest yield of 1.2 t/ha was obtained during the winter season in black gram using drip fertigation with 100% recommended water-soluble fertilizer doses. This particular treatment led to a remarkable 37% increase in yield compared to surface application (Ramah 2019). Despite the benefits of micro-irrigation such as drip and sprinkler systems, including improved water and nutrient use efficiency (as there is no loss of nutrients or water through deep percolation or leaching), increased plant growth and development, higher yields, improved quality, and flexibility in scheduling water application because of narrow row spacing and higher population per unit area in case of pulses (except in case of pigeon pea), its adoption is inadequate and not popular (Praharaj et al. 2016). A study conducted at ICAR-IIPR in Kanpur reported a significant impact of a single drip-fertigation schedule during the most critical stage of pigeon pea, leading to substantial improvements in crop and water productivity as well as enhanced resource use efficiency (Praharaj and Kumar 2012).

Unlike cereals and oilseeds requiring frequent irrigations, pulses are mostly grown under rainfed upland conditions and require less water. Under water-limited rainfed situations, deficit or critical stage-based irrigation strategies significantly enhance nutrient use efficiency in pulses by optimizing water and nutrient availability during the most critical phases of crop development. These strategies involve tailoring irrigation to match the specific growth stages of pulse crops when water demand is highest. By applying water precisely when needed, nutrient uptake is improved as the roots can access nutrients more efficiently in adequately moist soil. This approach minimizes nutrient leaching and runoff, reducing nutrient losses from the root zone. Certain pulse crops like red gram, black gram, and green gram are cultivated as irrigated crops during the summer season and typically require 3–4 irrigations at critical stages such as germination, flowering, and pod formation (Kannan 2019). Implementing deficit irrigation also results in an 8% improvement in water productivity for chickpeas in arid conditions (Jabow et al. 2015). The practice of deficit irrigation demonstrates that subjecting chickpeas to water stress during the vegetative stage can increase yield by 34% and enhance water productivity by 16% compared to full irrigation (Hirich et al. 2014). In situations where water is limited, applying 50% uniform deficit irrigation to green gram can achieve the highest water productivity while conserving 46% of water (Ambachew et al. 2014).

8.4 Conclusion

In conclusion, enhancing nutrient use efficiency in pulse crops is of paramount importance for ensuring global food security and environmental sustainability. Pulse crops, known for their ability to fix atmospheric nitrogen and enrich the soil with essential nutrients, play a crucial role in modern agriculture. Diversified farming systems, which include intercropping, crop rotation, and agroecological diversity, offer promising avenues for optimizing nutrient utilization in pulses. These practices not only improve nutrient resource utilization but also enhance soil fertility, reducing the risks of soil erosion and depletion. Agronomic interventions such as crop rotation, tillage practices, sowing window optimization, and precision water management significantly contribute to nutrient use efficiency. Additionally, varietal selection and microbial inoculation influence nutrient uptake and resilience to nutrient deficiencies, while balanced nutrient management and soil testing ensure that pulse crops receive the right nutrients in the right amounts. Finally, effective crop residue management not only recycles essential nutrients but also promotes sustainable agricultural practices. Altogether, these strategies outlined in this chapter provide a holistic approach to enhance nutrient use efficiency in pulse crops, ensuring their continued role in providing nutrient-rich food and preserving ecosystem health.

References

Ali M, Gupta S (2012) Carrying capacity of Indian agriculture: pulse crops. Curr Sci 102:874–881

Ali M, Kumar N, Ghosh PK (2012) Milestones on agronomic research in pulses in India. Indian J Agron 57(3rd IAC Special Issue):52–57

Ambachew S, Alamirew T, Melese A (2014) Performance of mungbean under deficit irrigation application in the semi-arid highlands of Ethiopia. Agric Water Manag 136:68–74

Babu S, Mohapatra KP, Das A, Yadav GS, Tahasildar M, Singh R, Panwar AS, Yadav V, Chandra P (2020) Designing energy-efficient, economically sustainable, and environmentally safe cropping system for the rainfed maize–fallow land of the eastern Himalayas. Sci Total Environ 722:137874. https://doi.org/10.1016/j.scitotenv.2020.137874

Betencourt E, Duputel M, Colomb B, Desclaux D, Hinsinger P (2012) Intercropping promotes the ability of durum wheat and chickpea to increase rhizosphere phosphorus availability in a low P soil. Soil Biol Biochem 46:181–190

Bhooshan B, Singh VK (2014) Effect of planting method, irrigation schedule and weed management practice on the performance of fieldpea (*Pisum sativum* L. arvense). J Food Legumes 27(2):112–116

Campbell CA, Zentner RP, Selles F, Biederbeck VO, McConkey BG, Jefferson PJ (2000) Quantifying short-term effects of crop rotations on soil organic carbon in southwestern Saskatchewan. Can J Soil Sci 80:193–202

Connor DJ, Gupta RK, Hobbs PR, Sayre KD (2003) Bed planting in rice-wheat system. In: Addressing resource conservation issues in rice-wheat systems of South Asia (no. CIS-3591). CIMMYT

Dabhi BM, Patel JC, Solanki RM (1998) Response of summer green gram to irrigation methods and varying moisture regimes. Legum Res 21(2):96–100

Dhaliwal SS, Sharma V, Shukla AK, Verma V, Behera SK, Singh P, … Hossain A (2021) Comparative efficiency of mineral, chelated and nano forms of zinc and iron for improvement of zinc and iron in chickpea (*Cicer arietinum* L.) through biofortification. Agronomy 11(12):2436

Faraz A, Faizan M, Sami F, Siddiqui H, Pichtel J, Hayat S (2019) Nanoparticles: biosynthesis, translocation and role in plant metabolism. IET Nanobiotechnol 13(4):345–352

Ferreira MC, Andrade DDS, Chueire LMDO, Takemura SM, Hungria M (2000) Tillage method and crop rotation effects on the population sizes and diversity of bradyrhizobia nodulating soybean. Soil Biol Biochem 32(5):627–637

Gan Y, Hamel C, O'Donovan JT, Cutforth H, Zentner RP, Campbell CA, … Poppy L (2015) Diversifying crop rotations with pulses enhances system productivity. Sci Rep 5(1):14625

Gao Y, Duan A, Qiu X, Liu Z, Sun J, Zhang J, Wang H (2010) Distribution of roots and root length density in a maize/soybean strip intercropping system. Agric Water Manag 98(1):199–212

Gärdenäs AI, Hopmans JW, Hanson BR, Šimůnek J (2005) Two-dimensional modeling of nitrate leaching for various fertigation scenarios under micro-irrigation. Agric Water Manag 74(3):219–242

Garland G, Bünemann EK, Oberson A, Frossard E, Six J (2017) Plant-mediated rhizospheric interactions in maize-pigeon pea intercropping enhance soil aggregation and organic phosphorus storage. Plant Soil 415:37–55

Gudadhe N, Dhonde MB, Hirwe NA (2015) Effect of integrated nutrient management on soil properties under cotton-chickpea cropping sequence in vertisols of Deccan plateau of India. Indian J Agric Res 49(3):207–214

Hagin J, Lowengart A (1996) Fertigation for minimizing environmental pollution by fertilizers. In: Fertilizers and environment: proceedings of the international symposium "fertilizers and environment", held in Salamanca, Spain, 26–29, September, 1994. Springer Netherlands, pp 23–25

Hayat R, Ali S, Amara U, Khalid R, Ahmed I (2010) Soil beneficial bacteria and their role in plant growth promotion: a review. Ann Microbiol 60:579–598

Hirich A, Choukr-Allah R, Fahmi H, Abdellatif RAMI, Laajaj K, Jacobsen SE, El Omari H (2014) Using deficit irrigation to improve crop water productivity of sweet corn, chickpea, faba bean and quinoa. Rev Maroc Des Sci Agron et Vétérinaires 2(1):15–22

Iqbal MA (2019) Nano-fertilizers for sustainable crop production under changing climate: a global perspective. Sustain Crop Prod 8:1–13

Jabow MA, Ibrahim OH, Adam HS (2015) Yield and water productivity of chickpea (*Cicer arietinum* L.) as influenced by different irrigation regimes and varieties under semi desert climatic conditions of Sudan. Agric Sci 6(11):1299

Kannan V (2019) Mechanized sowing and sprinkler method of irrigation for summer irrigated blackgram. Madras Agric J 106(March (1–3)):1
Kebede E (2021) Contribution, utilization, and improvement of legumes-driven biological nitrogen fixation in agricultural systems. Front Sustain Food Syst 5:767998
Kokate A, Suryavanshi VP, Sawant D, Gawande R, Waghmare P (2022) Response of gram (*Cicer arietinum* L.) to STCR based fertilizer application. Pharma Innov J 11:1150–1152
Kumar N, Prakash V, Mina BL, Gopinath KA, Srivastva AK (2010a) Effect of sowing methods, growth retardant and intercropping on horse gram (Microtyloma uniflorum) productivity. Indian J Agric Sci 80(4):335–337
Kumar N, Srinivas K, Mina BL, Kumar M, Srivastva AK (2010b) System productivity, profitability and competition indices of horsegram intercropping under rainfed condition. J Food Legumes 23(3&4):196–200
Kumar N, Mina BL, Chandra S, Srivastva AK (2011) In-situ green manuring for enhancing productivity, profitability and sustainability of upland rice. Nutr Cycl Agroecosyst 90:369–377
Kumar N, Singh MK, Ghosh PK, Hazra KK, Venkatesh MS, Nadarajan N (2012) Resource conservation technology in pulse based cropping systems, Technical Bulletin 1/2012. Indian Institute of Pulses Research, Kanpur. 35 pages
Kumar N, Hazra KK, Singh MK, Venkatesh MS, Kumar L, Singh J, Nadarajan N (2013) Weed management techniques in pulse crops, Technical bulletin 10/2013. Indian Institute of Pulses Research, Kanpur, 47 pages
Kumar N, Singh MK, Praharaj CS, Singh U, Singh SS (2015) Performance of chickpea under different planting method, seed rate and irrigation level in indo-Gangetic Plains of India. J Food Legumes 28(1):40–44
Kumar N, Hazra KK, Nath CP, Praharaj CS, Singh U, Singh SS (2016) Pulses in irrigated ecosystem: problems and prospects. Indian J Agron 61(4th IAC Special Issue):S199–S213
Kumar N, Nath CP, Hazra KK, Das K, Venkatesh MS, Singh MK, Singh SS, Praharaj CS, Singh NP (2019a) Impact of zero-till residue management and crop diversification with legumes on soil aggregation and carbon sequestration. Soil Tillage Res 189(6):158–167. https://doi.org/10.1016/j.still.2019.02.001
Kumar A, Patel JS, Meena VS, Ramteke PW (2019b) Plant growth-promoting rhizobacteria: strategies to improve abiotic stresses under sustainable agriculture. J Plant Nutr 42(11–12):1402–1415
Kumar Y, Tiwari KN, Nayak RK, Rai A, Singh SP, Singh AN, … Raliya R (2020) Nanofertilizers for increasing nutrient use efficiency, yield and economic returns in important winter season crops of Uttar Pradesh. Ind J Fertil 16:772–786
Kumar N, Hazra KK, Nath CP, Praharaj CS, Dutta A, Das K, Sewak S (2022a) Long-term benefits of pulses in sustaining rice-based cropping systems, Technical bulletin 1/2022. ICAR-Indian Institute of Pulses Research, Kanpur, 42 pages
Kumar N, Hazra KK, Nath CP, Praharaj CS, Dutta A, Das K, Singh R, Singh B (2022b) Diversification of maize-wheat cropping system with pulses: long-term impacts on soil health and yield sustainability, Technical bulletin – 2/2022. ICAR-Indian Institute of Pulses Research, Kanpur, 42 pages
Kumar S, Gopinath KA, Sheoran S, Meena RS, Srinivasarao C, Bedwal S, … Praharaj CS (2023) Pulse-based cropping systems for soil health restoration, resources conservation, and nutritional and environmental security in rainfed agroecosystems. Front Microbiol 13:1041124
Lasisi A, Liu K (2023) A global meta-analysis of pulse crop effect on yield, resource use, and soil organic carbon in cereal-and oilseed-based cropping systems. Field Crop Res 294:108857
Mahgoub HA, Fouda A, Eid AM, Ewais EED, Hassan SED (2021) Biotechnological application of plant growth-promoting endophytic bacteria isolated from halophytic plants to ameliorate salinity tolerance of *Vicia faba* L. Plant Biotechnol Rep 15:819–843
Makaure BT, Aremu AO, Magadlela A (2023) Soil nutritional status drives the co-occurrence of nodular bacterial species and arbuscular mycorrhizal fungi modulating plant nutrition and growth of *Vigna unguiculata* L. (Walp) in grassland and savanna ecosystems in KwaZulu-Natal, South Africa. J Soil Sci Plant Nutr 23(1):204–219

Malviya D, Varma A, Singh UB, Singh S, Singh HV, Saxena AK (2022) Sulfur-oxidizing bacteria from coal mine enhance sulfur nutrition in pigeonpea (*Cajanus cajan* L.). Front Environ Sci 10:932402

Mukherjee B, Kumar Naskar M, Nath R, Atta K, Visha Kumari V, Banerjee P, … Hossain A (2023) Growth, nodulation, yield, nitrogen uptake, and economics of lentil as influenced by sowing time, tillage, and management practices. Front Sustain Food Syst 7:1151111

Nath CP, Dutta A, Hazra KK, Praharaj CS, Kumar N, Singh SS, … Das K (2023) Long-term impact of pulses and organic amendments inclusion in cropping system on soil physical and chemical properties. Sci Rep 13(1):6508

Pandey P, Bisht S, Sood A, Aeron A, Sharma GD, Maheshwari DK (2012) Consortium of plant-growth-promoting bacteria: future perspective in agriculture. In: Bacteria in agrobiology: plant probiotics. Springer, Berlin, Heidelberg, pp 185–200

Patra AK, Patil SK, Ramesh K, Srivastava S, Dey P (2017) Soil health, the missing link for sustainable pulse production in India. In: Sustainable pulse production: from less for more. Indian Soc Soil Sci, New Delhi, p 19

Poongothai S, Chitdeshwari T (2003) Response of blackgram to multi-micronutrients. Madras Agric J 90:442–443

Porpavai S, Devasenapathy P, Siddeswaran K, Jayaraj T (2011) Impact of various rice based cropping systems on soil fertility. J Cereals Oilseeds 2(3):43–46

Praharaj CS, Blaise D (2016) Intercropping: an approach for area expansion of pulses. Indian J Agron 61(4th IAC Special issue):S113–S121

Praharaj CS, Kumar N (2012) Efficient management of water and nutrients through drip-fertigation in long duration pigeonpea under Indian Plains. In: Third international agronomy congress on agronomy, environment and food security for 21st century, vol 3. IARI, New Delhi, pp 819–820

Praharaj CS, Singh U, Singh SS, Singh NP, Shivay YS (2016) Supplementary and life-saving irrigation for enhancing pulses production, productivity and water-use efficiency in India. Indian J Agron 61. (4th IAC Special issue):S249–S261

Rajeshkumar A, Venkataraman NS, Ramadass S, Ashokkumar N, Thirumeninathan S (2017) Effect of weed control and cropping system on weed population and productivity of maize grown sole or intercropped with pulses. J Crop Weed 13(3):150–155

Ramah K (2019) Enhancing the productivity of maize-blackgram cropping system through drip fertigation. Plants Environ 1(1):1–4

Saroch K, Sandal SK, Rana K (2015) Effect of irrigation scheduling and NK fertigation on productivity of garden peas (Pisum sativum var. hortense L.). J Agric Res 41(2):126–131

Schoeny A, Jumel S, Rouault F, Lemarchand E, Tivoli B (2010) Effect and underlying mechanisms of pea-cereal intercropping on the epidemic development of ascochyta blight. Eur J Plant Pathol 126(3):317–331

Sharma A, Guled MB (2012) Effect of set-furrow method of cultivation in pigeonpea+ greengram intercropping system in medium deep black soil under rainfed conditions. Karnataka J Agric Sci 25(1):18–24.

Singh YV, Dey P, Meena R, Varma SK (2016) Effect of soil test-based fertilizer application on yield and economics of chick pea in incectisol. Ann Plant Soil Res 18(4):409–412

Singh SS, Kumar N, Praharaj CS, Singh NP (2017) Agro-technologies for enhancing pulses production in rice fallows. ICAR-Indian Institute of Pulses Research, Kanpur. 36 pages

Singh YV, Singh SK, Dey P (2018a) Fertilizer requirement of lentil based on soil test crop response correlation approach in an Inceptisol. J Indian Soc Soil Sci 66(3):324–328

Singh YP, Singh S, Nanda P, Singh AK (2018b) Impact of establishment techniques and maturity duration of pigeon pea cultivars on yield, water productivity and properties of soil. Agric Res 7:271–279

Singh R, Babu S, Avsthe RK, Yadav GS, Das A, Mohapatra KP, Kumar A, Singh VK, Chandra P (2021) Crop productivity, soil health, and energy dynamics of Indian Himalayan intensified organic maize-based systems. Int Soil Water Conserv Res 9(2):260–270. https://doi.org/10.1016/j.iswcr.2020.11.003

Sirivedhin T, Gray KA (2006) Factors affecting denitrification rates in experimental wetlands: field and laboratory studies. Ecol Eng 26(2):167–181

Srinivasarao CH (2021) Programmes and policies for improving fertilizer use efficiency in agriculture. Indian J Fertil 17(3):226–254

Stahl A, Vollrath P, Samans B, Frisch M, Wittkop B, Snowdon RJ (2019) Effect of breeding on nitrogen use efficiency-associated traits in oilseed rape. J Exp Bot 70(6):1969–1986

Tang X, Zhang C, Yu Y, Shen J, van der Werf W, Zhang F (2021) Intercropping legumes and cereals increases phosphorus use efficiency; a meta-analysis. Plant Soil 460:89–104

Thombre SV, Goud VV, Jaybhaye JN, Hodole SS, Tupe AR (2019) Effect of sowing dates and varieties on nutrient uptake and yield of chickpea. J Pharmacogn Phytochem 8(5):806–808

Tsubo M, Walker S, Ogindo HO (2005) A simulation model of cereal-legume intercropping systems for semi-arid regions: I. model development. Field Crops Res 93(1):10–22

Uzokwe VN, Mlay DP, Masunga HR, Kanju E, Odeh IO, Onyeka J (2016) Combating viral mosaic disease of cassava in the Lake zone of Tanzania by intercropping with legumes. Crop Prot 84:69–80

Venkatesh MS, Hazra KK, Katiyar R (2014) Nutrient acquisition and recycling through pulses. Resource conservation technology in pulses. Scientific Publishers, Jodhpur, pp 190–198

Wissuwa M, Mazzola M, Picard C (2009) Novel approaches in plant breeding for rhizosphere-related traits. Plant Soil 321(1):409–430

Yadav GS, Subhash B, Das A, Datta M, Mohapatra KP, Singh R, Singh VK, Rathore SS, Chakraborty M (2021) Productivity, soil health, and carbon management index of Indian Himalayan intensified maize-based cropping systems under live mulch-based conservation tillage practices. Field Crop Res. https://doi.org/10.1016/j.fcr.2021.108080

Zhou Y, Coventry DR, Gupta VV, Fuentes D, Merchant A, Kaiser BN, … Denton MD (2020) The preceding root system drives the composition and function of the rhizosphere microbiome. Genome Biol 21(1):1–19

Chapter 9
Crop Diversification with Legume for Sustainable Production and Environmental Gain

Nilutpal Saikia, Kadagonda Nithinkumar, Shreyas Bagrecha, SK Asraful Ali, Twinkle Jena, Megha Kumari, Vijay Pooniya, Prabhu Govindasamy, Rishi Raj, Kamlesh K. Lakhena, Anamika Barman, Manoj Kumar Singh, and Subhash Babu

Abstract This chapter explores the crucial topic of crop diversification using legumes for the purpose of achieving sustainable production with environmental gain. Faced with the environmental drawbacks of monoculture, the integration of legumes proves revolutionary. Crop diversification, strategically achieved through practices like intercropping and rotation, not only optimizes yields but also mitigates risks associated with pests and market fluctuations. Legumes provide a beneficial influence on cropping systems by suppressing diseases, facilitating alternative pest management, and enhancing nitrogen retention; thanks to their nitrogen-fixing capabilities. The integration of legumes in the cropping system improves resilience and decreases the overall environmental impact of farming. This sustainable approach offers a viable alternative to address the increasing global food needs.

Keywords Environmental gain · Intercropping · Nitrogen-fixing · Monoculture

---

N. Saikia (✉) · K. Nithinkumar · S. K. A. Ali · T. Jena · M. Kumari · V. Pooniya · P. Govindasamy · R. Raj · K. K. Lakhena · A. Barman · S. Babu
Division of Agronomy, ICAR-Indian Agricultural Research Institute, New Delhi, India

S. Bagrecha
Agronomy Section, ICAR-National Dairy Research Institute, Karnal, India

M. K. Singh
Department of Agronomy, Institute of Agricultural Sciences, Banaras Hindu University, Varanasi, India

S. Babu et al. (eds.), *Agricultural Diversification for Sustainable Food Production*, Sustainability Sciences in Asia and Africa,
https://doi.org/10.1007/978-981-97-7517-0_9

9.1 Introduction

In the ever-evolving landscape of agriculture, the imperative to strike a balance between meeting the escalating global demand for food and mitigating the environmental impacts of intensive farming practices has never been more critical. As we confront the ramifications of climate change, shifting weather patterns, and a burgeoning global population, the traditional paradigms of monoculture are proving inadequate. "Sustainability" means maintaining a certain level without depleting natural resources, aligning with the goal of preserving soil health and fertility for future generations. Monoculture has been widely embraced in recent centuries as a reaction to the increasing need for food caused by population expansion (Tilman et al. 2002). Concurrently, it has resulted in notable environmental issues, such as contamination of soil and water, depletion of biodiversity, and release of greenhouse gases (GHGs). This highlights the need of establishing sustainable agriculture methods. Crop diversification is a technique that may occur in many forms and sizes. Cropping systems should be planned to optimize crop yields and simultaneously uphold or elevate yield stability across diverse climatic conditions (Liu et al. 2019; Babu et al. 2016). Although there is a lack of a uniform framework and definition for diversification (Hufnagel et al. 2020), research in this area is growing (Babu et al. 2020b; Alletto et al. 2022). IPES-Food (2016) defines it as the upkeep of "multiple sources of production" and the diversification of cultivated crops across farming landscapes through practices like intercropping and over time via crop rotation. Crop diversification can reduce reliance on synthetic inputs, minimize environmental impacts, and boost resilience in both crop production and farming systems, contributing to a transition to more sustainable agrifood systems (Kremen et al. 2012; Tamburini et al. 2020; Paas et al. 2021; Messéan et al. 2021; Babu et al. 2023b). In agricultural settings, it is suggested that diversifying agricultural production systems spatially, temporally, and genetically can enhance ecosystem functioning, services, and biodiversity. This contributes to advancing sustainability goals (Beillouin et al. 2021). Crop diversification aims to reduce farmers' reliance on a single crop for income, mitigating risks associated with pests, environmental events, and market fluctuations. By cultivating a variety of crops, farmers enhance biodiversity, strengthen agroecosystems, and reduce the likelihood of complete crop failure, providing alternative income sources. This approach supports food security, allows the sale of diverse agricultural products in national and international markets, and fosters self-reliance in food production, particularly in developing countries (Clements et al. 2011; Avasthe et al. 2020; Babu et al. 2022). Ecological advantages encompass minimized nutrient losses, enhanced resource efficiency, and heightened productivity and stability within the production system. Global demand for plant-based protein is rising (Wu et al. 2014), while current crop production's environmental impact is worsened by low diversity and high input intensity (Tilman et al. 2002). Introducing legumes-based diversified cropping systems, offering ecosystem services, and promoting more sustainable practices with reduced trade-offs between economic and environmental impacts (Watson et al. 2017; Reckling et al. 2016; Babu et al. 2023a; Figs. 9.1 and 9.2), potentially aiding climate change mitigation (Jensen et al. 2012). Long-term field experiments conducted in France, Scotland, and Sweden reveal that incorporating legumes into cropping systems enhances the performance of cereals in European agriculture (Reckling et al. 2022).

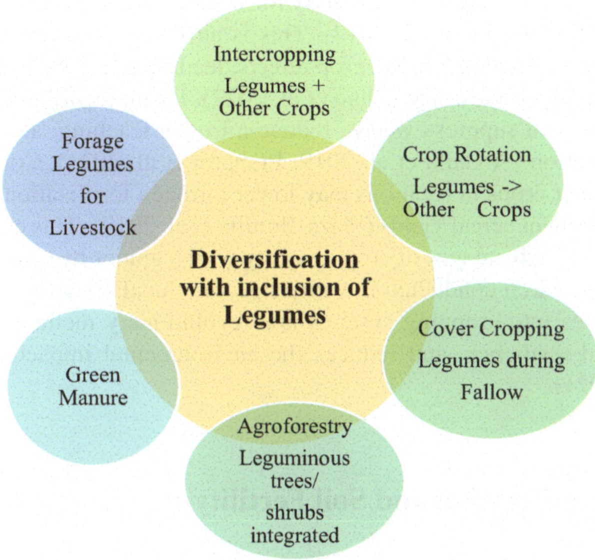

Fig. 9.1 Different ways of crop diversification with inclusion of legumes

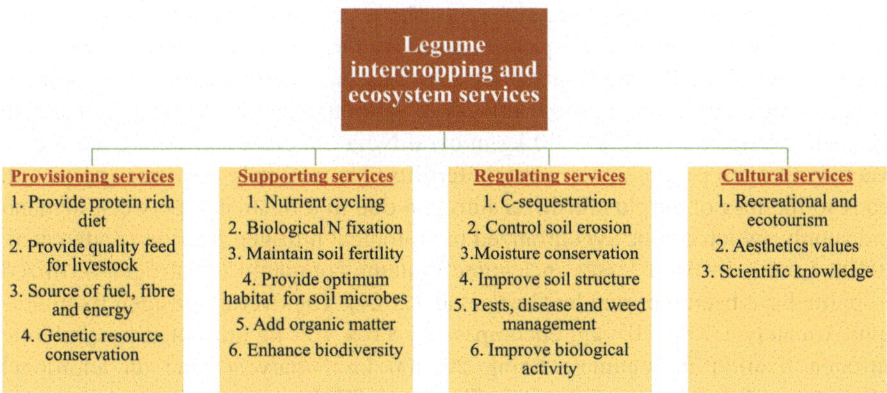

Fig. 9.2 Ecosystem services provided by legumes (Source: Kumawat et al. 2022)

Leguminous plants impact subsequent crops through break crop, nitrogen, and legume-specific effects (Watson et al. 2017; Das et al. 2022a, b). Break crops introduce diversity, reducing soil-borne diseases and enabling alternative pest control. Nitrogen effect involves the release of biologically fixed nitrogen, more pronounced in sandy soils (Jensen et al. 2004). Legume-specific effects enhance growth-promoting bacteria, benefiting succeeding crops. Nitrogen content in legume residues influences nitrate leaching and N_2O emissions, emphasizing the value of nitrogen-retaining cover crops, particularly for winter-fallow spring-sown scenarios (Pappa et al. 2011; Tuulos et al. 2014). This is attributed to a favorable pre-crop effect, including residual and break crop effects, leading to a 20% increase in main crop yields globally, on average (Zhao et al. 2022). Using leguminous cover crops in crop rotations can suppress weeds, reduce nitrogen leaching, and promote soil carbon sequestration (Abdalla et al. 2019; Praharaj et al. 2021). Combining grain legumes and cover crops in rotations may lower nitrogen fertilization rates without impacting subsequent cereal yields (Plaza-Bonilla et al. 2017). This chapter seeks to elucidate the many advantages of crop diversity with legumes by synthesizing ideas obtained from research conducted in various geographical situations. The incorporation of legumes into cropping systems is a revolutionary method that enhances crop yields and resilience, and reduces the environmental impact of agriculture (Singh et al. 2021).

9.2 Nitrogen Fixation and Soil Fertility

9.2.1 Nitrogen Fixation

Plants require nitrogen to the tune of 40–50% of their dry matter. The conversion of N_2 to NH_3 at normal temperature and atmospheric pressure is facilitated by nitrogenase, whereas the Haber-Bosch process relies on elevated temperature and pressure (Erismann et al. 2008). Biological nitrogen fixation (BNF) served as the primary source of nitrogen in agriculture (Smil 2004). Although certain soils can mineralize adequate nitrogen up to 300–400 kg mineral N/ha (Angus et al. 2015), additional reactive nitrogen is applied to maintain fertility. The untimely fertilizer application leads to the loss of surplus fertilizer nitrogen due to a mismatch between demand and supply, which can be synchronized in biological nitrogen fixation (Ladha et al. 2020). Lower nitrogen losses in legume systems compared to nitrogen-fertilized cropping have been reported by Costa et al. (2021). The overall nitrogen fixation is approximately 1.75×10^{11} kg, encompassing 8.0×10^{10} kg attributed to symbiotic nitrogen fixation in legumes, fixing 20–200 kg N/ha/year, and an additional 8.8×10^{10} kg from industrial fixation (Shah et al. 2021).

Nitrogen fixation constitutes dinitrogen formation in the form $N \equiv N$, which is carried out by nitrogenase activity. This process requires approximately 103 kJ/mol to break the triple bond and generates NH_3, which is utilized in the biochemical synthesis of amino acids and proteins, stored in microbial or plant biomass. These organic nitrogen sources contribute to the soil organic nitrogen (SON) pool, eventually released in an available form, such as nitrate (NO_3^-) through mineralization. The establishment of the legume-rhizobium symbiosis is initiated through a signal exchange between the host plant and its microsymbiont, forming root nodule primordia through cortical cell divisions. In most legumes, the development of infection threads in the root hair helps in the incorporation of bacteria into the nodule cells via endocytosis. Each bacterium, referred to as a symbiosome, further differentiates into nitrogen-fixing Bacteroides (Jones et al. 2007). Specificity is observed at both the species and genotypic levels (Wang et al. 2012) and is mediated by flavonoid compounds (Liu and Murray. 2016), rhizobial exopolysaccharides (Gibson et al. 2008), or the host's innate immunity mechanisms. Since nitrogenase is sensitive to oxygen, the formation of the oxygen diffusion barrier and the synthesis of the oxygen carrier protein leghemoglobin take place to facilitate its activity (Gordon et al. 2001). The process of N fixation is represented in Fig. 9.3.

BNF can help in replenishing the reservoirs of organic N in a plant–soil system. Average % nitrogen derived from atmosphere (% Ndfa) values ranged from 36% to 41% (common bean, green and black bean) to 77–87% (faba bean and pigeon pea), with the amounts of N fixed varying from 25 to 29 kg (cowpea and common bean) to 221 kg of N/ha per crop (Faba bean) across different geographic regions (Peoples et al. 2021). The magnitude of BNF depends on legume species, soil properties, climatic conditions, cropping system as well as soil management practices. The BNF fixing ability of commonly cultivated legumes across the world is shown in Table 9.1.

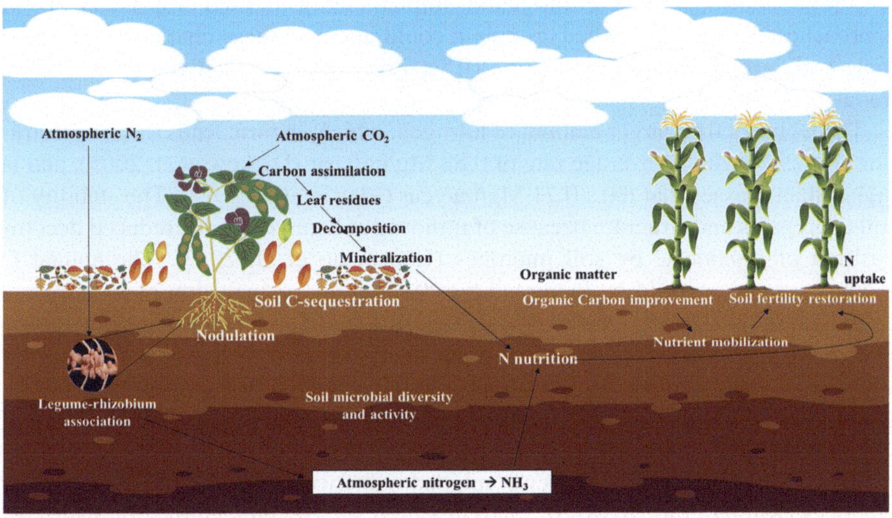

Fig. 9.3 Mechanism of nitrogen fixation in legumes

Table 9.1 BNF capacity of different legumes

Common name	*Scientific name*	BNF (kg/ha)	Cultivation area
Faba bean	*Vicia faba*	118.6–311	Greece, Italy
Pea	*Pisum sativum*	36.6–125.3	Canada, Greece
Common vetch	*Vicia sativa*	107–131	Switzerland
Grass pea	*Lathyrus sativus*	101–149	Switzerland
White lupin	*Lupinus albus*	53.1–64.1	Italy
Chickpea	*Cicer arietinum*	21.0–103.6	Canada
Lentil	*Lens culinaris*	23.0–86.8	Switzerland, Canada
Common bean	*Phaseolus vulgaris*	16.3–71.9	Canada
Cowpea	*Vigna unguiculata*	36–75	Brazil
Soybean	*Glycine max*	90–95	United States
Alfalfa	*Medicago sativa*	103–209	Canada, China
Egyptian clover	*Trifolium alexandrinum*	30–59	Switzerland
Red clover	*Trifolium pratense*	35.4–389	Denmark, United States

Source: Vasconcelos et al. (2020)

9.2.2 Soil Fertility

9.2.2.1 Role of Legume Crops in Soil Carbon Sequestration

A 2% rise in soil organic carbon (SOC) has the potential to boost wheat and maize yields by 10% and 23%, respectively (Oldfield et al. 2019). Yet, global SOC stocks in intensified agroecosystems have declined by 20–69% (Lal 2019), posing a risk to crop production and climate mitigation. Implementing focused crop management approaches, such as diversified cropping, could alleviate these challenges and promote soil carbon sequestration (Lal 2015a; Zhao et al. 2022; Babu et al. 2020a; Yadav et al. 2021).

The legumes fix a larger amount of nitrogen, which, in turn, leads to C sequestration (Fig. 9.3) with an average rate of 0.88 Mg/ha/year (Diekow et al. 2005), and is higher than grasses that fixes 0.71 Mg/ha/year (Martin et al. 2012). The stability of soil aggregates and structure because of turnover of legume residue reduced decomposition of organic C by soil microbes (Conceicao et al. 2013). The annual C sequestration rate could be increased by 20–75 g/m^2 by inclusion of leguminous crops (Liu et al. 2013). Various studies reported that perennial legumes increased SOC pool as compared to the annual crop (Young et al. 2009). Crop residues from leguminous plants are more effective in stabilizing soil carbon compared to nonlegume crops (Drinkwater et al. 1998). Leguminous plants, with their leaf-shedding ability and superior below-ground biomass production, contribute more to soil carbon sequestration, especially when combined with mixed farming and crop rotation practices (Kumar et al. 2018). In sandy-textured soil, legume cultivation increases soil organic carbon compared to oats (Avena sativa L.) cultivation on the same soil

Table 9.2 SOC gain through various legume tree species

Legumes	Soil depth (cm)	SOC increase (Mg/ha)	References
Medicago arborea		1.75	Alegre et al. (2004)
Medicago sativa	200	24.1	Guan et al. (2016)
Vicia sativa		10.0	Sisti et al. (2004)
Medicago sativa	100	0.33–1.56	Mortensen et al. (2004)
Acacia auriculiformis	0–10	3.41	Dreschel et al. (1991)
Cassia siamea	0–10	5.20	Dreschel et al. (1991)
Crotalaria grahamiana	0–20	1.69	Impala (2001)

(Hajduk et al. 2015). As per Singh et al. (2023), Napier grass + cowpea and tri-specific Napier hybrid + cluster bean helped in reducing the C-mineralization rates and potentially mineralizable C, implying a better sequestration efficiency. A meta-analysis study by Kumar et al. (2018) suggested that the legumes have the capacity to store 30% higher soil organic carbon (SOC) when compared to other crops and allow 7.21 g/kg DM sequestration of carbon, resulting in conservation of fossil fuel corresponding to nearly 277 kg/ha of CO_2 every year (Stagnari et al. 2017). The amount added to SOC pool by various legumes is presented in Table 9.2.

9.2.2.2 Role of Legume Vegetation in Improving the Soil N Pool

The depletion of nutrients from farming systems is acknowledged as a significant environmental issue, leading to air pollution and nutrient excess in water bodies. Integrating legumes into agriculture presents an opportunity to mitigate these losses while preserving or even improving agricultural productivity. When a portion of the biologically fixed nitrogen becomes available to the subsequent crop, the cultivation of legumes significantly reduces the need for chemical nitrogen fertilizers (Crozat and Fustec. 2006). BNF benefits extend beyond legumes to subsequent crops. The nitrogen provided to cereal crops results from the decomposition of legume-biomass residues, the senescence and decay of roots and nodules, and the release of nitrogen compounds through the exudation of living roots (Fustec et al. 2010). Nitrogen transfer between plants can happen among young plants, through mycorrhizal networks connecting plants, or via the exudation of nitrogen compounds by legumes (Moyer-Henry et al. 2006). Velvet bean intercropped with maize accumulates 160 kg N/ha in 12 weeks (Sanginga and Woomer 2009), and legume cover crops in potato systems reduce nitrogen losses from soil erosion, improving nitrogen use efficiency (Xing et al. 2011; Yadav et al. 2021). Legume roots decompose, releasing nitrogen into the soil for subsequent crops (Giller 2015). A mixed culture of maize and legumes, reported by Fu et al. (2023) is a nitrogen-saving alternative with higher productivity per unit of time and area compared to systems without legumes.

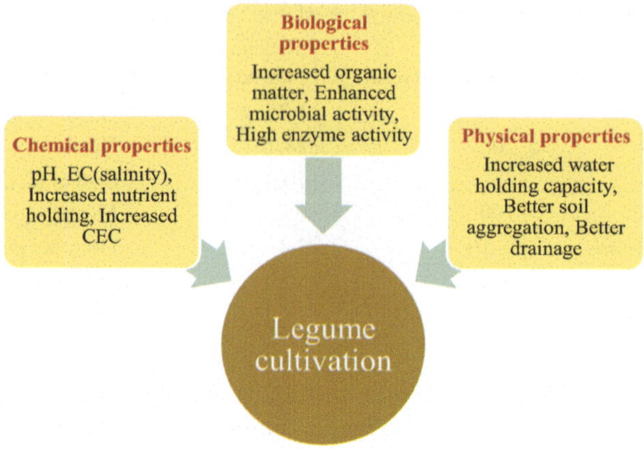

Fig. 9.4 Effect of legume cultivation on soil properties

9.2.2.3 Enhancing Soil Structure and Nutrient Availability

Legume cultivation leads to various changes in soil properties (soil physical, chemical, and biological), which is depicted in Fig. 9.4.

Soil Physical Properties

Aeration, runoff, erosion, water holding capacity, and infiltration rate influence various soil physical properties such as bulk density, porosity, aggregate stability, and texture (Juma and Van Averbeke 2005). Legume vegetation acts as a soil conditioner due to an increase in organic matter as a result of large biomass production capacity (Lal 2015b), and an increase in organic carbon stock by maintaining appropriate soil C/N ratio. In addition, leguminous cover crop protects the soil from erosion by physically modifying the soil profile through aggregation which results in higher infiltration of water (Srinivasarao et al. 2012; Meena et al. 2015). Incorporating legume crops into monoculture systems could enhance root development, water use efficiency, and soil water retention (Angus et al. 2015; Kazula et al. 2017).

Soil Chemical Properties

Soil cation exchange capacity (CEC), pH, nutrient levels, and soil organic carbon content are the primary determinants for the evaluation of soil fertility. Legume farming has dual effects on soil pH, acting as a natural buffer. Organic matter decomposition decreases pH through organic acid production, while the reduction of H^+ into H_2O and CO_2 raises pH (Buragohain et al. 2017; Mitran et al. 2018). Legume-based rotation alters the pH of the rhizospheric soil by root exudation of organic acids on the epidermal cell of root surfaces, which enhances P availability by dissolving the fixed phosphorus in soil (Varma et al. 2017). The hydrogen gas generated during nitrogen fixation stimulates the growth of other soil bacteria, contributing to enhanced soil health (Angus et al. 2015). Changes in pH due to root exudate and residue decomposition are recognized to aid in increasing the microbial biomass which helps in nutrient cycling and increases the availability of N, P, K, and trace elements to the succeeding plants (Benchaar et al. 2001; Webb et al. 2003). Ben-chuan et al. (2022) conducted an experiment on maize–soybean strip intercropping, maize–peanut strip intercropping, and reported that compared with the monoculture system, the N uptake, root length density, root surface area density, soil protease, urease, and nitrate reductase activities were significantly greater in intercropping than that of the corresponding monocultures. Introducing soybeans after maize/wheat for cropping diversification led to a one- to twofold improvement in the soil quality index compared to sole maize/wheat cultivation. This improvement was attributed to increased aggregate stability, fostering soil microbial growth, and carbon sequestration (Yan et al. 2023). In the topsoil under diversified cropping, there were notable increases in SOC, microbial biomass carbon, C-acquisition enzyme activity, and dissolved organic carbon concentrations within aggregates larger than 2 mm, compared to wheat/maize cultivation (Yan et al. 2023).

Soil Microbial Properties

Soil microorganisms provide a vital link between plant productivity and soil nutrient availability through nutrient cycling. Intercropping with legumes alters root distribution, architecture, and rhizosphere exudation, influencing the microbial community and promoting beneficial interactions with plants. Legumes actively participate in tripartite symbiotic relationships (Rhizobium–legume–Mycorrhiza), promoting the colonization of arbuscular mycorrhiza in agroecosystems with low inputs (Pivato et al. 2007). Legumes boost soil microbial biomass by creating a favorable microclimate, influencing soil properties, adjusting pH, and increasing root biomass. Microorganisms interacting with leguminous plants in the rhizospheric zone can enhance crop productivity (Graham and Vance 2000). Sesbania aculeata residue incorporation enhances soil microbial activities, vital for nutrient turnover and long-term soil productivity, leading to increased field crop yields (Pooniya et al. 2012).

9.2.2.4 Nutrient Availability and Uptake

Legumes, with their deep taproot system, can absorb nutrients from deeper soil layers, reducing the reliance on fertilizers (Sanderson et al. 2004). SOC and humus content, N and P availability and uptake were reported to be improved with legume-based cropping system (Stagnari et al. 2017). The activity in the root zone of the soil generally governs plant nutrient availability and uptake. The legumes by their nitrogen effect, which is due to nutrient provision from BNF and the break crop effect, which includes phosphorus mobilization, improvement of SOM, soil water retention, and availability contribute to the availability of different nutrients in legume-based cropping systems. The availability of phosphorus may be due to the exudation of organic acids that solubilize phosphorus, phosphatase enzyme, and the interaction between insoluble phosphorus and root surface (Ae and Shen 2002). Tang et al. (2021) found that intercropping cereal and legumes reduced phosphorus (P) fertilizer requirements by 21% compared to sole crops with the same yields. This highlights the potential of cereal-legume intercropping to enhance land-use efficiency and P uptake, improving P fertilizer use efficiency in agriculture. Incorporating stem-nodulating plants like *Sesbania rostrata* in the succeeding crop provides up to 6 tons of biomass, meeting 50–100% of the nitrogen requirement. Periodic green manuring with Sesbania aculeata, cowpea, and mungbean improves soil micronutrient availability (zinc, iron, manganese, and copper) for crops, compared to summer fallow (Pooniya and Shivay 2013).

9.3 Enhancing Yield and Quality

In the last two decades, the Indo-Gangetic Plain (IGP) in South Asia has faced declining yields and factor productivity in major cereal-based cropping systems. This has prompted a search for alternative crops to address the issues of diminishing productivity and ensure system sustainability. Incorporating legumes into cropping systems is one way to diversify farming operations and promote sustainable agriculture (Rodriguez et al. 2021). Legumes play a key role in maintaining sustainability as they lower the input by cutting down the N requirement. They have a unique ability to form symbiotic relationships with nitrogen-fixing bacteria in their root nodules. Further inclusion of legumes and soil nitrogen levels improve, benefiting subsequent crops and reducing the need for synthetic fertilizers.

9.4 Improved Crop Yields

Diversification with grain legumes offers a strategic approach to enhance agricultural yield. BNF and the N benefit from legume residues are the main factors influencing the legume pre-crop effect (Smith and Chalk 2020). Integrating perennial leys and grain legumes into cropping systems generally boosts the yield of subsequent crops compared to systems lacking legumes and perennial crops (Angus et al. 2015; Preissel et al. 2015; St-Martin et al. 2017). The pre-crop effect of the legume depended on the type of legume as different legumes have different BNF capacities and on the main crop species. However, the collective impacts of nitrogen (N) supply and non-N benefits (e.g., decreased pest and disease and better soil characteristics) frequently result in an increase in yields of crops after legumes. The inclusion

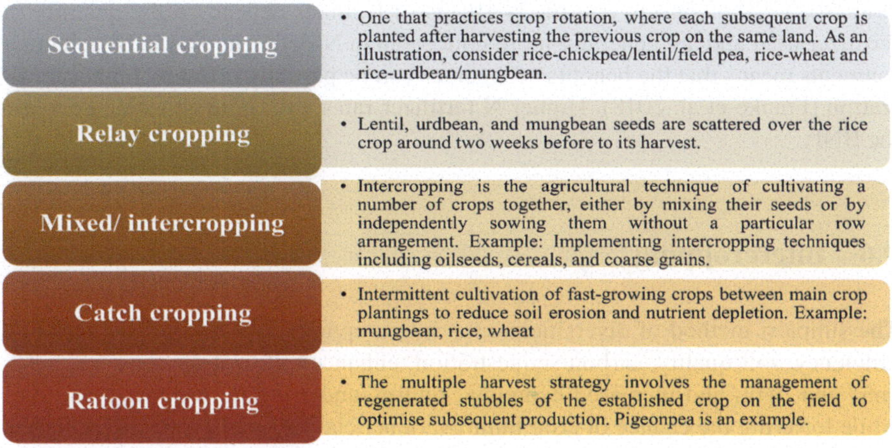

Fig. 9.5 Types of legume cropping (Source: Jangir et al. 2022)

of legume crops in cropping systems has been associated with a reported yield advantage of 15–25% in cereals (Kirkegaard et al. 2008). Zhao et al. (2022) conducted a meta-analysis of 60 legume-based cropping systems spanning 53 countries between 1959 and 2020. He reported an average of 20% increase in main crop yield from legumes. In main crops (e.g., rice, wheat, and maize) yield was the most stable. Nitrogen pre-crop effect had a negative correlation with both crop diversity and N fertilizer rate as these yield advantages decreased with increasing N fertilizer rates and crop diversity. Diversification through pulses could be done by crop rotation, intercropping, para/utera cultivation, etc. (Fig. 9.5). Under intercropping pulses can be incorporated via row intercropping, mixed intercropping, and relay intercropping. In Kenya's upper midland agro-ecological zones, sole potato yielded 11.4 t/ha, while intercropping with *Dolichos* and lima bean resulted in equivalent yields of 23.9 t/ha and 18.9 t/ha, respectively (Nyawade et al. 2020). Pierre et al. (2022) found that in rainfed conditions in the Yucatan Peninsula, intercropping maize with cowpea can maintain yields similar to maize monocropping, while also increasing the population of beneficial insects. Green leaf manuring with *Sesbania* and intercropping with cowpea improve soil fertility, subsequently boosting crop productivity (Bana et al. 2012).

9.5 Crop Rotation

Crop rotation is an age-old practice governing biodiversity in agricultural systems. Incorporating pulses into crop rotation systems offers a myriad of benefits that contribute to sustainable and productive agriculture. It improves soil health, deters pests and disease, and ultimately raises yield (Barbieri et al. 2019). Faba bean-wheat rotation recorded a 71.1% yield advantage as compared to wheat monocropping (Mesfin et al. 2023). The concept that the N effects of legumes depend on the quantity of fertilizer used and are generally more noticeable in N-deficient than in N-rich environments means that the benefits of legumes can be partially offset by higher fertilization (Franke et al. 2018). Higher N fertilizer rates will, in most cases, decrease the BNF.

9.6 Intercropping

The simplest method of diversification is intercropping, which improves land-use efficiency, soil quality, carbon sequestration, enhanced utilization of water, nutrients, and light, and increased economic returns under erratic weather conditions while lowering the chance of crop failures. Beneficial for small-scale and marginal farmers. Intercropping of pulses can be done with oilseeds (Indian mustard, linseed, safflower, soybean, etc.), cereals (maize, wheat, etc.), coarse grains (sorghum, pearl millet), and commercial crops (sugarcane, cotton, etc.) (Praharaj and Blaise. 2016; Praharaj et al. 2021). Maize + cowpea-wheat was the most profitable and productive intercrop land equivalent ratio (LER) = 1.71, sustainable value index (SVI) = 0.96, increasing net economic return to 43.63% with a Benefit:Cost ratio of 1.94 (Tripathi et al. 2021). Wheat + chickpea intercropping resulted in higher nutrient N and P content in both wheat and chickpea as well as higher system yield (Raza et al. 2016). Pearl millet intercropped with green gram exhibited higher equivalent yield, net returns, land equivalent ratio (LER), aggressively, competition ratio, monetary advantage index (MAI), and total crowding coefficient (Renu et al. 2018). Napier grass + cowpea and tri-specific Napier hybrid + cluster bean cropping significantly improved the yield, as well as the C and P fractions for the surface and subsurface soils (Singh et al. 2023). Nyawade et al. (2020) found that integrating lima bean and dolichos with potato in intercropping maintained a soil N balance of 4.1–6.6 kg N/ha, contrasting with −10.7 to −18.1 kg N/ha for sole potato and 2.9–22.3 kg N/ha for sole legumes. Nitrate leaching in sole potato cropping (4.2–46.6 kg N/ha) was alleviated by the deep root systems of legume intercrops. This suggests that legume intercropping can optimize soil N balance in smallholder potato farming systems by balancing nitrogen retention in biomass and soil.

9.7 Relay Cropping

The utera/paira is a sort of relay sowing in which pulses like lentils, lathyrus, urad bean, and mungbean seeds are broadcasted in a standing crop of rice 2 weeks before harvest. This method allows better utilization of the remaining moisture at rice harvesting time. It is widely practiced in Bihar, Eastern Uttar Pradesh, West Bengal, Chattisgarh, and Orissa. The variety of rice also affects the productivity of pulses. Sarkar et al. (2019) advocated spraying of 2% DAP for a higher yield of lathyrus under utera cultivation. In relay cropping lentil, grass pea, and pea had 46%, 37%, and 32% higher seed yields as compared to conventional techniques (Ali and Sarker 2017).

9.8 Weed Management

Effective weed control, through herbicide use and tillage, has led to declines in flowering plant species in agriculture (Swanton et al. 2006). Some legume crops can suppress weeds, but choosing the right species or crop mix is crucial. Intercropping legumes with cereals enhances competitiveness and reduces herbicide reliance (Poggio 2005). Reduced weed management, especially with no- or low-till practices, can limit weed populations and positively impact biodiversity, provided a well-planned legume-supported crop rotation is implemented. Cover crops can increase functional diversity and provide various ecosystem services, including weed management in grain agroecosystems without disrupting typical rotations or reducing yields (Rodriguez et al. 2021; Snapp et al. 2005). Legume crops can act as excellent living mulches, significantly boosting the overall competitiveness of intercrops against weeds (Naudin et al. 2011). In maize–legume intercrops (maize–bean, maize–cowpea), reduced light availability for weeds led to lower weed density and dry matter compared to sole crops (Bilalis et al. 2010). Introducing legume living mulches did not entirely eliminate weed growth but served as a relevant agroecological solution to reduce herbicide use, promoting sustainability, and diversifying weed control tactics in cropping systems.

9.9 Pest Management

Legume cropping fosters soil microbial growth by providing them with substrates, reducing the incidence of pests and phytopathogens (Bybee-Finley et al. 2016). Legume traits, like high nitrogen content and flowering, attract not only pollinators but also potential herbivore pests. Legumes serve as habitat and food for pest control agents like predatory insects (Géneau et al. 2012). Fauna diversity in legume-supported systems depends on chemical pest control, crop management, and rotation. Organic systems, without pesticides and mineral fertilizers, often show increased fauna diversity.

9.10 Nutritional Benefits

A significant number of people suffer from several forms of micronutrient deficiencies (World Bank. 2008). Legumes, second only to cereals in nutritional value, are vital for the diets of the economically weaker section in developing countries due to their rich protein content (Rani et al. 2019). Additionally, they serve as forage for animals and are used in various food products like doughnuts and bread. Proteins are essential for life, acting as catalysts in biological reactions and forming the main component of muscle tissue. Leafy plants, rich in proteins, play a vital role in photosynthesis. Feed and food proteins come from storage proteins in plant seeds and organs, supporting plant reproduction. Amino acids, the building blocks of proteins, contain about 16% nitrogen, a significant part of living organisms' nitrogen. Rhizobia bacteria on legume roots fix atmospheric nitrogen, making legumes a key source of reactive nitrogen, contributing to the nitrogen cycle and in protein provision, both in natural ecosystems and agricultural settings, particularly in situations where the application of nitrogen fertilizers is limited (Watson & Stoddard 2017). Legumes are nutrient-dense foods, are excellent sources of proteins, and generally contain a higher amount of lysine (130 mg/g in peas to 250 mg/g in pigeon peas) as compared to animal protein. They contain double the amount of protein found in cereals, ranging from 19.90 g in *Adzuki* beans to 36.17 g in lupin (per 100 g) (Table 9.3). However, they contain lower sulfur-containing amino acids (cystine and methionine), which can be compensated with the balanced uptake of cereals. Carbohydrate levels are low as compared to cereals (Dhull et al. 2022) averaging 60.76 g per 100 g (Table 9.3). Starch being its major constituent, is composed of

Table 9.3 Pulses approximate composition (per 100 g)

Bean/pulses	Energy (kcal/kJ)	Protein (g)	Total lipid/ fat (g)	Ash (g)	Carbohydrate (g)	Dietary fiber (g)
Adzuki bean	329/1377	19.90	0.53	3.26	62.90	12.70
Black bean	341/1427	21.60	1.42	3.60	62.36	15.50
Chickpeas	378/1582	20.47	6.04	2.85	62.95	12.20
Cowpea	343/1435	23.85	2.07	3.39	59.64	10.70
Cranberry bean	335/1402	23.00	1.23	3.31	60.00	24.70
Faba/broad bean	341/1427	26.12	1.53	3.08	58.29	25.00
Great northern bean	339/1418	21.90	1.14	3.93	62.40	20.20
Lentil	352/1473	24.63	1.06	2.71	63.35	10.70
Lima bean, large	338/1414	21.50	0.69	4.30	63.40	19.00
Lima bean, baby	335/1402	20.60	0.93	3.55	62.80	20.60
Lupin	371/1552	36.17	9.74	3.28	40.37	18.90
Moth bean	343/1435	22.94	1.61	4.26	61.52	–
Mung bean	347/1452	23.86	1.15	3.32	62.62	16.30
Navy bean	337/1410	22.33	1.50	3.32	60.75	15.30
Pigeon pea	343/1435	21.70	0.38	3.45	62.78	15.00
Pink bean	343/1435	21.00	1.13	3.66	64.20	12.70
Pinto bean	347/1452	21.42	1.23	3.46	62.55	15.50
Red kidney bean	337/1410	22.53	1.06	3.37	61.29	15.20
White bean	333/1393	23.40	0.85	4.20	60.30	15.20
Average	344/1438	23.10	1.86	3.50	60.76	15.55

Source: USDA (2022)

amylose and amylopectin. It contains a good amount of dietary fibers (5–37%) consisting of both soluble and insoluble fiber (Pittaway et al. 2007) and low fats averaging 1.86 g per 100 g. Insoluble fiber allows easier movement of substances in the digestive system by improving laxation and also boasts a low glycemic index (GI) owing to a higher concentration of ingestible carbohydrates, facilitating stable blood sugar levels, mitigating the risk of diabetes and stimulating prebiotics (*bifidobacteria* and *lactobacilli*) activity in the colon. Water-insoluble fibers and polyphenols are found in pulse grain hulls, while resistant starch, i.e., oligosaccharides and soluble fibers are found in cotyledons (Singh et al. 2017).

Minerals present in legumes (iron, zinc, calcium, selenium, phosphorus, potassium, and magnesium) act as a cofactor of various antioxidative enzymes, while adequate vitamins (pyridoxine, folate, thiamine, niacin, and riboflavin) are also present in it. Bioactive substances which are the non-nutrient components of food occurring in small amounts and have metabolic properties, imparting defense mechanisms in plants against stress and predators are present in the pulses like bioactive phytochemicals mainly polyphenols, including flavonoids, phenolic acids, lignans, and stilbenes (Singh et al. 2016) concentrated in the seed coat, which is removed by hulling. Phytosterol, another important bioactive phytochemical, is the sterol compound that reduces the level of blood cholesterol. It is found as part of the cell membrane of plants and also in the seeds of pea, lentil, chickpea, etc. Anti-nutritional factors which are also bioactive substances found in pulses constituting lectins, alkaloids, phytates, polyphenols, enzyme inhibitors, and oxalates have adverse effects on human health. It can, however, be reduced with the help of techniques like germination, autoclaving, soaking, dehulling, and cooking.

9.11 GHGs Emission

Current agricultural environmental challenges are climate change and heightened GHG emissions, notably N_2O. Agricultural soils contribute 75% of global N_2O emissions, with a warming potential 265 times that of CO_2 (Pachauri et al. 2014). Global nitrogen use efficiency varies from 21% to 41%, leaving 59–79% of unused nitrogen, contributing to environmental pollution (Omara et al. 2019). Research is exploring nitrogen-fixing legumes as a solution to reduce GHG emissions by decreasing reliance on energy-intensive synthetic fertilizers, known for significant carbon dioxide emissions. Legumes naturally fix nitrogen, cutting the reliance on energy-intensive fertilizers and reducing the associated carbon footprint. Stagnari et al. (2017) reported that legume-based systems emit 5–7 times fewer greenhouse gases, including nitrous oxide and carbon dioxide, per unit area than other field crop-based systems. It was found that maize–peanut intercropping could reduce soil N_2O emissions by 13% and CO_2 emissions by 23.4% compared with maize monoculture (Ma et al. 2022). In the context of maize–soybean intercropping, Chen et al. (2021) observed a reduction of 32.0% and 47.8% in soil N_2O emissions compared to monocultures of maize and soybean, respectively. Dyer et al. (2012) reported that maize–soybean intercropping reduced total soil CO_2 emissions by 18.1–20.4% compared to maize and soybean monocultures.

9.12 Resource Use Efficiency

More plant species diversity is believed to enhance ecosystem functionality by increasing the probability of plants having complementary resource requirements within the community (Barry et al. 2019). The rise in yield in intercropping may result from the complementarity effect, enabling more efficient utilization of resources (Hauggaard-Nielsen et al. 2012). Compared to traditional monoculture, integrated intercropping increased annual profit by 15.6–49.9%, net agricultural income by 39.2%, and reduced the environmental footprint by 17.3% (Yao et al. 2023). The integration of chickpeas with wheat enhanced both land-use efficiency and overall productivity (Banik et al. 2006). Adding mungbean to a maize–wheat system boosted productivity by 25% and net returns by 28% compared to the maize and wheat monoculture (Jat et al. 2018). In maize/soybean and maize/peanut intercropping systems, the land equivalent ratio (LER) based on yield exceeded or approached 1.0, indicating yield advantages compared to monocultures (Zhang et al. 2020). In India, rice–wheat–summer mungbean system outperformed over rice–wheat system, showing the highest net returns, system productivity, and energy efficiency (Pooniya et al. 2018).

9.13 Challenges and Considerations

Incorporating legumes into the cropping system requires careful planning. To optimize benefits and minimize risks, the preceding crop should efficiently absorb nitrogen, legume crops spaced for disease prevention, and the succeeding crop should rapidly use legume residual nitrogen. Management depends on climate and soil conditions, emphasizing site-specific designs (Döring. 2015). Agronomic challenges include crop establishment and maintaining one-third of legumes in grass-clover mixtures (Suter et al. 2015). Other disadvantages include delayed spring growth (Peyraud et al. 2009), reduced persistence compared to grass during grazing, the potential for livestock bloat, and difficulties in conservation as hay or silage (Phelan et al. 2015).

Policy intervention is needed as there is a lack of reliable information that exists among many farmers regarding suitable legume crops and their integration into farming systems. This is attributed to diminishing on-farm technical knowledge about legumes and limited progress in research. While legumes can be profitable for farmers, alternative crops often yield higher net revenue. Consequently, legumes are underutilized in farming systems, creating positive economic externalities. Addressing these calls for policies aimed at expanding the cultivation of legumes

(Kuhlman et al. 2017). The significant environmental benefits of growing legumes, coupled with their declining production, warrant policy intervention. Economic modeling can help design effective policies by simulating scenarios and assessing their impact on the environment and economy, comparing outcomes to a counterfactual without the proposed policy. In their study, Kuhlman et al. (2017) utilized the Common Agricultural Policy Regional Impact (CAPRI) model to simulate policy effects and compare them with the outcomes in the absence of policy action. Financial incentives are just one way to influence farmers. Advancements in legume research, applied to local conditions, can increase their appeal. Policies supporting research and collaboration with farmers can bridge the profitability gap between legumes and alternative crops.

9.14 Conclusion

Legumes actively improve soil health, especially in nitrogen dynamics and environmental sustainability. Grain legumes play a vital role in maintaining soil fertility and nutrition. Bridging awareness gaps and exploring integrated management practices are essential to understand their lasting positive impacts on soil properties and functions. The synthesis of research findings from diverse geographical contexts highlights the multifaceted advantages of crop diversification with legumes. Beyond optimizing yields and mitigating risks, this approach offers tangible environmental benefits, including reduced nutrient losses, improved resource efficiency, and increased stability in production systems. The relevance of legumes in boosting the efficiency of cropping systems is obvious in their capacity to inhibit diseases, offer alternative pest management, and allow nitrogen retention. The increasing worldwide need for plant-based protein has led to the recognition of legumes as a viable and sustainable alternative. This demonstrates that agricultural techniques may successfully balance economic objectives with environmental responsibility. Essentially, including legumes in cropping systems signifies a transformative transition toward a more resilient and ecologically aware approach to agriculture. By embracing crop diversification, we pave the way for a sustainable future, ensuring the continued productivity of our agricultural systems while safeguarding the health of our planet for future generations.

References

Abdalla M, Hastings A, Cheng K, Yue Q, Chadwick D, Espenberg M, Truu J, Rees RM, Smith P (2019) A critical review of the impacts of cover crops on nitrogen leaching, net greenhouse gas balance and crop productivity. Glob Chang Biol 25(8):2530–2543. https://doi.org/10.1111/gcb.14644

Ae NB, Shen RF (2002) Root cell-wall properties are proposed to contribute to phosphorus (P) mobilization by groundnut and pigeon pea. Plant Soil 245:95–103. https://doi.org/10.1023/A:1020669326602

Alegre J, Alonso-Bl'azquez N, de Andr'es EF, Tenorio JL, Ayerbe L (2004) Revegetation and reclamation of soils using wild leguminous shrubs in cold semiarid Mediterranean conditions: Litterfall and carbon and nitrogen returns under two aridity regimes. Plant Soil 263:203–212

Ali MO, Sarker A (2017) Pulses cultivation as relay cropping in the rice field: a technique for intensification of pulses production in the medium high to medium low lands of Bangladesh. J Food Microbiol Saf Hyg 2:2. https://doi.org/10.4172/2476-2059-C1-002

Alletto L, Celette F, Drexler D, Plaza-Bonilla D, Reckling M (2022) Editorial: crop diversifcation, a key pillar for the agroecological transition. Front Agron 4:950822. https://doi.org/10.3389/fagro.2022.950822

Angus JF, Kirkegaard JA, Hunt JR, Ryan MH, Ohlander L, Peoples MB (2015) Break crops and rotations for wheat. Crop Pasture Sci 66:523–552. https://doi.org/10.1071/CP14252

Avasthe RK, Babu S, Singh R, Yadav GS, Kumar A (2020) Productivity and profitability assessment of organically grown vegetables embedded in rice-based cropping sequences in Sikkim Himalayas, north East India. J Environ Biol 41:111–117

Babu S, Singh R, Avasthe RK, Yadav GS, Rajkhowa DJ (2016) Intensification of maize (*Zea mays*)–based cropping sequence in rainfed ecosystem of Sikkim Himalayas for improving system productivity, profitability, employment generation and energy-use efficiency under organic management condition. Ind J Agril Sci 86(6):778–784

Babu S, Mohapatra KP, Yadav GS, Lal R, Singh R, Avasthe RK, Das A, Chandra P, Gudade BA, Kumar A (2020a) Soil carbon dynamics in diverse organic land use systems in north eastern Himalayan ecosystem of India. Catena 194:104785. https://doi.org/10.1016/j.catena.2020.104785

Babu S, Mohapatra KP, Das A, Yadav GS, Tahasildar M, Singh R, Panwar AS, Yadav V, Chandra P (2020b) Designing energy-efficient, economically sustainable, and environmentally safe cropping system for the rainfed maize–fallow land of the eastern Himalayas. Sci Total Environ 722, 20 June 2020:137874. https://doi.org/10.1016/j.scitotenv.2020.137874

Babu S, Das A, Singh R, Mohapatra KP, Kumar S, Rathore SS, Yadav SK, Yadav P, Ansari MA, Panwar AS, Wani OA, Singh M, Ravishankar N, Layek J, Chandra P, Singh VK (2022) Designing an energy efficient, economically feasible, and environmentally robust integrated farming system model for sustainable food production in the Indian Himalayas. Sustain Food Technol. https://doi.org/10.1039/d2fb00016d

Babu S, Singh R, Avasthe RK, Kumar S, Rathore SS, Singh VK, Ansari MA, Valente D, Petrosillo I (2023a) Soil carbon dynamics under organic farming: impact of tillage and cropping diversity. Ecol Indic 147:109940. https://doi.org/10.1016/j.ecolind.2023.109940

Babu S, Singh R, Avasthe RK, Rathore SS, Kumar S, Das A, Layek J, Sharma V, Wani OA, Singh VK (2023b) Conservation tillage and diversified cropping enhance system productivity and ecofficiency and reduce greenhouse gas intensity in organic farming. Front Sustain Food Syst 7:1114617. https://doi.org/10.3389/fsufs.2023.1114617

Bana RS, Gautam RC, Rana KS (2012) Effect of different organic sources on productivity and quality of pearlmillet and their residual effect on wheat. Ann Agric Res 33(3):126–130

Banik P, Midya A, Sarkar BK, Ghose SS (2006) Wheat and chickpea intercropping systems in an additive series experiment: advantages and weed smothering. Eur J Agron 24:325–332

Barbieri P, Pellerin S, Seufert V (2019) Changes in crop rotations would impact food production in an organically farmed world. Nat Sustain 2:378–385

Barry KE, Mommer L, van Ruijven J, Wirth C, Wright AJ, Bai Y, Connolly J, De Deyn GB, De Kroon H, Isbell F, Milcu A (2019) The future of complementarity: disentangling causes from consequences. Trends Ecol Evol 34(2):167–180

Beillouin D, Ben-Ari T, Malézieux E, Seufert V, Makowski D (2021) Positive but variable effects of crop diversification on biodiversity and ecosystem services. Glob Chang Biol 27(19):4697–4710. https://doi.org/10.1111/gcb.15747

Benchaar C, Pomar C, Chiquette J (2001) Evaluation of dietary strategies to reduce methane production in ruminants: a modelling approach. Can J Anim Sci 81(4):563–574

Ben-chuan ZH, Ying ZH, Ping CH, Xiao-na ZH, Qing DU, Huan YA, Xiao-chun WA, Feng YA, Te XI, Long LI, Wen-yu YA (2022) Maize–legume intercropping promote N uptake through changing the root spatial distribution, legume nodulation capacity, and soil N availability. J Integr Agric 21(6):1755–1771

Bilalis D, Papastylianou P, Konstantas A, Patsiali S, Karkanis A, Efthimiadou A (2010) Weed-suppressive effects of maize–legume intercropping in organic farming. Int J Pest Manag 56(2):173–181

Buragohain S, Sharma B, Nath JD, Gogaoi N, Meena RS, Lal R (2017) Impact of ten years of biofertilizer use on soil quality and rice yield on an inceptisol in Assam. India Soil Res 56(1):49–58. https://doi.org/10.1071/SR17001

Bybee-Finley KA, Mirsky SB, Ryan MR (2016) Functional diversity in summer annual grass and legume intercrops in the northeastern United States. Crop Sci 56:2775–2790

Chen JS, Amin AS, Hamani AKM, Wang GS, Zhang YY, Liu K, Gao Y (2021) Effects of maize (zea mays l.) intercropping with legumes on nitrous oxide (N_2O) emissions. Appl Ecol Environ Res 19(5):3393

Clements R, Haggar J, Quezada A, Torres J (2011) Technologies for climate change adaptation—agriculture sector. In: Zhu X (Ed) UNEP Risø Centre, Roskilde, 2011. http://techaction.org

Conceicao PC, Diekow J, Bayer C (2013) Combined role of no-tillage and cropping systems in soil carbon stocks and stabilization. Soil Tillage Res 129:40–47. https://doi.org/10.1016/j.still.2013.01.006

Costa MP, Reckling M, Chadwick D, Rees RM, Saget S, Williams M, Styles D (2021) Legume-modified rotations deliver nutrition with lower environmental impact. Front Sustain Food Syst 5:656005. https://doi.org/10.3389/fsufs.2021.656005

Crozat Y, Fustec J (2006) Assessing the role of grain legumes in crop rotation: some agronomic concepts that can help. In: Grain legumes and the environment: how to assess benefits and impacts. Proceedings of the AEP workshop (ed). AEP, 18–19 November 2004, Zurich, Switzerland, pp 55–60

Das A, Babu S, Datta M, Kumar S, Singh R, Avasthe RK, Rathore SS, Yadav SK, Singh VK (2022a) Restoring soil carbon in marginal land of Indian Himalayas: impact of crop intensification and conservation tillage. J Environ Manag 318:115603. https://doi.org/10.1016/j.jenvman.2022.115603

Das A, Babu S, Singh R, Kumar S, Rathore SS, Singh VK, Datta M, Yadav SK, Wani OA, Yadav D (2022b) Impact of live mulch-based conservation tillage on soil properties and productivity of summer maize in Indian Himalayas. Sustainability 14:12078. https://doi.org/10.3390/su141912078

Dhull SB, Kinabo J, Uebersax MA (2022) Nutrient profile and effect of processing methods on the composition and functional properties of lentils (Lens culinaris Medik): a review. Legume Sci 5(1):156

Diekow J, Mileniczuk J, Knicker H, Bayer C, Dick DP, Kögel-Knabner I (2005) Soil C and N stocks affected by cropping systems and nitrogen fertilization in a southern Brasil Acrisol managed under no-tillage for 17 years. Soil Tillage Res 81:87–95

Döring TF (2015) Grain legume cropping systems in temperate climates. In: De Ron AM (ed) Grain legumes. Handbook of plant breeding. Springer Science+Business Media, New York, pp 401–434

Dreschel P, Glaser B, Zech W (1991) Effect of four multipurpose tree fallow in Central Togo. Agrofor Syst 16:193–202

Drinkwater LE, Wagoner P, Sarrantonio M (1998) Legume-based cropping systems have reduced carbon and nitrogen losses. Nature 396:262–265

Dyer L, Oelbermann M, Echarte L (2012) Soil carbon dioxide and nitrous oxide emissions during the growing season from temperate maize-soybean intercrops. J Plant Nutr Soil Sci 175:394–400

Erismann JW, Sutton MA, Galloway J, Klimont Z, Winiwarter W (2008) How a century of ammonia synthesis changed the world. Nat Geosci 1:636–639. https://doi.org/10.1038/ngeo325

Franke AC, Van den Brand GJ, Vanlauwe B, Giller KE (2018) Sustainable intensification through rotations with grain legumes in sub-Saha-ran Africa: a review. Agric Ecosyst Environ 261:172–118

Fu Z, Chen P, Zhang X, Du Q et al (2023) Maize-legume intercropping achieves yield advantages by improving leaf functions and dry matter partition. BMC Plant Biol 23:438

Fustec J, Lesuffleur F, Mahieu S, Cliquet JB (2010) Nitrogen rhizodeposition of legumes. A review. Agron Sustain Dev 30:57–66

Géneau CE, Wäckers FL, Luka H, Daniel C, Balmer O (2012) Selective flowers to enhance biological control of cabbage pests by parasitoids. Basic Appl Ecol 13(1):85–93

Gibson KE, Kobayashi, Walker GC (2008) Molecular determinants of a symbiotic chronic infection. Annu Rev Genet 42:413–441. https://doi.org/10.1146/annurev.genet.42.110807.091427

Giller KE (2015) Nitrogen fixation in tropical cropping systems, 2nd edn. CABI, Wallingford, p 423

Gordon AJ, Lea PJ, Rosenberg C, Trinchant JC (2001) Nodule formation and function. In: Lea PJ, Morot-Gaudry JF (eds) Plant nitrogen. SpringerVerlag, Berlin/Heidelberg/New York, pp 101–146

Graham PH, Vance CP (2000) Nitrogen fixation in perspective: an overview of research and extension needs. Field Crop Res 65:93–106

Guan XK, Turner NC, Song L, Gu YJ, Wang TC, Li FM (2016) Soil carbon sequestration by three perennial legume pastures is greater in deeper soil layers than in the surface soil. Bio Geosci 13:527–534. https://doi.org/10.5194/bg-13-527-2016

Hajduk E, Właśniewski S, Szpunar-Krok E (2015) Influence of legume crops on content of organic carbon in sandy soil. Soil Sci Annu 66:52–56

Hauggaard-Nielsen H, Mundus S, Jensen ES (2012) Grass-clover undersowing affects nitrogen dynamics in a grain legume–cereal arable cropping system. Field Crop Res 136:23–31

Hufnagel J, Reckling M, Ewert F (2020) Diverse approaches to crop diversifcation in agricultural research a review. Agron Sustain Dev 40(2):14. https://doi.org/10.1007/s13593-020-00617-4

Impala (2001) First annual report of IMPALA project, covering the period October 2000–December 2001, Nairobi, Kenya, 68pp

IPES-Food (2016) From uniformity to diversity: a paradigm shift from industrial agriculture to diversifed agroecological systems. International Panel of Experts on Sustainable Food systems

Jangir CK, Thakur A, Bijani H, Thakur P, Kumar S, Meena RS, Bedwal S, Rani K, Shukla UN, Meena AK, Dev P (2022) Residual nitrogen for succeeding crops in legume-based cropping system. In: Advances in legumes for sustainable intensification. Academic Press, Cambridge, pp 113–132

Jat RD, Jat HS, Nanwal RK, Yadav AK, Bana A, Choudhary KM, Kakraliya SK, Sutaliya JM, Sapkota TB, Jat ML (2018) Conservation agriculture and precision nutrient management practices in maize-wheat system: effects on crop and water productivity and economic profitability. Field Crop Res 222:111–120

Jensen CR, Joernsgaard B, Andersen MN, Christiansen JL, Mogensen VO, Friis P, Petersen CT (2004) The effect of lupins as compared with peas and oats on the yield of the subsequent winter barley crop. Eur J Agron 20:405–418

Jensen ES, Peoples MB, Boddey RM, Gresshoff PM, Hauggaard-Nielsen H, Alves BJR, Morrison MJ (2012) Legumes for mitigation of climate change and the provision of feedstock for biofuels and biorefineries. A review. Agron Sustain Dev 32:329–364

Jones KM, Kobayashi H, Davies BW, Taga ME, Walker GC (2007) How rhizobial symbionts invade plants: the Sinorhizobium–Medicago model. Nat Rev Microbiol 5:619–633. https://doi.org/10.1038/nrmicro1705

Juma KA, Van Averbeke W (2005) Response of Muxe to N and P availability in pots. In African crop science conference proceedings, vol 7, no. pt. 03 of 03, pp 1179–1182

Kazula MJ, Lauer JG, Arriaga FJ (2017) Crop rotation effect on selected physical and chemical properties of Wisconsin soils. J Soil Water Conserv 72:553–563

Kirkegaard J, Christen O, Krupinsky J, Layzell D (2008) Break crop benefits in temperate wheat production. Field Crop Res 107(3):185–195

Kremen C, Iles A, Bacon C (2012) Diversifed farming systems: an agroecological, systems-based alternative to modern industrial agriculture. Ecol Soc 17(4):44. https://doi.org/10.5751/ES-05103-170444

Kuhlman T, Helming J, Linderhof V (2017) Optimizing legume cropping: the policy questions. In: Legumes in cropping systems. CABI, Wallingford, pp 226–243

Kumar S, Meena RS, Lal R, Yadav GS, Mitran T, Meena BL, Dotaniya ML, EL-Sabagh A (2018) Role of legumes in soil carbon sequestration. In: Meena RS et al (eds) Legumes for soil health and sustainable management. Springer, Singapore, pp. 109–138. https://doi.org/10.1007/978-981-13-0253-4_4

Kumawat A, Bamboriya SD, Meena RS, Yadav D, Kumar A, Kumar S, Raj A, Pradhan G (2022) Legume-based inter-cropping to achieve the crop, soil, and environmental health security. In: Advances in legumes for sustainable intensification. Academic Press, pp 307–328

Ladha JK, Jat ML, Stirling CM, Chakraborty D, Pradhan P, Krupnik TJ, Sapkota TB, Pathak H, Rana DS, Tesfaye K, Gerard B (2020) Achieving the sustainable development goals in agriculture: the crucial role of nitrogen in cereal-based systems. Adv Agron 163:39–116

Lal R (2015a) Sequestering carbon and increasing productivity by conservation agriculture. J Soil Water Conserv 70:55A–62A

Lal R (2015b) Restoring soil quality to mitigate soil degradation. Sustainability 7(5):5875–5895 https://doi.org/10.3390/su7055875

Lal R (2019) Carbon cycling in global drylands. Curr Clim Chang Rep 5:221–232

Liu CW, Murray JD (2016) The role of flavonoids in nodulation host-range specificity: an update. Plan Theory 5:E33. https://doi.org/10.3390/plants5030033

Liu JB, Zhang YQ, Wu B, Qin SG, Jia X, Feng W (2013) Changes in soil carbon, nitrogen and phosphorus along a chronosequence of Caragana microphylla plantation, north western China. Pol J Environ Stud 23(2):385–391

Liu K, Johnson EN, Blackshaw RE, Hossain Z, Gan Y (2019) Improving the productivity and stability of oilseed cropping systems through crop diversification. Field Crop Res 237:65–73

Ma HY, Zhou J, Ge JY, Nie JW, Zhao J, Xue ZQ, Hu YG, Yang YD, Peixoto L, Zang HD, Zeng ZH (2022) Intercropping improves soil ecosystem multifunctionality through enhanced available nutrients but depends on regional factors. Plant Soil 480:71–84

Martin M, Dos R, Angersd A, Coraj E (2012) Carbohydrate composition and water-stable aggregation of an Oxisol as affected by crop sequence under no-till. Soil Sci Soc Am J 76:475–484. https://doi.org/10.2136/sssaj2011.0110

Meena RS, Meena VS, Meena SK, Verma JP (2015) The needs of healthy soils for a healthy world. J Clean Prod 102:560–565

Mesfin S, Gebresamuel G, Haile M, Zenebe A (2023) Potentials of legumes rotation on yield and nitrogen uptake of subsequent wheat crop in northern Ethiopia. Heliyon 9(6):e16126

Messéan A, Viguier L, Paresys L, Aubertot J-N, Canali S, Iannetta PJ, Eric KA, Keillor B, Kemper L, Muel F, Pancino B, Stilmant D, Watson C, Willer H, Zornoza R (2021) Enabling crop diversifcation to support transitions towards more sustainable European agrifood systems. Front Agric Sci Eng 8(3):474–480

Mitran T, Meena RS, Lal R, Layek J, Kumar S, Datta R (2018) Role of soil phosphorus on legume production. In: Meena et al (eds) Legumes for soil health and sustainable management. Springer. https://doi.org/10.1007/978-981-13-0253-4_15

Mortensen MC, Schuman GE, Ingram LJ (2004) Carbon sequestration in rangelands inter seeded with yellow-flowering alfalfa (Medicago sativa ssp. Falcate). Environ Manag 33:S475–S481

Moyer-Henry KA, Burton JW, Israel DW, Rufty TW (2006) Nitrogen transfer between plants: a 15N natural abundance study with crop and weed species. Plant Soil 282:7–20

Naudin C, Corre-Hellou G, Voisin A-S, OuryV SC, Crozat Y, Jeuffroy M-H (2011) Inhibition and recovery of symbiotic N2 fixation by peas (*P. Sativum* L.) in response to short-term nitrate exposure. Plant Soil 346:275–287. https://doi.org/10.1007/s11104-011-0817-8

Nyawade SO, Karanja NN, Gachene CK, Gitari HI, Schulte-Geldermann E, Parker M (2020) Optimizing soil nitrogen balance in a potato cropping system through legume intercropping. Nutr Cycl Agroecosyst 117:43–59

Oldfield EE, Bradford MA, Wood SA (2019) Global meta-analysis of the relationship between soil organic matter and crop yields. Soil 5(1):15–32

Omara P, Aula L, Oyebiyi F, Raun WR (2019) World cereal nitrogen use efficiency trends: review and current knowledge. Agrosyst Geosci Environ 2(1):1–8

Paas W, Coopmans I, Severini S, van Ittersum MK, Meuwissen MPM, Reidsma P (2021) Participatory assessment of sustainability and resilience of three specialized farming systems. Ecol Soc 26(2):2. https://doi.org/10.5751/ES-12200-260202

Pachauri, RK, Allen MR, Barros VR et al. (2014) Climate change 2014: synthesis report. Contribution of working groups I, II and III to the fifth assessment report of the intergovernmental panel on climate change. IPCC, p 151

Pappa VA, Rees RM, Walker RL, Baddeley JA, Watson CA (2011) Nitrous oxide emissions and nitrate leaching in an arable rotation resulting from the presence of an intercrop. Agric Ecosyst Environ 141:153–161

Peoples MB, Giller KE, Jensen ES, Herridge DF (2021) Quantifying country-toglobal scale nitrogen fixation for grain legumes: I. Plant reliance upon nitrogen fixation for soybean, groundnut and pulses. Plant Soil 469:1–14. https://doi.org/10.1007/s11104-021-05167-6

Peyraud JL, Le Gall A, Lüscher A (2009) Potential food production from forage legume-based systems in Europe: an overview. Irish J Agric Food Res 48:115–135

Phelan P, Moloney AP, McGeough EJ, Humphreys J, Bertilsson J, O'Kiely O'REG (2015) Forage legumes for grazing and conserving in ruminant production systems. Crit Rev Plant Sci 34:281–326

Pierre JF, Latournerie-Moreno L, Garruña R, Jacobsen KL, Laboski CA, Us-Santamaría R, Ruiz-Sánchez E (2022) Effect of maize–legume intercropping on maize physio-agronomic parameters and beneficial insect abundance. Sustainability 14(19):12385

Pittaway JK, Ahuja KDK, Robertson IK, Ball MJ (2007) Effects of a controlled diet supplemented with chickpea on serum lipids, glucose tolerance, satiety and bowel function. J Am Coll Nutr 26:334–340

Pivato B, Mazurier S, Lemanceau P, Siblot S, Berta G, Mougel C, Van Tuinen D (2007) Medicago species affect the community composition of arbuscular mycorrhizal fungi associated with roots. New Phytol 176:197–210. https://doi.org/10.1111/j.1469-8137.2007.02151.x

Plaza-Bonilla D, Nolot J-M, Raffaillac D, Justes E (2017) Innovative cropping systems to reduce N inputs and maintain wheat yields by inserting grain legumes and cover crops in southwestern France. Eur J Agron 82:331–341. https://doi.org/10.1016/j.eja.2016.05.010

Poggio SL (2005) Structure of weed communities occurring in monoculture and intercropping of field pea and barley. Agric Ecosyst Environ 109(1–2):48–58

Pooniya V, Shivay YS (2013) Enrichment of basmati rice grain and straw with zinc and nitrogen through ferti-fortification and summer green manuring crops under indo-Gangetic Plains of India. J Plant Nutr 36:91–117

Pooniya V, Shivay YS, Rana A, Nain L, Prasanna R (2012) Enhancing soil nutrient dynamics and productivity of basmati rice through residue incorporation and zinc fertilization. Eur J Agron 41:28–37

Pooniya V, Choudhary AK, Bana RS, Sawarnalaxmi K, Pankaj Rana DS, Puniya MM (2018) Influence of summer legume residue recycling and varietal diversification on productivity, energetics, and nutrient dynamics in basmati rice–wheat cropping system of western indo-Gangetic Plains. J Plant Nutr 41(12):1491–1506

Praharaj CS, Blaise D (2016) Intercropping: an approach for area expansion of pulses. Ind J Agron 61(4):113–121

Praharaj CS, Ali M, Kumar N, Dutta A, Singh U, Singh R (2021) Pulses for crop intensification and sustainable livelihood. Indian J Agron 66(5th IAC Special issue):S60–S72

Preissel S, Reckling M, Schläfke N, Zander P (2015) Magnitude and farm-economic value of grain legume pre-crop benefits in Europe: a review. Field Crop Res 175:64–79

Rani K et al (2019) Legumes for sustainable soil and crop management. In: Meena R, Kumar S, Bohra J, Jat M (eds) Sustainable management of soil and environment. Springer, Singapore. https://doi.org/10.1007/978-981-13-8832-3_6

Raza MA, Wang Z, Yasin HS, Gul H, Qin R, Rehman SU, Mahmood A, Iqbal Z, Ahmed Z, Juan SLC, Gitari XLH, Khalid MHB, Zhongming YFM (2016) Effect of crop combination on yield performance, nutrient uptake, and land use advantage of cereal/legume intercropping systems. Field Crop Res 304(2):109144

Reckling M, Bergkvist G, Watson CA, Stoddard FL, Zander PM, Walker RL, Pristeri A, Toncea I, Bachinger J (2016) Trade-offs between economic and environmental impacts of introducing legumes into cropping systems. Front Plant Sci 7:669

Reckling M, Albertsson J, Vermue A, Carlsson G, Watson CA, Justes E, Bergkvist G, Jensen ES, Topp CFE (2022) Diversifcation improves the performance of cereals in European cropping systems. Agron Sustain Dev 42(6):118. https://doi.org/10.1007/s13593-022-00850-z

Renu, Kumar A, Kumar P (2018) Performance of advance pearl millet hybrids and mungbean under sole cropping and intercropping systems under semi-arid environment. J Pharmacogn Phytochem 7:1671–1675

Rodriguez C, Mårtensson LMD, Jensen ES, Carlsson G (2021) Combining crop diversification practices can benefit cereal production in temperate climates. Agron Sustain Dev 41:48

Sanderson M, Skinner R, Barker D, Edwards G, Tracy B, Wedin D (2004) Plant species diversity and management of temperate forage and grazing land ecosystems. Crop Sci 44:1132–1114

Sanginga N, Woomer PL (2009) Integrated soil fertility management in Africa: principles, practices and development process. Tropical Soil Biology and Fertility Institute of the International Centre for Tropical Agriculture, Nairobi

Sarkar S, Samui I, Brahmachari K, Ray K, Ghosh A, Nanda MK (2019) Management practices for utera pulses in rice-fallow system under coastal saline zone of West Bengal. Ind Soc Coast Agric Res 37(2):98–103

Shah A, Nazari M, Antar M, Msimbira LA, Naamala J, Lyu D et al (2021) PGPR in agriculture: a sustainable approach to increasing climate change resilience. Front Sustain Food Syst 5:667546. https://doi.org/10.3389/fsufs.2021.667546

Singh JP, Kaur A, Singh N, Nim L, Shevkani K, Kaur H, Arora DS (2016) In vitro antioxidant and antimicrobial properties of jambolan (Syzygium cumini) fruit polyphenols. LWT Food Sci Technol 65:1025–1030

Singh B, Singh JP, Shevkani K, Singh N, Kaur A (2017) Bioactive constituents in pulses and their health benefits. J Food Sci Technol 54:858–870

Singh R, Babu S, Avasthe RK, Yadav GS, Das A, Mohapatra KP, Kumar A, Singh VK, Chandra P (2021) Crop productivity, soil health, and energy dynamics of Indian Himalayan intensified organic maize-based systems. Int Soil Water Conserv Res 9(2):260–270. https://doi.org/10.1016/j.iswcr.2020.11.003

Singh AK, Singh JB, Singh R, Kantwa SR, Jha PK, Ahamad S, Singh A, Ghosh A, Prasad M, Singh S, Singh S (2023) Understanding soil carbon and phosphorus dynamics under grass-legume intercropping in a semi-arid region. Agronomy 13(7):1692

Sisti CPJ, Dos Santos HP, Kochhann RA, Alves BJR, Urquiaga S, Boddey RM (2004) Change in carbon and nitrogen stocks in soil under 13 years of conventional or zero tillage in southern Brazil. Soil Tillage Res 76:39–58

Smil V (2004) Enriching the earth: Fritz Haber, Carl Bosch, and the transformation of world food production, p 358, MIT press

Smith CJ, Chalk PM (2020) Grain legumes in crop rotations under low and variable rainfall: are observed short-term N benefits sustainable? Plant Soil 453:271–279

Snapp SS, Swinton SM, Labarta R, Mutch D, Black JR, Leep R, Nyiraneza J, O'neil K (2005) Evaluating cover crops for benefits, costs and performance within cropping system niches. Agron J 97(1):322–332

Srinivasarao C, Venkateswarlu B, Lal R (2012) Long-term effects of soil fertility management on carbon sequestration in a rice-lentil cropping system of the indo-Gangetic plains. Soil Sci Soc Am J 76(1):167–178

Stagnari F, Maggio A, Galieni A, Pisante M (2017) Multiple benefits of legumes for agriculture sustainability: an overview. Chem Biol Technol Agric 4:2

St-Martin A, Vico G, Bergkvist G, Bommarco R (2017) Diverse crop-ping systems enhanced yield but did not improve yield stability in a 52-year long experiment. Agric Ecosyst Environ 247:337–342

Suter M, Connolly J, Finn JA, Loges R, Kirwan L, Sebastià MT, Lüscher A (2015) Nitrogen yield advantage from grass–legume mixtures is robust over a wide range of legume proportions and environmental conditions. Glob Chang Biol 21(6):2424–2438

Swanton CJ, Booth BD, Chandler K, Clements DR, Shrestha A (2006) Management in a modified no-tillage corn–soybean–wheat rotation influences weed population and community dynamics. Weed Sci 54(1):47–58

Tamburini G, Bommarco R, Wanger TC, Kremen C, van der Heijden MGA, Liebman M, Hallin S (2020) Agricultural diversifcation promotes multiple ecosystem services without compromising yield. Sci Adv 6(45):eaba1715. https://doi.org/10.1126/sciadv.aba1715

Tang X, Zhang C, Yu Y, Shen J, van der Werf W, Zhang F (2021) Intercropping legumes and cereals increases phosphorus use efficiency; a meta-analysis. Plant Soil 460:89–104

Tilman D, Cassman KG, Matson PA, Naylor R, Polasky S (2002) Agricultural sustainability and intensive production practices. Nature 418:671–677

Tripathi SC, Venkatesh K, Meena RP, Chander S, Singh GP (2021) Sustainable intensification of maize and wheat cropping system through pulse intercropping. Sci Rep 11:18805

Tuulos A, Yli-Halla M, Stoddard FL, Mäkelä PSA (2014) Winter turnip rape as a soil N scavenging catch crop in a cool humid climate. Agron Sustain Dev 35:359–366

USDA (US Dept of Agriculture) (2022) Food data central. https://fdc.nal.usda.gov/

Varma D, Meena RS, Kumar S, Kumari E (2017) Response of mungbean to NPK and lime under the conditions of Vindhyan region of Uttar Pradesh. Legume Res Int J 40(3):542–545

Vasconcelos MW, Grusak MA, Pinto E, Gomes A, Ferreira H, Balázs B et al (2020) The biology of legumes and their agronomic, economic, and social impact. In: The plant family Fabaceae. Springer, Singapore, pp 3–25. https://doi.org/10.1007/978-981-15-4752-2_1

Wang D, Yang S, Tang F, Zhu H (2012) Symbiosis specificity in the legume–rhizobial mutualism. Cell Microbiol 14:334–342. https://doi.org/10.1111/j.1462-5822.2011.01736

Watson CA, Stoddard FL (2017) Introduction-perspectives on legume production and use in European agriculture. In: Legumes in cropping systems. CABI, Wallingford UK, pp 1–17

Watson C, Reckling M, Preissel S, Bachinger J, Bergkvist G, Lindstrom K, Nemecek T, Topp C, Vanhatalo A, Zander Z, Murphy-Brokern D, Stoddard FL (2017) Grain legume production and use in European agricultural systems. Adv Agron 144:235–303

Webb J, Bellamy P, Loveland PJ, Goodlass G (2003) Crop residue returns and equilibrium soil organic carbon in England and Wales. Soil Sci Soc Am Journ 67:928–936

World Bank (2008) World development report 2008: agriculture for development. World Bank, Washington, DC

Wu G, Fanzo J, Miller D, Pingali P, Post M, Steiner JL, Thalacker-Mercer AE (2014) Production and supply of high-quality food protein for human consumption: sustainability, challenges, and innovations. Ann N Y Acad Sci 1321:1–19

Xing Z, Lien C, Herb W, Rees F, Monteith J, Lionel S (2011) A comparison of effects of one-pass and conventional potato hilling on water runoff and soil erosion under simulated rainfall. Can J Soil Sci 91:279–290

Yadav GS, Subhash B, Das A, Datta M, Mohapatra KP, Singh R, Singh VK, Rathore SS, Chakraborty M (2021) Productivity, soil health, and carbon management index of Indian Himalayan intensified maize-based cropping systems under live mulch-based conservation tillage practices. Field Crop Res 264:108080. https://doi.org/10.1016/j.fcr.2021.108080

Yan Z, Zhou J, Liu C, Jia R, Mganga KZ, Yang L, Yang Y, Peixoto L, Zang H, Zeng Z (2023) Legume-based crop diversification reinforces soil health and carbon storage driven by microbial biomass and aggregates. Soil Tillage Res 234:105848

Yao F, Wu Y, Liu X, Cao Y, Lv Y, Wei W, Xu W, Liu Z, Liang J, Wang Y (2023) Research progress and development trends of greenhouse gas emissions from Cereal–Legume intercropping systems. Agronomy 13(4):1085

Young RR, Wilson B, Harden S, Bernardi A (2009) Accumulation of soil carbon under zero tillage cropping and perennial vegetation on the Liverpool Plains, eastern Australia. Aust J Soil Res 47:273–285

Zhang J, Van der Heijden MGA, Zhang F, Bender SF (2020) Soil biodiversity and crop diversifcation are vital components of healthy soils and agricultural sustainability. Front Agr Sci Eng 7(3):236–242. https://doi.org/10.15302/j-fase-2020336

Zhao J, Chen J, Beillouin D, Lambers H, Yang Y, Smith P, Zeng Z, Olesen JE, Zang H (2022) Global systematic review with meta-analysis reveals yield advantage of legume-based rotations and its drivers. Nat Commun 13(1):4926

Chapter 10
Agroforestry for Sustainable Livelihood and Nutritional Security

Savita Dewangan, Alok Kumar Singh, Bipin Kumar Singh, and Saurbh Shukla

Abstract Agroforestry systems are present across agricultural landscapes, providing various products such as food, fuel wood, fiber, fodder, and more. Besides these goods, agroforestry delivers crucial ecosystem services that enhance livelihoods. By combining trees and/or shrubs with crops and/or livestock, agroforestry diversifies both agricultural and nonagricultural activities. This method, which involves integrating trees, plants, and livestock on the same land, offers a holistic approach to addressing urgent issues like food security, rural livelihoods, and environmental sustainability. Agroforestry is a well-managed land-use system where agricultural crops are grown alongside woody perennials on the same plot. While modern advancements have improved agroforestry techniques, it is essential to recognize and build upon traditional methods and indigenous knowledge. For agroforestry to be fully effective, collaboration among legislators, scientists, extension agents, farmers, and private sector businesses is necessary. Institutional and governmental frameworks must be established to promote adoption, provide incentives, and create a supportive environment. In conclusion, agroforestry is a promising strategy for improving rural livelihoods, achieving food security, and promoting environmental sustainability. Embracing agroforestry principles and practices will be crucial for

S. Dewangan (✉)
Department of Agronomy, Institute of Agricultural Sciences, Banaras Hindu University, Varanasi, Uttar Pradesh, India
e-mail: savita87bhu@bhu.ac.in

A. K. Singh
Rajiv Gandhi South Campus, Banaras Hindu University, Mirzapur, Uttar Pradesh, India

B. K. Singh
Department of Agroforestry, Acharya Narendra Deva University of Agriculture and Technology, Ayodhya, Uttar Pradesh, India

S. Shukla
Department of Silviculture and Agroforestry, Dr. Yashwant Singh Parmar University of Horticulture and Forestry, Solan, Himachal Pradesh, India

S. Babu et al. (eds.), *Agricultural Diversification for Sustainable Food Production*, Sustainability Sciences in Asia and Africa,
https://doi.org/10.1007/978-981-97-7517-0_10

ensuring a more resilient and equitable future as we face the challenges of climate change and increasing demands for food and resources.

Keywords Agroforestry · Environmental sustainability · Food security · Livelihood · Policy

10.1 Introduction

Worldwide, 3.83 billion individuals reside in households that depend on agri-food system-based livelihoods, supporting a workforce of nearly 1.23 billion people (Davis et al. 2023). In a world marked by growing environmental concerns and increased pressure on the earth's natural resources, the quest for sustainable and harmonious solutions to address the intricate balance between human livelihood and environmental security is of paramount importance (Rathore et al. 2022). This increased human population may require the production of more food grains with limited available resources to satisfy their nutritional needs (Foley et al. 2005; Sirohi et al. 2022; Yadav et al. 2023). Hunger and poverty represent critical global challenges, leading to malnutrition and increased reliance on food imports. To ensure food security, it is imperative to reform production, consumption, and distribution patterns (Gómez et al. 2022). With the limited availability of land, it is not possible to boost agricultural production without the diversification of land use. The contemporary challenge of climate variability and change constitutes a pivotal factor influencing livelihoods and food security (Okunlola et al. 2019; Kumar et al. 2022a). In this context, agroforestry emerged as a beacon of hope, offering a multifaceted and innovative approach to tackle the complex web of issues surrounding livelihood and environmental security (Pretty 2018; Castle et al. 2021; Rathore et al. 2022; Wani et al. 2023). The promotion of intensive farming practices in the preceding century, characterized by the utilization of external inputs, has led to a considerable carbon footprint and the degradation of agricultural ecosystems (Rigueiro-Rodri-Guez et al. 2009; Buttould 2013; Yadav et al. 2021; Das et al. 2022). Presently, there is an expectation for agricultural and forestry systems to maximize the provision of ecosystem services, encompassing those associated with provisioning, regulating, and cultural aspects, in accordance with the specifications outlined by the Common International Classification of Ecosystem Services (Haines-Young 2016). The potential of agroforestry as a land-use practice to resolve multiple issues viz. ecological, environmental, and socioeconomic has also been recognized by Intergovernmental Panel on Climate Change (IPCC) in its third assessment report (McCarthy et al. 2001; Kumar et al. 2022a). Agroforestry as a land-use system is a multifunctional, sustainable agricultural production system that besides meeting the production needs of farmers also provides ecosystem services and has the potential to mitigate climate change (Kumar et al. 2022b; Bhatia et al. 2023; Bishaw et al. 2022). Agroforestry represents a meticulously managed agricultural land utilization system that involves the concurrent cultivation of agricultural crops alongside woody perennials within the same land unit. This integrated approach aims to create a more productive, diverse, and sustainable land-use system that not only enhances landowner objectives but also fosters overall land health and profitability (Diwakar et al. 2021; Bhatia et al. 2023). The components of agroforestry, including both woody and non-woody elements, can be strategically arranged either spatially through practices such as intercropping annual crops with trees or temporally by implementing crop rotation with trees (Fouladbash and Currie 2015). The effectiveness of agroforestry as a land-use practice hinges on the judicious selection of crop combinations that not only meet the production needs of farmers but also prove to be sustainable by preserving the resource base (Nair 1993; Babu et al. 2020). Moreover, the adoption of agroforestry by farmers not only increases agricultural yields but also enhances diversification. This diversification becomes especially vital in extreme cases, such as droughts and floods, where trees exhibit greater resilience (Araya et al. 2023; Wani et al. 2023). Research indicated that more than 10% of tree cover is present on half of the world's cultivated lands, which amounts to approximately 1 billion hectares of agricultural land (Garrity 2012). The agroforestry systems have greater production, protection, and economic benefits over monocropping of trees and crops (Kumar et al. 2021). Agroforestry provides diverse agricultural products (fruits, grain, nuts, etc.), timber, and various NTFPs. For example, enset-coffee homegardens in southern Ethiopia, specifically in the Sidama and Gedeo communities, feature the *Ensete ventricosum* plant, a perennial that acts as a primary food source for approximately 15 million people in the area (Abebe 2013). The pearl millet was found to be three times greater beneath the canopy of trees compared to locations located at a distance of five tree-crown radii (Roupsard et al. 2020). They provide a range of services viz. biodiversity conservation, soil health improvement, carbon sequestration, water quality and quantity, microclimatic regulation, pollination services, livestock forage, income diversification, and rural employment (Bhardwaj et al. 2023). The carbon sequestration potential of agroforestry systems ranged from 0.29–15.2 Mg C/ha/year and 30–300 Mg C/ha/year in aboveground biomass and in soil (up to 1 m depth), respectively (Nair et al. 2009). In the next 50 years, practicing sustainable agroforestry can store carbon up to 2.2 Pg (Lorenz and Lal 2014). The use of woody perennials in agricultural land-use systems increases soil microbial diversity and activity, accumulates soil organic C quickly based on the quantity of litter and rate of decomposition (Aguiar et al. 2010), reduces runoff and soil's hydraulic properties, and thus enhances the nutrient cycling (Udawatta et al. 2017). The carbon stock in agroforestry systems varied across the regions due the various reasons like type of plants, age, soil and climatic conditions, etc. (Table 10.1). The World Forestry Congress, New Delhi recognizes the role of agroforestry in the eradication of hunger, increasing biodiversity, and mitigation and adaptation to climate change (Khasa et al. 2017). Agroforestry is officially recognized as a prospective land-use strategy for mitigating climate change. This acknowledgment is explicitly articulated in Article 3.3 of the Kyoto Protocol, where afforestation and reforestation, incorporating agroforestry practices, are identified as mechanisms for the reduction of greenhouse gas emissions (Singh et al. 2023).

Table 10.1 Biomass and soil carbon stocks in different agroforestry systems

Agroforestry system	Study area	Biomass carbon stock (Mg/ha)	Soil carbon stock (Mg/ha)	Depth of soil C stock determination (cm)	References
Homegarden agroforestry	Ethiopia	27.4	130.37	0–60	Betemariyam et al. (2020)
Coffee-based agroforestry	Ethiopia	63.1	131.86	0–60	Betemariyam et al. (2020)
Horti-Silvipastoral	India	29.94	–	–	Saleem et al. (2023)
Parklands	Northern Ethiopia	7.79	71.69	0–60	Manaye et al. (2021)
Woodlots	Northern Ethiopia	31.12	96.95	0–60	Manaye et al. (2021)
Dispersed trees	Mexico	12.6–34.8	144.3–182.5	0–30	Aryal et al. (2022)
Parkland agroforestry	Ethiopia	8.34	51.31	0–40	Tsedeke et al. (2021)
Rhamnus prinoides based agroforestry	Northern Ethiopia	13.82	121.9	0–60	Gebremeskel et al. (2021)
Danielliaoliveri agroforestry parklands	Cameroon	85.44–236.55	–	–	Dangai et al. (2021)
Agri-horticulture	Northern Himalayas	8.05–81.68	43.67–151.7	–	Panwar et al. (2022)
Agri-horticulture	Eastern Himalayas	–	52.8–71.6	0–100	Yadav et al. (2021)
Alder + large cardamom	Northern Himalayas	–	140.7	0–45	Babu et al. (2020)

10.2 Agroforestry as a Diversification Strategy

The term “diversification” originates from the word “diverge,” signifying the action of moving or extending in a direction distinct from a shared point (Jha et al. 2009). Agricultural diversification refers to the strategy of expanding range of crops, livestock, or agricultural activities within a farming system to reduce risk, improve sustainability, and enhance overall resilience (Deogharia 2018). The diversification of agriculture is necessary to boost income and employment opportunities, alleviate disparities across different regions and time periods, and mitigate the degradation of natural resources (Joshi et al. 2004; Das et al. 2022). The need for diversification arises because intensive agriculture and prevalent cropping systems have led to degradation of soil and water resources and a decline in agricultural productivity. Agroforestry is currently recognized as an important land-use system around the globe that has many solutions to global challenges (Verma et al. 2023a).

10.2.1 Integrating Trees, Crops, and Livestock

Incorporating woody vegetation within agricultural, forage, and livestock systems enhances resilience to interannual variability by promoting crop diversification and optimizing resource utilization (Olson et al. 2000). Trees can provide timber, food, fruit, fuelwood, fodder, fiber, and various NTFP's. Trees can also provide a number of ecosystem services such as erosion control, water conservation, enhanced soil fertility, biodiversity conservation, and livelihood support (Verchot et al. 2007). The study has shown that 70%, 50%, and 90% of wood demand in Kenya, India, and, Bangladesh is met from the farm-grown trees respectively (Singh et al. 2016). Agroforestry also involves livestock as a component of agroforestry as agroforestry makes provisions for fodder and shelter for livestock. The farmers with fodder trees on their farms have the advantage of year-round availability of high-quality forage. Franzel et al. (2014) reported that over 200,000 smallholder farmers in East Africa are estimated to have fodder trees on their farms. Fodder banks and hedgerow intercropping are the leading agroforestry practices that integrate fodder trees, though fodder production was reported in almost all agroforestry systems. Feeding experiments show that fodder trees increased dry matter uptake, and improved animal growth, milk production, and high-quality manure. Thus, fodder trees and shrubs provide increased income from the sale of animal products, forage, etc. Many benefits are expected from the diversification through agroforestry

- Adoption of high-value crops will increase profitability
- Withstand price fluctuation
- Increased food security
- Reduce the risk of pest attack
- Price fluctuations can be reduced through diversified products
- More income resources
- Fodder production of livestock
- Lower environmental pollution
- Livelihood security

10.2.2 Coping up with Shocks and Building Sustainable Livelihoods

Climate change will have a negative impact on agriculture as it leads to an increase in incidences of natural hazards such as drought, and floods (Porter et al. 2014). To limit global warming to 1.5 °C by 2050, it is necessary to increase one billion ha of forest (IPCC 2018). Agricultural crops are inherently susceptible to the heightened risk of crop failure resulting from extreme weather events (Simelton et al. 2015). The intentional integration of trees within agricultural landscapes is anticipated to facilitate carbon sequestration while concurrently providing shielding effects to adjacent crops. This protective role encompasses mitigating the impact of elevated temperatures, minimizing wind-induced stress, and preventing soil erosion through the shading influence of trees, the establishment of windbreaks, and the binding action on soil, respectively (Nair 1993; Verma et al. 2021). Furthermore, this integrated approach serves to diminish the risk of crop failure, as trees exhibit a comparatively shorter recovery period following most natural disasters (Simelton et al. 2015). Globally the rainfed areas constitute about 79% of croplands (FAO 2021) and are home to millions of poor farmers whose whole livelihood is affected by the continuous drought (Palsaniya et al. 2011). Inclusion of goats in agroforestry systems and construction of farm ponds in drought-prone areas can enhance the net returns by 36% and 226% when managed under intensive and semi-intensive systems and thus increases the resilience and income of farmers which ultimately contributes to better livelihood conditions of resource-poor farmers (Palsaniya et al. 2023). Farmers in Kenya heavily depend on fruit trees during periods of floods when other crops fail to yield returns, and they resort to selling fuels to ensure food security during droughts (Thorlakson and Neufeldt 2012). The resilience of these fruit trees lies in their deep root systems, allowing them to efficiently utilize larger volumes of water and nutrients, thereby maintaining production even in the face of droughts and floods (Verchot et al. 2007). This resilience sets them apart from shallow-rooted cash crops, making them a reliable source of income for farmers during challenging climatic conditions. One of the key livelihood benefits associated with agroforestry, particularly during floods and droughts, is the sale of fruits. This income derived from fruit sales plays a crucial role in sustaining the basic needs of farmers. The findings underscore the significance of agroforestry as an alternative livelihood strategy, particularly during times of environmental stress. By offering a source of income during floods and droughts, agroforestry emerges as a valuable asset in enhancing the resilience of livelihoods. The concept of agroforestry as a supporting livelihood during floods and droughts implies that it can contribute to building household resilience. This involves better-preparing households to manage the impacts of shocks, navigate uncertainties, and adapt to changing environmental conditions. The multifaceted benefits of agroforestry position it as a holistic approach to agricultural practices, not only providing economic stability but also contributing to environmental sustainability and adaptation to a variable climate (Verma et al. 2023b).

10.3 Historical Foundation and Evaluation of Agroforestry

10.3.1 Early Practices and Indigenous Knowledge

Agroforestry is as old as agriculture. It was an ancient practice in many regions of the world to retain indigenous trees on the farms or boundaries of fields. Agroforestry has been a traditional practice spanning generations across various regions worldwide, employing indigenous technological knowledge and expertise. Archaeological excavations provide support for the early domestication of trees in the vicinity of South Asian settlements. This evidence can be traced back to the Mesolithic period (10,000–4000 BC), during which people are known to have consumed the fruits of 63 different plant species, such as bael (*Aegle marmelos*), figs (*Ficus glomerata*), gooseberry (*Emblica officinalis*), mahua (*Madhuca indica*), mango (*Mangifera indica*), and jujube (*Ziziphus mauritiana*) in various forms. These plants were cultivated in close proximity to human habitats, as documented by Randhawas, in 1980. The significance of several frequently encountered trees, including khejri (*Prosopis cineraria*), peepal (*Ficus religiosa*), palas or dhak (*Butea monosperma*), and varana (*Crataevaroxburghii*), in the cultural traditions of India has been documented in ancient texts such as the Rig Veda, Atharva Veda, and other Indian scriptures (Mann and Saxena 1980). The farmers also retain the trees for the collection of fodder for livestock and for fuelwood. In the Indian subcontinent rearing of silkworms and lac insects on host trees based on the traditional indigenous knowledge of society was also documented by Dhyani and Handa (2013). In Europe, the forest lands were cleared, the slash was burned, and agricultural crops were sown on the cleared patch of land for varying periods based on the fertility of soil, abandoning the cultivation for several years and repeating the same cycle. This practice was widely recognized as slash-and-burn cultivation or shifting cultivation or swidden agriculture. In the Philippines, a form of shifting cultivation aims to retain certain trees in agricultural land to provide shade for reduction of excessive exposure of soil to the sun. The supplementary purpose of retaining trees in Hanuno cultivation is that woody trees are the source of food, fodder, timber, medicines, and services (soil conservation, windbreaks, water conservation, and nutrient cycling) (Conklin 1957). Similarly, in Africa, certain crops like maize, yams, and pumpkins were grown under scattered trees (Forde 1937). These systems besides providing material output to the owner combat problems like erosion and nutrient losses. In Ethiopia, indigenous populations employ agroforestry techniques for multifaceted objectives, encompassing the acquisition of fuelwood, fruits, shading, fodder, construction resources, and soil enrichment. These practices serve to yield a spectrum of products and income streams, thereby enhancing the overall diversification of resources and economic sustenance (Rossier and Lake 2014). It was common practice during ancient times to plant fruit trees, vegetables, and multipurpose trees in vicinity of home, which evolved to known as homegarden agroforestry. The intercropping of annuals with perennials and animal-based agroforestry systems are also prevalent. A survey of 900 households during 2004 reveals that 90% of households grow fruit trees on their own field to meet their livelihood needs. Consequently, indigenous communities have accrued expertise in the integration of trees and livestock within their agricultural systems to meet their subsistence requirements and bolster their financial resilience.

10.3.2 Modernization of Agroforestry

Agroforestry, as a scientific discipline, started to grow in the late 1980s and 1990s. The first step in its evolvement was marked by U Pan Hle (a Karen of Myanmar Forest Division), who established a teak plantation using a technique of taungya and later on, presented to Dietrich Brandis (Blanford 1958). This method was then widely spread to other parts of the world (Hailey 1957; Raghvan 1960). There were several crop modeling approaches widely recognized to determine the potential of agricultural land-use systems. However, these modeling approaches are not developed to simulate the agroforestry systems (Mead and Willey 1980). This might be due to the low importance given to agroforestry among modern agricultural land uses because of the complex interactions between the crops and trees. In recent times, some agroforestry simulation models have developed but these are incapable of determining the dynamism in growth and yield of component crops (Rahman et al. 2023). These simulation models are given below in Table 10.2.

10.3.3 Current Global Concerns Deriving Adoption of Agroforestry

Current agricultural practices increasingly advocate for agroforestry due to its sustainable, multifaceted benefits. Agroforestry enhances soil health, conserves biodiversity, and mitigates climate change by acting as a carbon sink. It offers adaptive strategies against climate impacts, such as shade for crops and windbreaks. Diversifying income streams through timber and non-timber products, agroforestry contributes to economic stability (Yadav et al. 2021). The combination of trees and crops ensures diverse agricultural products, promoting food security. Improved water use efficiency, community empowerment, and government support further drive the adoption of agroforestry. With global recognition and ongoing research validating its effectiveness, agroforestry emerges as a holistic solution, aligning with the principles of sustainability and resilience in modern agriculture (Das et al. 2022). There are various concerns at a global level that promote the theory of agroforestry for achieving sustainability. Climate change, marked by extreme weather events, altering precipitation patterns, and rising temperatures, poses threats to agricultural productivity and food security. Agroforestry emerges as a solution by actively participating in climate change mitigation through carbon sequestration via tree growth. The diverse vegetation in agroforestry systems further acts as a safeguard, enhancing resilience against climate uncertainties and buffering against extreme weather events (Zhang et al. 2024). Biodiversity loss, fueled by deforestation and intensive agricultural practices, stands as a significant concern compromising ecosystem health and resilience. Agroforestry addresses this by integrating trees with diverse crops, creating habitat diversity that supports various plant and animal species. This integration not only promotes biodiversity but also enhances

Table 10.2 Simulation models for agroforestry systems

S. No.	Model name	Objective
1.	ALWAYS (alternative land use with agroforestry systems) model	Climate regulation and carbon storage, soil carbon, and nitrogen
2.	APISM next generation, agroforestry model	Biomass production, yield, N losses, soil organic matter, soil erosion
3.	CO_2 fix	Carbon sequestration, carbon balance, and soil organic carbon
4.	Coffee agroforestry model	Biomass, yield, carbon, nitrogen, water balance, leaf area, and climate
5.	Hi-sAFe (sAFe tree + STIC crop model)	Biomass, yield, carbon sequestration, N losses, soil erosion, environment, and climate
6.	Yield-SAFE model	Biomass, yield, carbon sequestration, soil erosion, N and P losses and Green House Gases (GHG) emissions
7.	SCUAF (Soil Changes Under Agroforestry Model)	Soil organic carbon, soil erosion, N losses, total soil carbon, organic C, nitrogen and phosphorus, GHG's
8.	WaNulCAS (Water, Nutrient and Light Capture in Agroforestry Systems)	Biomass and yield prediction, soil organic matter, carbon sequestration, water, nitrogen, climate, and soil erosion
9.	Farm-SAFE (Work with Hi-SAFE and Yield SAFE for economic analysis for agroforestry)	Economics of system, carbon sequestration, soil erosion, N and P losses, and GHG's emissions

Source: Rahman et al. (2023)

ecosystem resilience, contributing substantially to overall biodiversity conservation. Unsustainable land-use practices leading to soil erosion, loss of fertility, and degradation are prominent concerns impacting agricultural productivity. Agroforestry plays a pivotal role in combating land degradation by offering ground cover, reducing soil erosion, enhancing soil fertility through nutrient cycling, and promoting sustainable land management practices.

The challenge of meeting future food demand due to a growing global population and changing dietary patterns is another critical issue. Agroforestry emerges as a solution by enhancing food security through the diversification of agricultural production. The integration of trees with crops provides a variety of food products, ensuring more stable yields and increasing resilience to climate-related shocks. Increasing water scarcity, exacerbated by climate change and inefficient water management, threatens agricultural water supply. Agroforestry systems prove more water-efficient than monoculture crops, with tree shade reducing evaporation and deep root systems improving water retention in the soil. This efficient water management is evidenced by the effective utilization of approximately 25% of water transpired by trees during the dry season, showcasing their capacity to harness off-season rainfall. This mutualistic interaction between trees and annual crops expands the potential for soil moisture uptake, optimizing soil resource utilization more efficiently than observed in monocultural systems. In the context of rural livelihoods and socioeconomic equity, agroforestry addresses challenges faced by rural communities related to poverty, lack of economic opportunities, and social inequities. It provides avenues for diversified income sources, improved livelihoods, and community empowerment. Smallholder farmers particularly benefit from engaging in sustainable agroforestry practices that enhance both economic and social well-being. Lastly, the imperative for a transition to a green economy that balances economic growth with environmental sustainability is a key concern. Agroforestry aligns with the principles of sustainable development, contributing to environmental conservation while supporting economic growth. It serves as a model for sustainable land use that effectively integrates ecological and economic goals.

10.4 Current Agroforestry Practices

Agroforestry practices are essential to rural development and sustainable land usage. These systems cover a range of methodologies, each suited to the ecological conditions, socioeconomic requirements, and cultural settings of the particular area. In India, silvopastoral systems, agrisilviculture systems, and agri silvopastoral systems are the three major types of agroforestry practices. In silvopastoral systems, trees are integrated with the grazing of livestock. By cycling nutrients, this approach improves soil fertility in addition to giving animals a supply of fodder (Nair 1993). By using tree cover for shade and reducing heat stress in cattle, silvopastoral systems provide a dual benefit in areas with limited grazing pasture. Certain tree species, such as Prosopis and Acacia, are specifically chosen because of their capacity to fix nitrogen, improve soil fertility, and produce nutritious fodder. The placement of trees enhances microclimate conditions in addition to providing shade for cattle, which reduces heat stress. The production of forage is greatly aided by silvopastoral systems, which produce a varied and nutrient-rich feed basis. These systems not only benefit animals but also encourage environmental resilience, nitrogen cycling, and soil conservation. Timber, non-timber forest products, and a variety of marketable goods provide farmers with many sources of income, which makes silvopastoral systems a sustainable and profitable land-use plan that balances livelihood and ecological goals.

10.4.1 Agrisilvicultural Systems

Agrisilvicultural systems combine trees with crop production, offering advantages such as increased yield, improved soil structure, and enhanced biodiversity. The arrangement of trees in these systems is carefully planned to avoid competition for resources with crops. For example, Eucalyptus species are often used in agrisilvicultural systems as they have a relatively narrow canopy and release fewer allelopathic substances that could hinder crop growth (Nair 1993). Carefully adopting tree species, like eucalyptus, ensures compatibility with crops in these systems, reducing competition for resources. The layout of the space is carefully designed to maximize the dispersion of nutrients, water, and sunlight, promoting mutually beneficial relationships. Along with improving total land productivity, this integration also has other advantages including better soil structure and more biodiversity. Increased yields, better adaptation to climate variability, and economic diversification through timber and non-timber forest products are experienced by farmers who implement agrisilvicultural systems. This strategy offers a complex solution for sustainable land use and rural development in a wide range of agroclimatic zones, demonstrating a strategic interaction between tree and crop components.

10.4.2 Agro Silvopastoral Systems

Agro silvopastoral systems represent a comprehensive approach, integrating trees with both crops and livestock. These systems were developed with intended spatial arrangements that consider the requirements of both crops and grazing animals in order to maximize the efficiency of resource utilization. In addition to giving animals shade and feed, the trees act as nutrient cycle agents, enhancing the land's fertility. This complex integration supports biodiversity, environmental resilience, and soil protection in addition to raising agricultural productivity overall. Agrisilvipastoral systems provide farmers with a variety of revenue streams, such as crop yields, timber, and livestock products. This strategy is both economically and sustainably feasible, and it effectively addresses the complex issues facing contemporary agriculture and rural development. The adoption of agroforestry practices in India is influenced by a combination of ecological, economic, and social factors. Farmers are increasingly recognizing the potential of agroforestry to improve land productivity while maintaining environmental sustainability. Government initiatives and support programs further encourage the adoption of agroforestry, promoting practices that align with the diverse agroclimatic zones across the country. In conclusion, the rich tapestry of agroforestry practices in India reflects a dynamic interplay between traditional knowledge and modern agricultural techniques. Silvopastoral, agrisilvicultural, and agrosilvopastoral systems represent strategic approaches to address the complex challenges of resource scarcity, climate change, and rural livelihood improvement. As India continues to grapple with the demands of a growing population and changing climate, agroforestry stands out as a resilient and sustainable land-use strategy that contributes to the well-being of both the environment and the people.

10.5 Advantages of Agroforestry

10.5.1 Increasing Agricultural Productivity

Agroforestry plays a crucial role in increasing agricultural productivity by harnessing the synergies between trees, crops, and livestock. Numerous advantages of this integrated strategy support increased yields, better soil fertility, and all-around sustainable land management.

10.5.1.1 Nutrient Cycling and Soil Fertility

One of the fundamental contributions of agroforestry to increased agricultural production lies in nutrient cycling and soil fertility enhancement (Patel et al. 2022). Agroforestry systems are often home to nitrogen-fixing trees, which are essential because they transform atmospheric nitrogen into a form that plants can utilize. These trees' roots are habitat to microorganisms that fix nitrogen, which enriches the soil with this essential element. Furthermore, the tree's leaves, twigs, and other organic materials add to the soil's organic content, enhancing its structure and ability to hold nutrients.

10.5.1.2 Biodiversity and Pest Management

Agroforestry systems increase biodiversity, making an ecosystem more robust and dynamic. A wide range of insects, birds, and other helpful creatures are drawn to the diversity of plant species. By acting as a biological barrier against pests, these naturally occurring predators lessen the need for chemical pesticides. Agroforestry systems create a complex equilibrium that improves crop production by strengthening insect resistance and resilience.

10.5.1.3 Microclimate Regulation

Microclimate regulation is significantly impacted by the presence of trees in agroforestry systems. Because trees provide shade, there are less extremes in temperature, which improves the growing conditions for crops. This temperature control is especially useful in areas with severe weather since it prolongs growing seasons and creates the ideal environment for higher agricultural yields.

10.5.1.4 Water Use Efficiency

Agroforestry contributes to improved water use efficiency through various mechanisms. The canopy of trees serves as a natural barrier, reducing evaporation from the soil surface and minimizing water loss. The root systems of trees also contribute to soil structure, enhancing water retention capacity. In water-scarce regions, agroforestry becomes a sustainable approach to optimize the use of available water resources, leading to increased agricultural productivity.

10.5.1.5 Enhanced Crop Yields

The spatial arrangement of trees in agroforestry systems is carefully designed to optimize resource use and minimize competition with crops (Verma et al. 2023c). This thoughtful planning leads to improved photosynthesis, reduced nutrient competition, and enhanced overall crop yields. Agroforestry has been demonstrated to increase the productivity of various crops, ranging from staple food crops to cash crops, addressing the growing demand for food production to feed a burgeoning global population.

10.5.1.6 Soil Erosion Control

Soil erosion poses a significant threat to agricultural productivity, particularly in areas with sloping terrain or high rainfall. Agroforestry systems act as natural buffers against soil erosion. The root systems of trees stabilize the soil, preventing it from being washed away during heavy rains. This soil conservation aspect of agroforestry preserves the integrity of agricultural land, ensuring its long-term productivity.

10.5.1.7 Agroforestry for Livestock Integration

Agroforestry provides opportunities for integrating livestock into farming systems, further enhancing agricultural productivity. Trees can offer shade for livestock, reducing heat stress and improving overall animal welfare. Additionally, the forage produced by trees can supplement the diet of grazing animals, contributing to improved livestock health and productivity.

10.5.2 Diversifying Income Sources

10.5.2.1 Income from Woody Components

Tree products such as timber, fruit, fuelwood, fodder, gums and resins, tans and dyes, essential oils, medicines, and other NTFPs are the source of income generation from agroforestry (Muthuri et al. 2023). Some products provide a steady flow of income to farmers while others provide at an interval. The woody perennials are raw material for charcoal making which serves as an energy source for urban livelihoods (Iiyama et al. 2014). A survey of 22 countries in Africa showed that nearly 12%, 13%, and 19% of annual gross income of farmers of Ethiopia, Tanzania, and Uganda come from farm-grown trees (Miller et al. 2017). On-farm firewood production constitutes a significant revenue stream for farmers in East Africa. Concurrently, the cultivation of *Senegalia senegal*, yielding gum Arabic, emerges as a pivotal economic driver for farmers in Sudan.

10.5.2.2 Income from Sequestered Carbon

Agroforestry offers to the farmers, a dual benefit by not only contributing to environmental sustainability but also creating a viable revenue stream through participating in ecosystem service markets. Although the carbon credit market is young, utilizing this concept in agroforestry projects has the potential to improve socioeconomic status of farmers and increase food security (Allwardt 2011). The carbon stored in the above- and belowground biomass can be sold in the form of credits to the purchasing agencies (Brown et al. 2011). An international program "The International Small Group Tree Planting Program" functioning in Uganda, Tanzania, and Kenya since 2005 quantifies the carbon stored in the land use and sells sequestered carbon through VCS (Verra) Standard. Under Clean Development Mechanism (CDM), a project named Humbo Community-based Natural Regeneration Project (Brown et al. 2011), and Kenya Agricultural Carbon Project (Nyberg et al. 2020) also provide evidence of income generation from the sale of carbon credits. Kumar et al. (2020) found poplar and wheat-based agroforestry systems to sequester more carbon, produce higher carbon credits, and higher benefit-cost ratio over open farming practices. According to the present estimation provided by the World Agroforestry Center, the carbon market is anticipated to surpass 1 trillion US dollars by the year 2025, implying that substantial funds could potentially be generated through interventions in agroforestry (WAC 2010).

10.5.2.3 Income from Sale of Agricultural Crops

As a sustainable land-use technique, agroforestry has a big impact on the revenue that is obtained by selling agricultural products. With this integrated strategy, trees are deliberately grown alongside traditional crops to create a dynamic system that increases farmers' overall profits. The optimization of resource consumption is one of the main ways that agroforestry helps to generate money. In agroforestry systems, trees are essential for improving soil quality and cycling nutrients, which creates an environment that is beneficial to crop growth. Higher yields as a result increase the quantity of products that are available for sale. Studies show that agroforestry practices boost crop nutrient availability and have a well-documented favorable impact on soil fertility and structure (Nair et al. 2010).

Agroforestry systems also improve the microclimate, which makes the environment better for growing crops. The presence of trees improves soil moisture retention, lowers wind speed, and controls temperature. These improvements to the microclimate help crops withstand unfavorable weather, which lowers the possibility of output losses (Verma et al. 2023c). As a result of this, agroforestry farmers are in a better position to continually provide the market with high-quality crops, ensuring a consistent source of revenue. In agroforestry systems, crop diversification is another important component that affects income. By integrating trees with conventional crops, one can produce a variety of goods and reduce their dependence on a single source. In addition to distributing the financial risk, this diversification gives farmers access to new markets. Agroforestry systems that include fruit or nut-bearing trees, for instance, can generate income sources in addition to conventional crop sales. Additionally, constant and secure revenue is a benefit of agroforestry's long-term sustainability. These systems' enhanced ecological resilience guarantees sustained productivity, giving farmers a steady stream of income over time. According to research, well-designed agroforestry systems can produce yields that are sustained or even higher than those of monoculture methods, which can have long-term financial benefits.

10.5.2.4 Other Income Sources

Livestock are also an integral part of many agroforestry systems. Farmers may get additional income from the sale of livestock and livestock products (Hughes et al. 2020). Agroforestry facilitates the integration of livestock by providing essential components such as fodder, shade, and shelter. Fodder provision is a common attribute across various agroforestry systems, encompassing specialized configurations like fodder banks, live hedges of fodder trees, and hedgerows explicitly designed for fodder production (Muthuri et al. 2023). As estimated by Dawson et al. (2014), a substantial number, exceeding 200,000 smallholder farmers in East Africa, incorporate fodder trees into their agricultural landscapes, significantly contributing to the sustainability of livelihoods. The increased biodiversity and presence of nectar-rich trees provide shelter for bees and in turn farmers may get income from the sale of honey and waxes from bee-keeping.

10.5.3 Enhancing Resilience to Climate Change

The agriculture sector is the prime source of livelihood needs in less developed nations and these nations are adversely affected by the land pressures caused by the high population which threatens food security (Mbow et al. 2014). The intensification of agriculture to boost agricultural productivity poses a threat to environmental health (Krausmann et al. 2013; Wani et al. 2022; Kumar et al. 2022a). To simultaneously increase the productivity per unit area, certain alternative land-use systems based on diversification of components are required (Gockowski and Asten 2012; Williamson 2022). Agroforestry might be an ideal land-use system that can enhance agricultural productivity, store carbon for a longer duration, and serve as a strategy for climate change mitigation (Unruh et al. 1993; Kumar and Nair 2012). The estimated worldwide capacity for greenhouse gas (GHG) sequestration in agriculture falls within a range from 1500 to 4300 million metric tons of carbon dioxide equivalent per year (Mt CO_2e/yr). Approximately, 70% of this capacity is located in developing countries, with 90% of the potential stemming from activities related to restoring soil carbon and preventing net soil carbon emissions (Mbow et al. 2014). Climate-smart agriculture through agroforestry also supported by FAO is based on four betters: better production, better nutrition, better environment, and better life for all promotes agroforestry practices for resilience to climate change (Araya et al. 2023). Agroforestry offers numerous biophysical benefits including a reduction in mean air temperature (Gomes et al. 2020) and crop transpiration (Verchot et al. 2007; Lin 2010) through shading effect on crops. The deep tap-rooted trees can withdraw water from deeper soil layers, thus conserving water to be taken by agricultural crops (Padovan et al. 2015). They reduce soil loss during heavy rainfall (Lal et al. 1991). Agroforestry systems that have medium-sized trees, reduce the stresses to heat, drought, and nitrogen by 20–35% when they increase two times over a determined period (Reyes et al. 2021). Research conducted across seven West African countries has revealed that the widespread implementation of agroforestry practices has the potential to capture 135 metric tons of carbon annually over a 22 years period. This amount is equivalent to 166% of the combined carbon emissions from fossil fuels and deforestation within these countries (Chemura et al. 2022).

10.5.4 *Ecological Benefits*

10.5.4.1 Soil Fertility

The persistent decline in soil fertility, attributed to various causative factors, poses a significant impediment to sustainable agriculture (Altieri and Rogé 2009). Of particular concern is topsoil erosion, an adverse manifestation of soil degradation exacerbated by the prolonged removal of surface litter and crop residues. Agroforestry emerges as a potent strategy to ameliorate soil fertility, primarily attributable to the augmentation of soil organic matter and the facilitation of biological nitrogen fixation by leguminous trees (Lehmann et al. 1998). The presence of woody perennials on agricultural lands fosters a more intricate nutrient cycling dynamic than monoculture systems, concurrently enhancing soil nutrient content and organic matter while ameliorating structural properties. By virtue of their water extraction capabilities and the prevention of nutrient leaching, trees play a pivotal role in nutrient recovery, soil moisture conservation, and the enhancement of soil organic matter within agroforestry systems (Duguma and Hager 2011). Agroforestry is a second soil fertility paradigm (Sanchez 1999). The nitrogen-fixing tree species has a pivotal role in any land-use system to enhance soil fertility, thereby stimulating the growth and productivity of co-existing species. In line with this, mixed plantings that include nitrogen-fixing tropical tree species have been observed to yield greater aboveground biomass compared to monoculture stands in similar environmental conditions (Bauhus et al. 2004a, b; Forrester et al. 2006). Three of the most promising low-cost options to increase soil fertility through agroforestry practices include improved tree fallow systems, biomass transfer techniques, and mixed intercropping methods (Niang et al. 1996; Sanchez et al. 1997; Kiptot 2008). Enhanced tree fallows involve the intentional planting of rapidly growing leguminous trees or shrubs in crop rotation. Biomass transfer is a technology where shrub/tree biomass, whether grown on or off the farm, is harvested and integrated into the soil as green manure during crop planting. Mixed intercropping entails the strategic planting of nitrogen-fixing trees that can withstand continuous and substantial pruning alongside crops like maize (*Zea mays* L.). These technologies have the potential to assist farmers in enhancing soil quality and increasing their incomes, thereby contributing to improved food security. Kaba and Abunyewa (2021) found that *Gliricidiasepium* can be used as green manure in cocoa plantations to meet the nitrogen demand reducing the need for external input and subsequently increasing farmer's profits. Young gliricidia trees can be pruned at 4 months interval while old trees can be pruned at an interval of 8 months. So, introducing this leguminous tree in cocoa plantations represents an important tool for meeting nitrogen demand and enhancing soil fertility.

10.5.4.2 Biodiversity

Agroforestry systems exhibit greater diversity compared to monocultures of agricultural crops or forest plantations, as articulated by Sanchez (1995). Agroforestry systems have the capacity to augment biodiversity conservation in agricultural environments by increasing structural variability, and diverse environmental conditions (Haggar et al. 2019). This increased biodiversity in agroforestry might be due to increased access to food, habitat, shelter, favorable microclimate, and improved soil-water-plant relationship (Udawatta et al. 2017). Ng'endo and Bhagwat (2019) found that agroforestry has 60% more species diversity than forest ecosystem. Homegardens, a prevalent agroforestry system in tropical regions, are characterized by notable floristic diversity and are sometimes regarded as mimicking natural forests (Ewel 1999). Kumar and Nair (2004) provided estimates indicating that species richness in home gardens ranges from 27 (Sri Lanka) to 602 (West Java). The diversity and population density of insects are influenced by the various tree-crop combinations and their spatial distribution. Microhabitats at the edges of windbreaks contribute to increased insect density and diversity (Brandle et al. 2004), while also serving as habitats for wild animals, thereby augmenting animal diversity. Gallina et al. (1996) conducted an estimation, revealing that shade trees in coffee-based agroecosystems in Mexico provide shelter, food, and protection to mammals. They further emphasized that a reduction in the complexity of the system leads to a substantial loss of 45% and 43% in ecological richness and diversity within the respective guilds. In semiarid Tehuacan Valley, agroforestry systems demonstrated the preservation of 59% of plant species and 94% of genetic variation in dominant cacti compared to forest ecosystems. However, these systems exhibited limited capacity to conserve rare species (Moreno-Calles et al. 2010). Agroforestry practices promote biodiversity conservation in the following ways:

- Serves as habitat-wide species of plants, animals, and microorganisms.
- Promotes genetic diversity.
- Reduce the conversion rate of natural habitat by providing a more sustainable alternative over conventional agriculture.
- Promotes biological diversity through supporting ecosystem services viz. erosion control, water recharge, and conservation.

Agroforestry for the purpose of biodiversity conservation necessitates meticulous design and management within the ecosystem, calling for less intensive cultural practices. The key design features that agroforestry systems require to promote heightened biodiversity include:

- High floristic diversity
- High structural diversity
- Low management intensity
- Long rotation periods
- Cultural practices viz., thinning, pruning, and prescribed fire
- Site specificity of system (should be close to natural forests)

The implementation of agroforestry systems on degraded wastelands characterized by low biodiversity yields enhanced biodiversity in improved land-use scenarios. However, it is imperative to note that deforestation undertaken for the establishment of agroforestry can exert adverse effects on biological diversity, as highlighted by Jose (2012).

10.5.4.3 Carbon Sequestration

Carbon sequestration, as defined by Jose (2009), involves the extraction of carbon from the atmosphere and its deposition into carbon sinks such as oceans, soil, and vegetation through either biological or physical mechanisms. The incorporation of woody perennials into agricultural systems has the potential to augment the quantity of sequestered carbon compared to monocultures of field crops or pastures, as articulated by Sharrow and Ismail (2004). The basic principle of carbon sequestration through agroforestry is based on the basic ecological processes of photosynthesis, respiration, and decomposition (Nair and Nair 2002). The total carbon sequestered by the ecosystem is the difference between carbon acquired by photosynthesis and carbon released by the components. The accumulated carbon in the biomass can be transferred to the ground (more than half) through litter deposition, root growth, and exudations. Agroforestry can be sequestered of carbon in both above (plant biomass) and below ground (soil) and hence contribute to climate change. The incorporation of trees into agricultural systems is acknowledged as a substantial reservoir of atmospheric carbon, attributable to their rapid growth and heightened productivity (Chauhan et al. 2010). This dual function not only augments the quantity of carbon stored in agricultural lands but also facilitates the concurrent cultivation of food crops (Kursten 2000). The average global carbon storage capacity of agroforestry systems in biomass is 21.4 Mg C/ha (Zomer et al. 2016) with humid areas having the highest potential (Montagnini and Nair 2004). Agroforestry systems have the capability to sequester more carbon as compared to land use without trees as trees produce more biomass and can reach deeper in to the soil and attains more height as compared to agricultural crops (Howlett et al. 2011; Kumar et al. 2023). Based on tree growth rates, wood production, and assuming a 1:2 ratio of tree-stem biomass to carbon content (i.e., 50% of stem wood consists of carbon), research estimates that agroforestry practices can store approximately 9, 21, 50, and 63 megagrams (Mg) of carbon per hectare in semiarid, subhumid, humid, and temperate regions, respectively. The higher carbon storage levels observed in temperate regions are attributed to longer cutting cycles, leading to more extended long-term storage.

Enhanced carbon sequestration can be achieved through the utilization of trees with higher rotation cycles or the conversion of harvested woody biomass into more enduring products, as outlined by Jose (2009). Within the agroforestry domain, an aggregate land area spanning 585–1215 million hectares across Africa, Asia, and America possesses the potential to sequester 1.1–2.2 petagrams (Pg) of carbon over a 50-year timeframe (Dixon and Krankina 1995). The Intergovernmental Panel on Climate Change (IPCC) in 2000 projected that through the refinement of existing agroforestry practices, an additional 12,000 megagrams of carbon per year could be sequestered by 2010, escalating to 17,000 megagrams per year by 2040. Furthermore, the carbon sequestered within agroforestry systems holds the potential for monetization in carbon credit markets, where such opportunities exist, according to Jose (2009).

10.5.5 Institutional and Policy Framework for Agroforestry

Despite the recognized potential of agroforestry to positively impact land use across various scales, including productivity, protection, and socioeconomic aspects, it is observed that agroforestry is not accorded the same level of promotion and support as traditional agriculture or forestry within existing policy frameworks and institutional structures (Rodriguez-Rigueiro et al. 2021). The absence of adequate policies and institutional support can significantly influence the adoption of agroforestry practices and impact their sustainability and environmental effectiveness (Pantera et al. 2021). Notably, recent efforts have been made in certain countries, such as the USA, France, and India, to systematically promote and adopt agroforestry practices (Pantera et al. 2021). The successful adoption of agroforestry innovations through policy implementation hinges upon addressing three primary research challenges. These include the intricate issues of land and tree tenure, providing support to farmers during the time lag phase (defined as the duration within which trees on farmlands remain unproductive), and the comprehensive valuation of environmental externalities, as emphasized by Sanchez (1995). The complexities are further compounded by the insufficient understanding of site-specific crop combinations and the willingness of farmers to embrace the intensification of agricultural practices. This lack of knowledge poses a formidable barrier to the effective promotion of agroforestry policies, as highlighted in the work of Santiago-Freijanes et al. (2021). It is crucial to recognize the role of policies in stimulating innovation within the agroforestry sector, achieved through the dissemination of knowledge, bottom-up stakeholder involvement, and the implementation of multi-actor approaches. The robust engagement of numerous non-governmental organizations (NGOs) and international institutions, including the World Agroforestry Center, the Association for Temperate Agroforestry, and the European Agroforestry Foundation, along with the concerted efforts of researchers and farmers, serves to underscore the significance of agroforestry land use in the pursuit of sustainable livelihood objectives (Santiago-Freijanes et al. 2021).

10.5.6 Promoting Adoption and Scaling

The parklands are a traditional agroforestry system of Sahel region which consists of scattered trees throughout agricultural landscapes. During the late twentieth century, it was felt that numbers of young trees were significantly reduced over time (Boffa 1999) as farmers had been removing young trees because of uncertainty of ownership and use rights to the trees. Thereafter some NGOs ("Serving in Mission" and "CARE International") along with the Government of Niger take initiative and demonstrated the importance of farmer-managed natural regeneration agroforestry practice and played a significant role in the promotion of farmer-managed natural regeneration agroforestry in Southern Sahel region (Reij et al. 2009). The fertilizer tree system which consists of growing nitrogen-fixing trees in agricultural lands was widely promoted by the government to replenish the soil fertility and for increasing crop production (Kwesiga et al. 2003). This system was widely promoted during the early 2000s in response to economic burden on the government following heavily subsidized fertilizer support to maize production system. Some government projects such as ZIAP (Zambia Integrated Agroforestry Project), AFSP (Agroforestry Food Security Project), and TLC (Total Land Care) led to increased awareness and adoption of system by farmers of Africa. e.g., tobacco farmers had to keep 10% of land under tree cover to cater to the wood demand of the crop for curing and drying. The HASHI project promoted the regeneration of *ngitili* (a defined area protected from grazing and other activities to serve as a fodder source during the dry season) with new agroforestry technologies, including planting of exotics. The success of the HASHI project was reflected as are under *ngitili* was increased from 600 ha to nearly 78,000 ha of land (Pye-Smith 2010). The Government of Nepal has initiated a proactive measure by formulating a National Policy on Agroforestry, making it the second country, following India, to officially endorse a National Agroforestry Policy. This strategic policy framework recognizes agroforestry as a promising land-use system capable of addressing challenges associated with agricultural land issues stemming from labor shortages (NAFP 2019). The poplar-based agroforestry system is widely acknowledged in Northern India. In this system, wheat and barley are cultivated as intercrops within a rotation period of 8 years alongside poplar cultivation. The introduction of poplar to India in 1969 from Australia prompted the Uttar Pradesh government to initiate the cultivation of wheat, sugarcane, and mustard in conjunction with poplar (Zomer et al. 2007). This endeavor gained momentum in the 1980s when a World Bank-funded project aimed to promote the expansion of agroforestry on both individual farms and communal lands. Prior to the 1970s, the demand for raw materials by wood-based industries was met through state-owned government forests. However, a ban on green felling in state forests, enacted by a Supreme Court decision in 1996, coupled with an escalating demand for wood and wood-based products, led industries to seek raw materials from farmers. WIMCO (Western India Match Company) Ltd. played a notable role in advocating and extending poplar-based agroforestry. The collaborative initiative between WIMCO and NABARD in 1984 marked a significant advancement, wherein farmers were provided credit facilities to offer financial support for the adoption of agroforestry practices. Silvipastoralism and Sustainable Management International Congress held in Spain in 2004, highlighted the importance of Silvipastoral agroforestry and its role in enhancing rural economy and called for its promotion through capacity building and transfer of technology (Mosquera-Losada et al. 2006).

10.5.7 Public and Private Sector Cooperation

Public and private sector cooperation in agroforestry is essential for harnessing the full potential of this sustainable land-use practice. This collaboration brings together the strengths and resources of both sectors, facilitating comprehensive and impactful initiatives that address ecological, economic, and social dimensions. The following points highlight the significance and key aspects of public and private sector cooperation in agroforestry:

(i) **Policy Framework and Regulation:** Public sector involvement is crucial in establishing a conducive policy framework and regulatory environment for agroforestry. Governments can develop and implement policies that incentivize and support agroforestry practices. This may include providing subsidies, tax incentives, and regulatory frameworks that promote sustainable agroforestry practices.

(ii) **Research and Development:** Public institutions often play a leading role in research and development activities. Collaborations between public research institutions and private enterprises can drive innovation in agroforestry. Research initiatives can focus on developing new tree varieties, sustainable agroforestry practices, and technologies that enhance productivity while minimizing environmental impact.

(iii) **Extension Services and Capacity Building:** Public extension services are instrumental in disseminating knowledge and best practices in agroforestry to farmers. Public-private partnerships can enhance extension services by integrating private sector expertise and resources. This collaboration facilitates training programs, workshops, and capacity-building initiatives to empower farmers with the skills needed for successful agroforestry implementation.

(iv) **Funding and Investment:** Both public and private sectors can contribute financial resources to promote agroforestry initiatives. Governments can allocate funds for agroforestry programs, while private investors, including agribusinesses and impact investors, can provide capital for agroforestry projects. This collaboration attracts funding at various stages, from research and pilot projects to large-scale implementation.

(v) **Technology Transfer and Adoption:** Private sector involvement is crucial for the development and dissemination of technologies that support agroforestry. This may include the provision of agroforestry inputs, machinery, and precision agriculture technologies. Collaborations between public research institutions and private companies facilitate the efficient transfer and adoption of these technologies by farmers.

(vi) **Value Chain Integration:** Private enterprises, including agribusinesses and forest-based industries, can actively participate in agroforestry value chains. This integration ensures that the products derived from agroforestry, such as timber, fruits, nuts, and non-timber forest products, find viable markets. Private sector involvement in value chains contributes to increased income for farmers and enhances the economic viability of agroforestry.

(vii) **Certification and Quality Standards:** Private sector engagement in agroforestry can contribute to the development and adherence to certification and quality standards. This is particularly important for products like sustainably harvested timber and organic fruits. Private companies can work with certification bodies to ensure that agroforestry products meet recognized environmental and social standards, enhancing market access.

(viii) **Public-Private Partnerships (PPPs):** Formal collaborations between the public and private sectors through public-private partnerships (PPPs) can be instrumental in implementing large-scale agroforestry projects. These partnerships leverage the strengths of both sectors, combining public resources and expertise with the efficiency and innovation of private enterprises. PPPs can involve joint ventures, co-financing arrangements, and shared responsibility for project implementation.

(ix) **Market Access and Linkages:** Private sector involvement facilitates better market access for agroforestry products. Companies can establish market linkages, negotiate fair prices, and create demand for sustainably produced agroforestry goods. This connection to markets provides farmers with a reliable income stream, encouraging the adoption and continuation of agroforestry practices.

(x) **Climate Change Mitigation and Corporate Social Responsibility (CSR):** Private sector engagement in agroforestry aligns with corporate social responsibility (CSR) goals and climate change mitigation strategies. Companies can invest in agroforestry projects as part of their commitment to sustainability. These initiatives contribute to carbon sequestration, biodiversity conservation, and community development, showcasing the positive impact of the private sector on environmental and social outcomes.

(xi) **Community Engagement and Social Inclusion:** Both public and private sectors must actively engage with local communities for successful agroforestry initiatives. This collaboration ensures that projects are culturally sensitive, address local needs, and provide socioeconomic benefits to communities. Private enterprises can implement inclusive business models that prioritize the well-being of local populations.

(xii) **Risk Sharing and Insurance:** Agroforestry projects involve inherent risks, including climate-related uncertainties and market fluctuations. Public-private collaboration can facilitate the development of risk-sharing mechanisms and insurance products tailored for agroforestry. These financial instruments help mitigate risks for farmers and investors, encouraging greater participation in agroforestry ventures. The collaboration between the public and private sectors in agroforestry is integral to realizing its full potential for sustainable development. This synergy leverages the strengths of each sector, creating a conducive environment for policy support, research and development, technology transfer, funding, and market access. As the world faces the challenges of climate change, deforestation, and food security, effective public and private sector cooperation in agroforestry becomes increasingly imperative for building resilient and sustainable agricultural systems.

10.6 Conclusion

Agroforestry is a dynamic method of managing land use that combines cattle, crops, and trees to improve sustainability, resilience, and productivity. One of the most important things to remember from our conversation is how important agroforestry is to solving a number of global issues, from improving rural livelihoods and food security to preserving the environment and reducing the effects of climate change. Agroforestry promotes soil fertility, biodiversity conservation, and carbon sequestration in addition to increasing yields and diversifying revenue streams by utilizing the synergies between agricultural production and ecological services. Agroforestry also demonstrates the value of traditional knowledge and methods from indigenous peoples for sustainable land management. Agroforestry techniques have advanced as a result of modernity, but it is important to acknowledge and build upon the rich heritage of traditional methods and indigenous knowledge. Further, to fully utilize agroforestry, a number of parties must work together, including legislators, scientists, extension agents, farmers, and businesses in the private sector. For agroforestry methods to flourish, institutional and governmental frameworks that encourage adoption, offer incentives, and foster a supportive atmosphere must be designed and put into place. In summary, agroforestry presents a viable route to enhancing rural livelihoods, attaining food security, and advancing environmental sustainability. Adopting the tenets and methods of agroforestry will be essential to ensuring a more resilient and just future for future generations as we negotiate the challenges posed by a changing climate and rising food and resource demands.

References

Abebe T (2013) Determinants of crop diversity and composition in Enset-coffee agroforestry homegardens of Southern Ethiopia. J Agric Rural Dev Trop Subtrop 114(1):29–38

Aguiar ACF, Bicudo SJ, Costa Sobrinho JRS, Martins ALS, Coelho KP, Moura EG (2010) Nutrient recycling and physical indicators of an alley cropping system in a sandy loam soil in the pre-Amazon region of Brazil. Nutr Cycl Agroecosyst 86:189–198. https://doi.org/10.1007/s10705-009-9283-6

Allwardt J (2011) Carbon credit payment options for agroforestry projects in Africa (No. 1097-2016-88565)

Altieri MA, Rogé P (2009) The ecological role and enhancement of biodiversity in agriculture. In: Agriculture, biodiversity and markets. Routledge, pp 15–32

Araya YN, Emmott A, Rawes W, Zuza EJ (2023) Promoting climate-smart sustainable agroforestry to tackle social and environmental challenges: the case of macadamia agroforestry in Malawi. J Agric Food Res 14:100846. https://doi.org/10.1016/j.jafr.2023.100846

Aryal DR, Danilo EM, Susana L, César NT, Alejandra L, José AJ, Perez-Sanchez E (2022) Silvopastoral systems and remnant forests enhance carbon storage in livestock-dominated landscapes in Mexico. Sci Rep 12(1):16769. https://doi.org/10.1038/s41598-022-21089-4

Babu S, Mohapatra KP, Yadav GS, Lal R, Singh R, Avasthe RK, Das A, Chandra P, Gudade BA, Kumar A (2020) Soil carbon dynamics in diverse organic land use systems in north eastern Himalayan ecosystem of India. Catena 194:104785. https://doi.org/10.1016/j.catena

Bauhus J, Van Winden AP, Nicotra AB (2004a) Aboveground interactions and productivity in mixed-species plantations of *Acacia mearnsii* and *Eucalyptus globulus*. Canadian J Forest Res 34(3):686–694. https://doi.org/10.1139/x03-243

Bauhus J, Vor T, Bartsch N, Cowling A (2004b) The effects of gaps and liming on forest floor decomposition and soil C and N dynamics in a *Fagus sylvatica* forest. Can J For Res 34(3):509–518. https://doi.org/10.1139/x03-218

Betemariyam M, Negash M, Worku A (2020) Comparative analysis of carbon stocks in home garden and adjacent coffee-based agroforestry systems in Ethiopia. Small-Scale For 19:319–334. https://doi.org/10.1007/s11842-020-09439-4

Bhardwaj DR, Salve A, Kumar J (2023) Biomass production and carbon storage potential of agroforestry land use systems in high hills of north-western Himalaya: an approach towards natural based climatic solution, vol 14. Bioref, Biomass Conv, p 18079. https://doi.org/10.1007/s13399-023-03952-0

Bhatia A, Sharma K, Thakur M, Kumar S, Sharma R, Kumar A (2023) Ecological and economical assessment of Harar–soybean agroforestry system in subtropical conditions of Himachal Pradesh, India. Agrofor Syst 98:151. https://doi.org/10.1007/s10457-023-00896-2

Bishaw B, Soolanayakanahally R, Karki U, Hagan E (2022) Agroforestry for sustainable production and resilient landscapes. Agrofor Syst 96(3):447–451. https://doi.org/10.1007/s10457-022-00737-8

Blanford HR (1958) Highlights of 100 years of forestry in Burma. Empire For Rev 37(1):33–42

Boffa JM (1999) Agroforestry parklands in sub-Saharan Africa (No. 34). Food and Agriculture Organization

Brandle JR, Hodges L, Zhou XH (2004) Windbreaks in north American agricultural systems. In: Nair PKR, Rao MR, Buck LE (eds) New vistas in agroforestry. advances in agroforestry, vol 1. Springer, Dordrecht. https://doi.org/10.1007/978-94-017-2424-1_5

Brown DR, Dettmann P, Rinaudo T, Tefra H, Tofu A (2011) Poverty alleviation and environmental restoration using the clean development mechanism: a case study from Humbo, Ethiopia. Environ Manag 48:322–333. https://doi.org/10.1007/s00267-010-9590-3

Buttould, G. (2013). Advancing agroforestry on the policy agenda. http://www.fao.org/3/a-i3182e.pdf. Accessed 1 Sept 2017

Castle SE, Miller DC, Ordonez PJ, Baylis K, Hughes K (2021) The impacts of agroforestry interventions on agricultural productivity, ecosystem services, and human well-being in low-and middle-income countries: a systematic review. Campbell Syst Rev 17(2):e1167. https://doi.org/10.1002/cl2.1167

Chauhan SK, Sharma SC, Beri V, Ritu V, Yadav A, Gupta N (2010) Yield and carbon sequestration potential of wheat (*Triticum aestivum*) and poplar (*Populus deltoides*) based Agri-silvicultural system. Indian J Agric Sci 80(2):129–135

Chemura A, Kutywayo D, Hikwa D, Gornott C (2022) Climate change and cocoyam (*Colocasia esculenta* (L.) Schott) production: assessing impacts and potential adaptation strategies in Zimbabwe. Mitig Adapt Strateg Glob Chang 27:42. https://doi.org/10.1007/s11027-022-10014-9

Conklin HC (1957) Hanunoo agriculture. A report on an integral system of shifting cultivation in The Philippines, vol 2. Food and Agricultural Organization of the United Nations, Rome

Dangai Y, Hamawa Y, Tsobou R, Oumarou HZ (2021) Carbon stocks in Danielliaoliveri agroforestry parklands in the Sudano-sahelian zone of Cameroon. Environ Chall 5:100397. https://doi.org/10.1016/j.envc.2021.100397

Das A, Babu S, Singh R, Kumar S, Rathore SS, Singh VK, Datta M, Yadav SK, Wani OA, Yadav D (2022) Impact of live mulch-based conservation tillage on soil properties and productivity of summer maize in Indian Himalayas. Sustainability 14:12078. https://doi.org/10.3390/su141912078

Davis B, Mane E, Gurbuzer LY, Caivano G, Piedrahita N, Schneider K, Azhar N, Benali M, Chaudhary N, Rivera R, Ambikapathi R, Winters P (2023) Estimating global and country-level employment in agrifood systems, FAO statistics working paper series, no. 23–34. Rome, FAO. https://doi.org/10.4060/cc4337en

Dawson IK, Carsan S, Franzel S, Kindt R, van Breugel P, Graudal L, Jamnadass R (2014) Agroforestry, livestock, fodder production and climate change adaptation and mitigation in East Africa: issues and options. World Agroforestry Center, Nairobi. https://doi.org/10.5716/WP14050

Deogharia PC (2018) Diversification of agriculture: a review. J Econ Soc Dev 15(1):46–59

Dhyani SK, Handa AK (2013) Agroforestry in India and its potential for ecosystem services. In: Agroforestry systems in India: livelihood security & ecosystem services. Springer India, New Delhi, pp 345–365. https://doi.org/10.1007/978-81-322-1662-9_11

Diwakar SK, Zaidi SFA, Kumar S, Kumar A, Avasthe RK, Singh R, Babu S, Gudade BA, Verma G, Kumar N (2021) Effect of land use systems on soil health in eastern region of Uttar Pradesh. Indian J Agric Sci 91(4):647–650. https://doi.org/10.56093/ijas.v91i4.112743

Dixon RK, Krankina ON (1995) Can the terrestrial biosphere be managed to conserve and sequester carbon? In: Carbon sequestration in the biosphere: processes and prospects. Berlin/Heidelberg, Springer, pp 153–179

Duguma LA, Hager H (2011) Farmers' assessment of the social and ecological values of land uses in Central Highland Ethiopia. Environ Manag 47:969–982. https://doi.org/10.1007/s00267-011-9657-9

Ewel JJ (1999) Natural systems as models for the design of sustainable systems of land use. Agrofor Syst 45(1–3):1–21. https://doi.org/10.1023/A:1006219721151

FAO (2021) FAO statistical year book 2021. https://www.fao.org/3/cb4477en/online/cb4477en.html#

Foley JA, DeFries R, Asner GP, Barford C, Bonan G, Carpenter SR, Chapin FS, Coe MT et al (2005) Global consequences of land use. Science 309:570–574. https://doi.org/10.1126/science.1111772

Forde CD (1937) Land and labour in a Cross River village, Southern Nigeria. Geogr J 90(1):24–47. https://doi.org/10.2307/1788245

Forrester DI, Bauhus J, Cowie AL (2006) Carbon allocation in a mixed-species plantation of *Eucalyptus globulus* and *Acacia mearnsii*. For Ecol Manag 233:275–284. https://doi.org/10.1016/j.foreco.2006.05.018

Fouladbash L, Currie WS (2015) Agroforestry in Liberia: household practices, perceptions and livelihood benefits. Agrofor Syst 89:247–266. https://doi.org/10.1007/s10457-014-9763-9

Franzel S, Carsan S, Lukuyu B, Sinja J, Wambugu C (2014) Fodder trees for improving livestock productivity and smallholder livelihoods in Africa. Curr Opin Environ Sustain 6:98–103. https://doi.org/10.1016/j.cosust.2013.11.008

Gallina S, Mandujano S, González-Romero A (1996) Conservation of mammalian biodiversity in coffee plantations of Central Veracruz, Mexico. Agrofor Syst 33:13–27. https://doi.org/10.1007/BF00122886

Garrity D (2012) Agroforestry and the future of global land use. In: Nair P, Garrity D (eds) Agroforestry – the future of global land use. Advances in agroforestry, vol 9. Springer, Dordrecht. https://doi.org/10.1007/978-94-007-4676-3_6

Gebremeskel D, Birhane E, Rannestad MM, Gebre S, Tesfay G (2021) Biomass and soil carbon stocks of *Rhamnus prinoides* based agroforestry practice with varied density in the drylands of northern Ethiopia. Agrofor Syst 95(7):1275–1293. https://doi.org/10.1007/s10457-021-00608-8

Gockowski J, Asten PV (2012) Agricultural intensification as a climate change and food security strategy for sub-Saharan Africa. In: Wollenberg E, Nihart A, Tapio-Bostro ML, Grieg-Gran M (eds) Climate change mitigation and agriculture. ICRAF-CIAT, London/New York, pp 382–390

Gomes LC, Bianchi FJJA, Cardoso IM, Fernandes RBA, Fernandes Filho EI, Schulte RPO (2020) Agroforestry systems can mitigate the impacts of climate change on coffee production: a spatially explicit assessment in Brazil. Agric Ecosyst Environ 294:106858. https://doi.org/10.1016/j.agee.2020.106858

Gómez MU, Bueno AL, León AC, Uribe Bernal JI, Hernández Aguirre SA (2022) Traditional agroforestry systems: a methodological proposal for its analysis, intervention, and development. Agrofor Syst 1-13:491. https://doi.org/10.1007/s10457-021-00692-w

Haggar J, Pons D, Saenz L, Vides M (2019) Contribution of agroforestry systems to sustaining biodiversity in fragmented forest landscapes. Agric Ecosyst Environ 283:106567. https://doi.org/10.1016/j.agee.2019.06.006

Hailey (1957) The differing faces of Africa. Foreign Aff 36(1):143–153. https://doi.org/10.2307/20029269

Haines-Young R (2016) Report of results of a survey to assess the use of CICES. Support to EEA tasks under the EU MAES process. Negotiated procedure no EEA/NSS/16/002.

Howlett DS, Mosquera-Losada MR, Nair PR, Nair VD, Rigueiro-Rodríguez A (2011) Soil carbon storage in silvopastoral systems and a treeless pasture in northwestern Spain. J Environ Qual 40(3):825–832. https://doi.org/10.2134/jeq2010.0145

Hughes K, Morgan S, Baylis K, Oduol J, Smith-Dumont E, Vågen TG, Kegode H (2020) Assessing the downstream socioeconomic impacts of agroforestry in Kenya. World Dev 128:104835. https://doi.org/10.1016/j.worlddev.2019.104835

Iiyama M, Neufeldt H, Dobie P, Njenga M, Ndegwa G, Jamnadass R (2014) The potential of agroforestry in the provision of sustainable woodfuel in sub-Saharan Africa. Curr Opin Environ Sustain 6:138–147. https://doi.org/10.1016/j.cosust.2013.12.003

IPCC (2018) Global warming of 1.5. In: Masson-Delmotte V et al (eds) An IPCC special report on the impacts of global warming of 1.5C above pre-industrial levels and related global greenhouse gas emission pathways, in the context of strengthening the global response to the threat of climate change, sustainable development, and efforts to eradicate poverty. Switzerland

Jha B, Kumar N, Mohanty B (2009) Pattern of agricultural diversification in India. In: Working paper series no. E/302/2009. Institute of Economic Growth, University of Delhi Enclave North Campus, New Delhi

McCarthy JJ, Canziani OF, Leary NA, Dokken DJ, White KS (2001) Climate change 2001: impacts, adaptation and vulnerability, vol 2001. Cambridge University Press, Cambridge, UK, p 967. (Report of Working Group II)

Jose S (2009) Agroforestry for ecosystem services and environmental benefits: an overview. Agrofor Syst 76:1–10. https://doi.org/10.1007/s10457-009-9229-7

Jose S (2012) Agroforestry for conserving and enhancing biodiversity. Agrofor Syst 85:1–8. https://doi.org/10.1007/s10457-012-9517-5

Joshi PK, Gulati A, Birthal PS, Tewari L (2004) Agriculture diversification in South Asia: patterns, determinants and policy implications. Econ Polit Wkly 39(24):2457–2467

Kaba JS, Abunyewa AA (2021) New aboveground biomass and nitrogen yield in different ages of Giliricidia (*Gliricidia sepium* Jacq.) trees under different pruning intensities in moist semi-deciduous forest zone of Ghana. Agrofor Syst 95(5):835–842. https://doi.org/10.1007/s10457-019-00414-3

Khasa DP, Olivier A, Atangana AR, Bonneville J (2017) Two decades of agroforestry training, education and research at universities Laval, Quebec, Canada. Agrofor Syst 91:825–833. https://doi.org/10.1007/s10457-015-9871-1

Kiptot E (2008) Adoption dynamics of *Tithonia diversifolia* for soil fertility management in pilot villages of western Kenya. Exp Agric 44(4):473–484. https://doi.org/10.1017/S0014479708006704

Krausmann F, Erb KH, Gingrich S, Haberl H, Bondeau A, Gaube V, … Searchinger TD (2013) Global human appropriation of net primary production doubled in the 20th century. Proc Natl Acad Sci 110(25):10324–10329. https://doi.org/10.1073/pnas.1211349110

Kumar BM, Nair PKR (2004) The enigma of tropical homegardens. In: Nair PKR, Rao MR, Buck LE (eds) New Vistas in Agroforestry. Advances in Agroforestry, vol 1. Springer, Dordrecht. https://doi.org/10.1007/978-94-017-2424-1_10

Kumar BM, Nair PKR (2012) Carbon sequestration potential of agroforestry systems opportunities and challenges. Springer. https://doi.org/10.1007/978-94-007-1630-8

Kumar A, Singh V, Shabnam S, Oraon PR (2020) Carbon emission, sequestration, credit and economics of wheat under poplar-based agroforestry system. Carbon Manag 11(6):673–679. https://doi.org/10.1080/17583004.2020.1840875

Kumar R, Singh JK, Singh AK, Minz SD, Kumar NM (2021) Boron management in green gram (*Vigna radiata* L. Wilczek) under custard apple (*Annona squamosa* L.) based agri-horti system in semiarid region. Ann Arid Zone 60(3&4):01–05

Kumar SS, Mir SA, Wani OA, Babu S, Yeasin M, Bhat MA, Hussain N, Wani AA, Kumar R, Yadav D, Dar SR (2022a) Land-use systems regulate carbon geochemistry in the temperate Himalayas, India. J Environ Manag. https://doi.org/10.1016/j.jenvman.2022.115811

Kumar SS, Wani OA, Mir SA, Babu S, Sharma V, Chisti MH, Baba ZA, Sofi PA, Wani F, Dar SR, Singh R, Yadav D (2022b) Soil carbon dynamics in the temperate Himalayas: impact of land use management. Front Environ Sci 10:1009660. https://doi.org/10.3389/fenvs.2022.1009660

Kumar R, Kumar R, Karmakar S, Kumar A, Singh AK, Kumar A, Singh J (2023) Impact of amide fertilizer on carbon sequestration under the agroforestry system in the eastern plateau region of India. Sustain For 15(12):9775. https://doi.org/10.3390/su15129775

Kursten E (2000) Fuelwood production in agroforetry systems for sustainable land use and CO_2 mitigation. Ecol Eng 16:569–572. https://doi.org/10.1016/S0925-8574(00)00054-9

Kwesiga F, Akinnifesi FK, Mafongoya PL, McDermott MH, Agumya A (2003) Agroforestry research and development in southern Africa during the 1990s: review and challenges ahead. Agrofor Syst 59(3):173–186. https://doi.org/10.1023/B:AGFO.0000005222.68054.38

Lal R, Regnier E, Eckert DJ, Edwards WM, Hammond R (1991) Expectations of cover crops for sustainable agriculture. In: Hargrove W (ed) Cover crops for clean water. Soil and Water Conservation Society Publication, pp 1–11

Lehmann J, Peter I, Steglich C, Gebauer G, Huwe B, Zech W (1998) Below-ground interactions in dryland agroforestry. Forest Ecol Manag 111:157–169. https://doi.org/10.1016/S0378-1127(98)00322-3

Lin BB (2010) The role of agroforestry in reducing water loss through soil evaporation and crop transpiration in coffee agroecosystems. Agric For Meteorol 150(4):510–518. https://doi.org/10.1016/j.agrformet.2009.11.010

Lorenz K, Lal R (2014) Soil organic carbon sequestration in agroforestry systems. A review. Agron Sustain Dev 34:443–454. https://doi.org/10.1007/s13593-014-0212-y

Manaye A, Tesfamariam B, Tesfaye M, Worku A, Gufi Y (2021) Tree diversity and carbon stocks in agroforestry systems in northern Ethiopia. Carbon Balance Manag 16(1):14. https://doi.org/10.1186/s13021-021-00174-7

Mann HS, Saxena SK (1980) Khejri (*Prosopis cineraria*) in the Indian desert: its role in agroforestry. CAZRI, Jodhpur

Mbow C, Van Noordwijk M, Luedeling E, Neufeldt H, Minang PA, Kowero G (2014) Agroforestry solutions to address food security and climate change challenges in Africa. Curr Opin Environ Sustain 6:61–67. https://doi.org/10.1016/j.cosust.2013.10.014

Mead R, Willey R (1980) The concept of a 'land equivalent ratio' and advantages in yields from intercropping. Exp Agric 16(3):217–228. https://doi.org/10.1017/S0014479700010978

Miller DC, Ordonez PJ, Baylis K, Hughes K, Rana P (2017) Protocol for an evidence and gap map. The impacts of agroforestry on agricultural productivity, ecosystem services, and human well-being in low-and middle-income countries: an evidence and gap map, vol 13, Campbell collaboration, p 1. Ardmore, US

Montagnini F, Nair PKR (2004) Carbon sequestration: an underexploited environmental benefit of agroforestry systems. Agrofor Syst 61:281–295. https://doi.org/10.1007/978-94-017-2424-1_20

Moreno-Calles A, Casas A, Blancas J, Torres I, Masera O, Caballero J, Rangel-Landa S (2010) Agroforestry systems and biodiversity conservation in arid zones: the case of the Tehuacán Valley, Central México. Agrofor Syst 80:315–331. https://doi.org/10.1007/s10457-010-9349-0

Mosquera-Losada MR, Fernández-Núñez E, Rigueiro-Rodríguez A (2006) Pasture, tree and soil evolution in silvopastoral systems of Atlantic Europe. For Ecol Manag 232(1–3):135–145. https://doi.org/10.1016/j.foreco.2006.05.057

Muthuri CW, Kuyah S, Njenga M, Kuria A, Öborn I, van Noordwijk M (2023) Agroforestry's contribution to livelihoods and carbon sequestration in East Africa: a systematic review. Trees Forests People 14:100432. https://doi.org/10.1016/j.tfp.2023.100432

NAFP National Agroforestry Policy (2019) Ministry of Agriculture and livestock development (MoALD), Kathmandu

Nair P (1993) An introduction to agroforestry. Kluwer Academic Publisher, Dordrecht

Nair PKR, Nair VD (2002) Carbon storage in north American agroforestry systems. In: The potential of US forest soils to sequester carbon and mitigate the greenhouse effect. CRC Press, pp 333–346

Nair PKR, Kumar BM, Nair VD (2009) Agroforestry as a strategy for carbon sequestration. J Plant Nutr Soil Sci 172(1):10–23. https://doi.org/10.1002/jpln.200800030

Nair PKR, Nair VD, Kumar BM, Showalter JM (2010) Carbon sequestration in agroforestry systems. Adv Agron 108:237–307. https://doi.org/10.1016/S0065-2113(10)08005-3

Ng'endo M, Bhagwat SA (2019) The role of biodiversity in agroforestry and other types of smallholder farming. In: Agroforestry for sustainable agriculture. Burleigh Dodds Science Publishing, pp 301–316

Niang A, Amadalo B, Gathumbi S, Obonyo C (1996) Maize yield response to green manure application from selected shrubs and tree species in western Kenya: a preliminary assessment. East African Agric For J 62(1–2):199–207. https://doi.org/10.1080/00128325.1996.11663302

Nyberg Y, Musee C, Wachiye E, Jonsson M, Wetterlind J, Öborn I (2020) Effects of agroforestry and other sustainable practices in the Kenya agricultural carbon project (KACP). Land 9(10):389. https://doi.org/10.3390/land9100389

Okunlola JO, Oke DO, Adekunle VAJ, Owolabi KE (2019) Effects of climate change and coping strategies among crop farmers in south west, Nigeria. Agrofor Syst 93:1399–1408. https://doi.org/10.1007/s10457-018-0263-1

Olson RK, Schoeneberger MM, Aschmann SG (2000) An ecological foundation for temperate agroforestry. In: Garrett HE, Rietveld WJ, Fisher RF (eds) North American agroforestry: an integrated science and practice. American Society of Agronomy, Inc, Madison, pp 31–62

Padovan MP, Cortez VJ, Navarrete LF, Navarrete ED, Deffner AC, Centeno LG, Munguia R, Barrios M, Vilchez-Mendoza JS, VegaJarquin C, Costa AN, Brook RM, Rapidel B (2015) Root distribution and water use in coffee shaded with *Tabebuia rosea* Bertol and *Simarouba glauca* DC. Compared to full sun coffee in suboptimal environmental conditions. Agrofor Syst 89:857–868. https://doi.org/10.1007/s10457-015-9820-z

Palsaniya DR, Singh R, Yadav RS, Tewari RK, Dhyani SK (2011) Now it is water all the way in Garhkundar-Dabar watershed of droughtprone semi-arid Bundelkhand, India. Curr Sci 100:1287–1288

Palsaniya DR, Kumar S, Das MM, Rai SK, Kumar TK, Kumar S, Kumar P (2023) Rain water harvesting, agroforestry and goat-based intensification for livelihood resilience in drought prone

rainfed smallholder farming system: a case for semi-arid tropics. Agrofor Syst 97:1405–1419. https://doi.org/10.1007/s10457-023-00863-x
Pantera A, Mosquera-Losada MR, Herzog F, Den Herder M (2021) Agroforestry and the environment. Agrofor Syst 95(5):767–774. https://doi.org/10.1007/s10457-021-00640-8
Panwar P, Mahalingappa DG, Kaushal R, Bhardwaj DR, Chakravarty S, Shukla G, ... Gurung T (2022) Biomass production and carbon sequestration potential of different agroforestry systems in India: a critical review. Forests 13(8):1274. https://doi.org/10.3390/f13081274.
Patel AK, Bohra JS, Dewangan S, Singh AK (2022) Effect of fertility levels on sesame (*Sesamum indicum*) varieties under Agri-horti system. Indian J Agric Sci 92(8):1015–1017. https://doi.org/10.56093/ijas.v92i8.98465
Porter JR, Xie L, Challinor AJ et al (2014) Food security and food production systems. In: Field CB, Barros VR, Dokken DJ, Mach KJ, Mastrandrea MD, Bilir TE, Chatterjee M, Ebi KL, Estrada YO, Genova RC, Girma B, Kissel ES, Levy AM, MacCracken S, Mastrandrea PR, White LL (eds) Climate change 2014: impacts, adaptation, and vulnerability. Part A: global and sectoral aspects. Contribution of working group II to the fifth assessment report of the intergovernmental panel on climate change. Cambridge University Press, Cambridge, UK, pp 485–533. http://www.ipcc.ch/report/ar5/wg2/
Pretty J (2018) Intensification for redesigned and sustainable agricultural systems. Science 362:362. https://doi.org/10.1126/science.aav0294
Pye-Smith C (2010) The fruits of success: a programme to domesticate west and Central Africa's wild fruit trees is raising incomes, improving health, and stimulating the rural economy. World Agroforestry Centre, Nairobi
Raghvan MS (1960) Genesis and history of the Kumari system of cultivation. In: Proceedings of the ninth silvicultural conference 1956. Dehradun, Forestry Research Institute (FRI)
Rahman MH, Ahrends HE, Raza A, Gaiser T (2023) Current approaches for modeling ecosystem services and biodiversity in agroforestry systems: challenges and ways forward. Front Glob Change 5:1032442. https://doi.org/10.3389/ffgc.2022.1032442
Rathore SS, Babu S, El-Sappah AH, Shekhawat K, Singh VK, Singh RK, Upadhyay PK, Singh R (2022) Integrated agroforestry systems improve soil carbon storage, water productivity, and economic returns in the marginal land of the semi-arid region. Saudi J Biol Sci 29:103427. https://doi.org/10.1016/j.sjbs.2022.103427
Reij C, Tappan G, Smale M (2009) Agroenvironmental transformation in the Sahel. Food policy, IFPRI discussion paper. http://www.ifpri.org/sites/default/files/publications/ifpridp00914.pdf
Reyes F, Gosme M, Wolz KJ, Lecomte I, Dupraz C (2021) Alley cropping mitigates the impacts of climate change on a wheat crop in a mediterranean environment: a biophysical model based assessment. Agriculture 11:356. https://doi.org/10.3390/agriculture11040356
Rigueiro-Rodri-Guez A, McAdam J, Mosquera-Losada MR (2009) Agroforestry in Europe: current status and future prospects advances in agroforestry, vol 9. Springer, Dordrech
Rodríguez-Rigueiro FJ, Santiago-Freijanes JJ, Mosquera-Losada MR, Castro M, Silva-Losada P, Pisanelli A, ... & Ferreiro-Domínguez N (2021) Silvopasture policy promotion in European Mediterranean areas. PLoS One, 16(1), e0245846. https://doi.org/10.1371/journal.pone.0245846h
Rossier C, Lake F (2014) Indigenous traditional ecological knowledge in agroforestry. USDA National Agroforestry Center
Roupsard O, Audebert A, Ndour AP, Clermont-Dauphin C, Agbohessou Y, Sanou J et al (2020) How far does the tree affect the crop in agroforestry? Newspatial analysis methods in a *Faidherbia* parkland. Agric Ecosyst Environ 296:106928. https://doi.org/10.1016/j.agee.2020.106928
Saleem I, Mugloo JA, Pala NA, Bhat GM, Masoodi TH, Mughal AH, Mehraj B (2023) Biomass production, carbon stock and sequestration potential of prominent agroforestry systems in north-western Himalaya. India Front Glob Change 6:1192382. https://doi.org/10.3389/ffgc
Sanchez PA (1995) Science in agroforestry. In: Sinclair FL (eds) Agroforestry: Science, Policy and Practice. Forestry Sciences, vol 47. Springer, Dordrecht. https://doi.org/10.1007/978-94-017-0681-0_1

Sanchez P (1999) Improved fallows come of age in the tropics. Agrofor Syst 47:3–12. https://doi.org/10.1023/A:1006287702265

Sanchez PA, Shepherd KD, Soule MJ, Place FM, Buresh RJ, Izac AMN, Woomer PL (1997) Soil fertility replenishment in Africa: an investment in natural resource capital. In: Buresh RJ, Sanchez PA, Calhoun F (eds) Replenishing soil fertility in Africa. Soil Science Society of America American Society of Agronomy, Madison, pp 1–46. https://doi.org/10.2136/sssaspecpub51.c1

Santiago-Freijanes JJ, Mosquera-Losada MR, Rois-Díaz M, Ferreiro-Domínguez N, Pantera A, Aldrey JA, Rigueiro-Rodríguez A (2021) Global and European policies to foster agricultural sustainability: agroforestry. Agrofor Syst 95(5):775–790. https://doi.org/10.1007/s10457-018-0215-9

Sharrow SH, Ismail S (2004) Carbon and nitrogen storage in agroforests, tree plantations, and pastures in western Oregon, USA. Agrofor Syst 60:123–130. https://doi.org/10.1023/B:AGFO.0000013267.87896.41

Simelton E, Dam BV, Catacutan D (2015) Trees and agroforestry for coping with extreme weather events: experiences from northern and central Viet Nam. Agrofor Syst 89(6):1065–1082. https://doi.org/10.1007/s10457-015-9835-5

Singh B, Kaur N, Gill RIS, Dagar JC (2016) Agroforestry for crop diversification and carbon sequestration. Agrofor Res Dev 99:101–148

Singh AK, Singh BK, Kumar NM, Karada MS, Agnihotri D, Singh R (2023) Agroforestry as a mechanism for reforestation: scenarios within REDD. Biol Forum Int J 15(7):370–377

Sirohi C, Dhillon RS, Chavan SB, Handa AK, Balyan P, Bhardwaj KK, Ahlawat KS (2022) Development of poplar-based alley crop system for fodder production and soil improvements in semi-arid tropics. Agrofor Syst 96(4):731–745. https://doi.org/10.1007/s10457-022-00735-w

Thorlakson T, Neufeldt H (2012) Reducing subsistence farmers' vulnerability to climate change: evaluating the potential contributions of agroforestry in western Kenya. Agric Food Secur 1:15. https://doi.org/10.1186/2048-7010-1-15

Tsedeke RE, Dawud SM, Tafere SM (2021) Assessment of carbon stock potential of parkland agroforestry practice: the case of Minjar Shenkora; north Shewa, Ethiopia. Environ Syst Res 10:1–11. https://doi.org/10.1186/s40068-020-00211-3

Udawatta RP, Gantzer CJ, Jose S (2017) Agroforestry practices and soil ecosystem services. In: Soil Health and Intensification of Agroecosytems, pp 305–333. https://doi.org/10.1016/B978-0-12-805317-1.00014-2

Unruh JD, Houghton RA, Lefebvre PA (1993) Carbon storage in agroforestry: an estimate for sub-Saharan Africa. Clim Res 3:39–52

Verchot LV, Noordwijk MV, Kandji S, Tomich T (2007) Climate change: linking adaptation and mitigation through agroforestry. Mitig Adapt Strateg Glob Chang 12:901–918. https://doi.org/10.1007/s11027-007-9105-6

Verma K, Sharma P, Kumar D, Vishwakarma SP, Meena NK (2021) Strategies sustainable management of agroforestry in climate change mitigation and adaptation. Int J Curr Microb Applie Sci 10(01):2439–2449. https://doi.org/10.20546/ijcmas.2021.1001.282

Verma K, Sharma P, Bhardwaj DR, Kumar R, Kumar NM, Singh AK (2023a) Land and environmental management through agriculture, forestry and other land use (AFOLU) system. Land Environ Manag Forest:247–271. https://doi.org/10.1002/9781119910527.ch10

Verma T, Bhardwaj DR, Sharma U, Sharma P, Kumar D, Kumar A (2023b) Agroforestry systems in the mid-hills of the north-western Himalaya: a sustainable pathway to improved soil health and climate resilience. J Environ Manag 348:119264. https://doi.org/10.1016/j.jenvman.2023.119264

Verma K, Prasad SK, Singh MK, Sharma P (2023c) Response of alley-cropped pearl millet (*Pennisetum glaucum*) to nitrogen and zinc schedules under semi-arid regions. Indian. J Agron 68(1):105–109. https://doi.org/10.59797/ija.v68i1.213

Wani OA, Kumar S, Hussain N, Wani AA, Babu S, Alam P, Rashid M, Popescu SM, Mansoor S (2022) Multi-scale processes influencing global carbon storage and land-carbon-climate nexus: a critical review. Pedosphere 33:250–267. https://doi.org/10.1016/j.pedsph.2022.07.002

Wani OA, Sharma V, Kumar SS, Babu S, Sharma KR, Rathore SS, Marwaha S, Ganai NA, Dar SR, Yeasin M, Singh R, Tomar J (2023) Climate plays a dominant role over land management in governing soil carbon dynamics in North Western Himalayas. J Environ Manag 338:117740. https://doi.org/10.1016/j.jenvman.2023.117740

Williamson JD (2022) Field hand labor in the cybershire imaginary and urban agrarian questions beyond the metabolic rift. University of California, Berkeley, pp 1–24

World Agroforestry Centre (2010) Medium-term plan 2011–2013. Consultative group on International Agriculture Research, Montpellier, p 148

Yadav GS, Kandpal BK, Das A, Babu S, Mohapatra KP, Devi G, Chandra P, Singh R, Barman KK (2021) Impact of 28-year-old agroforestry systems on soil carbon dynamics in eastern Himalayas. J Environ Manag 283:111978. https://doi.org/10.1016/j.jenvman

Yadav D, Singh D, Babu S, Madegowda M, Singh DV, Mandal D, Rathore AC, Sharma VK, Singhal V, Kumawat A, Yadav DK, Yadav RK, Kumar S (2023) Intensified cropping reduces soil erosion and improves rainfall partitioning and soil properties in the marginal land of the Indian Himalayas. Int Soil Water Cons Res 10:002. https://doi.org/10.1016/j.iswcr

Zhang L, Yadav V, Zhu H, Shi Y, Hu X, Wang X, Zhou X, Babu S, Kang Y (2024) Climate drivers of litterfall biomass dynamics in three types of forest stands on the loess plateau. Ecol Indic 112088. https://doi.org/10.1016/j.ecolind.2024

Zomer RJ, Bossio DA, Trabucco A, Yuanjie L, Gupta DC, Singh VP (2007) Trees and water: smallholder agroforestry on irrigated lands in northern India, vol 122. IWMI

Zomer RJ, Neufeldt H, Xu J, Ahrends A, Bossio D, Trabucco A, Wang M (2016) Global tree cover and biomass carbon on agricultural land: the contribution of agroforestry to global and national carbon budgets. Sci Rep 6(1):29987. https://doi.org/10.1038/srep29987

Chapter 11
Food–Water–Energy Nexus in Designing Sustainable Food Production Systems

S. K. Naik, S. S. Mali, M. K. Dhakar, Reshma Shinde, and S. Mondal

Abstract Scope of horizontal expansion of agriculture is limited due to land, water, and energy shortages and therefore, an additional food demand must be met by vertical expansion, i.e., by increasing cropping intensity and productivity of existing agricultural land. Uncertainty of monsoon rain is a serious issue in attaining secured yield. Adoption of different resource conservation technologies (RCTs) could be instrumental in attaining sustainable productivity and higher cropping intensity. Declining soil health, irregularities in monsoonal rain, lack of irrigation facilities, low residual soil moisture, prevalence of long-duration rice variety, and free grazing/poor socioeconomic conditions of farmers are major issues leading to unsustainability of mono-cropped rice. In scenario of global climate change, low carbon emitting and low energy input farming systems are given paramount importance. The conservation agriculture-based systems having higher system productivity, better C-sequestration potential, lower C-footprint, and higher energy and water productivity have a very good potential for sustaining soil health and crop yield. The fertigation dose and maintaining optimal plant population under polythene mulched drip irrigation improves yield and increases nutrient and water-use efficiencies of many vegetable crops like tomato, cabbage, cauliflower, broccoli, and sweet corn in the uplands. Growing biomass-yielding plants in the alley area of orchards and recycling of harvested biomass in the plant basin of fruit trees have been found to improve soil fertility and plant growth of number of fruit trees. A multitier cropping system helps to maximize the use of resources like sunshine, land, water, and air by producing two or more crops of varying heights on a single plot of ground. Multitier cropping systems based on mango, litchi, and guava were shown to be suited for the highland environment for sustainable food production. Energy and carbon budgeting of different production systems are extremely reliant on the type of tillage oper-

S. K. Naik (✉) · S. S. Mali · M. K. Dhakar · R. Shinde
ICAR Research Complex for Eastern Region, Farming System Research Centre for Hill and Plateau Region, Ranchi, Jharkhand, India

S. Mondal
ICAR-Central Citrus Research Institute, Nagpur, India

S. Babu et al. (eds.), *Agricultural Diversification for Sustainable Food Production*, Sustainability Sciences in Asia and Africa,
https://doi.org/10.1007/978-981-97-7517-0_11

ation, crop residue, and choice of crops. This calls for precise and accurate assessments of energy and carbon budget of diversified cropping systems. The interaction among the food, water, and energy systems emerges as an integrated tool for designing better cropping systems, energy, and carbon-efficient cropping systems in the region for sustainable food production systems.

Keywords Carbon · Circular economy · Energy · Sustainable agriculture

11.1 Food–Water–Energy Nexus

Food, water, and energy are crucial resources and key elements of life and human welfare. The nexus among the key elements means that water, energy, and food production are closely connected. The food–water–energy nexus is generally interpreted as the interactions among the different subsystems within the nexus system. The food–water nexus is presented as the interdependencies between food and water, as they are coupled in their supply, distribution, processing, and use (Yadav et al. 2021). Similarly, food–water–energy nexus are intimately related, as water is needed for production processes both for food and energy (Pan et al. 2024). Energy is required for water abstraction, distribution, and treatment. Further, food can be used to generate energy in the form of biofuels. The extreme weather events cause draught, which draws more water for the production process and ultimately affects food and energy security. As the securities of these three subsystems become severe, the failure in one subsystem may exert pressure on the other two subsystems, and thus requiring a holistic management among these subsystems to make the food–water–energy nexus more vibrant. The food–water–energy nexus considers how food, water, and energy subsystems operate and aims to maximize synergies (mutually beneficial outcomes) and minimize trade-offs (nonoptimal outcomes) and improve resource use efficiency. Further, the nexus focused to strengthen and improve management outcomes to enhance food, water, and energy security.

At present, all the elements of the nexus are under increasing stress. The global water demand will increase by 55% by 2050 so 40% of the global population will be living under water-stressed conditions (OECD 2012). The water demand for energy production will increase by about 20% (IEA 2012). About 800 million people are currently hungry and by 2050 food production will have to increase by about 50% due to the rising world population (WHO 2017). The climate change will further aggravate the water resources under stress because of higher temperatures, resulting in shifts in precipitation patterns and snow cover, and a likely increase in the frequency of flooding and droughts (EEA 2018), which will have an impact on global food production. Water is a vital element for raising different kinds of fruits, vegetables, cereals, and animals along with the products that we use as food. Water for food production accounts for 70% of water withdrawals. Globally, the progress of agricultural practices is influenced by the accessibility of water assets. Water is required for food production, especially for irrigation and crop processing. Nowadays, the relationship between water and food is vulnerable. First, changing water supply patterns that affect the reliability of agriculture and second growing competition for limited water resources to meet the planned addition in food needs (Ferroukhi et al. 2015). Several food security concerns are influenced by accessibility to water with acceptable quality. In order to produce food and further downstream effects during the preparation of food, it is necessary to prepare and consume sufficient water with acceptable quality. The various water-related risks to food security are (i) increased variability in water availability due to climate change results in changes in supply of food products, leading to higher price volatility; (ii) impact of water quality on food production and consumption, which results in utilization of poor-quality water along different stages of food supply chain have negative impacts, which includes soil degradation and accumulation of contaminants within the food chain.

The various food-related risks to water security are (i) the use of external inputs for agriculture and food production can lead to water pollution affecting downstream activities and aquatic life, (ii) increased agricultural land leasing could lead to expanded use of local water resources with negative local socioeconomic impacts, and (iii) water resource overutilization due to food security.

Water and energy are multifaceted problems with many parts changing their supply and demand and both are important for economic growth. Water plays an important role in almost every stage of energy development, consisting of the exploitation, production, and processing of fossil fuels, electricity generation, extraction, treatment, and processing of water from energy-related measures. At the same time, energy is needed to lift, move, distribute, and purify water (Pereira-Cardenal et al. 2016). The water–energy nexus studies focus on energy generation from crops evaluated by water footprint and sustainable development. Energy is a critical element in food production and for mechanization, land preparation, fertilizer production and application, irrigation, packaging, food processing, and storage (Yadav et al. 2020).

A sustainable food production system refers to the agriculture system that delivers healthy food to meet the current food requirements, while at the same time maintaining healthy and sustainable ecosystems that are capable of providing food for generations to come, with controlled and minimum negative impact on the environment (Allen and Prosperi 2016; Rathore et al. 2022; Babu et al. 2023). It is a system that encourages local production and knowledge, providing nutritious and healthy food, which is available, accessible, and affordable to all and at the same time protecting farmers, consumers, and communities.

11.2 Factors Governing the Sustainability of Food Production Systems

At present the sustainability of our food security is challenged, as 30–50% of the food is lost or wasted in different stages of the food systems with consequent higher consumption of animal foodstuffs and larger environmental footprints.

Some of the factors responsible for making the food production system more sustainable.

Efficient Use of Natural Resources

- Reduction in the use of fossil fuels and optimization of water use in production.
- Optimum land use and reduction in the conversion of land for agriculture.
- Designing the water and energy-efficient food production.
- Recycling of manure from animal production for industrial uses to recover energy and nutrients.

Protection of the Quality of Natural Resources

- Optimum use of fertilizers and pesticides to avoid pollution of soils and waterways.
- Reduction in greenhouse gas emissions to help maintain air quality.
- Restoration of organic matter content of soils by minimizing soil loss.
- Farming practices promote biodiversity for safeguarding land, water, and energy resources.
- Reduction in the pollution of coastal areas from fertilizers and litter for the aquatic animals.
- Prioritizing resource use efficiency, sustainable food production systems can improve economic viability for farmers, mitigate environmental degradation, and ensure long-term resilience in the face of global challenges such as climate change and population growth.

Technology and Innovations

Technology and innovations in agriculture are paramount for the advancement of sustainable food production systems, offering tools and practices that enhance efficiency, productivity, and environmental stewardship. Precision agriculture technologies, such as satellite imagery and sensors, enable precise resource management, optimizing the use of water, fertilizers, and pesticides.

Promotion of Environmentally Efficient Food Packaging

- Optimization of packaging use
- Material should be recycled and have lower environmental impact

Reduction in Food Waste and Food Losses

- Extension of shelf life and use of functional packaging to reduce spoilage and waste of perishable foods.
- Minimization of food waste by diverting excess food to food recycling plants and its further use as animal feed.

- The food losses and wastes as well as by-products from animals and vegetables may be subjected to recover the energy as biofuel and promotion as preparation of organic manure through efficient composting process.
- Farming methods of increased mechanization can reduce food loss.
- Structural infrastructure like transport and storage systems and processing facilities will reduce food loss.
- Social education on reduce, reuse, and recycle of food waste may contribute to increased sustainable consumption behavior.

Building Stronger Local and Circular Food Systems
The circular food systems help to maintain the valuable natural resources, minerals, and nutrients in the loop. For example, wasteater can be treated to extract important minerals like phosphates. Food loss and waste can be composted so that valuable nutrients return to the soil instead of being thrown away. The organic waste can be used for bioenergy to power homes and agribusiness.

Education and Awareness
Education and awareness are pivotal for the success of sustainable food production systems as they empower farmers, consumers, and stakeholders to make informed decisions that promote environmentally conscious and socially responsible practices. Furthermore, educating policymakers, researchers, and the general public about the importance of sustainable food production fosters a collective commitment to creating and implementing policies that prioritize long-term environmental health, social equity, and resilience in agriculture.

11.3 Sustainable Food Production Systems

11.3.1 Micro-irrigation

Micro-irrigation is the most widely used irrigation method for crops in many regions of the world. Amid rising water scarcity, micro-irrigation technology as a tool to improve water-use efficiency and farmers' welfare has occupied a prominent place in policy discourse in India (Barman et al. 2001). These technologies are promoted showcasing the prominent gains in terms of water saving in irrigated agriculture, increased income, and enhanced food and nutritional security. Micro-irrigation technologies result in a significant productivity improvement and, hence, economic gain over the traditional method of surface irrigation (Mali et al. 2019; Naik et al. 2022). The most important determinants of micro-irrigation adoption include access to groundwater, the prevailing cropping pattern, level of education, financial resources, the social stratum of the household, and the wealth or poverty status of the farmer. The states of Jharkhand, Bihar, Assam, West Bengal, and Chhattisgarh offer a lot of potential for bringing newer areas under micro-irrigation (Table 11.1).

Table 11.1 Adoption potential (Mha) of micro-irrigation in the states of eastern region of India

State	Sprinkler	Drip	Total
Assam	0.08	0.05	0.13
Bihar	2.92	1.01	3.93
Chhattisgarh	0.53	0.4	0.93
Jharkhand	0.13	0.04	0.17
Odisha	0.33	0.29	0.62
West Bengal	2.05	0.95	3.00
Total	6.04	2.74	8.78

Table 11.2 Drip and sprinkler irrigated area in eastern region of India

State	Drip (ha)	Sprinkler (ha)	Micro-irrigation (ha)	Percent of NIA (%)
Assam	2374	11,320	13,694	3.74
Bihar	12,488	106,979	119,467	3.84
Chhattisgarh	27,504	316,456	343,960	21.98
Jharkhand	25,081	17,298	42,379	18.35
Odisha	26,134	105,095	131,229	12.02
West Bengal	10,329	78,182	88,511	2.85

Source: Ministry of Agriculture and Farmers Welfare, Government of India

From the perspective of climate-smart agriculture, micro-irrigation can play a stellar role in saving water and energy. According to FAO micro-irrigation, especially the small holders' low-cost micro-irrigation gives a window to maneuver and break the nexus to an extent. Several researchers have highlighted the benefits of the technology. The studies revealed that micro-irrigation systems can lead to water saving of about 40–70% over surface irrigation methods and increased yield up to 100% for some crops. In addition, saline water can also be used in this system and the salt is accumulated only at the surface of the periphery of wetting zone and hence does not affect crop growth.

The technology of modern drip irrigation developed in the 1960s marked a significant step in the history of irrigation science. Many recent studies have proved the advantages of drip irrigation over surface methods of irrigation in terms of water saving, improved economic returns, and increased yields and water-use efficiency. Present status of area under the drip and sprinkler irrigation system is given in Table 11.2.

Drip or trickle irrigation is the most efficient and widely used fertigation method for crops in many regions of the world (Mali et al. 2017). Micro-irrigation (sprinkler and drip) technologies enhance water application efficiency in agriculture and reduce loss of water. Drip systems apply water and minerals evenly across crops, thus helping to reduce wastage and increase crop yields. It involves the application of water at a determined rate to the root zone of crops. This significantly reduces water run loss through deep percolation or evaporation. Required minerals can also be added to the water, thus allowing for increased efficiency, minimized

Table 11.3 Performance of different crops under drip irrigation

Crop	Fertilizer dose/Application	Planting geometry	Spacing (cm)	Yield (t/ha)	Water productivity (kg/m)
Tomato	120:60:60 kg NPK/ha (higher dose during late reproductive stage) (13th–18th week after transplanting)	Triangular pair row system	50 (PP) × 40 (RR)	81	34.8
Chili	120:60:60 kg NPK/ha (higher dose during early reproductive stage) (8th–13th WAT)	Triangular pair row system	30 (PP) × 40 (RR)	14.5	5.2
Broccoli	150:60:60 kg NPK/ha (higher dose during early reproductive stage) (7th–11th WAT)	Triangular pair row system	30 (PP) × 40 (RR)	31.3	15.8
Sweet corn	150:60:60 kg NPK/ha (higher dose during late reproductive stage) (10th–12th week after transplanting)	Rectangular pair row system	15 (PP) × 50 (RR)	21.3	7.4

Source: Jha et al. (2015)

oversaturation, and improved productivity. Many states are now taking lead in promoting and installing micro-irrigation systems in India. Adoption of mini-sprinkler led to more effective consumptive use of water and water saving to the tune of 57.14% as compared to flood irrigation (Kadam et al. 2014). Performance of drip irrigation systems in different crops is detailed in Table 11.3.

The major benefits of micro-irrigation technology are increase in yield, water-use efficiency, reduction in the cost of water, fertilizers, and manures. This technology offers higher benefits like irrigation efficiency (50–90%), fertilizer (28.8%), and energy (30.5%) (Kumar and van Dam 2013). Micro-irrigation is an intervention to address various issues of agricultural growth and hence it is considered leveraging technology for sustainable food production.

11.3.2 Conservation Agriculture

Rice–wheat (RW) system, the largest cropping system in the world, occupies an area of ~24 m ha spread over Asia and 13.5 m ha in South Asia. The traditional system of growing rice with puddling helps in reducing the water loss through percolation and weed management; it led to several problems in post-rice crops, threatening sustainability of ecosystem as follows:

- Erratic stand establishment of post-rice crops owing to poor contact of the seed with soil.

- Delayed sowing of wheat due to late harvesting of long-duration transplanted rice, resulting in lower wheat yield.
- Overexploitation of the groundwater.
- Soil compaction with consequent increase in bulk density resulting in restricted root growth.
- Weed shift and development of herbicide-resistant weeds.
- Sharp decline in soil organic matter.
- Multi-nutrient deficiencies.

To counteract some of these problems, CA is now being promoted which consists of combining the three principles: (a) soils are not disturbed (by more than 15 cm in width or more than 25%, whichever is less, of the cropped area) or tilled periodically; (b) more than 30% of the soil has to be covered with organic residues at planting; and (c) crop rotation has to be implemented involving at least three different crops. Conservation agriculture embodies a set of practices that prioritize environmental health, soil conservation, and long-term agricultural resilience. Conservation agriculture has emanated as environmentally sustainable management alternative to tillage-intensive practices of crop production (FAO 2011; Edralin et al. 2016). It helps farmers to maintain and boost yields and increase profits while reversing land degradation, protecting the environment, and responding to growing challenges of climate change. It enhances biodiversity (Kumar et al. 2023) and natural biological processes above and below ground surface, which contribute to increased water and nutrient use efficiency for sustainable crop production.

11.3.2.1 Influence of CA on Soil Properties and Crop Yield

The main sources of C input into soil through plants are roots–shoots, root exudates, and root-borne organic substances released into the rhizosphere during plant growth as well as root hairs and fine roots sloughed by root elongation. The elimination of monocultures and incorporation of plant species into rotations that return greater amounts of residue to the soil are often associated with greater SOC stock in CA systems (Yadav et al. 2018; Kumar et al. 2022a). The reduction in tillage and the preservation of organic matter on the field contribute to increased carbon content in the soil, mitigating greenhouse gas emissions and bolstering the soil's ability to act as a carbon sink (Yadav et al. 2019). The practice of CA (legume crop was taken along with rice and wheat with CA) in eastern IGP registered the highest total organic carbon (TOC) stock of 47.71 Mg C/ha and resulted in significant increase of 14.57% relative to Farmer's practice in 0–30 cm soil depth after 7 years of field trial. Further, the system productivity was enhanced by 56.1% in the CA-based production system (Samal et al. 2017). The legume cover crops, which contain carbon compounds likely more resistant to microbial metabolism, could also increase the complexity and diversity of soil carbon, making it more stable (Singh et al. 2016, 2021a). The rice–fallow ecosystems of eastern India on three crop establishment methods: (1) zero-till-direct-seeded rice (ZTDSR), (2) conventional-till direct-seeded rice (CTDSR), and (3) transplanted puddled rice (TPR), and two residue management practices (i) with residue, and (ii) without residue in combination with five potential winter season crops, i.e., chickpea (*Cicer arietinum* L.), lentil (*Lens culinaris* L.), safflower (*Carthamus tinctorius* L.), linseed (*Linum usitatissimum* L.), and mustard (*Brassica juncea* L.) revealed that ZTDSR treatment reduced carbon-footprint (C-footprint) by 2.8% compared to TPR-based production systems. The CA-based production system enhanced the system productivity by 46.5% due to diversification of rice–fallow systems with oilseed and pulse crops (Kumar et al. 2022b). Pulse-based crop rotations (rice–lentil and rice–chickpea) under ZTDSR with surface crop residue yielded 21.5% higher system net returns as compared to rice–oilseed production systems. The rice–chickpea, rice–lentil, and rice–safflower production systems in combination with ZTDSR along with residue retention can be viable production systems to lower C-footprint.

Conservation agriculture mitigates soil degradation and erosion, but it also enhances water-use efficiency, making it particularly valuable in regions prone to drought. The CA-based crop management practices in the eastern plateau and hill region of India highlighted the importance of including second and third crops in the rice–fallow ecosystems (Naik et al. 2017). This practice enhanced the very labile C in all of the soil depths and stabilized the SOC as a non-labile recalcitrant fraction in 0–0.30 m soil depth. Further, the CA practice showed a lower carbon-footprint, higher carbon sustainability index, and carbon efficiency (Babu et al. 2023a, b). A cropping system having direct-seeded rice (DSR)–mustard–black gram under conservation agriculture was adjudged as the best practice in terms of C-sequestration, C-stabilization, biological activity with higher system productivity (23.7%), energy use efficiency (53.5%), and reduced carbon footprints (26.6%). The CA-based system enhanced the water productivity of rice, mustard, and black gram by 7.0%, 23.6%, and 14.1% over CP, respectively, and overall system productivity by 23.7% (Naik et al. 2023).

Conservation agriculture has been identified as one of the technological options to meet the global challenges of increasing food production and conserving environment, thereby improving food and nutritional security and alleviating poverty.

11.3.3 Integrated Farming Systems

To ensure food security while safeguarding the environment, there is need for resilient agricultural practices like Integrated Farming Systems (IFS), bio-intensive cropping, conservation agriculture, and organic farming to promote resource recycling and profitability (Shyam et al. 2023). The IFS, a complex matrix of soil, plants, animals, and other inputs, integrates various farm enterprises. It offers a diversified and resilient approach to agriculture, combining crops, animal husbandry, fisheries, forestry, sericulture, horticulture, and poultry for optimal resource utilization (Fatima et al. 2023). Diversified agricultural systems enhance resilience and address challenges posed by land degradation and climate change, contributing to global food security. By adopting a climate-resilient integrated approach, IFS provides a stable and sustainable production system, minimizing risks and building resilience to climate change (Babu et al. 2023). Integrated Farming Systems (IFS) play critical roles in the food–water–energy nexus by contributing to the design of sustainable food production systems. The food–water–energy nexus represents the interconnected relationship between food production, water resources, and energy consumption. To ensure long-term sustainability, it is essential to optimize the utilization of these resources and minimize their environmental impacts through integrated farming system.

11.3.3.1 Resource Optimization

Water Management Integrated farming systems often incorporate practices such as crop rotation, intercropping, and agroforestry, which use water more effectively, reducing water waste and enhancing water-use efficiency (Rathore et al. 2022). Trees in agroforestry systems help in water retention and reduce soil erosion (Yadav et al. 2021). Efficient irrigation practices, such as drip irrigation combined with crop diversification, are often part of IFS, optimizing water use in agriculture.

Energy Efficiency IFS focus on optimizing resource use, including energy. It promotes practices like reduced tillage, the use of organic matter, and diversified cropping systems, which can reduce energy inputs in agriculture. This contributes to energy conservation and sustainability in food production. By integrating various components, IFS can enhance energy efficiency (Babu et al. 2023). For example, integration of livestock will help in plowing fields and transportation, livestock waste can be utilized for biogas production, used for cooking and lighting, which indeed reduces the reliance on external energy sources (Fatima et al. 2023). Integration of renewable energy sources, such as solar panels or biogas units powered by organic waste, aligns with the goal of sustainable energy use in farming.

Nutrient Cycling Nutrient cycling in Integrated Farming Systems (IFS) minimizes reliance on synthetic fertilizers, fostering a sustainable and eco-friendly farming approach. IFS encourages nutrient management by producing rich sources of nutrients, such as farmyard manure and vermicompost, from agricultural waste (Das et al. 2019). Livestock manure fertilizes crops, while crop residues serve as animal feed, promoting efficient nutrient cycling and soil fertility (Naik et al. 2018). The integration of livestock and fisheries with crops is found to be beneficial for increasing nutrient use efficiency, nutrient recycling, and higher soil microbial activity in soil. IFS practices like crop diversification and cover cropping reduce the demand for synthetic fertilizers, decrease nutrient runoff, and enhance soil health, ultimately improving water quality.

11.3.3.2 Diversified Food Pproduction

IFS involves the cultivation of a variety of crops, including cereals, legumes, and cash crops, as well as livestock. This diversification helps in using resources more efficiently and reduces the risk of resource depletion in mono-cropping systems. Diversified farming systems are less susceptible to pests and diseases, enhancing food security (Singh et al. 2021b). These systems provide a balanced diet and reduce the need for energy-intensive food processing. The diversification of cropping systems with vegetables, tuber, pulses, and fruit crops provides nutritional and food security (Babu et al. 2019; Pan et al. 2022).

11.3.3.3 Climate Resilience

The diversified nature of Integrated Farming Systems (IFS) acts as a protective measure against climate variability (Babu et al. 2019; Das et al. 2021). Varied crops and livestock species exhibit different responses to climatic conditions, thereby reducing overall risks to the farming system. The inclusion of agroforestry components in IFS contributes to carbon sequestration, addressing concerns related to climate change. In a study by Salton et al. (2014), net greenhouse gas (GHG) emissions were observed as positive in conventional systems and negative in IFS. This trend was primarily attributed to higher soil carbon sequestration in the IFS system, counterbalancing N_2O emissions.

Efficient and resilient crop and livestock production systems in the face of climate change can be achieved through various management practices. Practices such as nutrient management through composting and crop residues, as well as the use of legumes for nitrogen fixation, contribute to increased resilience of crops to changing climates and reduce GHG emissions. IFS is recognized as a promising strategy to enhance agricultural production, rehabilitate degraded pastures, and mitigate GHG emissions (Gil et al. 2015). Barbosa et al. (2015) noted IFS as a viable strategy to reduce GHG emissions and nutrient loss through improved nutrient recycling and the utilization of crop residues as animal feed. Thus, IFS help reduce environmental pollution by decreasing greenhouse gas emissions, soil erosion, and runoff, ultimately contributing to soil and water resource sustainability and resilience in agriculture (Table 11.4).

11.3.3.4 Biodiversity Conservations

IFS play a crucial role in biodiversity conservation by integrating crops, trees, shrubs, and livestock, creating diverse habitats that promote biodiversity in agricultural landscapes. This diversity enhances ecosystem services such as pollination, natural pest control, risk reduction of crop failures, and maintenance of soil fertility, and contributes to ecosystem resilience. Additionally, IFS stimulates soil microbial biodiversity through the addition of compost, manure, or duck droppings in

Table 11.4 GHG emission from different IFS models in India

Location	IFS Model Components	GHG emission (kg CO_2 eq/ha)
Palampur	Crop + Dairy + Horticulture + Fodder + Vermicompost + Boundry Plantations + Kitchen Gardening	−1787
Johrat	Crop + Dairy + Horticulture + Poultry + Fishery + duckery + Goatery + Apairy +Vermicompost + Biogas + Liquid manure + FYM	−3175
Kalyani	Crop + Dairy + Horticulture + Fishery + Vermicompost + Biogas Gardening	−4517
Raipur	Crop + Dairy + Horticulture + Poultry + Fishery + Goat + Mushroom + Vermicompost + Boundary Plantations + Kitchen Gardening	−7713
Telangana	Crop + Livestock + Hortipasture	−27,036

Source: Ravisankar et al. (2019) and Paramesh et al. (2022)

rice–fish–duck culture. Nayak et al. (2018) observed structural variations in soil microbial diversity due to nutrient recycling (organic manures) with the production of planktons and macro-benthos in rice–fish–duck, rice–duck, and rice–fish systems over conventional rice production systems.

11.3.3.5 Economic Benefits

Integrated Farming Systems (IFS) provide farmers with opportunities to diversify, significantly contributing to rural development. As per Paramesh et al. (2022), IFS boosts farm profitability, increasing net income by 265% compared to single enterprises. Noteworthy examples include a 14% rise in net income in the Rice–rice–Azolla/Calotropis + Fish system (Shanmugasundaram et al. 1995) and a remarkable 1838% increase in the crops + pigeon + buffalo + agroforestry + farm pond system (Shekinah et al. 2005) compared to monocrop/single enterprise. Jayanthi et al. (2003) demonstrated a 25% increase in economic returns through crop integration with fish and poultry in Tamil Nadu's lowland conditions. Furthermore, Rautaray et al. (2005) highlighted the rice–fish system, integrating vegetables, fruits, ornamental plants, and agroforestry, as having the potential to produce 2.8 times higher income compared to rice alone in Assam's lowland ecologies. The higher product diversification in IFS, especially from livestock like dairy and poultry, holds the potential to generate daily income for small and marginal farmers (Babu et al. 2023).

11.3.3.6 Food and Nutritional Security

Integrated Farming Systems (IFS) play a crucial role in ensuring food and nutritional security. These systems efficiently utilize land and time by cultivating short-duration vegetable crops, pulses, and fodder for livestock. For example, in Kerala,

0.2 ha homestead farming integrated with livestock, supports a family of four with a year-round supply of vegetables, milk, and eggs (John 2014). Devendra and Thomass (2002) highlight the significance of IFS for small and marginal farmers, emphasizing its role in meeting protein requirements through livestock products. IFS, which include crop components like legumes, vegetables, oilseed, fruit crops, or agroforestry, present a promising approach to achieve food and nutritional security by optimizing available resources (Paramesh et al. 2022).

Brief integrated farming systems are crucial components in the design of sustainable food production systems within the food–water–energy nexus. They enable more efficient resource utilization, reduce environmental impacts, enhance food security, and contribute to the long-term sustainability of agriculture. These systems play a pivotal role in addressing the global challenges of food production while preserving essential natural resources.

11.3.4 Multitier Cropping Systems

11.3.4.1 Fruit-based Multitier Cropping System

Multitier cropping systems have been developed over long periods of time based on agroecological conditions, plant diversity, farmer resources, and needs. The eastern region of India is characterized by rainfed agriculture constrained with soil acidity, high rate of soil erosion, and poor water-holding capacity of soil. The lowlands and the medium uplands are used for rainfed rice production, whereas the uplands are seldom utilized for production of agronomical crops. These uplands are either used as open pasture or for social forestry as per the conditions. Further, due to the lower productivity of rainfed crops, growing agricultural crops is unprofitable under upland conditions. Fruit tree-based multitier production system can offer suitable option for profitable utilization of the uplands. The technology on fruit-based multitier cropping system comprises planting of fruit trees with large canopies (Mango, litchi, Aonla, Jackfruit) at a spacing of 10 × 10 m as the main crop, planting of precocious bearing fruit species with dwarf canopy (guava, custard apple, lime, lemon, and peach) at a spacing of 5 × 5 m between rows and between plants in the same field as filler crop and growing of intercrops in the interspaces. The system accommodates 100 large size trees and 300 small-sized trees per ha. During the initial years, any kind of inter crops can be grown. After the plants of the main crop attain full canopy size, the filler plants can be removed and shade-tolerant intercrops like turmeric, ginger, and elephant foot yam can be grown. A productivity level of 12.0 Mg/ha of paddy equivalent yield can be obtained from a 10-year-old fruit-based multitier system under the eastern plateau region of India. The mango-based, litchi-based, and aonla-based multitier cropping systems have B:C ratios of 2.40, 4.86, and 2.51, respectively.

11.3.4.2 Silvi–Horti–Agricultural System

Silvi–horti–agricultural system is an approach to increase land productivity in low-external-input, resource-limited situations of rainfed uplands. The light-demanding intercrops are planted initially, but later due to shading effect of perennials, either perennials were harvested before their rotation age or shade-tolerant intercrops are recommended. The silvi–horti–agricultural system is a continuous income-generating system, which not only helps the farmers for assured income in the events of weather vagaries but also protects the land from degradation and enhances the soil quality (Ali et al. 2024). The silvi–horti–agricultural system comprises integration of mahogany and mango as the main crops with the filler crops aonla and peach and annual crops rice and finger millet (ragi). The perennial woody plants of mango and mahogany were planted at a distance of 20 × 20 m from row to row and 10 × 10 m from plant to plant (5 × 5 m between mango and mahogany within row). The filler crops (Peach and Aonla) were planted at a distance of 5 × 5 m from row to row and 10 × 10 m from plant to plant. Intercrop components of the system, i.e., *kharif* crops (rice and ragi) were taken in association with perennial woody plants and horticultural plants. The highest paddy equivalent yield was recorded under Mango + Mahogany + Peach + Rice followed by Mango + Mahogany + Peach + Ragi system. Both the systems have potential to achieve optimal yields for light-demanding cereals in these systems on a continuous basis. The mango yield varied from 29.75 kg/plant to 36 kg/ plant under different systems. The filler crop peach yield (50 kg/ha) was found higher under Mango + Mahogany + Peach + Rice system. Based on the economics multi-story cropping system Mango + Mahogany + P each + Finger millet and Mango + Mahogany + Peach + Rice system gave net income of Rs. 80,000 and 60,000, respectively from one-hectare area. The total soil organic carbon stock was also found highest of more than 50.00 Mg/ha under the multitier systems having peach as filler trees in the entire soil depth of 0–60 cm. The very labile, labile, less labile, and non-labile carbon stocks were highest at 21.1, 14.9, 11.0, and 13.2 Mg/ha, respectively, in mango + mahogany + peach + rice multitier systems.

11.3.5 RCT-based Production Systems

Various resource conservation-based production systems play a crucial role in the broader framework of sustainable production. These systems are designed to optimize the use of natural resources, promote environmental health, and enhance the overall sustainability of agricultural practices. Here are several types of resource conservation-based production systems within the context of sustainable production.

11.3.5.1 Precision Agriculture

Precision agriculture employs advanced technologies to optimize resource use, enhance crop yields, and minimize environmental impact. Through the integration of global positioning systems (GPS), sensors and data analytics, precision agriculture enables farmers to tailor their practices with unprecedented accuracy (Finger et al. 2019). This approach facilitates precise application of water, fertilizers, and pesticides, reducing waste and mitigating environmental pollution. By utilizing real-time data on soil conditions and crop health, farmers can make informed decisions, leading to more efficient use of resources and improved overall productivity. Precision agriculture not only enhances economic viability for farmers by minimizing input costs but also contributes to the long-term sustainability of agriculture through the promotion of resource efficiency, conservation, and resilience in the face of evolving environmental challenges.

11.3.5.2 Water-Efficient Irrigation Systems

Water-efficient irrigation systems play a pivotal role in promoting sustainable crop production by optimizing water use and addressing the growing challenges of water scarcity and climate change. Technologies such as drip irrigation and sprinkler systems deliver water directly to the root zones of plants, minimizing evaporation and runoff, and ensuring precise application (Abdelraouf et al. 2020). These systems enhance water-use efficiency, conserving this precious resource and mitigating the environmental impact associated with conventional irrigation methods. Moreover, adopting smart irrigation technologies that incorporate soil moisture sensors and weather data enables farmers to tailor irrigation schedules based on real-time conditions, preventing overwatering and reducing energy consumption. By integrating water-efficient irrigation systems into agricultural practices, not only do farmers enhance crop yields, but they also contribute to sustainable water management, supporting both environmental conservation and the long-term viability of global food production systems.

11.3.5.3 Integrated Pest Management (IPM)

By combining biological, cultural, mechanical, and chemical control methods, IPM aims to manage pest populations effectively while minimizing the impact on the environment and human health. This strategy promotes the use of natural enemies, such as predators and parasites, to control pests, reducing reliance on chemical pesticides. Crop rotation, companion planting, and the use of resistant crop varieties are cultural practices integrated into IPM to disrupt pest life cycles. Regular monitoring and early detection of pests, facilitated by technology like sensors and data analytics, enable timely and targeted interventions, preventing pest outbreaks. By emphasizing preventive measures and considering the ecological context, IPM not only protects crop yields but also contributes to biodiversity conservation and the overall health of agroecosystems. In this way, IPM aligns with the principles of sustainable agriculture, promoting resilient and environmentally friendly crop production systems.

11.3.5.4 Climate-Smart Agriculture

This approach integrates adaptive and mitigative measures to enhance agricultural resilience, minimize greenhouse gas emissions, and optimize resource use. By employing practices such as drought-resistant crop varieties, efficient water management, and precision farming, climate-smart agriculture enhances the ability of farming systems to withstand extreme weather events and shifting climatic patterns (Imran et al. 2019). Climate smart agriculture (CSA) involves the selection of climate-resilient crop varieties, altered planting schedules, and the incorporation of innovative technologies like precision agriculture and controlled-environment agriculture. Additionally, practices such as agroforestry, cover cropping, and conservation tillage contribute to soil health, water conservation, and carbon sequestration. CSA not only helps farmers adapt to the impacts of climate change, such as altered precipitation patterns and increased frequency of extreme weather events but also contributes to global efforts to mitigate climate change by reducing the environmental footprint of agriculture. By fostering a flexible and sustainable approach to crop production, climate-smart agriculture addresses the challenges posed by a changing climate while promoting food security and environmental stewardship.

11.4 Way Forward in Sustainable Food Production Systems

(a) The climate-smart agricultural practices and sustainable land management can simultaneously reduce the GHG emissions from agriculture and increase resilience to climate-related challenges.
(b) Prioritizing production efficiency through sustainable agricultural practices and encouraging the adoption of renewable energy in farming operations for sustainable food production.
(c) Conventional agricultural practices should be replaced by practices that work with nature to protect long-term farm productivity and improve human and animal wellbeing.
(d) Emerging innovation and farming techniques to produce the new food system with low environmental impacts.
(e) Reducing food losses and waste has a great opportunity in enhancing the sustainability of the food system and simultaneously improve food security and nutrition.

(f) Understanding the nexus among food–energy–water governs the ability to meet the current global challenges.

11.5 Conclusion

The global population will face a major challenge in future from the increasing demand for food fueled by global population growth and changing dietary habits with unsustainable food production systems. The progress toward healthy food systems for all can be accelerated through the national and local work undertaken to meet the sustainable development goals. Depending on the local context, both innovative and improved sustainable food-producing practices can be implemented for effective resilience to climate challenges. However, when promoting sustainable food systems, it is important to consider the interaction among food, water, and energy. The food–water–energy nexus has gained momentum among the research communities for the security of food, water, and energy due to future uncertainties. The various sustainable food production systems comprising micro-irrigation, conservation agriculture, integrated farming system, multitier cropping system, and resource conservation technologies have the ability to secure the subsystems of food, water, and energy and thus providing the holistic management of the subsystems to make the food–water–energy nexus more vibrant.

References

Abdelraouf RE, El-Shawadfy MA, Ghoname AA, Ragab R (2020) Improving crop production and water productivity using a new field drip irrigation design. Plant Arch 20(1):3553–3564

Ali A, Das B, Dhakar MK, Naik SK, Patel VB, Mishra GP, Sarkar PK, Shinde R, Jha AK, Bhatt BP (2024) Enhancing soil health and fruit yield through Tephrosia biomass mulching in rainfed guava orchards. Sci Rep 14(1):13913

Allen T, Prosperi P (2016) Modeling sustainable food systems. Environ Manag 57(5):956–975

Babu S, Das A, Mohapatra KP, Yadav GS, Singh R, Tahashildar M, Devi MT, Das S, Panwar AS, Prakash N (2019) Pond dyke utilization: an innovative means for enhancing productivity and income under integrated farming system in north east hill region of India. Indian J Agric Sci 89(1):117–122. https://doi.org/10.56093/ijas.v89i1.86190

Babu S, Das A, Singh R, Mohapatra KP, Kumar S, Rathore SS, Yadav SK, Yadav P, Ansari MA, Panwar AS, Wani OA (2023) Designing an energy efficient, economically feasible, and environmentally robust integrated farming system model for sustainable food production in the Indian Himalayas. Sustain Food Technol 1(1):126–142. https://doi.org/10.1039/D2FB00016D

Babu S, Singh R, Avasthe RK, Kumar S, Rathore SS, Singh VK, Ansari MA, Valente D, Petrosillo I (2023a) Soil carbon dynamics under organic farming: Impact of tillage and cropping diversity. Ecol Indic 147:109940. https://doi.org/10.1016/j.ecolind.2023.109940

Babu S, Singh R, Avasthe RK, Rathore SS, Kumar S, Das A, Layek J, Sharma V, Wani OA, Singh VK (2023b) Conservation tillage and diversified cropping enhance system productivity and ecoefficiency and reduce greenhouse gas intensity in organic farming. Front Sustain Food Syst 7:1114617. https://doi.org/10.3389/fsufs.2023.1114617

Barbosa FA, Soares Filho BS, Merry FD, de Oliveira AH, Costa WLS, Coe MT, da Silveira BE, Maciel TC, Sheepers LC, de Oliveira AR (2015) Cenários para apecuária de corte amazônica. IGC/UFMG, pp 1–154

Barman A, Singh VK, Rathore SS, Babu S, Sharma VK (2001) Deficit-irrigation management and agri-horti system compatibility using drip for enhancing water-use efficiency and profitability in Indian mustard (*Brassica juncea*). Indian J Agron 66(3):378–381. https://doi.org/10.59797/ija.v66i3.2864

Das A, Layek J, Ramkrushna G, Babu S, Devi MT, Dey U, Suting D, Yadav GS, Lyngdoh DBD, Prakash N (2019) Integrated organic farming system: an innovative approach for enhancing productivity and income of farmers in North Eastern Hill Region of India. Indian J Agric Sci 89(8):1267–1272. https://doi.org/10.56093/ijas.v89i8.92843

Das A, Datta D, Samajdar T, Idapuganti RG, Islam M, Choudhury BU, Mohapatra KP, Layek J, Babu S, Yadav GS (2021) Livelihood security of small holder farmers in eastern Himalayas, India: Pond based integrated farming system a sustainable approach. Curr Res Environ Sustain. https://doi.org/10.1016/j.crsust.2021.100076

Devendra C, Thomas D (2002) Smallholder farming systems in Asia. Agric Syst 71(1–2):17–25

Edralin DI, Sigua GC, Reyes MR (2016) Dynamics of soil carbon, nitrogen and soil respiration in farmer's field with conservation agriculture, Siem Reap, Cambodia. Int J Plant Soil Sci 11(1):1–13

European Environment Agency (EEA) (2018) Climate impacts on water resources. EEA, Kongens Nytorv 6, 1050 Copenhagen K, Denmark

FAO (2011) Save and grow, a policy maker's guide to the sustainable intensification of smallholder crop production. FAO, Rome. http://www.fao.org/docrep/014/i2215e/i2215e.pdf. Accessed 1 Nov 2011

Fatima A, Singh VK, Babu S, Singh RK, Upadhyay PK, Rathore SS, Kumar B, Hasanain M, Parween H (2023) Food production potential and environmental sustainability of different farming system models in northwest India. Front Sustain Food Syst. https://doi.org/10.3389/fsufs.2023.959464

Ferroukhi R, Nagpal D, Lopez-Pena A, Hodges T, Mohtar RH, Daher B, Mohtar S, Keulertz M (2015) Renewable energy in the water, energy & food nexus. IRENA, Abu Dhabi, pp 1–125

Finger R, Swinton SM, El Benni N, Walter A (2019) Precision farming at the nexus of agricultural production and the environment. Ann Rev Resour Econ 11:313–335

Gil J, Siebold M, Berger T (2015) Adoption and development of integrated crop–livestock–Forestry systems in Mato Grosso, Brazil. Agric Ecosyst Environ 199:394–406. https://doi.org/10.1016/j.agee.2014.10.008

IEA (2012) World energy outlook 2012. IEA, Paris

Imran MA, Ali A, Ashfaq M, Hassan S, Culas R, Ma C (2019) Impact of climate smart agriculture (CSA) through sustainable irrigation management on resource use efficiency: a sustainable production alternative for cotton. Land Use Policy 88:104113

Jayanthi C, Balusamy M, Chinnusamy C, Mythili S (2003) Integrated nutrient supply system of linked components in low land integrated farming system. Indian J Agron 48(4):41–46

Jha BK, Mali SS, Naik SK, Kumar A, Singh AK (2015) Optimal planting geometry and growth stage based fertigation in vegetable crops. Technical bulletin no R-56/Ranchi-25. Research Centre Ranchi, ICAR-Research Complex for Eastern Region, Patna, India

John J (2014) Homestead farming in Kerala: a multi-faceted land-usesystem. Rev Agrar Stud 4:80–105

Kadam CS, Thanki JD, Gudadhe NN (2014) Response of chickpea to irrigation methods, fertilisers and biofertiliser under south Gujarat condition. Indian J Fert 10(4):20–24

Kumar MD, van Dam JC (2013) Drivers of change in agricultural water productivity and its improvement at basin scale in developing economies. Water Int 38(3):312–325

Kumar R, Mishra JS, Naik SK, Mondal S, Meena RS, Kumar S, Dubey AK, Makarana G, Jha BK, Mali SS, Biswas AK (2022a) Impact of crop establishment and residue management on soil properties and productivity in rice-fallow ecosystems in India. Land Degrad Dev 33(5):798–812

Kumar R, Mishra JS, Mali SS, Mondal S, Meena RS, Lal R, Jha BK, Naik SK, Biswas AK, Hans H, Sundaram PK (2022b) Comprehensive environmental impact assessment for designing carbon-cum-energy efficient, cleaner and eco-friendly production system for rice-fallow agro-ecosystems of South Asia. J Clean Prod 331:129973

Kumar R, Choudhary JS, Naik SK, Mondal S, Mishra JS, Poonia SP, Kumar S, Hans H, Kumar S, Das A, Kumar V (2023) Influence of conservation agriculture-based production systems on bacterial diversity and soil quality in rice-wheat-greengram cropping system in eastern Indo-Gangetic Plains of India. Front Microbiol 14:1181317

Mali SS, Jha BK, Singh R, Meena M (2017) Bitter gourd response to surface and subsurface drip irrigation under different fertigation levels. Irrig Drain 66:615–625

Mali SS, Naik SK, Jha BK, Singh AK, Bhatt BP (2019) Planting geometry and growth stage linked fertigation patterns: impact on yield, nutrient uptake and water productivity of chilli pepper in hot and sub-humid climate. Sci Hortic 249:289–298

Naik SK, Maurya S, Bhatt BP (2017) Soil organic carbon stocks and fractions in different orchards of eastern plateau and hill region of India. Agrofor Syst 91:541–552

Naik SK, Maurya S, Mukherjee D, Singh AK, Bhatt BP (2018) Rates of decomposition and nutrient mineralization of leaf litter from different orchards under hot and dry sub-humid climate. Arch Agron Soil Sci 64(4):560–573

Naik SK, Jha BK, Singh AK (2022) Drip fertigated planting systems with polythene mulching on cauliflower–eggplant cropping systems in hot and subhumid climate: impact on soil health and crop yield. Commun Soil Sci Plant Anal 53(10):1261–1276

Naik SK, Mali SS, Jha BK, Kumar R, Mondal S, Mishra JS, Singh AK, Biswas AK, Choudhary AK, Choudhary JS, Hans H (2023) Intensification of rice-fallow agroecosystem of south Asia with oilseeds and pulses: impacts on system productivity, soil carbon dynamics and energetics. Sustainability 15(2):1054

Nayak PK, Nayak AK, Panda BB, Lal B, Gautam P, Poonam A, Shahid M, Tripathi R, Kumar U, Mohapatra SD, Jambhulkar NN (2018) Ecological mechanism and diversity in rice based integrated farming system. Ecol Indic 91:359–375. https://doi.org/10.1016/j.ecolind.2018.04.025

OECD (2012) Environmental outlook. OECD, Paris

Pan RS, Kumar R, Bhatt BP, Mishra JS, Naik SK, Shinde R, Mali SS, Sarkar PK, Kumawat N, Singh AK (2022) Production potential and soil health of diversified production system of hill and plateau region of eastern India. Indian J Agric Sci 92:101–104

Pan RS, Mali SS, Kumar R, Naik SK, Upadhyay PK, Shinde R, Jha BK, Jeet P, Das A (2024) An evaluation of energy and carbon budgets in diverse cropping systems for sustainable diversification of rainfed uplands in India's eastern hill and plateau region. Front Sustain Food Syst 8:1340638

Paramesh V, Ravisankar N, Behera U, Arunachalam V, Kumar P, Solomon R, Dhar Misra S, Mohan Kumar R, Prusty AK, Jacob D, Panwar AS, Mayenkar T, Reddy VK, Rajkumar S (2022) Integrated farming system approaches to achieve food and nutritional security for enhancing profitability, employment, and climate resilience in India. Food Energy Secur 11:e321. https://doi.org/10.1002/fes3.321

Pereira-Cardenal SJ, Mo B, Gjelsvik A, Riegels ND, Arnbjerg-Nielsen K, Bauer-Gottwein P (2016) Joint optimization of regional water-power systems. Adv Water Resour 92:200–207

Rathore SS, Babu S, El-Sappah AH, Shekhawat K, Singh VK, Singh RK, Upadhyay PK, Singh R (2022) Integrated agroforestry systems improve soil carbon storage, water productivity, and economic returns in the marginal land of the semi-arid region. Saudi J Biol Sci. https://doi.org/10.1016/j.sjbs.2022.103427

Rautaray SK, Das PC, Sinhababu DP, Singh BN (2005) Increasing farm income through rice-fish based integrated farming system in rainfed lowlands of Assam. Indian J Agric Sci 75(2):79–82

Ravisankar N, Singh P, Mishra RP, Prusty AK, Shamim M, Singh R, Tripathi D, Mohan B (2019) Annual report of AICRP on integrated farming system. Indian Council of Agricultural Research, New Delhi

Salton JC, Mercante FM, Tomazi M, Zanatta JA, Concenço G, Silva WM, Retore M (2014) Integrated crop-livestock system in tropical Brazil: toward a sustainable production system. Agric Ecosyst Environ 190:70–79. https://doi.org/10.1016/j.agee.2013.09.023

Samal SK, Rao KK, Poonia SP, Kumar R, Mishra JS, Prakash V, Mondal S, Dwivedi SK, Bhatt BP, Naik SK, Choubey AK, Kumar V, Malik RK, Mc Donald A (2017) Evaluation of long-term conservation agriculture and crop intensification in rice-wheat rotation of Indo-Gangetic plains of South Asia: carbon dynamics and productivity. Eur J Agron 90:198–208

Shanmugasundaram VS, Baluswamy M, Rangaswamy A (1995) Integrated farming system research in Tamil Nadu. J Farm Syst Res Develop 1(1):1–9

Shekinah DE, Jayanthi C, Sankaran N (2005) Physical indicators of sustainability – a farming systems approach for the small farmer in the rainfed vertisols of the western zone of Tamil Nadu. J Sustain Agric 25(3):43–65

Shyam CS, Shekhawat K, Rathore SS, Babu S, Singh RK, Upadhyay PK, Dass A, Fatima A, Kumar S, Sanketh GD, Singh VK (2023) Development of integrated farming system model – a step towards achieving biodiverse, resilient and productive green economy in agriculture for small holdings in India. Agronomy 13:955. https://doi.org/10.3390/agronomy13040955

Singh R, Babu S, Avasthe RK, Yadav GS, Rajkhowa DJ (2016) Productivity, economic profitability and energy dynamics of rice (*Oryza sativa* L.) under diverse tillage and nutrient management practices in rice-vegetable pea cropping system of Sikkim Himalayas. Indian J Agric Sci 86(3):326–330

Singh R, Babu S, Avasthe RK, Meena RS, Yadav GS, Das A, Mohapatra KP, Rathore SS, Kumar A, Singh C (2021a) Conservation tillage and organic nutrients management improve soil properties, productivity, and economics of a maize-vegetable pea system in the Eastern Himalayas. Land Degrad Dev:32, 1–18. https://doi.org/10.1002/ldr.4066

Singh R, Babu S, Avsthe RK, Yadav GS, Das A, Mohapatra KP, Kumar A, Singh VK, Chandra P (2021b) Crop productivity, soil health, and energy dynamics of Indian Himalayan intensified organic maize-based systems. Int Soil Water Conserv Res 9(2):260–270. https://doi.org/10.1016/j.iswcr.2020.11.003

WHO (2017) The state of food security and nutrition in the world. https://www.unicef.org/reports/state-food-security-and-nutrition-world-2017

Yadav GS, Das A, Lal R, Subhash B, Meena RS, Saha P, Singh R, Datta M (2018) Energy budget and carbon footprint in a no-till and mulch-based rice–mustard cropping system. J Clean Prod 191:144–157

Yadav GS, Babu S, Das A, Bhowmik SN, Datta M, Singh R (2019) Soil carbon dynamics and productivity of rice–rice system under conservation tillage in submerged and unsubmerged ecologies of eastern Indian Himalaya. Carbon Manag 10(1):51–62. https://doi.org/10.1080/17583004.2018.1545518

Yadav GS, Das A, Subhash B, Mohapatra KP, Lal R, Rajkhowa DJ (2020) Potential of conservation tillage and altered land configuration to improve soil properties, carbon sequestration and productivity of maize-based cropping system in eastern Himalayas, India. Int Soil and Water Conserv Res. https://doi.org/10.1016/j.iswcr.2020.12.003

Yadav GS, Das A, Kandpal BK, Babu S, Lal R, Datta M, Das B, Singh R, Singh VK, Mohapatra KP, Chakraborty M (2021) The food-energy-water-carbon nexus in a maize-maize-mustard cropping sequence of the Indian Himalayas: An impact of tillage-cum-live mulching. Renew Sust Energ Rev 151:111602. https://doi.org/10.1016/j.rser.2021.111602

Chapter 12
Harnessing Below and Above-ground Interaction in Diversified Cropping for Sustainable Food Production

Kadagonda Nithinkumar, Nilutpal Saikia, Shreyas Bagrecha, Alapati Nymisha, and S. K. Asraful Ali

Abstract Adoption of intensive agricultural practices is essential to ensure food availability and security, addressing the demands of a teeming global population amid limited land resources. Unfortunately, these practices have numerous ecological and environmental consequences. Diversified cropping is an important strategy to limit the existing agricultural challenges and promote sustainable agriculture, maintaining environmental quality and agroecosystem services. The plants and microorganisms interact with each other in the natural ecosystem, which succeeds as rhizosphere and phyllosphere in belowground and aboveground interaction, respectively. Understanding the aboveground and belowground and their interactions in diversified cropping systems is the most important criterion. About 25% soil microbial community is found below ground, which represents the biggest component of the earth's biodiversity and plays a vital role in nutrient cycling, carbon storage, and soil structure formation. The soil contains roots of the plant, the region which is the rhizosphere, consisting of number of diverse organisms like rhizobacteria and arbuscular mycorrhizal fungi (AMF). Similarly, aboveground interactions in diversified cropping systems such as allelopathy and its role, companion planting strategy, pest and disease management through diversity, pollinators, and their importance, thus, ultimately improving crop yields and soil quality. Hence, the integration of below- and aboveground interaction scenarios of the diversified cropping system to improve soil health, increase food production, and increase farmers' profitability. Therefore, globally, harnessing the below- and aboveground interactions in

K. Nithinkumar (✉) · N. Saikia · S. K. A. Ali
Division of Agronomy, ICAR – Indian Agricultural Research Institute, New Delhi, India

S. Bagrecha
Agronomy Section, ICAR – National Dairy Research Institute, Karnal, India

A. Nymisha
Division of Soil Science and Agricultural Chemistry, ICAR – Indian Agricultural Research Institute, New Delhi, India

S. Babu et al. (eds.), *Agricultural Diversification for Sustainable Food Production*, Sustainability Sciences in Asia and Africa,
https://doi.org/10.1007/978-981-97-7517-0_12

the above system could be an effective approach to achieving a more sustainable food production system.

Keywords Phyllosphere · Companion planting strategy · Soil microbial diversity · Rhizosphere

12.1 Introduction

Galloping population, climate change, and rapid urbanization exert tremendous pressure on agriculture and natural resources (Yadav et al. 2017). Therefore, one of the most important questions facing civilization is how to sustainably supply the rising needs for food, feed, biomass, and various ecosystem services (Yadav et al. 2021a). The projected growth in global population to 11 billion people by 2100 will necessitate a 70% increase in food production between 2005–2007 and 2050 (UNO 2017; Babu et al. 2022). Due to intensive agricultural techniques, food production expanded significantly in the second half of the twentieth century; however, on the other hand, this had numerous unfavorable effects on the ecosystem, as the loss of biodiversity, soil degradation, and pesticide contamination of the land, water, air, and food (Yadav et al. 2021a, b), which could be harmful to the health of both human and the environment (Yadav et al. 2020; Das et al. 2022). However, the adoption of intensive agricultural practices leads to several negative consequences like loss of biodiversity (Tian et al. 2020), nutrient leaching (Goemann et al. 2021), emission of greenhouse gases, buildup of insect pests, diseases, and degradation of soil health (Singh et al. 2021a; Yadav et al. 2023). Hence, there are several efforts to minimize the detrimental effects on the environment of intensive agricultural practices are mostly dedicated to reducing external (Wittwer et al. 2017) and various cropping strategies and management techniques like crop diversification by crop rotation, conservation tillage (Raj et al. 2023), and cover crops, enhancing the soil health and agricultural productivity on a sustainable basis (Rathore et al. 2022).

Crop diversification plays a direct and important role in maintaining ecosystem sustainability (Babu et al. 2020a) and it can potentially be achieved through crop rotation and multiple and intercropping systems (Babu et al. 2020b). An agroecological technique known as intercropping, which is growing two or more crops concurrently in a specific arrangement within the same field (Brooker et al. 2015) and is extensively adopted by smallholders, particularly in underdeveloped countries (Zhang and Li 2003). It often provides higher agricultural productivity than sole cropping (Huang et al. 2017) mainly due to efficient utilization of the resources (Singh et al. 2021b; Babu et al. 2023). However, the introduction of crop diversification enhances the diversity within agricultural systems through the utilization of spatial elements (intercropping), temporal aspects (crop rotations and multi-cropping), and genetic variations (genotype mixture) (Messéan et al. 2021). Thus, this technique may positively impact crop growth and production and the sustainability of ecosystems (Beillouin et al. 2021). Both temporally and spatially diverse crops are a robust strategy for reducing external input dependency due to their ability to reduce pest pressure. Thus, it stands as a central tenet of agroecology (Jabran et al. 2015).

Crop diversification includes two primary forms, i.e., (i) horizontal and (ii) vertical. Horizontal diversification entails cultivators cultivating a diverse array of crops, including fruits, medicinal plants, cut flowers, and vegetables. On the other hand, vertical diversification involves farmers engaging in additional value chain activities or alternative enterprises, in conjunction with crop cultivation (Ijaz et al. 2019).

A diversified cropping system significantly influences the below and aboveground interactions. In the aboveground, crops contend with both advantageous and antagonistic organisms, influencing the economic dynamics of agricultural practices (Holland et al. 2016). Herbivores and foliar pathogens pose threats to crops, while insects contribute to their well-being through pollination. Soil organisms (belowground) can significantly impact soil health and crop development. For instance, arbuscular mycorrhizal fungi (AMF) (Bowles et al. 2017), diazotrophic bacteria (Nitrogen fixers) (Canfield et al. 2010), and rhizobacteria (Raaijmakers et al. 2009) that promote plant growth (Haas and Défago 2005) and may strengthen a plant's resistance to stress and pests, improve the uptake of plant resources, and encourage growth of the plant and nutrition (Raaijmakers et al. 2009).

12.2 Belowground and Aboveground Interactions: The Concept

The boundary between the soil and atmosphere establishes a spatial differentiation in interspecific interactions within an intercropping system, distinguishing between aboveground and belowground dynamics. Several studies have shown that the implementation of root barriers effectively segregates belowground interactions from their aboveground counterparts (Chen et al. 2015; Das et al. 2022). In comparison to planting a single crop, the belowground interaction has a greater influence on an intercropped system's productivity than the aboveground interaction (Li et al. 2014). On the other hand, opposing perspectives assert that aboveground interactions have a greater impact on the benefits of intercropping compared to belowground interactions (Fig. 12.1) (Yang et al. 2017). The impacts of both belowground and aboveground interspecific relations on productivity exhibit variability contingent upon the combinations of crop species (Mariotti et al. 2009; Zhiipao et al. 2023) and are additionally influenced by the availability of natural resources (Cahill 1999). Therefore, to maximize the benefits of intercropping, it is crucial to gain a comprehensive insight into the role of aboveground and belowground interactions in crop growth.

Fig. 12.1 Above and below- ground interaction under different intensified systems

The primary benefits from belowground interactions are largely attributed to the enhancement of composition of soil microbial community, soil nutrient content, and soil enzymatic activity (Li et al. 2018; Babu et al. 2020a; Ansari et al. 2023), or enhancement of symbiotic N_2 fixation by legume crops and iron (Fe) nutrition (Inal et al. 2007; Ansari et al. 2023). Likewise, the palpable aspects of aboveground interspecific interactions can be heightened by the photosynthetic performance of crops, leading to increased resource utilization efficiency under various production system (Yadav et al. 2023, 2021b). The intercropping advantages on grain yield of maize primarily arise from higher maize production (average 95%), which is fueled by belowground interactions and aboveground interspecific competition for light (Jiao et al. 2021). Grasslands in aboveground environments play a paramount in supporting pollinators and communities of natural enemies, thereby improving the delivery of biological control and pollination services in nearby locations (Werling et al. 2014). Furthermore, it has been shown that grasslands can sometimes even match forests in providing essential belowground ecosystem services like carbon sequestration, regulation of water, and several others (Bardgett et al. 2021). Due to cutting, grazing, and senescence, the carbon stored in aboveground biomass has a limited shelf life. On the other hand, a significant fibrous root system found in belowground biomass frequently sequesters 60–80% of the carbon (Kizeková et al. 2019; Wani et al. 2023). Interactions between belowground and aboveground components in crop diversification are predominantly influenced by key ecological principles such as biodiversity, cycling of nutrients, and health of the soil (Yadav et al. 2023; Babu et al. 2020a).

12.2.1 Nutrient Cycling

Crop diversification is a key strategy for uplifting the soil's inherent capacity of providing nutrients and agricultural production, leveraging legume crops' ability to fix atmospheric nitrogen (Yadav et al. 2019; Babu et al. 2023, 2020b). Leguminous crops play a distinctive part in the nitrogen (N) cycle globally by harnessing nitrogen from the environment and enriching the soil with it (Yadav et al. 2023; Singh et al. 2023). As a result, reducing synthetic nitrogen fertilizer is necessary not only for legume crops but also for subsequent crops, enhancing the activity of soil microbes and structure (Matse et al. 2020; Rathore et al. 2022). Legumes contribute to enhancing soil fertility by forming symbiotic associations with microbes, like rhizobia, it facilitates the fixation of nitrogen from the environment, known as biological nitrogen fixation (BNF), and enhances the availability of nitrogen to the legume host and other crops (Kebede 2021). Many soil microorganisms are involved in BNF that can be influenced by belowground interactions (Fig. 12.2). In the biosphere, the biological fixation of atmospheric N_2 stands as a crucial process, second most in significance only after photosynthesis, vital for sustaining life on the globe (Ladha et al. 2022). BNF has a beneficial impact on soil N availability, resulting in a 9.7–20.5% rise in the residual nitrogen concentration of rice fields (Yu et al. 2014). Hence, above phenomenon was observed in rice crop rotation with legumes such as beans and vetch. Approximately 7–11% of the total nitrogen fixed by legumes is assigned to their roots and nodules, which translates to nearly 11–14 kg of locked nitrogen per thousand kg of belowground plant matter, constituting nearly half part of the overall aboveground plant biomass (Carranca et al. 2015). Every year, leguminous groups of crops collectively fix an impressive ~21.45 teragrams (Tg) nitrogen, with soybean contributing a significant 16.44 Tg, and food legumes, including chickpeas, common beans, peas, faba beans, cowpeas, and lentils accounting for 2.95 Tg of nitrogen fixation (Herridge et al. 2008). Similarly, nitrogen obtained from legume

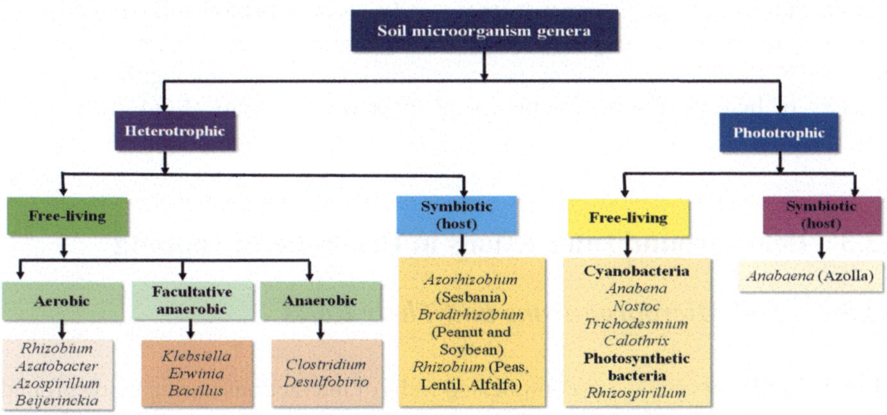

Fig. 12.2 Important soil microorganism genera involved in BNF and associated host plant species

rhizodeposits plays a significant role in boosting residual soil nitrogen content, contributing to an increase of 35–44%. These rhizodeposits also make up a substantial portion of belowground nitrogen in plants when they reach maturity, ranging from 79% to 85% (Mayer et al. 2003). In addition, Jensen (1996) demonstrated that around half part of the total belowground N through crops can be traced back to the root exudates of legumes. Therefore, the inclusion of legumes in crop diversification will significantly enhance nutrient availability and improve the soil's health.

12.2.2 Soil Biodiversity and Health

Soil health pertains to the propensity of living soil to show its biological functionalities within the boundaries of natural or managed ecosystems, sustaining plant and animal productivity (Bender et al. 2016). Soil microorganism, including bacteria, fungi, protozoa, nematodes, earthworms, and plant roots contributes more than 25% of belowground biomass of all described species (Bach et al. 2020). Crop diversification is recognized as a significant catalyst for achieving sustainable intensification. Increasing dependence on ecosystem services and biodiversity is how ecologically increasing agriculture achieves its benefits (Kleijn et al. 2019) encompassing soil management and their biota (Zhang et al. 2020). Soils serve as habitats for a rich organism diversity, encompassing invertebrates, microbes, and vertebrates, being one of the most biodiverse places on Earth, it supports as many as a thousand species per cubic meter of soil (Bardgett and van der Putten 2014). Soil microbiota, plant roots, and their associated microbiome engage in various interactions, collectively performing functions like preventing pests and diseases, storing carbon, improving nutrient availability, and water-holding capacity, and refining soil structure (Kumar et al. 2022; Das et al. 2022). However, high biodiversity in the soil enhances ecosystem steadiness and adaptability, specific ways may be required to meet the distinct needs of agricultural crop production, like promoting efficient nutrient utilization, high plant productivity, and resistance to biotic pressure. Species diversity enhances not only the effectiveness of resource use efficiency but also promotes higher yields, reducing the crop failure risk by enhancing the buffering capacity of heterogeneous systems under diversified systems (FAO 2016; Babu et al. 2020a, b).

12.3 Belowground Interactions in Diversified Cropping

12.3.1 Soil Microorganisms and Their Roles

Microbes and roots engage in complicated webs of interactions during belowground diversification, some of which are helpful, some of which are destructive, and some of which are neutral. According to Bakker et al. (2013), the microbial population in the rhizosphere may be broadly classified as helpful, detrimental, or neutral. The production of root exudates helps to boost the bacterial population in the rhizosphere, which in turn encourages the proliferation of bacteria (Bais et al. 2006). Subterranean interactions with beneficial microbes, such as Plant growth-promoting fungi (PGPFs), Plant growth-promoting rhizobacteria (PGPRs), and AMFs, alter the soil microbiome's function and structure. In their fight against parasitic arthropods and plant nematodes, among other diseases, these microorganisms indirectly impact soil ecosystems, which are vital for the generation of biomass aboveground (de Vries and Wallenstein 2017). Rhizosphere microbes secrete phytohormones (e.g., abscisic acid, salicylic acid, and jasmonic acid), signaling molecules (e.g., 1-aminocyclopropane-1-carboxylate (ACC) deaminase), phenolic compounds, and enzymes that protect plants from free radical damage (e.g., peroxidase, polyphenol oxidase, superoxide dismutase, and catalase). Collectively, these compounds shield plants against a wide range of biotic and abiotic dangers (Sharma and Sharma 2021).

12.3.2 Ecological Interactions of Microorganisms Within the Rhizosphere

A diverse ecosystem's rhizosphere plays a crucial role in its belowground diversity. It is linked to enhanced microbial activity, which depends on a range of factors, including biotic and abiotic, including the soil physicochemical properties and the messages that plants and other creatures emit via their biochemistry (Jain et al. 2020). The word "rhizosphere" was originally used by Hiltner (1904) to refer to a zone of maximal microbial activity. The belowground mass of a plant influences the development of this particular soil zone (Berendsen et al. 2012). Plant roots release various chemicals called "root exudates," fulfilling imperative roles of the attractor as well as repellers (Bais et al. 2006). Despite their repellent potential, microbes are drawn to these exudates, fostering the development of the population of microbes in the rhizosphere (Fig. 12.3). As experimentally established in Arabidopsis (Badri et al. 2013) and potatoes (Hoysted et al. 2018), the soil community relies on root exudates, which are essential for the development and upkeep of the soil food chain. Furthermore, roots release volatile organic molecules, such as (E)-β-caryophyllene, which are hydrophobic and diffuse in the natural soil setting and exhibit robust antifungal properties against a range of species of soil fungi (Massalha et al. 2017). Since the rhizosphere ecosystem relies on roots for energy, they are essential to the establishment of the microbial population. The discharged energy via root channels is then used by microorganisms to flow through, where it is transformed into microbial biomolecules that support aboveground interactions by producing volatile chemicals and belowground productivity. The set belowground communities are essential to ecological processes, particularly amid the current challenges posed by changes in the climate (Yang et al. 2018), and maintain close interconnections with aboveground communities (Jia et al. 2020).

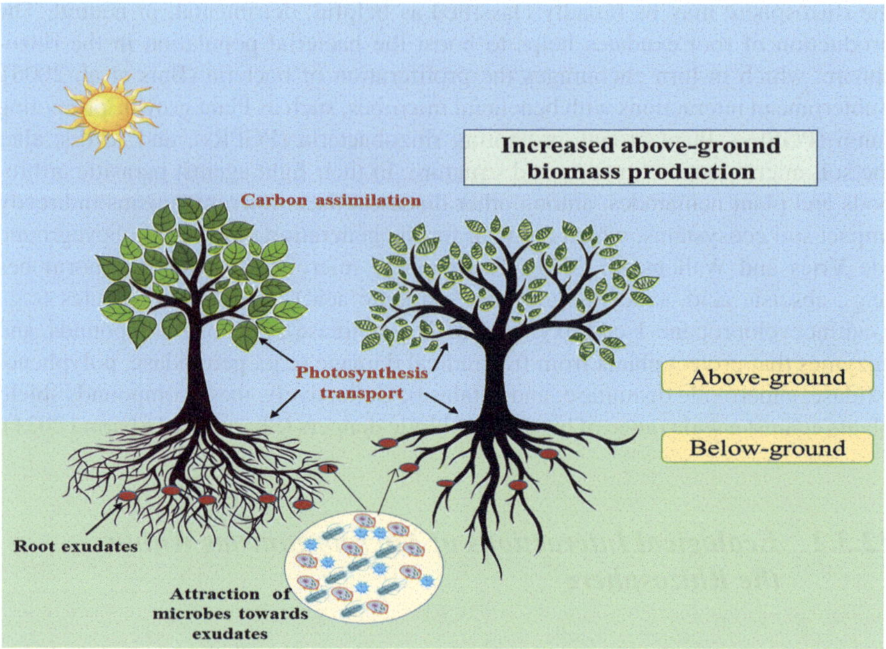

Fig. 12.3 Microbial community associated with rhizosphere that allows aboveground production

12.3.3 Rhizobacteria

In legume-based cropping systems, rhizobacteria engage in several imperative roles and diverse microorganisms exist in the rhizosphere, bacteria are the most dominant (Kaymak 2010). Plant growth-promoting rhizobacteria (PGPR) consists of a group of bacterial genera that heavily influence the development of plants as a whole. These genera include *Pseudomonas, Bacillus, Azospirillum, Variovorax, Klebsiella, Enterobacter, Burkholderia, Azotobacter*, and *Serratia*. According to Nadeem et al. (2010), multiple direct and indirect mechanisms have been identified by which PGPR enhances plant development, like phosphorus solubilization, biological nitrogen fixation, production of siderophores, phytohormones, organic acids, and protective enzymes like chitinase, ACC, deaminase, and glucanase (Glick et al. 2007; Berg 2009). Rhizobacteria exhibit variations in their nitrogen-fixation capacity, allowing for the selection of effective strains that fix large amounts of nitrogen even under stressful conditions (Hungria et al. 2013). Endophytic rhizobacteria inoculants can effectively colonize roots, establishing a conducive environment for proliferation and functioning. In a similar vein, nonsymbiotic endophytes inhabit plant tissues' intercellular gaps, which are rich in inorganic minerals, amino acids, and carbohydrates (Bacon and Hinton 2006). Intercropping of groundnut and maize proves effective in improving the iron nutrition of groundnut through the secretion of phytosiderophores by maize root, altering the microbial properties and biogeochemical of the rhizosphere (Zuo et al. 2000). Additionally, according to Guo et al. (2014), this intercropping method increases the expression of iron transporter gene as AhNRAMP1 and AhDMT1 in the leaf of groundnut, as well as iron absorption gene as AhFRO1 and AhYSL1 in the root of groundnut. This improves nitrogen fixation in the groundnut symbiotic relationship (Li et al. 2018).

12.3.4 Mycorrhizal Fungi

The terrestrial ecosystem consists majority of crops, which participate in relationships with coevolved microbial symbionts in the soil called arbuscular mycorrhizal (AM), a fungus, which associates with an estimated 80% of terrestrial trees and plants forming such associations suitable for their hosts (Frew et al. 2019). These fungi create colonies in the roots of plants, making it easier for the plants to get nutrients. In return, the plants give the fungi carbon (C) in the form of lipids and sugars (Keymer and Gutjahr 2018). Plants with mycorrhizal associations exhibited a 30% increase in aboveground biomass compared to plants without AM fungi. Moreover, herbivory on roots resulted in a 19% increase in aboveground biomass for nonmycorrhizal plants and a 27% increase for mycorrhizal plants. Consequently, plants associated with AM fungi demonstrated a 35% higher total plant biomass (Frew et al. 2019).

12.4 Aboveground Interactions in Diversified Cropping

12.4.1 Crop Rotation Effect on Soil Health

Diversified cropping systems are recognized as pivotal solutions to address sustainability concerns, offering a significant avenue to enhance agricultural resource utilization efficiency and crop productivity. Within diversified cropping strategies, crop rotation emerges as a crucial approach exerting a substantial impact on the soil features (Babu et al. 2020b, 2023; Yadav et al. 2023), increasing the fertility of the soil, and improving resource utilization efficiency (Ouda et al. 2018; Yadav et al. 2024). In contemporary agriculture, incorporating crop rotations involving vegetables or pulses holds the potential to augment the production of essential dietary micronutrients, including vitamins and, minerals addressing issues related to hidden hunger (Biswakarma et al. 2023; Dwivedi et al. 2017). In addition, introducing high-value crops, may enhance financial profitability (S'anchez et al. 2022). Diversified rotation of crops also plays a vital function in addressing environmental and sustainability challenges, with rotations involving water-use-efficient species and legumes shown to mitigate greenhouse gas emissions (Liu et al. 2022) and reduce declines in groundwater tables (Yang et al. 2021). Moreover, a strategy to restore farmland biodiversity is to implement diversified crop rotations (Sietz et al. 2022). Significant improvements in soil health indicators, encompassing physical (bulk density, gravimetric water content, and soil aggregate stability), chemical (soil pH, soil organic carbon, available nitrogen, phosphorus, potassium, micronutrients, and cation exchange capacity), and biological properties (soil microorganisms, alkaline phosphatase, and urease activity) are observed in diversified crop rotations (Ali et al. 2015) (Fig. 12.4). These parameters are essential for increasing agricultural productivity and soil health. For instance, a diversified crop rotation involving potatoes, faba beans, and common vetch resulted in a 15.0% and 12.7% increase in tuber yield, respectively, compared to potatoes grown after winter wheat (Malihe et al. 2015). Similarly, the legumes included in crop rotations contribute to increased crop biomass, grain yield, and soil health (Alemayehu et al. 2020; Babu et al. 2020b). Moreover, crop rotation can also impact the soil microbe's population. Li et al. (2023) observed that potato crop rotation with oats and fodder maize significantly changed the proportions of the bacterial and fungal population. Continuous potato cropping decreases the soil pH due to the accumulation of phenolic substances in the soil (Lv et al. 2023). Leguminous crops have a good influence on the physical properties of soil like aggregation and stability, which are important for controlling soil activities. This effect is ascribed to leaf drop, deep roots, root exudates, and increased rhizosphere activity. Therefore, crop rotation is a crucial technology to enhance soil health and agricultural production in both above- and belowground diversified cropping systems.

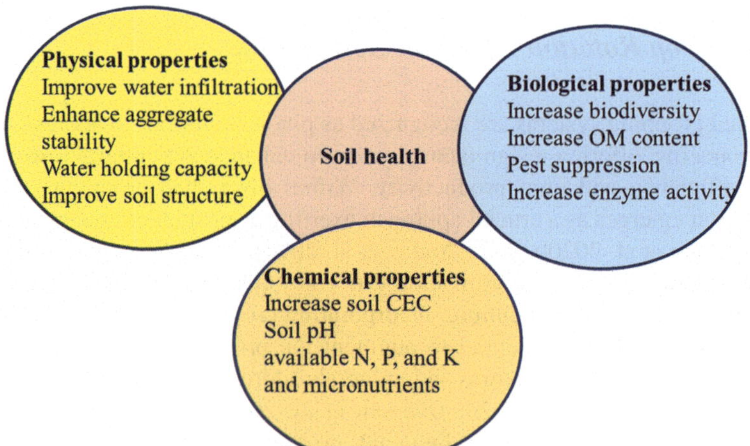

Fig. 12.4 Crop rotation effect on soil health

12.4.2 Pest and Disease Management Through Crop Diversification

Diversification of crops stands as a crucial technology for bolstering the populations of natural enemies of pests and mitigating incidents of pest and disease outbreaks within cropping systems. This strategy hinders pest populations through three primary mechanisms: improving crop resistance to pests, changing how arthropod pests identify and choose their major host agricultural plants, and destroying the spatiotemporal continuity of resources (Jaworski et al. 2023). The influence of crop diversification on pests' significant capacity to identify and choose their host plant, particularly through modifications in chemical and visual signals specific to their host crop (Cook et al. 2007). Many arthropod pests utilize visual cues from crops to select and detect their host plant (Vander Kooi et al. 2021). For example, aphids base their host plant selection on color, indicative of nutritional status (Doring and Kirchner 2022), whereas changes to the visual cues of virus-infected plants that are hosts, influence the choice of hosts for whiteflies (Zhang et al. 2022). Arthropod pests use sights to locate host crops, although variables like distance may play a role (Lund et al. 2019) and olfactory signals composed of organic substances that are volatile (Bouwmeester et al. 2019).

Various pest control strategies are implemented in crop diversification, including intercropping and push-pull technology (Han et al. 2022). In push-pull technology, an intercrop that repels pests is typically used for the push component, while attractive plants that lower insect fitness relative to the main crop are needed for the trap plants (Badenes-Perez 2018). For example, Brachiara was planted as a border trap crop and *Desmodium intortum* was interplanted with maize fields (Midega et al. 2018). This led to 83% decrease in the population of *Spodoptera frugiperda* larvae per plant of maize and an 87% reduction in crop damage (Midega et al. 2018). Furthermore, crop diversification contributes to enhancing crop defenses against pest attacks, typically categorized under resistance or tolerance strategies (Mitchell et al. 2016). Through chemical or physical barriers (such as thickened cuticles) or physical hindrances, resistance features work to prevent or lessen pest harm. Tolerance characteristics, in contrast, protect against pest damage by slowing the rate of growth after tissue injury or by showing delayed phenology in comparison to the life cycles of arthropod pests (Cook et al. 2007).

12.4.3 Allelopathy and Its Role

Weeds pose a significant threat to crop production in aboveground interactions, causing substantial reductions in crop growth, development, and yield. On average, weed interference leads to a ~34% reduction in crop yield (Oerke 2006). Potential yield reductions due to weeds are reported as 23% in wheat, 37% in rice, 40% in maize, 37% in soybeans, 36% in cotton, and 30% in potatoes (Oerke 2006; Jabran et al. 2015). The continuous herbicides used for weed control resulted in the development of herbicide resistance in weed biotypes. Introducing allelopathic crops into diversified cropping systems proves effective in controlling weeds and enhancing crop yields. Allelopathy is a method that can be used to address issues with contamination of the environment and pesticide resistance. Various crops exhibit allelopathic properties through the release of allelochemicals. Important allelopathic crops like rice, wheat, sorghum, rye, sunflower, and rape seed (Table 12.1). Allelochemicals released by these crops control weeds and stimulate underground microbial activity. Introducing allelopathic crops into crop rotation, mulching, cover crops, and intercropping allows the allelochemicals released through roots or crop residue decomposition, effectively suppressing weeds and other pests. Intercropping allelopathic crops with other plants, for instance, sesame with soybean and sorghum in cotton, significantly reduced the 70–96% of purple nutsedge density and 71–97% of dry biomass (Amosse et al. 2013). Similarly, planting cowpea and maize intercropped on different ridges increased land use efficiency and decreased weed intensity by around 50% (Saudy 2015). The efficacy of relay-intercropping legumes alongside wheat for weed reduction was also investigated in another research (Amosse et al. 2013). As per Fernandez-Aparicio et al. (2010), intensity of *Orobanche crenata* Forsk. was reduced when broad beans, peas, and berseem were intercropped. Crop diversification strategies like cover crops, mixed cropping, intercropping, crop rotation, and mulches with allelopathic potential not only suppress weeds but also conserve soil moisture, improve the fertility status of the soil, and enhance agricultural productivity. Wheat, canola, red clover, buckwheat, rape seed, brown mustard, cowpea, oats, mustards, annual ryegrass, hairy vetch, and black mustard, are a few significant cover crops that have the potential to be allelopathic (Altieri and Toledo 2011). Growing lucerne in wheat production fields as a mulch increased soil nitrogen content due to the fixation of atmospheric nitrogen, resulting in high wheat grain yields and weed suppression (Wiens et al. 2006).

Table 12.1 Allelochemicals produce of different crops

Crop	Allelochemical	References
Rye	2(3H)-benzoxazolinone (BOA) and benzoxazinones [2,4-dihydroxy-1,4(2H)-benzoxazin-3-one (DIBOA)]	Tabaglio et al. (2013)
Sorghum	Hydrophobic p-benzoquinone (sorgoleone), phenols, and acyanogenic glycoside (dhurrin)	Weston et al. (2013)
Brassica	Glucosinolate (isothiocyanates)	Petersen et al. (2001)
Sunflower	Phenolic acids	Alsaadawi et al. (2012)
Rice	Vanillic, p-hydroxybenzoic, o-hydroxy phenylacetic, ferulic, and syringic acid	Chung et al. (2006)
Wheat	Polyphenols and hydroxamic acids	Chung et al. (2006)

12.5 Monitoring and Assessing Belowground and Aboveground Interactions

The above and belowground interactions are influenced by the physical, chemical, and biological properties of soil and were ascertained using techniques shown in Tables 12.2 and 12.3.

Table 12.2 Monitoring and assessing belowground interactions

Parameters	Method/reagent/instrument	References
Soil texture	Hydrometer method	Bouyoucos (1962)
Soil pH	1:2.5 soil–water suspension	Jackson (1973)
Electrical conductivity	1:2.0 soil–water suspension	Jackson (1973)
Organic carbon	Wet oxidation	Walkley and Black (1934)
Available N	Alkaline permanganate method	Subbiah and Asija (1956)
Available P	Olsen's method	Olsen et al. (1954)
Available K	1 N NH_4OAc at pH 7.0	Hanway and Heidal (1952)
Available micronutrients	DTPA extraction	Lindsay and Norvell (1978)
Dehydrogenase activity	2,3,5-triphenyl tetrazolium chloride test	Klein et al. (1971)
Microbial biomass carbon	Fumigation–extraction method	Jenkinson and Powlson (1976)
Microbial biomass phosphorus	Fumigation–extraction method	Brookes et al. (1982)
Acid phosphatase	Colorimetry	Tabatabai and Bremner (1969)
Alkaline phosphatase	Colorimetry	Tabatabai and Bremner (1969)

Table 12.3 Monitoring and assessing aboveground interactions

Parameters	Method/reagent/instrument	References
Plant N	H_2SO_4 digestion	Kjeldahl (1883)
Plant P	Di-acid digestion–Vanadomolybdate yellow color method	Jackson (1973)
Plant K	Flame photometric method	Isaac and Kerber (1971)
Micronutrient content in plant	Atomic absorption spectrophotometer	Lindsay and Norvell (1978)
Greenness of plant	Normalized difference vegetative index (NDVI)), soil plant analysis development (SPAD), leaf color chart (LCC)	

12.6 Integrating Belowground and Aboveground Interactions

12.6.1 Holistic Approaches to Diversified Cropping Systems

Holistic approaches to diversified cropping systems involve integrated and sustainable farming practices that consider the entire ecosystem, foster biodiversity, and enhance overall farm resilience (Fig. 12.5). The goal of these approaches is to optimize natural resources utilization, minimize environmental impact, and enhance the long-term productivity of the soil. Crop diversification is becoming more and more well known as a sustainable strategy for raising or preserving productivity and promoting agricultural growth (Cong et al. 2021). Temporal and spatial diversification methods like crop rotation and intercropping have been extensively demonstrated to enhance grain production (Zhao et al. 2020), addressing challenges associated with continuous single crops (monocropping) and providing various ecological advantages. The cover crops incorporation into crop rotations has been shown to significantly increase 15.5% of organic carbon of soil (Jian et al. 2020), and extended intercropping techniques have the potential to increase the 4% of total nitrogen and

Fig. 12.5 Holistic approach of belowground and aboveground interaction in diversified cropping system

11% of soil organic carbon (Cong et al. 2015). Microbial diversity and activity are closely related to the positive effects of a diverse plant community on soil organic matter and nutrient availability (Jia et al. 2021). Furthermore, both intercropping and crop rotations contribute to the proliferation of functional microbes such as AMF and beneficial rhizobia and thereby improving nutrient use efficiency (Wang et al. 2019) and reducing diseases by disrupting pathogen life cycles (Li et al. 2019). For example, maize + soybean intercropping considerably reduced red crown rot (Gao et al. 2014) and soybean root rot (Chang et al. 2020) as compared to monoculture.

Crop diversity has the potential to release a wider range of belowground exudates that enhance microbial diversity. This phenomenon resulted in shifts in the makeup and function of the soil microbiome (Denier et al. 2022; Yadav et al. 2024), influencing trophic interaction networks that encompass both antagonistic and facilitative relationships among different community individuals. Ecological network functions may be better understood by analyzing the arrangement and distribution of these linkages (Thébault and Fontaine 2010), offering a new viewpoint on the benefits of diverse cropping systems over monocultures. Consequently, deciphering the framework behind the advantages of diversifying crops requires a thorough grasp of the pattern of soil microbial community composition and activities.

12.6.2 *Effect of Above and Belowground Interaction on Food Production*

The interaction between aboveground and belowground play significantly influences crop production by enhancing crop growth and yield while improving soil fertility, ultimately impacting food grain production. Practices such as intercropping, mixed cropping, and crop diversification contribute to increased productivity by effectively utilizing land, water, light, and nutrients (Jiao et al. 2021). Soil microorganism and their interactions in the belowground are key to the breakdown of organic matter, liberating vital nutrients, and enhancing the overall properties of the soil (Bardgett and van der Putten 2014). Plant roots are key players in underground interactions as they release root exudates into the soil, potentially altering the behavior and diversity among soil microbes (Philippot et al. 2013). These connections can enhance nutrient availability, water-holding capacity, soil health, and overall ecosystem functioning. Aboveground interactions also impact soil fertility and microbial properties. Crop diversification increases food production in above- and belowground interaction (Table 12.4) (Kallenbach et al. 2016). Diversified crop rotation with monoculture wheat–maize with cash crops like sweet potato and legumes like groundnut and soybean in Northern China plains increased equivalent yield by up to 38% (Yang et al. 2024). Moreover, aboveground and belowground interactions can influence pest and disease dynamics in agricultural systems. Certain fungi and bacteria in the soil may suppress plant diseases, lowering the need for synthetic pesticides and fostering an eco-friendly environment conducive to soil

Table 12.4 Percent increase in food production due to crop diversification of above- and belowground interaction

Cropping system		Region	Percent increase in grain yield (%)	References
Intercropping	Maize + groundnut	Hebei Province, China	25%	Jiao et al. (2021)
Relay cropping	Rice–lentil, grass pea and pea	Bangladesh	46%, 37%, and 32%, respectively	Ali and Sarker (2017)
Crop rotation	Dekeko–wheat	Alaje, Ethiopia	52.3%	Mesfin et al. (2023)
Crop rotation	Field pea–wheat	Alaje, Ethiopia	40.6%	Mesfin et al. (2023)
Crop rotation	Lentil–wheat	Alaje, Ethiopia	34.5%	Mesfin et al. (2023)
Crop rotation	Faba bean–wheat	Alaje, Ethiopia	71.1%	Mesfin et al. (2023)

biodiversity (Birkhofer et al. 2008). A deeper understanding and management of aboveground and belowground interactions are crucial for sustainable agriculture and food production. By nurturing these relationships, soil fertility, soil health, and climate resilience can be enhanced, ultimately contributing to global food security.

12.6.3 Benefits of Harmonizing Belowground and Aboveground Interactions

It is generally agreed upon that optimal crop diversification has many advantages for both farmers and the environment. Crop yields and soil health are known to improve with increased crop diversification because, over a variety of spatial dimensions, they increase the diversity of soil chemical nutrients, soil physical structures, and functional microorganisms (Maron et al. 2011). However, the nature of this relationship may be influenced by factors such as the host specificity of certain soil-borne diseases and species redundancy (Zhu et al. 2000). Aforementioned, Bainard et al. (2017) recorded the positive impact of increased crop diversity on reducing soil-borne diseases might not be universal and they revealed that the inclusion of more pulse crops in rotations led to a significant improve in the pathogen index, potentially attributed to a rise in pulse-specific pathogens.

12.6.4 Challenges and Barriers to Integration

The diversification of crop species within a region can be impacted by numerous factors, including the experiences of farmers, rainfall patterns, the demographics of households, and community-level characteristics (Degrande et al. 2006). In some cases, adding additional crop species to a diversified cropping system might result in serious pest incursions nearby (Hall 2003). Additionally, limited access to domestic and foreign markets, which is impacted by government regulations, subsidies, input costs and availability, infrastructure limitations for agricultural produce storage and transportation, and other factors, can also be detrimental to the success of novel crop species in cropping systems. When crop selection is not in line with market demand, farmers also run the danger of losing money. In the absence of sufficient demand, drought-tolerant crop types, for instance, could not fetch as high a price (Clements et al. 2011). Factors pertaining to the environment, society, and the market all play an important part in shaping crop diversification. Government policies should consider providing subsidies to farmers embracing crop diversification, and funding for research organizations should be prioritized to study and promote crop diversification initiatives.

12.7 Future Directions and Research Needs

The complex relationships, between plants and microorganisms, are still not fully understood so it is important to research these plant and microbes' interactions; hence, this is crucial for sustainable agriculture. We need to pay attention to the communities and their roles in disease control, nutrient cycling, and soil health. It is essential to uncover how different crops influence the diversity and function of soil microbes in the root zone. Additionally, we should explore how different crops interact with aboveground crop diversification considering factors like canopy structure, allelopathy (chemical interactions between plants), and competition for resources. Similarly, we need to focus on how crop diversification contributes to carbon sequestration and this includes studying root secretions matter input and activity to carbon storage. Moreover, we must investigate whether the benefits of cropping can be scaled up to agroecosystems. This involves examining its impact, on biodiversity, ecosystem services, and resilience against stressors. Develop adaptive management strategies for diversified cropping systems, considering the dynamic interactions between crops, microbes, and the environment. This includes assessing how these systems respond to climate variability and changing environmental conditions.

12.8 Conclusion

Addressing the challenges of increasing global population, climate change, and urbanization requires a sustainable approach to agriculture. Intensive farming practices have led to environmental issues, necessitating strategies like crop diversification. Horizontal and vertical diversification methods offer benefits, enhancing soil health, nutrient cycling, and overall sustainability. Intercropping, a key aspect of diversification, optimizes resource use and promotes belowground and aboveground interactions. Understanding these interactions is crucial for optimizing crop growth. Furthermore, integrating legumes in diversification enhances the fixation of nitrogen, enhancing soil fertility while decreasing the use of synthetic fertilizers. Diversified cropping systems, encompassing both belowground and aboveground interactions, offer multifaceted benefits. By interacting with beneficial microbes, crop diversification improves soil health. It mitigates pests and diseases, improves nutrient cycling, and fosters ecosystem resilience. Practices like intercropping, crop rotation, and incorporating allelopathic crops play pivotal roles. However, challenges, including market dynamics and potential pest invasions, need addressing. Adopting holistic, sustainable farming practices, and supportive government policies are crucial for successfully integrating diverse crops, promoting environmental health, and ensuring economic viability for farmers. Ultimately, embracing diversified cropping systems is pivotal for achieving long-term economic, biological, and environmental sustainability in agriculture.

References

Alemayehu G, Shibabaw A, Adgo E, Asch F, Freyer B (2020) Crop rotation and organic matter application restore soil health and productivity of degraded highland crop farms in Northwest Ethiopia. Cogent Food Agric 6(1):1831124

Ali MO, Sarker A (2017) Pulses cultivation as relay cropping in the rice field: a technique for intensification of pulses production in the medium-high to medium-low lands of Bangladesh. J Food Microbio Safety Hygiene 2:2. https://doi.org/10.4172/2476-2059-C1-002

Ali W, Jan A, Abbas A (2015) Residual effect of preceding legumes and nitrogen levels on subsequent maize. Int J Res Agron Agric Res 7:78–85

Alsaadawi IS, Sarbout AK, Al-Shamma LM (2012) Differential allelopathic potential of sunflower (*Helianthus annuus* L.) genotypes on weeds and wheat (*Triticum aestivum* L.) crop. Arch Agron Soil Sci 58:1139–1148

Altieri MA, Toledo VM (2011) The agroecological revolution in Latin America: rescuing nature, ensuring food sovereignty and empowering peasants. J Peasant Stud 38(3):587–612

Amosse C, Jeuffroy MH, Celette F, David C (2013) Relay-intercropped forage legumes help to control weeds in organic grain production. Eur J Agron 49:158–167

Ansari MA, Ravisankar N, Ansari MH, Babu S, Layek J, Panwar AS (2023) Integrating conservation agriculture with intensive crop diversification in the maize-based organic system: impact on sustaining food and nutritional security. Front Nutr 10:1137247. https://doi.org/10.3389/fnut.2023.1137247

Babu S, Mohapatra KP, Das A, Yadav GS, Tahasildar M, Singh R, Panwar AS, Yadav V, Chandra P (2020a) Designing energy-efficient, economically sustainable and environmentally safe cropping system for the rainfed maize–fallow land of the eastern Himalayas. Sci Total Environ 722. https://doi.org/10.1016/j.scitotenv.2020.137874

Babu S, Singh R, Avasthe RK, Yadav GS, Mohapatra KP, Selvan T, Das A, Singh VK, Valente D, Petrosillo I (2020b) Soil carbon dynamics in Indian Himalayan intensified organic rice-based cropping sequences. Ecol Indic 114:106292. https://doi.org/10.1016/j.ecolind.2020.10629

Babu S, Singh R, Yadav D, Rathore SS, Raj R, Avasthe R, Yadav SK, Das A, Yadav V, Yadav B, Shekhawat K, Upadhyay PK, Yadav DK, Singh VK (2022) Nanofertilizers for agricultural and environmental sustainability. Chemosphere 292:133451. https://doi.org/10.1016/j.chemosphere.2021.133451

Babu S, Singh R, Avasthe RK, Kumar S, Rathore SS, Singh VK, Ansari MA, Valente D, Petrosillo I (2023) Soil carbon dynamics under organic farming: impact of tillage and cropping diversity. Ecol Indic. https://doi.org/10.1016/j.ecolind.2023.109940

Bach EM, Ramirez KS, Fraser TD, Wall DH (2020) Soil biodiversity integrates solutions for a sustainable future. Sustainability 12(7):2662. https://doi.org/10.3390/su12072662

Bacon CW, Hinton DM (2006) Bacterial endophytes: the endophytic niche, its occupants, and its utility. In: Gnanamanickam SS (ed) Plant-associated bacteria. Springer, Dordrecht, pp 155–194

Badenes-Perez FR (2018) Trap crops and insectary plants in the order Brassicales. Ann Entomol Soc Am 112:318–329

Badri DV, Chaparro JM, Zhang RF, Shen Q, Vivanco JM (2013) Application of natural blends of phytochemicals derived from the root exudates of Arabidopsis to the soil reveal that phenolic-related compounds predominantly modulate the soil microbiome. J Biol Chem 288:4502–4512

Bainard LD, Chagnon PL, Cade-Menun BJ, Lamb EG, LaForge K, Schellenberg M, Hamel C (2017) Plant communities and soil properties mediate agricultural land use impacts on arbuscular mycorrhizal fungi in the mixed prairie ecoregion of the north American Great Plains. Agric Ecosyst Environ 249:187–195

Bais HP, Weir TL, Perry LG (2006) The role of root exudates in rhizosphere interactions with plants and other organisms. Annu Rev Plant Biol 57(1):233–266

Bakker PA, Berendsen RL, Doornbos RF (2013) The rhizosphere revisited: root microbiomics. Front Plant Sci 4:165

Bardgett RD, Van Der Putten WH (2014) Belowground biodiversity and ecosystem functioning. Nature 515(7528):505–511

Bardgett RD, Bullock JM, Lavorel S, Manning P, Schaffner U, Ostle N, Chomel M, Durigan G, Fry EL, Johnson D, Lavallee JM, Le Provost G, Luo S, Png K, Sankaran M, Hou X, Zhou H, Ma L, Ren W, Shi H (2021) Combatting global grassland degradation. Nat Rev Earth Environ 2:720–735

Beillouin D, Ben-Ari T, Malézieux E, Seufert V, Makowski D (2021) Positive but variable effects of crop diversification on biodiversity and ecosystem services. Glob Chang Biol 27(19):4697–4710. https://doi.org/10.1111/gcb.15747

Bender SF, Wagg C, van der Heijden MGA (2016) An underground revolution: biodiversity and soil ecological engineering for agricultural sustainability. Trends Ecol Evol 31(6):440–452

Berendsen RL, Pieterse CM, Bakker PA (2012) The rhizosphere microbiome and plant health. Trends Plant Sci 17(8):478–486

Berg G (2009) Plant-microbe interactions promoting plant growth and health: perspectives for controlled use of microorganisms in agriculture. Appl Microbiol Biotechnol 84(1):11–18

Birkhofer K, Bezemer TM, Bloem J, Bonkowski M, Christensen S, Dubois D, Ekelund F, Fließbach A, Gunst L, Hedlund K, Mäder P (2008) Long-term organic farming fosters below and aboveground biota: implications for soil quality, biological control and productivity. Soil Biol Biochem 40(9):2297–2308

Biswakarma N, Pooniya V, Zhiipao RR, Kumar D, Shivay YS, Das TK, Roy D, Das B, Choudhary AK, Swarnalakshmi K, Govindasamy P, Lakhena KK, Das K, Lama A, Jat RD, Babu S, Khan SA, Behara B (2023) Identification of a resource-efficient integrated crop management practice for the rice-wheat rotations in south Asian indo-Gangetic Plains. Agric Ecosyst Environ 357:108675. https://doi.org/10.1016/j.agee.2023.108675

Bouwmeester H, Schuurink RC, Bleeker PM, Schiest LF (2019) The role of volatiles in plant communication. Plant J 100:892–907

Bouyoucos GJ (1962) Hydrometer method improved for making particle size analysis of soils. Agron J 54:464–465

Bowles TM, Jackson LE, Loeher M, Cavagnaro TR (2017) Ecological intensification and arbuscular mycorrhizas: a meta-analysis of tillage and cover crop effects. J Appl Ecol 54:1785–1793

Brooker RW, Bennett AE, Cong WE, Daniell TJ, George TS, Hallett PD, Hawes C, Iannetta PPM, Jones HG, Li AJ, Karley L, McKenzie BM, Pakeman RJ, Paterson E, Schob C, Shen JB, Squire G, Watson CA, Zhang CC, Zhang FS, Zhang JL, White PJ (2015) Improving intercropping: a synthesis of research in agronomy, plant physiology and ecology. New Phytol 206:07–117

Brookes PC, Powlson DS, Jenkinson DS (1982) Measurement of microbial biomass phosphorus in soil. Soil Biol Biochem 14(4):319–329

Cahill JF (1999) Fertilization effects on interactions between above- and below-ground competition in an old field. Ecology 80:466–480

Canfield DE, Glazer AN, Falkowski PG (2010) The evolution and future of Earth's nitrogen cycle. Science 330:192–196

Carranca C, Torres MO, Madeira M (2015) Underestimated role of legume roots for soil N fertility. Agron Sustain Dev 35:1095–1102

Chang X, Yan L, Naeem M, Khaskheli MI, Zhang H, Gong G, Zhang M, Song C, Yang W, Liu T (2020) Maize/soybean relay strip intercropping reduces the occurrence of Fusarium root rot and changes the diversity of the pathogenic Fusarium species. Pathogens 9:211

Chen GD, Chai Q, Huang GB, Yu AZ, Feng FX, Mu YP, Kong XF, Huang P (2015) Belowground interspecies interaction enhances productivity and water use efficiency in maize-pea intercropping systems. Crop Sci 55:420–428

Chung IM, Kim JT, Kim SH (2006) Evaluation of allelopathic potential and quantification of momilactone A, B from rice hull extracts and assessment of inhibitory bioactivity on paddy field weeds. J Agric Food Chem 54:2527–2536

Clements R, Haggar J, Quezada A, Torres J (2011) In: Zhu X (ed) Technologies for climate change adaptation – agriculture sector. UNEP Risø Centre, Roskilde, August 2011, ISBN: 978-87-550-3927-8.

Cong WF, Hoffland E, Li L, Six J, Sun JH, Bao XG, Zhang FS, Van Der Werf W (2015) Intercropping enhances soil carbon and nitrogen. Glob Chang Biol 21(4):1715–1726. https://doi.org/10.1111/gcb.12738

Cong W-F, Zhang C, Li C, Wang G, Zhang F (2021) Designing diversified cropping systems in China: theory, approaches and implementation. Front Agric Sci Eng 8:362–372

Cook SM, Khan ZR, Pickett JA (2007) The use of push-pull strategies in integrated pest management. Annu Rev Entomol 52:375–400

Das A, Babu Subhash, Datta M. Kumar Sanjeev, Singh R, Avasthe RK, Rathore SS, Yadav SK, Singh VK (2022) Restoring soil carbon in marginal land of Indian Himalayas: impact of crop intensification and conservation tillage. J Environ Manag 318, 115603. https://doi.org/10.1016/j.jenvman.2022.115603,

de Vries FT, Wallenstein MD (2017) Below-ground connections underlying above-ground food production: a framework for optimising ecological connections in the rhizosphere. J Ecol 105(4):913–920

Degrande A, Schreckenberg K, Mbosso C, Anegbeh P, Okafor V, Kanmegne J (2006) Farmers' fruit tree-growing strategies in the humid forest zone of Cameroon and Nigeria. Agrofor Syst 67(2):159–175

Denier J, Faucon M-P, Dulaurent A-M, Guidet J, Kervroëdan L, Lamerre J, Houben D (2022) Earthworm communities and microbial metabolic activity and diversity under conventional, feed and biogas cropping systems as affected by tillage practices. Appl Soil Ecol 169:104232

Doring TF, Kirchner SM (2022) A model for colour preference behavior of spring migrant aphids. Phi Trans Roy Soc B 377:20210283

Dwivedi SL, Lammerts van Bueren ET, Ceccarelli S, Grando S, Upadhyaya HD, Ortiz R (2017) Diversifying food systems in the pursuit of sustainable food production and healthy diets. Trends Plant Sci 22:842–856

FAO (2016) Pulses contribute to food security. http://www.fao.org/3/a-c0063e.pdf

Fernandez-Aparicio M, Emeran AA, Rubiales D (2010) Inter-cropping with berseem clover (*Trifolium alexandrinum*) reduces infection by *Orobanche crenata* in legumes. Crop Prot 29:867–871

Frew A, Powell JR, Johnson SN (2019) Aboveground resource allocation in response to root herbivory as affected by the arbuscular mycorrhizal symbiosis. Plant Soil 447:463–473

Gao X, Wu M, Xu R, Wang X, Pan R, Kim H-J, Liao H (2014) Root interactions in a maize/soybean intercropping system control soybean soil-borne disease, red crown rot. PLoS One 9:e95031

Glick BR, Todorovic B, Czarny J, Cheng Z, Duan J, McConkey B (2007) Promotion of plant growth by bacterial ACC-deaminase. Crit Rev Plant Sci 26:227–242

Goemann HM, Gay JD, Mueller RC, Brookshire ENJ, Miller P, Poulter B, Peyton BM (2021) Aboveground and belowground responses to cyanobacterial biofertilizer supplement in a semi-arid, perennial bioenergy cropping system. GCB Bioenergy 13(12):1908–1923

Guo XT, Xiong HC, Shen HY, Qiu W, Ji CQ, Zhang ZJ, Zuo YM (2014) Dynamics in the rhizosphere and iron-uptake gene expression in peanut induced by intercropping with maize: role in improving iron nutrition in peanut, plant. Physiol Biochemist 76:36–43

Haas D, Défago G (2005) Biological control of soil-borne pathogens by fluorescent pseudomonads. Nat Rev Microbiol 3:307–319

Hall J (2003) Environment: aliens plant species invade southern Africa. Global Info Network 27:1–2

Han P, Lavoir AV, Rodriguez-Saona C, Desneux N (2022) Bottom-up forces in agroecosystems and their potential impact on arthropod pest management. Annu Rev Entomol 67:239–259

Hanway JJ, Heidal H (1952) Soil analysis methods as used in Iowa State College Soil Testing Laboratory. Iowa State College of Agriculture bulletin 57:1–31

Herridge D, Peoples M, Boddey R (2008) Global inputs on biological nitrogen fixation in agricultural systems. Plant Soil 311:1–18

Hiltner L (1904) Recent experiments and new problems in soil bacteriology, with especial reference to green manuring and fallows. Arb Deu Landwirtschafts Gesellschaft 98:59–78

Holland JM, Bianchi FJ, Entling MH, Moonen AC, Smith BM, Jeanneret P (2016) Structure, function and management of semi-natural habitats for conservation biological control: a review of European studies. Pest Manag Sci 72:1638–1651

Hoysted GA, Bell CA, Lilley CJ, Urwin PE (2018) Aphid colonization affects potato root exudate composition and the hatching of a soil borne pathogen. Front Plant Sci 9:1278

Huang LQ, Gou X, Li X, Zhang C, VanderWerf W, Zhang FS (2017) Plant growth patterns in a tripartite strip relay inter crop are shaped by a symmetric above ground competition. Field Crop Res 201:41–51

Hungria M, Nogueira MA, Araujo RS (2013) Co-inoculation of soybeans and common beans with rhizobia and *azospirilla*: strategies to improve sustainability. Biol Fertil Soils 49:791–801

Ijaz M, Nawaz A, Ul-Allah S, Rizwan MS, Ullah A, Hussain M, Sher A, Ahmad S (2019) Crop diversification and food security. In: Hasanuzzaman M (ed) Agronomic crops, vol 1. Springer, Singapore, pp 607–621. https://doi.org/10.1007/978-981-32-9151-5_26

Inal A, Gunes A, Zhang FS, Cakmak I (2007) Peanut/maize-intercropping induced changes in rhizosphere and nutrient concentrations in shoots. Plant Physiol Biochem 45:350–356

Isaac RA, Kerber JD (1971) Atomic absorption and flame photometry: techniques and uses in soil, plant, and water analysis. In: Instrumental methods for analysis of soils and plant tissue. SSSA, Madison, pp 17–37

Jabran K, Mahajan G, Sardana V, Chauhan BS (2015) Allelopathy for weed control in agricultural systems. Crop Prot 72:57–65

Jackson ML (1973) Soil chemical analysis. Prentice Hall of India Pvt, Ltd, New Delhi

Jain A, Chakraborty J, Das S (2020) Underlying mechanism of plant–microbe crosstalk in shaping microbial ecology of the rhizosphere. Acta Physiol Plant 42(1):1–13

Jaworski CC, Thomine E, Rusch A, Lavoir AV, Wang S, Desneux N (2023) Crop diversification to promote arthropod pest management: a review. Agriculture. Communications 100004:100004

Jenkinson DS, Powlson DS (1976) The effects of biocidal treatments on metabolism in soil—V: a method for measuring soil biomass. Soil Biol Biochem 8:209–213

Jensen ES (1996) Rhizodeposition of N by pea and barley and its effect on soil N dynamics. Soil Biol Biochem 28:65–71

Jia Y, Shi Z, Chen Z, Walder F, Tian C, Feng G (2020) Soil moisture threshold in controlling above-and belowground community stability in a temperate desert of Central Asia. Sci Total Environ 703:134650

Jia Y, Zhai G, Zhu S, Liu X, Schmid B, Wang Z, Ma K, Feng X (2021) Plant and microbial pathways driving plant diversity effects on soil carbon accumulation in subtropical forest. Soil Biol Biochem 161:108375

Jian J, Du X, Reiter MS, Stewart RD (2020) A meta-analysis of global cropland soil carbon changes due to cover cropping. Soil Biol Biochem 143:107735. https://doi.org/10.1016/j.soilbio.2020.107735

Jiao N, Wang J, Ma C, Zhang C, Guo D, Zhang F, Jensen ES (2021) The importance of aboveground and belowground interspecific interactions in determining crop growth and advantages of peanut/maize intercropping. Crop J 9(6):1460–1469

Kallenbach CM, Frey SD, Grandy AS (2016) Direct evidence for microbial-derived soil organic matter formation and its ecophysiological controls. Nat Commun 7(1):13630

Kaymak HC (2010) Potential of PGPR in agricultural innovations. In: Maheshwari D (ed) Plant growth and health promoting bacteria. Microbiology monographs, vol 18. Springer, Berlin/Heidelberg

Kebede E (2021) Contribution, utilization, and improvement of legumes-driven biological nitrogen fixation in agricultural systems. Front Sustain Food Syst 5:767–998

Keymer A, Gutjahr C (2018) Cross-kingdom lipid transfer in arbuscular mycorrhiza symbiosis and beyond. Curr Opin Plant Biol 44:137–144

Kizeková M, Kanianska R, Jančová Ľ, Čunderlík J, Dugátová Z, Makovníková J (2019) Above and below ground biomass and carbon stock in permanent grasslands of Slovakia. Agric (Pol'nohospodárstvo) 65(4):155–163

Kjeldahl J (1883) A new method for the determination of nitrogen in organic matter. Ztschr Anal Chem 22:366

Kleijn D, Bommarco R, Fijen TPM, Garibaldi LA, Potts SG, van der Putten WH (2019) Ecological intensification: bridging the gap between science and practice. Trends Ecol Evol 34(2):154–166

Klein DA, Loh TC, Goulding RL (1971) A rapid procedure to evaluate the dehydrogenase activity of soils low in organic matter. Soil Biol Biochem 3:385–387

Kumar S, Sindhu SS, Kumar R (2022) Biofertilizers: an ecofriendly technology for nutrient recycling and environmental sustainability. Curr Res Microbial Sci 3:100094

Ladha JK, Peoples MB, Reddy PM, Biswas JC, Bennett A, Jat ML, Krupnik TJ (2022) Biological nitrogen fixation and prospects for ecological intensification in cereal-based cropping systems. Field Crop Res 283:108541

Li L, Tilman D, Lambers H, Zhang FS (2014) Plant diversity and overyielding: insights from belowground facilitation of intercropping in agriculture. New Phytol 203(1):63–69

Li Q, Chen J, Wu LK, Luo XM, Li N, Lin S, Lin WX (2018) Below ground interactions impact the soil bacterial community, soil fertility, and crop yield in maize/peanut intercropping systems. Int J Mol Sci 19:622

Li X, Jousset A, de Boer W, Carrión VJ, Zhang T, Wang X, Kuramae EE (2019) Legacy of land use history determines reprogramming of plant physiology by soil microbiome. ISME J 13:738–751

Li QM, Zhang D, Zhang JZ, Zhou ZJ, Pan Y, Yang ZH, Zhang LF (2023) Crop rotations increased soil ecosystem multifunctionality by improving keystone taxa and soil properties in potatoes. Front Microbiol 14:1034761

Lindsay WL, Norvell WA (1978) Development of a DTPA soil test for zinc, iron, manganese, and copper. Soil Sci Soc Am J 42:421–428

Liu C, Daniel PB, Jeffrey ACH, Kutcher R, Beckie HJ, Wang L, Floc'h JB, Hamel C, KHM S, Li LL (2022) Diversifying crop rotations enhances agroecosystem services and resilience. Adv Agron 173:299–335

Lund M, Brainard DC, Szendrei Z (2019) Cue hierarchy for host plants election in Pieris rapae. Entomol Exp Appl 167:330–340

Lv Q, Chi B, He N, Zhang D, Dai J, Zhang Y, Dong H (2023) Cotton-based rotation, intercropping, and alternate intercropping increase yields by improving root–shoot relations. Agronomy 13(2):413

Malihe M, Mohammad F, Alireza K, Mohsen N, Mehdi T (2015) Farmyard manure application of potato with three level of nitrogen fertilizer. Int J Farm Sci 4:536–540

Mariotti M, Masoni A, Ercoli L, Arduini I (2009) Above- and below-ground competition between barley, wheat, lupin and vetch in a cereal and legume inter cropping system. Grass Forage Sci 64:401–412

Maron JL, Marler M, Klironomos JN, Cleveland CC (2011) Soil fungal pathogens and the relationship between plant diversity and productivity. Ecol Lett 14:36–41

Massalha H, Korenblum E, Tholl D, Aharoni A (2017) Small molecules below-ground: the role of specialized metabolites in the rhizosphere. Plant J 90(4):788–807

Matse DT, Huang C, Huang Y, Yen M (2020) Effects of coinoculation of rhizobium with plant growth promoting rhizobacteria on the nitrogen fixation and nutrient uptake of *Trifolium repens* in low phosphorus soil. J Plant Nutr 43:739–752

Mayer J, Buegger F, Jensen ES, Schloter M, Hess J (2003) Estimating N rhizodeposition of grain legumes using a 15N in situ stem labeling method. Soil Biol Biochem 35:21–28

Mesfin S, Gebresamuel G, Haile M, Zenebe A (2023) Potentials of legumes rotation on yield and nitrogen uptake of subsequent wheat crop in northern Ethiopia. Heliyon 9(6):e16126

Messéan A, Viguier L, Paresys L, Aubertot JN, Canali S, Iannetta PP, Justes E, Karley A, Keillor B, Kemper L (2021) Enabling crop diversification to support transitions towards more sustainable European agrifood systems, vol 8. Front Agric Sci Eng, pp 474–480

Midega CAO, Pittchar JO, Pickett JA, Hailu GW, Khan ZR (2018) A climate-adapted push-pull system effectively controls fall armyworm (*Spodoptera frugiperda)* (J E Smith), in maize in East Africa. Crop Prot 105:10–15

Mitchell C, Brennan RM, Graham J, Karley AJ (2016) Plant defense against herbivorous pests: exploiting resistance and tolerance traits for sustainable crop protection. Front Plant Sci 7:191972

Nadeem SM, Zahir ZA, Naveed M, Asghar HN, Arshad M (2010) Rhizobacteria capable of producing ACC-deaminase may mitigate salt stress in wheat. J Soil Sci Soc Am 74(2):533–542

Oerke E-C (2006) Crop losses to pests. J Agric Sci 144:31–43

Olsen SR, Cole CV, Watandbe FS, Dean LA (1954) Estimation of available phosphorus in soil by extraction with sodium bicarbonate. J Chem Inf Model 53:1689–1699

Ouda S, Zohry AE-H, Noreldin T (2018) Crop rotation: an approach to secure future food. Springer International Publishing, Cham. https://doi.org/10.1007/978-3-030-05351-2

Petersen J, Belz R, Walker F, Hurle K (2001) Weed suppression by release of isothiocyanates from turnip-rape mulch. Agron J 93:37–43

Philippot L, Raaijmakers JM, Lemanceau P, Van Der Putten WH (2013) Going back to the roots: the microbial ecology of the rhizosphere. Nat Rev Microbiol 11(11):789–799

Raaijmakers JM, Paulitz TC, Steinberg C, Alabouvette A, MoënneLoccoz Y (2009) The rhizosphere: a playground and battlefield for soilborne pathogens and beneficial microorganisms. Plant Soil 321:341–361

Raj R, Das TK, Chakraborty D, Bhattacharyya R, Babu S, Govindasamy P, Kumar V, Ekka U, Sen S, Ghosh S, Roy A, Sharma T (2023) Soil physical environment and active carbon pool in rice–wheat system of South Asia: impact of long-term conservation agriculture practices. Environ Technol Innov 29:102966. https://doi.org/10.1016/j.eti.2022.102966

Rathore SS, Babu S, Shekhawat K, Singh R, Yadav SK, Singh VK, Singh C (2022) Designing energy cum carbon-efficient environmentally clean production system for achieving green economy in agriculture. Sustain Energy Technol Assess 52:102190. https://doi.org/10.1016/j.seta.2022.102190

S'anchez AC, Kamau HN, Grazioli F, Jones SK (2022) Financial profitability of diversified farming systems: a global meta-analysis. Ecol Econ 201:107595

Saudy HS (2015) Maize–cowpea intercropping as an ecological approach for nitrogen-use rationalization and weed suppression. Arch Agron Soil Sci 61(1):1–14. https://doi.org/10.1080/03650340.2014.920499

Sharma IP, Sharma AK (2021) Above and below-ground involvement in cyclic energy transformation that helps in the establishment of rhizosphere microbial communities. Symbiosis 85(1):21–30

Sietz D, Klimek S, Dauber J (2022) Tailored pathways toward revived farmland biodiversity can inspire agroecological action and policy to transform agriculture. Commun Earth Environ 3:211

Singh R, Babu S, Avasthe RK, Yadav GS, Das A, Mohapatra KP, Kumar A, Singh VK, Chandra P (2021a) Crop productivity, soil health, and energy dynamics of Indian Himalayan intensified organic maize-based systems. Int Soil Water Conserv Res 9(2):260–270. https://doi.org/10.1016/j.iswcr.2020.11.003

Singh P, Adhale P, Guleria A, Vaidya MK (2021b) Is crop diversification vulnerable to climate, agricultural and socio-economic factors in Himachal Pradesh, India? Current Science, pp 707–711

Singh R, Kumar A, Babu S, Avasthe R, Das A, Rathore SS, Kumar S, Singh C, Sharma V, Bhupenchandra I (2023) Development of organic nutrients management system for profitable and soil-supportive French bean (*Phaseolus vulgaris* L.) farming in north eastern Himalayas, India. Front Sustain Food Syst 7:1115521. https://doi.org/10.3389/fsufs.2023.1115521

Subbiah BV, Asija GL (1956) A rapid procedure for the determination of available nitrogen in soil. Curr Sci 25:259–260

Tabaglio V, Marocco A, Schulz M (2013) Allelopathic cover crop of rye for integrated weed control in sustainable agroecosystems. Ital J Agron 8:e5

Tabatabai MA, Bremner JM (1969) Use of p-nitrophenyl phosphate for assay of soil phosphatase activity. Soil Biol Biochem 1(4):301–307

Thébault E, Fontaine C (2010) Stability of ecological communities and the architecture of mutualistic and trophic networks. Science 329:853–856

Tian H, Xu R, Canadell JG, Thompson RL, Winiwarter W, Suntharalingam P, Davidson EA, Ciais P, Jackson RB, Janssens-Maenhout G, Prather MJ, Regnier P, Pan N, Pan S, Peters GP, Shi H, Tubiello FN, Zaehle S, Zhou F, Yao Y (2020) A comprehensive quantification of global nitrous oxide sources and sinks. Nature 586(7828):248–256

UNO (United Nations Organisation) 2017. World population prospects (2017). www.esa.un.org/unpd/wpp/Download/Standard/Population/

Vander Kooi CJ, Stavenga DG, Arikawa K, Belusic G, Kelber A (2021) Evolution of insect color vision: from spectral sensitivity to visual ecology. Annu Rev Entomol 66:435–461

Walkley A, Black IA (1934) An examination of the digestrates. Method for determining soil organic matter and propose modification of the chronic acid titration method. Soil Sci 37:29–38

Wang G, Schultz P, Tipton A, Zhang J, Zhang F, Bever JD (2019) Soil microbiome mediates positive plant diversity-productivity relationships in late successional grassland species. Ecol Lett 22:1221–1232

Wani OA, Sharma V, Kumar SS, Babu S, Sharma KR, Rathore SS, Marwaha S, Ganai NA, Dar SR, Yeasin M, Raghavendra S, Tomar J (2023) Climate plays a dominant role over land management in governing soil carbon dynamics in North Western Himalayas. J Environ Manag 338:117740. https://doi.org/10.1016/j.jenvman.2023.117740

Werling BP, Dickson TL, Isaacs R, Gaines H, Gratton C, Gross KL, Liere H, Malmstrom CM, Meehan TD, Ruan L, Robertson BA, Robertson GP, Schmidt TM, Schrotenboer AC, Teal TK, Wilson JK, Landis DA (2014) Perennial grasslands enhance biodiversity and multiple ecosystem services in bioenergy landscapes. Proc Natl Acad Sci USA 111:1652–1657

Weston LA, Alsaadawi IS, Baerson SR (2013) Sorghum allelopathyd from ecosystem to molecule. J Chem Ecol 39:142e153

Wiens MJ, Entz MH, Martin RC, Hammermeister AM (2006) Agronomic benefits of alfalfa mulch applied to organically managed spring wheat. Can J Plant Sci 86(1):121–131

Wittwer RA, Dorn B, Jossi W, van der Heijden MG (2017) Cover crops support ecological intensification of arable cropping systems. Sci Rep 7(1):41911

Yadav GS, Datta R, Imran Pathan S, Lal R, Meena RS, Babu S, Das A, Bhowmik SN, Datta M, Saha P, Mishra PK (2017) Effects of conservation tillage and nutrient management practices on soil fertility and productivity of rice (*Oryza sativa* L.)–rice system in north eastern region of India. Sustainability 9(10):1816

Yadav GS, Das A, Lal R, Babu S, Datta M, Meena RS, Patil SB, Singh R (2019) Impact of no-till and mulching on soil carbon sequestration under rice (*Oryza sativa* L.)–rapeseed (*Brassica campestris* L. var. rapeseed) cropping system in hilly agro-ecosystem of the eastern Himalayas, India. Agric Ecosyst Environ 275:81–92. https://doi.org/10.1016/j.agee.2019.02.001

Yadav GS, Babu S, Das A, Mohapatra KP, Singh R, Avasthe RK, Roy S (2020) No till and mulching enhance energy use efficiency and reduce carbon footprint of a direct-seeded upland rice production system. J Clean Prod 271:122700. https://doi.org/10.1016/j.jclepro.122700

Yadav GS, Babu S, Das A, Datta M, Mohapatra KP, Singh R, Singh VK, Rathore SS, Chakraborty M (2021a) Productivity, soil health, and carbon management index of Indian Himalayan intensified maize-based cropping systems under live mulch-based conservation tillage practices. Field Crop Res 264:108080. https://doi.org/10.1016/j.fcr.2021.108080

Yadav GS, Das A, Kandpal BK, Babu S, Rattan L, Datta M, Das B, Singh R, Singh VK, Mohapatra KP, Chakraborty M (2021b) The food-energy-water-carbon nexus in a maize-maize-mustard cropping sequence of the Indian Himalayas: an impact of tillage-cum-live mulching. Renew Sust Energ Rev 151:111602. https://doi.org/10.1016/j.rser.2021.111602

Yadav D, Singh D, Babu S, Madegowda M, Singh DV, Mandal D, Rathore AC, Sharma VK, Singhal V, Kumawat A, Yadav DK, Yadav RK, Kumar S (2023) Intensified cropping reduces soil erosion and improves rainfall partitioning and soil properties in the marginal land of the Indian Himalayas. International Soil and Water Conservation Research 12:521. https://doi.org/10.1016/j.iswcr.2023.10.002

Yadav D, Singh D, Madhu M, Singh DV, Islam S, Kumar G, Gupta AK, Mandal D, Babu S, Barh A, Kumawat A, Dubey R, Pati K, Bhargavi B, Yadav DK (2024) Interplanted crops and rainfall variability regulate soil erosion in rainfed maize of the Indian Himalayas. Land Degrad Dev 35:4271. https://doi.org/10.1002/ldr.5221

Yang F, Liao D, Wu X, Gao R, Fan Y, Raza MA, Wang X, Yong T, Liu W, Liu J, Du J, Shu K, Yang W (2017) Effect of above ground and below ground interactions on the intercrop yields in maize-soybean relay inter cropping systems. Field Crop Res 203:16–23

Yang G, Wagg C, Veresoglou SD, Hempel S, Rillig MC (2018) How soil biota drive ecosystem stability. Trends Plant Sci 23(12):1057–1067

Yang X, Steenhuis TS, Davis KF, Werf W, Ritsema CJ, Pacenka S, Zhang F, Siddique KHM, Du T (2021) Diversified crop rotations enhance groundwater and economic sustainability of food production. Food Energy Secur 10(4), e311

Yang X, Xiong J, Du T, Ju X, Gan Y, Li S, Butterbach-Bahl K (2024) Diversifying crop rotation increases food production, reduces net greenhouse gas emissions and improves soil health. Nat Commun 15(1):198

Yu Y, Xue L, Yang L (2014) Winter legumes in rice crop rotations reduces nitrogen loss, and improves rice yield and soil nitrogen supply. Agron Sustain Dev 34:633–640

Zhang FS, Li L (2003) Using competitive and facilitative interact ions in intercropping systems enhances crop productivity and nutrient-use efficiency. Plant Soil 248:305–312

Zhang J, Van Der Heijden MG, Zhang F, Bender SF (2020) Soil biodiversity and crop diversification are vital components of healthy soils and agricultural sustainability. Front Agric Sci Eng 7(3):236

Zhang Z, Zhang B, He H, Yan M, Li J, Yan F (2022) Changes in visual and olfactory cues in virus-infected host plants alter the behavior of *Bemisia tabaci*. Front Ecol Evol 2022:10

Zhao J, Yang Y, Zhang K, Jeong J, Zeng Z, Zang H (2020) Does crop rotation yield more in China? A meta-analysis. Field Crop Res 245:107659

Zhiipao RR, Pooniya V, Kumar D, Biswakarma N, Shivay YS, Dass A, Kumar Bainsla N, Lakhena KK, Pandey RK, Burman A, Bhatia A, Jat RD, Govindasamy P, Swarnalakshmi K, Das K, Choudhary RL, Babu S (2023) Above and below-ground growth, accumulated dry matter and nitrogen remobilization of wheat (*Triticum aestivum* L.) genotypes grown in PVC tubes under well- and deficit-watered conditions. Front Plant Sci 14:1087343. https://doi.org/10.3389/fpls.2023.1087343

Zhu Y, Chen H, Fan J, Wang Y, Li Y, Chen J, Fan J, Yang S, Hu L, Leung H, Mew TW, Teng PS, Wang Z, Mundt CC (2000) Genetic diversity and disease control in rice. Nature 406:718–722

Zuo YM, Cao YP, Li XL, Zhang FS (2000) Studies on the improvement in iron nutrition of peanut by inter cropping with maize on a calcareous soil. Plant Soil 220:13–25

Chapter 13
Crop Diversification Opportunities in Rainfed Farming: Approaches and Strategies

G. Ravindra Chary, Vinod Kumar Singh, V. Ramamurthy, K. A. Gopinath, B. Bhargavi, Raghavendra Singh, and Subhash Babu

Abstract Rainfed agriculture in India still constitutes 48% of the net cropped area and contributes approximately 40% of the nation's food supply. The impacts of climate change and climate variability are felt across Indian agriculture, with rainfed agriculture being particularly vulnerable. The primary challenge is to sustain unirrigated and partially irrigated operational holdings by promoting sustainable intensification, ensuring stable productivity and profitability, and improving the livelihoods of small and marginal farmers. Crop diversification in rainfed regions is essential to enhance resource use efficiency, increase land productivity and profitability per unit area, mitigate various risks, create employment opportunities, improve farmers' livelihoods, and ensure food and nutritional security. Traditionally, mixed or intercropping systems prevail in core rainfed areas, providing resilience against adverse weather conditions during the growing season. Therefore, adopting climate-resilient crop diversification strategies in rainfed areas is crucial for maximizing income and ensuring sustainable agricultural systems.

Keywords Crop intensification · Low input · Nutritional security · Pulses

---

G. R. Chary · V. K. Singh · K. A. Gopinath · B. Bhargavi
ICAR-Central Research Institute for Dryland Agriculture, Hyderabad, India

V. Ramamurthy (✉)
ICAR-National Bureau of Soil Survey & Land Use Planning, Regional Centre, Bangalore, India

R. Singh
ICAR-Indian Institute of Farming Systems Research, Modipuram, Meerut, Uttar Pradesh, India

S. Babu
Division of Agronomy, ICAR-Indian Agricultural Research Institute, New Delhi, India

S. Babu et al. (eds.), *Agricultural Diversification for Sustainable Food Production*, Sustainability Sciences in Asia and Africa,
https://doi.org/10.1007/978-981-97-7517-0_13

13.1 Introduction

Rainfed agriculture in India accounts for 48% of the net crop sown area. It contributes about 40% of the country's food basket (rice: 37%; nutri-cereals: 82%; pulses: 77%; oilseeds: 69%). The need for crop intensification in rainfed areas is to enhance resource use efficiency, land productivity, and profitability per unit area, minimize various risks, increase employment opportunities, enhance the livelihoods of the farmers, and address food and nutritional security. Rainfed agriculture is totally dependent on the south-west monsoon and thus, is synonymous with risk due to erratic monsoon. Climate change and climate variability impact on Indian agriculture in general and more pronounced in rainfed agriculture. A district-level climatic analysis in the country revealed spatial shifts of climate zones in about 27% of the geographical area in the country, i.e., a substantial increase in the arid region in Gujarat, and a decrease in the arid region in Haryana, an increase in semiarid region in Madhya Pradesh, Tamil Nadu, and Uttar Pradesh due to shift of climate from dry subhumid to semiarid. Likewise, the moist subhumid pockets in Chhattisgarh, Orissa, Jharkhand, Madhya Pradesh, and Maharashtra states shifted to dry subhumid to a larger extent. These climate shifts in rainfed areas will have larger implications for crop planning, water resources assessment, and prioritizing drought-proofing programs. Out of the total operational land holdings in the country, 58.14 m operational land holdings are unirrigated out of which 85% account for marginal and small operational holdings while the total operational holdings under partly irrigated operational holdings are 17.21 m out of which 80% marginal and small operational holdings. The challenge is to sustain these unirrigated and partially irrigated operational holdings with respect to sustainable intensification, stabilized productivity and profitability, and enhanced livelihoods of the small and marginal farmers. Although the average productivity levels have increased to 1.2 tons per hectare at present in rainfed areas, large yield gaps still remain in several rained crops. With current gains in the existing rainfed production systems, there is a need for more profitable land-use systems. While evolving strategies for crop intensification, due attention must be given to bridging present yield gaps, and regional imbalances in terms of natural resources and technology intake capacity of farmers.

The basic resource that determines the success of rainfed agriculture is water availability. The imbalance between the rainwater input and water withdrawals from surface and groundwater sources is likely to widen further. It is estimated that even after achieving full irrigation potential, nearly 40% of the total cultivated area of the country will remain rainfed. The National Water Mission institutionalized under the National Action Plan for Climate Change has set the target to improve water use efficiency by at least 20%. With the expected productivity levels and crop intensification, the demand for nutrients is bound to increase severalfold. The soil organic carbon is 5 g/kg in soils in rainfed areas, whereas the desired level is 11 g/kg. Although about 80 million tons of crop residues are produced annually in rainfed areas, their recycling is not done due to competitive uses and burning. There is a need to develop and demonstrate proven agroecology-specific nature-based solutions for improving soil quality in rainfed areas. Farm mechanization in rained areas is very low due to small and marginal holdings and other biophysical and socioeconomic settings. About 80% of existing farm mechanization is skewed toward land preparation, whereas low for intercultural, weed management, and harvest operations. Farm mechanization for conservation tillage practices to be standardized for crop intensification in rainfed production systems. The projected domestic demand for different crop groups shows that rice and wheat may be in surplus, whereas other cereals will be in acute shortage. Out of the 59 million tons shortfall, most of the produce will constitute maize that would go for animal/poultry feed. Also, the deficit would be primarily for oilseeds, fruits, vegetables, and pulses. Hence, the challenge would be to enhance the productivity levels of these crops by developing/identifying climate-resilient varieties along with matching natural resource management (NRM) technologies. Further, as rice and wheat are going to be surplus, we need to follow a two-pronged strategy, i.e., to increase their productivity by bridging the yield gaps and shifting some of the areas under these crops to oilseeds, millets, and pulses. Currently, there is an imbalance between natural resource endowment and cropping patterns in the country. It is an irony that areas with less rainfall are net exporters of agricultural produce to areas with sufficient rainfall and untapped groundwater potential. Crop diversification is needed for (i) risk reduction and as adaptation strategies in resilient rainfed agriculture, (ii) aligning cropping pattern with natural resource status, (iii) food and nutritional security, (iv) self-sufficiency of pulses and oilseeds and to reduce dependence on import, (v) aligning production and consumption pattern to reduce logistic effort and wastage at the state level, (vi) income enhancement of farmers, and (vii) to address environmental concerns such as resource degradation (Vittal et al. 2006, Vittal and Ravindra Chary 2007; Ravindra Chary et al. 2016a, b; Bhargavi et al. 2019).

13.2 Approaches

Crop diversification in rainfed farming could be horizontal and or vertical diversification. The following three broad approaches viz. crop alignment as per moisture availability, agroecology potential crop zoning, and diversifying within the farm are suggested that are likely to enable crop diversification at various scales.

13.2.1 Crop Alignment as per Moisture Availability

Under the changing climate scenario, many conventional practices and cropping systems are becoming redundant and ineffective, and thus need revalidation and modification in accordance with changing climate and soil-site conditions; this also calls for critical examination of important modifiers of cropping systems, such as soil types, rainfall patterns, length of growing season, and temperature regimes so that available farm resources are effectively used. Accordingly, the length of growing period (LGP)/moisture availability period was found to be a better parameter for crop planning (Velayutham 1999). The LGP is the period when the moisture and temperature regimes are suitable for crop growth, and it is computed as the sum of the periods when moisture is more than 0.5 PET plus the time taken to utilize stored soil moisture. Therefore, planning cropping systems using LGP (Table 13.1) has the potential to use available renewable resources efficiently and reduce the chances of crop failure. For example, an LGP of 210 days in deep black soils of Vidharbha could suitably be used for growing a short-duration soybean (*kharif*) followed by chickpea (*rabi*) instead of a single long-duration crop. While short-duration drought tolerant pulses such as moth bean and cereals of 10–12 weeks' duration such as pearl millet and minor millets could be cultivated in arid regions with low seasonal rainfall (<500 mm) and length of growing period (< 90 days).

13.2.2 Agroecology-Specific Potential Crop Zoning

The cropping pattern in rainfed areas is largely driven by management (accumulated knowledge), monsoon (southwest), and often with market influence. Traditionally, mixed or intercropping dominated in core rained areas, which provide risk resilience during aberrant weather conditions in crop growing season. The

Table 13.1 Potential cropping systems and agricultural drought vulnerability based on rainfall and soil types

Mean annual rainfall (mm)	Major soil order	Growing season (Moisture availability period) (weeks)	Suitable cropping system	Agricultural drought (frequency)
350–650	Alfisols, shallow Vertisols, Aridisols, and Entisols	15	Single rainy season	Severe drought (Once in <5 seasons)
350–650	Deep Aridisols and Inceptisols	20	Either rainy or post-rainy season crop	Moderate drought (Once in 5–10 seasons)
350–650	Deep Vertisols	20	Post-rainy season crop	Moderate drought (Once in 5–10 seasons)
650–800	Alfisols, Vertisols, Inceptisols	20–30	Intercropping	Less prone to drought (Once in 10–20 seasons)
800–1100	Deep Vertisols, Alfisols, and Entisols	30	Double cropping	Less prone to drought (Once in 10–20 seasons)
>1100	Deep Alfisols, Oxisols	30+	Double cropping	Nil to less prone to drought (once in >20 seasons)

Source: Velayutham (1999), modified from CRIDA 1997

Table 13.2 Strategies for potential zones of the base crop

Potential zone of the base rainfed crop	Strategies
Highly potential zones	Technological interventions (soil, water, crop, energy-based) for higher productivity, profitability and stability of the rainfed crop-based system Sustained, quality, and adequate quantity seed production of the base crop Cost-effective and energy-efficient total farm mechanization of the base crop Development of the value chain, weather indices-based insurance of the base crop Developing rainfed crop-based integrated farming systems
Moderately potential zones	Rainfed crop-based crop diversification/intensification Strengthening traditional rained farming systems/agroforestry systems
Marginally potential and non-potential zones	Replacing base crop/crop substitution with alternate crops/cropping systems and agroforestry systems

recent trend of a shift in climate and the impact of rainfall variability in a region/agroclimatic zone in crop growing season impacts productivity, profitability, and stability of rainfed crop production systems and also results in poor soil quality. This calls for concerted efforts in efficient agroecology-specific potential crop zoning refers to the specific regions/areas of crops and cropping sequences that are bio-physically suitable and also have high productivity and high spread. Efficient crop zones have similar geographic settings in terms of soils, landforms, rainfall, temperature, length of growing period, and irrigation potentials, are suitable for specific crop and cropping sequences, and have the potential to respond similarly to similar kinds of management practices. The potential crop zoning helps in developing strategies for various potential zones of the base crop and is given in Table 13.2.

13.2.2.1 Methodology for Delineation of Potential Zones of Crops

Potential crop zoning involves the development of soil-site characterization, biophysical suitability evaluation, and linking of biophysical suitable maps to the relative spread and productivity of reference crops and cropping sequences. The methodology for delineating potential crop zones is described hereunder (Fig. 13.1; Ramamurthy et al. 2016).

13.2.3 Diversifying Within the Farm

Crop diversification in rainfed farming has to be diversifying within farms in integrated farming system mode, i.e., integration of crop–trees–fodder–animal systems. There is a need for evolving rainfed integrating farming systems by strengthening

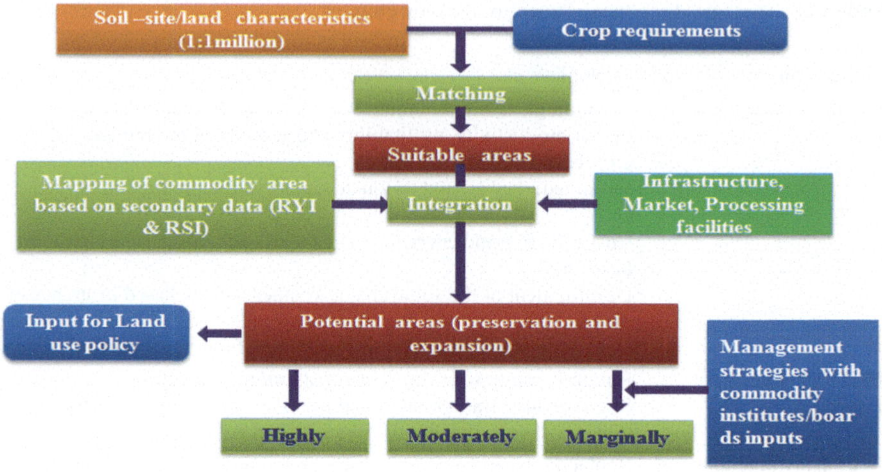

Fig. 13.1 Methodology for delineation of potential zones of crops

Table 13.3 Suggested strategies for strengthening traditional rained farming systems

Rainfall zone (mean annual rainfall)	Strengthening predominant traditional rained farming systems	Agroecology components along with efficient in situ and ex situ rainwater management practices
< 500 mm	Livestock-crop based	Small ruminants, nutritious cereals/millets
500–750 mm	Crop-horticulture-livestock based	Small/large ruminants, predominant rainfed crops, and dryland horticulture
750–1000 mm	Crop-horticulture-livestock-poultry based	Predominant rainfed crops, dryland horticulture, agri-horti systems, rainfed vegetable crops, small/large ruminants, improved breeds of poultry
> 1000 mm	Multiple enterprises based on multiple water use	Predominant rainfed crops, lowland rice with water-saving technologies, dryland horticulture, vegetable crops, other high-value crops, agri-horti systems, small/large ruminants, improved breeds of poultry, fish and other income-generating enterprises like seed production, apiary, and mushroom cultivation

Source: Ravindra Chary et al. (2022)

predominant traditional rained farming systems that enhance resource use efficiency, risk resilience, food, and nutritional security, staggered employment, and income (Table 13.3). Agroecology-specific alternate land-use systems/agroforestry systems based on land capability have to be promoted for risk resilience and staggered income, biomass production, and ecosystem services. Boundary plantations with perennial tree species have to be promoted for forage, greenleaf manure, mulching, and ecosystem services at the individual farm level.

13.3 Key Strategies for Crop Diversification in Rainfed Farming

The crop diversification in rainfed farming has to be in alignment with the biophysical suitability, i.e., soil-site suitability, including moisture availability period. The key strategies include technological, institutional, infrastructure, and policy interventions and are presented briefly below.

13.3.1 Technological Interventions

13.3.1.1 Real-Time Contingency Planning Under Delayed Onset of Monsoon

Introduction of alternate crops under delayed onset of monsoon reduces the risk of total crop failure and also provides alternative means of generating income, as different crops will respond to climate scenarios in different ways. The choice of crops and varieties is more relevant under highly complex rainfed production systems and areas frequented with weather vagaries. As a general rule, rainfed crops are sown early with the onset of monsoon to realize higher yields. Beyond the sowing window, the choice of alternate crops depends on the farming situation, soil, rainfall, and cropping pattern in the location and extent of delay in the onset of monsoon. For example, pulses and oilseeds are preferred over cereals due to less water requirement and hence can be grown under delayed *kharif* sowing. Further, cluster bean, moth bean, and horse gram are suitable for low rainfall areas. During *rabi* season, under residual soil moisture, chickpeas and lentils are preferred over peas and french bean. Similarly, oilseed crops viz. groundnut, castor, sesame, and niger perform well under rainfed conditions during the *kharif* season. In the rapeseed-mustard group of crops, taramira is the best choice for light-textured soil with low moisture storage capacity, followed by Indian mustard. Among the *kharif* cereals, coarse cereals–millet, finger millet, and sorghum are better choices over maize and rice. Similarly, in the *rabi* season, barley does well under conserved soil moisture than wheat. Among the millets, Setaria is the most suitable crop for the late sown condition without affecting productivity.

The onset of monsoon is critical for taking up timely sowing of normal crops during *kharif* in rainfed areas. If the onset is delayed beyond 4 weeks, accordingly, the farmer has to select alternate crops, to produce for his domestic needs and income, though sometimes yield levels are not so high (Srinivasa Rao et al. 2016). Beyond the sowing window, the choice of alternate crops or cultivars depends on the farming situation, soil, rainfall, and cropping pattern in the location and extent of delay in the onset of monsoon. The contingent crops under delayed monsoon situations are suggested in Table 13.4.

Table 13.4 Contingent crops to cope with delayed onset of monsoon

State	Agroclimatic zone/ Climate/soil type; Recommendation domain (Districts)	Major rainfed production systems/ crops	Suggested alternate crops/intercropping systems with delay in onset of monsoon		
			4 weeks	6 weeks	8 weeks
Karnataka	Northern Dry zone/ Semiarid (Hot dry)/ Vertisols; Vijayapura, Bagalkote, Gadag, Bellary, Koppal, parts of Dharwad, Belagavi, and Raichur districts	Rabi Sorghum	Pigeon pea, castor, moth bean, and cowpea	Pearl millet, castor, cowpea, Setaria	Sunflower, pigeon pea, green gram, castor, cowpea
Maharashtra	Scarcity zone/Semiarid (Hot dry)/Vertisols; Solapur, Ahmednagar, Sangli, Pune, Dhule, Jalgaon, Nandurbar, Satara districts	Rabi Sorghum	Pigeon pea, castor, moth bean, horse gram, cowpea	Pearl millet, castor, Setaria spp., cowpea, sorghum fodder	Sunflower, pigeon pea, green gram, black gram, castor, pearl millet, cowpea
Uttar Pradesh	South-western semiarid zone/Semiarid (Hot dry)/Inceptisols; Agra, Mathura, Mainpuri, Aligarh, Hathras, Firozabad, Etah, and Kanshiram Nagar districts	Pearl millet	Cluster bean, green gram, black gram	Pearl millet for fodder, green gram, black gram,	–
Gujarat	Northern Gujarat zone/ Semiarid/Arid (Hot dry)/Entisols; Banaskantha, Aravali, Sabarkantha, Patan, Mehsana, Gandhinagar, and Kachchh districts	Pearl millet	Black gram, groundnut, sesame, cluster bean, castor+ green gram	Black gram, pigeon pea, cluster bean, fodder sorghum, castor,	Black gram, pigeon pea, fodder sorghum, castor, fennel
	North Saurashtra zone/ Semiarid (Hot dry)/ Vertisols; Jamnagar, Devbhoomi Dwaraka Morbi, Surendranagar, Amreli, Bhavnagar, and Rajkot districts	Groundnut	Castor, sorghum, green gram, black gram, sesame	Green gram, castor, sorghum, sesame, pigeon pea	Sesame, castor, sorghum, green gram, black gram, pearl millet
Karnataka	Eastern, Southern, and Central Dry zones/ Semiarid (Hot moist)/ Alfisols; Mysore, Tumkur, Kolar, Ramanagara, Bengaluru (Rural and Urban), Chamarajanagar, Hassan, Chikkaballapur and Mandya districts	Finger millet	Groundnut, pigeon pea, cowpea,	Little millet, foxtail millet	Little millet, foxtail millet, cowpea, sunflower, field bean

(continued)

Table 13.4 (continued)

State	Agroclimatic zone/ Climate/soil type; Recommendation domain (Districts)	Major rainfed production systems/ crops	Suggested alternate crops/intercropping systems with delay in onset of monsoon		
			4 weeks	6 weeks	8 weeks
Madhya Pradesh	Malwa zone/Semiarid (Hot moist)Vertisols; Mandsaur, Ratlam, Ujjain, Rajgarh, Shajapur, Indore, Dewas, Dhar, Jhabua, and Sehore districts	Soybean	Maize fodder, sesame,	Sunflower, maize for fodder	Maize fodder
	Keymore Plateau and Satpura Hill Zone/ Subhumid (Hot dry)/ Vertisols; Rewa, Sidhi, Singroli, Satna, Panna, Katni, Jabalpur and Seoni districts	Soybean	Pigeon pea, green gram, black gram	Sesame, black gram, green gram	Green gram, black gram
	Central Maharashtra Plateau zone/Semiarid (Hot moist)/Vertisols; Parbhani, Hingoli, Nanded, Beed, Latur Osmanabad, Aurangabad, and Jalna districts	Cotton	Green gram, black gram, sorghum, pearl millet, pigeon pea	Sesame, castor, green gram, black gram, sorghum, pearl millet, pigeon pea	Sesame, castor, green gram, black gram, sorghum, pearl millet, pigeon pea

Source: Ravindra Chary et al. (2022)

13.3.1.2 Crop Intensification with Intercropping Systems in Mono-cropped Areas

Intercropping systems enhance resource use efficiency overall system productivity and income per unit area to the smallholders. Further, in no-cropped or mono-cropped areas, crop intensification with intercropping systems provides better conditions for food and nutritional security and enables farmers to grow surplus products or sell at market and thus help to obtain increased income to meet other needs related to household well-being. Diversifying from the monoculture of traditional staples can have important nutritional benefits for farmers. Diversification can also manage price risk, on the assumption that not all products will suffer low market prices at the same time and increase the profitability of the farming community. Pigeon pea-based intercropping systems with cereals, pulses, oilseeds, and cotton are most promising across rainfed agroecologies. Pigeon pea area is also expanding due to its cultivation as an intercrop in Bt cotton (1:4/1:6 row ratio) in the southern states of India, like Andhra Pradesh, Telangana, Maharashtra, Karnataka, and Tamil Nadu, especially under rainfed conditions. Similarly, in the southern plateau, intercropping of cowpea and horse gram with Bt cotton was found advantageous. Under rainfed ecosystems, short-duration chickpea as a preferred intercrop is considered over others since it is most compatible with many winter cereals and oilseeds. The rainfed cereals, pulses, oilseeds, and cotton-based production systems, state and agroclimatic zone-wise recommendation domain-wise intercropping systems are suggested in Tables 13.5, 13.6, 13.7, and 13.8.

Table 13.5 Suggested rainfed cereal crops-based intercropping systems

State	Agroclimatic zone/Climate/Soil type/ Recommendation domain (Districts)	Major rainfed crop-based production system	Efficient Intercropping systems
Chattisgarh	Bastar Plateau Zone/Subhumid (Hot moist)/Inceptisols; Bastar, Narayanpur, Kondagaon, Bijapur, Sukma, Dantewadaand Kanker districts	Rainfed rice	Upland rice + pigeon pea (8:1); Finger millet + pigeon pea (6:1) Maize + pigeon pea (4:1); Maize + cowpea (4:1); Maize + Okra (4:1)
Haryana	South-Western Dry zone/Arid (Hyper)/Inceptisols; Sirsa, Fatehabad, Hisar, Bhiwani, Mahendergarh, Rewari, Jhajjar, Nuh, Palwal, Jind, Gurugramand and Panchkula districts	Pearl millet	Pearl millet + green gram/ cluster bean/cowpea (8:4)
Karnataka	Eastern, Southern, and Central Dry zones/Semiarid (Hot moist)/Alfisols; Mysore, Tumkur, Kolar, Ramanagara, Bengaluru (Rural and Urban), Chamarajanagar, Hassan, Chikkaballapur, and Mandya districts	Finger millet	Finger millet + pigeon pea (8:2); Maize+ pigeon pea (4:1)
	Northern Dry zone/Semiarid (Hot dry)/Vertisols; Vijayapura, Bagalkote, Gadag, Bellary, Koppal, parts of Dharwad, Belagavi, and Raichur districts	*Rabi* sorghum	*Rabi* sorghum + chickpea (2:1) Pearl millet + pigeonpea/ castor (2:1); Pearl millet + bunch groundnut (4:2); Sorghum + chickpea(2:1); Foxtail millet + niger (2:4)
Maharashtra	Western Vidarbha zone/Semiarid (Hot moist)/Vertisols; Akola, Washim, Buldana, Yavatmal, Amravati, Wardha, and Nagpur districts	Cotton	Pearl millet + pigeon pea (1:1)
	Central Maharashtra Plateau zone/ Semiarid (Hot moist)/Vertisols; Parbhani, Hingoli, Nanded, Beed, Latur Osmanabad, Aurangabad, and Jalna districts	Cotton	Sorghum + pigeon pea (4:2); Maize–soybean (2:2); Maize + green gram (1:1)
	Scarcity zone/Semiaarid (Hot dry)/ Vertisols; Solapur, Ahmednagar, Sangli, Pune, Dhule, Jalgaon, Nandurbar, Satara districts	*Rabi* sorghum	Sorghum + cowpea (4:4/6:6); Pearl millet + sunflower (2:1/2:2)

(continued)

Table 13.5 (continued)

State	Agroclimatic zone/Climate/Soil type/ Recommendation domain (Districts)	Major rainfed crop-based production system	Efficient Intercropping systems
Madhya Pradesh	Malwa zone/Semiarid (Hot moist) Vertisols; Mandsaur, Ratlam, Ujjain, Rajgarh, Shajapur, Indore, Dewas, Dhar, Jhabua, and Sehore districts	Soybean	Maize + soybean (2:4)
	Keymore Plateau and Satpura Hill zone/Subhumid (Hot dry)/Vertisols; Rewa, Sidhi, Singroli, Satna, Panna, Katni, Jabalpur and Seoni districts	Soybean	Wheat + chickpea (2:1)
Odisha	Northeastern Ghat zone/Subhumid (Hot moist)/Inceptisols; Kandhamal, Gajapati, Rayagada, Ganjam, and part Koraput districts	Rainfed Rice	Maize + cowpea/pigeon pea/runner bean (2:2); Maize + bitter gourd/ cucumber/ridge gourd (2:1)
Rajasthan	Southern zone/Semiarid(Hot dry)/ Vertisols; Bhilwara, Chittaurgarh, Rajsamand, Udaipur and part of Sirohi districts	Maize	Maize + black gram (2:2); Maize + pigeon pea/castor (1:1); Sorghum + green gram (2:1)
Uttar Pradesh	South-western semiarid zone/ Semiarid (Hot dry)/Inceptisols; Agra, Mathura, Mainpuri, Aligarh, Hathras, Firozabad, Etah, and Kanshiram Nagar districts	Pearl millet	Pearl millet + pigeon pea/ green gram/black gram (2:1)
	Eastern plain and Vindhayan zone/ Subhumid (Hot dry)/Inceptisols; Varanasi, Chandauli, Sant Ravidas Nagar, Jaunpur, Ghazipur, Mirzapur, and Sonbhadra districts	Rice	Maize + black gram (1:2); Maize + okra (1:1)

Source: Amrutsagar et al. (2022); Prasad et al. (2012); Ravindra Chary et al. (2022); Shirahatti et al. (2022); Singh et al. (2022)

Table 13.6 Suggested rainfed pulse crops-based intercropping systems

State/	Agroclimatic zone/Climate/Soil type/Recommendation domain	Major rainfed crop-based production system	Efficient intercropping systems
Gujarat	Northern Gujarat/Semiarid/Arid (Hot dry)/Entisols; Banaskantha, Aravali, Sabarkantha, Patan, Mehsana, Gandhinagar and Kachchh districts	Pearl millet	Green gram + castor as a relay crop (2:1)
Jharkhand	Western Plateau zone/Subhumid (Hot moist)/Inceptisols; Palamu, Garhwa, Latehar, Chatra, Lohardaga, Gumla and Simdega districts	Rainfed Rice	Pigeon pea+ okra/maize/ sorghum (1:1); Pigeon pea + groundnut (1:2); Chickpea + linseed (4:2); Chickpea + safflower (2:1)

(continued)

Table 13.6 (continued)

State/	Agroclimatic zone/Climate/Soil type/Recommendation domain	Major rainfed crop-based production system	Efficient intercropping systems
Karnataka	Eastern, Southern, and Central Dry zones/Semiarid (Hot moist)/ Alfisols; Mysore, Tumkur, Kolar, Ramanagara, Bengaluru (Rural and Urban), Chamarajanagar, Hassan, Chikkaballapur, and Mandya districts	Finger millet	Pigeon pea + field bean/ cowpea (1:1)
	Northern Dry zone/Semiarid (Hot dry)/Vertisols; Vijayapura, Bagalkote, Gadag, Bellary, Koppal, parts of Dharwad, Belagavi, and Raichur districts	*Rabi* sorghum	Pigeon pea + bunch groundnut (4:2)
Maharashtra	Western Vidarbha zone/Semiaarid (Hot moist)/Vertisols; Akola, Washim, Buldana, Yavatmal, Amravati, Wardha, and Nagpur districts	Cotton	Pigeon pea + Sesame (2:4)
	Scarcity zone/Semiarid (Hot dry)/ Vertisols; Solapur, Ahmednagar, Sangli, Pune, Dhule, Jalgaon, Nandurbar, Satara districts	*Rabi* sorghum	Pigeon pea + sunflower/pearl millet (1:2); Pigeon pea + soybean (2:4 or 1:6); Pigeon pea+ black gram/green gram/ bajra fodder (1:3); Sorghum + cowpea (4:4 and 6:6); Chickpea + safflower (2:4)
Madhya Pradesh	Keymore Plateau and Satpura Hill Zone/Subhumid (Hot dry)/ Vertisols; Rewa, Sidhi, Singroli, Satna, Panna, Katni, Jabalpur and Seoni districts	Soybean	Chickpea + linseed (4:2)
Rajasthan	Southern zone/Semiarid (Hot dry)/ Vertisols; Bhilwara, Chittaurgarh, Rajsamand, Udaipur and part of Sirohi districts	Maize	Pigeon pea + groundnut (1:1); Chickpea + mustard (Intra-cropping) across the sowing of chickpea at 2.7 m interval
Uttar Pradesh	South-western semiarid zone/ Semiarid (Hot dry)/Inceptisols; Agra, Mathura, Mainpuri, Aligarh, Hathras, Firozabad, Etah, and Kanshiram Nagar districts	Pearl millet	Cluster bean + pearl millet/ sesame (6:1); Pigeon pea + green gram (2:2); Chickpea + mustard (4–5:1)
	North Eastern Plain Zone/ Subhumid (Hot dry)/Inceptisols; Faizabad, Sultanpur, Basti, Gonda, and Ambedkarnagar districts	Rainfed rice	Pigeon pea + maize/green gram/black gram (1:1); Chickpea + mustard (4:2)
	Eastern plain and Vindhayan zone/ Subhumid (Hot dry)/Inceptisols; Varanasi, Chandauli, Sant Ravidas Nagar, Jaunpur, Ghazipur, Mirzapur, and Sonbhadra districts	Rainfed Rice	Pigeon pea + rice/sesame/ okra (1:1); Pigeon pea+black gram (1:2); Chickpea + mustard (4:1); Chickpea +linseed (2:1)

Source: Ravindra Chary et al. (2022)

Table 13.7 Suggested rainfed oilseeds crops-based intercropping systems

State	Agroclimatic zone/Climate/Soil type/ Recommendation domain	Major rainfed crop-based production system	Efficient intercropping systems
Assam	North Bank Plains zone/Humid (Hot)/ Alfisols; Dhemaji, Lakhimpur, Biswanath, Sonitpur, Darrang, and Udalguri districts	Rainfed rice	Sesame + black gram (1:1)
Andhra Pradesh	Scarcity zone/Semiarid/Arid (Hot)/ Alfisols; Ananthapuramu, Kurnool, YSR Kadapa and Chittoor districts	Groundnut	Groundnut+ pigeon pea (8:1); Groundnut+castor (15:1)
Gujarat	North Saurashtra zone/Semiarid (Hot dry)/Vertisols; Jamnagar, Devbhoomi Dwarka Morbi, Surendranagar, Amreli, Bhavnagar, and Rajkot) districts	Groundnut	Groundnut + castor (3:1); Groundnut + cotton (2:1) Castor + cowpea/green gram (1:1); Castor + green gram (2:1)
	Northern Gujarat/Semiarid/Arid (Hot dry)/Entisols; Banaskantha, Aravali, Sabarkantha, Patan, Mehsana, Gandhinagar, and Kachchh districts	Pearl millet	Castor + green gram/ cowpea/cluster bean/ sesame (1:1)
Karnataka	Eastern, Southern, and Central Dry zones/Semiarid(Hot moist)/Alfisols; Mysore, Tumkur, Kolar, Ramanagara, Bengaluru (Rural and Urban), Chamarajanagar, Hassan, Chikkaballapur, and Mandya districts	Finger millet	Groundnut + pigeon pea (8:2)
	Northern Dry zone/Semiarid (Hot dry)/ Vertisols; Vijayapura, Bagalkote, Gadag, Bellary, Koppal, parts of Dharwad, Belagavi, and Raichur districts	*Rabi* sorghum	Safflower + chickpea(2:4 or 1:3)
Maharashtra	Western Vidarbha zone/Semiaarid (Hot moist)/Vertisols; Akola, Washim, Buldana, Yavatmal, Amravati, Wardha, and Nagpur districts	Cotton	Soybean + pigeon pea (4:2)
	Central Maharashtra Plateau zone/ Semiarid (Hot moist)/Vertisols; Parbhani, Hingoli, Nanded, Beed, Latur Osmanabad, Aurangabad, and Jalna districts	Cotton	Soybean + pigeon pea (4:2)
Madhya Pradesh	Malwa zone/Semiarid (Hot moist) Vertisols; Mandsaur, Ratlam, Ujjain, Rajgarh, Shajapur, Indore, Dewas, Dhar, Jhabua, and Sehore districts	Soybean	Sunflower + pigeon pea
	Keymore Plateau and Satpura Hill Zone/Subhumid (Hot dry)/Vertisols; Rewa, Sidhi, Singroli, Satna, Panna, Katni, Jabalpur, and Seoni districts	Soybean	Soybean+pigeon pea (4:2)

(continued)

Table 13.7 (continued)

State	Agroclimatic zone/Climate/Soil type/ Recommendation domain	Major rainfed crop-based production system	Efficient intercropping systems
Rajasthan	Southern zone/Semiarid(Hot dry)/ Vertisols; Bhilwara, Chittaurgarh, Rajsamand, Udaipur, and part of Sirohi districts	Maize	Castor + green gram (1:2)
Uttar Pradesh	North Eastern Plain Zone/Subhumid (Hot dry)/Inceptisols; Faizabad, Sultanpur, Basti, Gonda, and Ambedkarnagar districts	Rainfed rice	Linseed + chickpea/ lentil (2:4)
	Eastern plain and Vindhayan zone/ Subhumid (Hot dry)/Inceptisols; Varanasi, Chandauli, Sant Ravidas Nagar, Jaunpur, Ghazipur, Mirzapur, and Sonbhadra districts	Rainfed rice	Linseed + mustard (4/6:2)

Source: Ravindra Chary et al. (2022)

Table 13.8 Suggested rainfed cotton-based intercropping systems

State	Agroclimatic zone/Climate/Soil type/Recommendation domain	Major rainfed crop-based production system	Efficient intercropping systems
Gujarat	North Saurashtra zone/Semiarid (Hot dry)/Vertisols; Jamnagar, Devbhoomi Dwarka Morbi, Surendranagar, Amreli, Bhavnagar, and Rajkot) districts	Groundnut	Cotton + sesame/cowpea/ green gram/groundnut/ cowpea
Maharashtra	Western Vidarbha zone/Semiarid (Hot moist)/Vertisols; Akola, Washim, Buldana, Yavatmal, Amravati, Wardha, and Nagpur districts)	Cotton	Cotton + green gram/ cowpea/cluster bean (1:1); Cotton + Soybean (4:10)–safflower; Cotton + Soybean (6:6)–mustard; Cotton + soybean + pigeon pea + Soybean (3:1:1:1)/(3:2:2:2)
	Central Maharashtra Plateau zone/Semiarid (Hot moist)/ Vertisols; Parbhani, Hingoli, Nanded, Beed, Latur Osmanabad, Aurangabad, and Jalna districts	Cotton	Cotton + pigeon pea (6:1); Cotton + green gram (1:1)

Source: Ravindra Chary et al. (2022)

13.3.1.3 Single Cropping to Double Cropping in Rainfed Areas

The double-cropping system aims to make optimum use of land by permitting the production of an extra crop cultivated in winter/*rabi* after the *kharif* season. In general, double cropping becomes feasible when there is an on-time onset of monsoon rains that helps to successfully establish the first crop, then persist for an extended period to allow the planting and maturation of the second crop. To develop feasible and sustainable double-cropping systems, production factors such as length of growing season, cropping sequence, crop compatibility, biological complementarity, and planting time must be considered. Sequence/double cropping necessitates more thorough planning, which includes choosing hybrids or cultivars that will mature early enough to allow for next season's field operations and the establishment of the winter/*rabi* crops. Reliable legumes or cereals crops with short growing periods like chickpeas and wheat should be preferred to be planted in the post-rain season (September/October) relying on residual moisture in major soil types of *rainfed* regions of India. The use of early maturing cultivars of the main crop in *kharif* facilitates timely planting of succeeding winter crops to ensure optimum production as irrigation is the dominant constraint during the *rabi* for *rainfed* farming. It is necessary to look for the best combination and compatibility of crops, and the inclusion of legumes in double-cropping systems makes it essential, which can fix and incorporate nitrogen into the soil and improve soil structure, avoiding the formation of hardpan, and promoting better aeration with soil sustainability. In rainfed areas, there is untapped potential for crop intensification with double and relay cropping in various bimodal and or high rainfall zones. However, this is possible only with rainwater harvesting in farm ponds/percolation tanks/other rainwater harvesting structures and efficient utilization through micro-irrigation systems. This preparedness will also enable adjusting sowing windows for double cropping, taking up normal sowings and critical/life-saving irrigation to mitigate dry spells during *Kharif* season, pre-sowing/supplemental irrigation for early/rabi cropping and enhance overall resource use efficiency and system productivity and profitability. Further, crop intensification in rainfed areas shall lead to climate adaptation, doubling farmers' income, livelihood security of small farm holders, and also contribute toward food and nutritional security at the household level. In Bihar, under *Dholi* condition, paira cropping of lentils had a potential production to the extent of 2.5 times higher than sequential planting (lentils sown only after the harvest of rice). The state and agro-climatic zone and recommendation domain-wise double-cropping systems are suggested in Table 13.9.

Table 13.9 Suggested double-cropping systems

State	Agroclimatic zone/Climate/Soil type/Recommendation domain	Major rainfed crop-based production system	Double-cropping system
Andhra Pradesh	Scarcity zone/Semiarid/Arid(Hot)/Alfisols; Ananthapuramu, Kurnool, YSR Kadapa, and Chittoor districts	Groundnut	Groundnut-horse gram/fodder sorghum
Chhattisgarh	Bastar Plateau Zone/Subhumid (Hot moist)/Inceptisols; Bastar, Narayanpur, Kondagaon, Bijapur, Sukma, Dantewada, and Kanker districts	Rice	Rice-chick pea/field pea/mustard/linseed

(continued)

Table 13.9 (continued)

State	Agroclimatic zone/Climate/Soil type/Recommendation domain	Major rainfed crop-based production system	Double-cropping system
Jharkhand	Western Plateau zone/Subhumid(Hot moist)/Inceptisols; Palamu, Garhwa, Latehar, Chatra, Lohardaga, Gumla, and Simdega districts	Rice	Rice–wheat/chickpea/lentil/linseed/safflower
Karnataka	Eastern, Southern, and Central Dry zones/Semiarid(Hot moist)/Alfisols; Mysore, Tumkur, Kolar, Ramanagara, Bengaluru (Rural and Urban), Chamarajanagar, Hassan, Chikkaballapur, and Mandya districts	Finger millet	Field bean–finger millet; Cowpea–finger millet; Sesame–finger millet Fodder bajra–cowpea/chilies
	Northern Dry zone/Semiarid (Hot dry)/Vertisols; Vijayapura, Bagalkote, Gadag, Bellary, Koppal, parts of Dharwad, Belagavi, and Raichur districts	Rabi sorghum	Green gram/sunflower–*rabi* sorghum/chickpea/safflower; Onion–*rabi* sorghum/chickpea; Cucumber–sunflower/*rabi* sorghum/chickpea
Maharashtra	Western Vidarbha zone/Semi arid (Hot moist)/Vertisols; Akola, Washim, Buldana, Yavatmal, Amravati, Wardha, and Nagpur districts)	Cotton	Soybean–chickpea; Green gram/black gram–chickpea/safflower; Cotton and soybean (4:10)–safflower
	Central Maharashtra Plateau zone/Semiarid (Hot moist)/Vertisols; Parbhani, Hingoli, Nanded, Beed, Latur Osmanabad, Aurangabad, and Jalna districts	Cotton	Green gram–*rabi* sorghum; Soybean–chickpea/linseed; Sorghum–safflower
	Scarcity zone/Semiaarid (Hot dry)/Vertisols; Solapur, Ahmednagar, Sangli, Pune, Dhule, Jalgaon, Nandurbar, Satara districts	*Rabi* sorghum	Black gram–*rabi* sorghum/chickpea/safflower; Green gram–*rabi* sorghum/chickpea/safflower; Cowpea–*rabi* sorghum/chickpea/safflower
Madhya Pradesh	Malwa zone/Semiarid (Hot moist)/Vertisols; Mandsaur, Ratlam, Ujjain, Rajgarh, Shajapur, Indore, Dewas, Dhar, Jhabua, and Sehore districts	Soybean	Soybean–chickpea/safflower; Maize–chickpea/safflower; Black gram–chickpea/safflower
	Keymore Plateau and Satpura Hill zone/Subhumid (Hot dry)/Vertisols; Rewa, Sidhi, Singroli, Satna, Panna, Katni, Jabalpur, and Seoni districts	Soybean	Rice–wheat/chickpea/lentil; Soybean–wheat/chickpea
Meghalaya	Hot and humid/Ri Bhoi, East Khasi, Garo hills	Maize	Maize–French bean, Maize–Soybean, Maize–urdbean, Maize–mungbean, Maize–toria

(continued)

Table 13.9 (continued)

State	Agroclimatic zone/Climate/Soil type/Recommendation domain	Major rainfed crop-based production system	Double-cropping system
Odisha	Northeastern Ghat zone/Subhumid (Hot moist)/Inceptisols; Kandhamal, Gajapati, Rayagada, Ganjam, and part Koraput districts	Rice	Upland rice–green gram/horse gram; Maize–potato/horse gram/mustard
Rajasthan	Southern zone/Semiarid (Hot dry)/Vertisols; Bhilwara, Chittaurgarh, Rajsamand, Udaipur, and part of Sirohi districts	Maize	Maize–chickpea; Sorghum–mustard; Sorghum + cowpea (green fodder)–Taramira; Black gram–mustard; Maize–Chickpea + mustard
Sikkim	Hot and Humid/East Sikkim, West Sikkim, and South Sikkim	Maize	Maize + cowpea–pea Maize + cowpea–rajmash Maize–black gram–buckwheat
Telangana	Northern Telangana zone/Subhumid (Hot moist)/Vertisols; Adilabad, Nizamabad, Asifabad, Mancherial, Nirmal, Kamareddy, and Jagtial districts	Soybean	Soybean–chickpea
Tripura	Hot and Humid/Dhalai, South Tripura	Maize	Maize–maize–mustard Maize + cowpea–maize–mustard Maize–maize–lentil Maize–pea
Uttar Pradesh	South-western semiarid zone/Semiarid (Hot dry)/Inceptisols; Agra, Mathura, Mainpuri, Aligarh, Hathras, Firozabad, Etah, and Kanshiram Nagar districts	Pearl millet	Green gram/black gram (green manuring after first picking)–mustard; Pearl millet + cowpea (fodder)–mustard
	Eastern plain and Vindhayan zone/Subhumid (Hot dry)/Inceptisols; Varanasi, Chandauli, Sant Ravidas Nagar, Jaunpur, Ghazipur, Mirzapur, and Sonbhadra districts	Rainfed rice	Rice–chickpea/lentil/mustard

Source: Amrutsagar et al. (2022); Babu et al. (2016); Prasad et al. (2012); Ravindra Chary et al. (2022); Shirahatti et al. (2022); Singh et al. (2021a, b, 2022); Babu et al. (2020a, b, c, 2023); Yadav et al. (2021a, b)

Cropping System Intensification in Rice Fallows in the Eastern Plateau Region of India Rice fallows are the lowland rainfed rice-growing areas that remain uncropped after *kharif* rice. These are mostly predominant in Eastern India and account for about 9.7 million ha of total rice–fallow area (11.7 million ha) in the country. Winter pulses such as grasspea, mung bean, black gram, lentil, and chickpea, and winter oilseeds such as linseed and mustard have scope to grow in rice fallows. Black gram and green gram, being shorter in duration, are recommended for crop intensification under rice–fallow regions of southern and eastern regions and Lathyrus, lentil, and chickpea in eastern and central regions of the country. Relay cropping of lentil and lathyrus is very common in Eastern India and also in Chhattisgarh. In this system, seeds of lentil/lathyrus are broadcasted in standing rice 7–10 days before their harvesting. Adoption of soil moisture conservation practices and advancement of sowing time following *utera* (relay) cultivation may improve winter crop productivity.

Cropping System Intensification in Maize Fallows in the North-Eastern Himalayan Region of India Maize–fallow is a dominating production system in upland terrains of the North-Eastern Hill part of India. The sustainability of maize–fallow system is highly questioned due to poor soil health and related environmental issues. The maize monoculture has been the primary reason for soil degradation/loss in the region. Continuous maize cultivation on hilly terrains has deteriorated the soil's physical, chemical, and biological properties and accelerated carbon emissions in the environment (Yadav et al. 2019, 2020; Babu et al. 2020a; Das et al. 2022a, b). Furthermore, intense rainfall during the pre-*kharif* to post-*kharif* season (March–October) further accelerated the nutrients and soil erosion losses in hilly terrains of the NEH region of India. Hill slope soils are low in C and nutrients, and prone to erosion (Yadav et al. 2019; Singh et al. 2021b).

13.3.1.4 Double-Cropping to Triple-Cropping Systems

The potential areas for enabling double- to triple-cropping systems could be possible in rainfed areas with harvested rainwater in farm ponds and micro-irrigation systems in high rainfall zones such as north-eastern regions and eastern regions with dependable groundwater. Triple-cropping systems with harvested rainwater in the north bank plains zone of Assam are suggested in Table 13.10.

The above suggested technological interventions for crop diversifications are agroecology-specific and have to be implemented/disseminated/demonstrated in the delineated potential crop zones.

13.3.2 Institutional Interventions

- Delineation of potential crop zones for major rainfed crops for enabling crop diversification in diverse rainfed agroecologies.
- Comprehensive interventions should be implemented in selected clusters to demonstrate a crop diversification system in an integrated approach. Model demonstration units of at least 50 ha on crop diversification as learning sites/field schools and capacity-building sites in these clusters need to be established. Further, investments are needed for building infrastructure for the production of inputs, storage, and marketing of produce, value addition facilities, and training facilities.
- Institutional interventions such as seed systems, custom hiring centers, nutrient banks, agro-advisories, market intelligence, and postharvest value addition need

Table 13.10 Suggested triple-cropping systems

State	Agroclimatic zone/Climate/ Soil type/Recommendation domain	Major rainfed crop-based production system	Triple-cropping system
Assam	North Bank Plains zone/ Humid (Hot)/Alfisols; Dhemaji, Lakhimpur, Biswanath, Sonitpur, Darrang, and Udalguri districts	Rainfed rice	Maize–toria–green gram/ black gram; Maize–linseed–green gram/ black gram; Maize–buckwheat–green gram/black gram; Maize–niger–green gram/ black gram; Maize–rajma–green gram/ black gram; Maize–potato–green gram/ black gram; Direct Seeded *Ahu* rice–green gram–toria/ rajma/potato
Sikkim	Hot and Humid/East Sikkim, South Sikkim	Rainfed rice	Rice–pea–sesbania Rice–broccoli–sesbania Rice–pea–maize/ (green cobs)
Tripura	Hot and Humid/Dhalai, South Tripura	Rainfed rice	Rice–mustard Rice–pea Rice–rice

Source: Ravindra Chary et al. (2022); Babu et al. (2020a, c); Yadav et al. (2021a, b)

to be established in these clusters. Thus, there would be a need for a strategic mix of better technology adoption, institutional innovations, and incentive systems to enhance the productivity of rainfed cereals.

- Measures for cropping systems should include value chain, financial inclusion, insurance, and concessional credit.
- Capacity building of line department officials, farmer producer organizations, and other stakeholders.
- Information sharing should focus on last-mile delivery of early warning advisories and drought mitigation measures.
- Awareness through mass media and social media campaigns should promote inclusive access to schemes and entitlements, governance of natural resources, and adoption of good practices for drought mitigation.

13.3.3 Infrastructure Interventions

- A Model Custom hiring, training, and maintenance center has to be established. This is to enable innovative custom service on a rental model by institutionalization for farm machinery/implements for small and marginal farmers. Performance

testing and certification at designated testing centers have to be ensured. There is a need for strengthening training programs for different stakeholders on the operation, repair, and maintenance of agricultural machinery, tractors, power tillers, combines, and other machinery.
- Establishing processing and value addition units at strategic places production areas for in public–private partnership (PPP) mode.
- Helping farmers in marketing their processed products (forward linkages).
- Skill development, particularly farm women in primary and secondary processing.
- The marketing intelligence and creation of market infrastructure to realize higher prices would enhance output price.
- Improving the holding capacity of produce by farmers in warehouses.

13.3.4 Policy Interventions

- Map ongoing central and state schemes that can create convergence through supplementation of resources toward scaling out crop intensification programs in rainfed areas.
- Scaling out crop intensification technologies in a saturation mode in identified clusters (subdistricts/villages) as an area approach or on several farm holdings is suggested. In these clusters, the convergence of various central sectors/sponsored or state schemes/programs of various ministries/departments is needed for implementing NRM, crop, farm mechanization, and inputs-based interventions.
- Incentivization for the adoption of crop intensification technologies along with an insurance policy.
- A national scheme on crop intensification in rainfed areas is suggested to create agroeconomic zones in rainfed areas toward sustainable intensification, food and nutritional security, environmental safety, and enhancing the livelihoods of a large number of small and marginal farmers in rainfed areas.

13.4 Way Forward

- Mapping potential sites for surplus runoff and rainwater harvesting in farm ponds.
- Implementation and popularization of agroecology-specific improved soil and water management technologies for increasing area under supplemental irrigation.
- Creating water assets such as farm ponds based on agroecology-specific runoff potentials and feasibility for lifesaving/supplemental/pre-sowing.
- Popularization of farm pond technology package (selection of ideal site, digging, harvesting, lining, minimizing evaporation losses, lifting pump, and micro-irrigation system), including efficient utilization of stored water for higher water productivity.

- Desilting tanks to increase the volume of water for irrigation of crops and groundwater stabilization.
- Promotion of less water and energy-intensive pulse-based cropping systems in rainfed Indo-Gangetic plain (IGP) areas in view of poor soil quality and poor groundwater resources.
- Judicious exploitation of groundwater potential for crop intensification in the eastern region.
- Crop intensification with pulses, oilseeds, and millets in low productive upland rice and other crop areas and also with short-duration *rabi* pigeon pea in Peninsular India.
- Identifying/developing climate-resilient varieties of rainfed crops suitable to delayed onset of monsoon and to cope with drought, floods, heat waves, and cold wave.
- Production and timely supply of adequate breeder seed of rainfed crops for further multiplication with strict enforcement of quality control assuring supply of quality seed.
- Wider adoption of bio-fertilizers and biopesticides, crop residue recycling, INM and SSNM.
- Implementation and popularization of seed-to-seed smart mechanization in major rained crops.

References

Amrutsagar VM, Upadhye SK, Ranshur NJ, Ravindra Chary G, Khadtare SV, Kathmale DK (2022) Overview of dryland agriculture research and achievements in scarcity zone of Maharashtra. Indian J Dryland Agric Res Dev 37(2):60–68

Babu S, Raghavendra S, Avasthe RK, Yadav GS, Rajkhowa DJ (2016) Intensification of maize (Zea mays)–based cropping sequence in rainfed ecosystem of Sikkim Himalayas for improving system productivity, profitability, employment generation and energy-use efficiency under organic management condition. Indian J Agric Sci 86(6):778–784

Babu S, Mohapatra KP, Das A, Yadav GS, Tahasildar M, Singh R, Panwar AS, Yadav V, Chandra P (2020a) Designing energy-efficient, economically sustainable and environmentally safe cropping system for the rainfed maize–fallow land of the Eastern Himalayas. Sci Total Environ. https://doi.org/10.1016/j.scitotenv.2020.137874

Babu S, Singh R, Avasthe RK, Yadav GS, Mohapatra KP, Selvan T, Das A, Singh VK, Valente D, Petrosillo I (2020b) Soil carbon dynamics in Indian Himalayan intensified organic rice-based cropping sequences. Ecol Indic. https://doi.org/10.1016/j.ecolind.2020.106292

Babu S, Mohapatra KP, Yadav GS, Lal R, Singh R, Avasthe RK, Das A, Chandra P, Gudade BA, Kumar A (2020c) Soil carbon dynamics in diverse organic land use systems in North Eastern Himalayan ecosystem of India. Catena 194:104785. https://doi.org/10.1016/j.catena.2020.104785

Babu S, Raghavendra S, Avasthe RK, Rathore SS, Kumar S, Das A, Layek J, Sharma V, Wani OA, Singh VK (2023) Conservation tillage and diversified cropping enhance system productivity and ecoefficiency and reduce greenhouse gas intensity in organic farming. Front Sustain Food Syst 7:1114617. https://doi.org/10.3389/fsufs.2023.1114617

Bhargavi B, Behera UK, Rana KS, Singh R (2019) Productivity, resource-use efficiency and profitability of high-value crops embedded diversified cropping systems. Indian J Agric Sci 89(5):821–827

Das A, Babu S, Datta M, Kumar S, Raghavendra S, Avasthe RK, Rathore SS, Yadav SK, Singh VK (2022a) Restoring soil carbon in marginal land of Indian Himalayas: impact of crop intensification and conservation tillage. J Environ Manag 318:115603. https://doi.org/10.1016/j.jenvman.2022.115603

Das A, Babu S, Singh R, Kumar S, Rathore SS, Singh VK, Datta M, Yadav SK, Wani OA, Yadav D (2022b) Impact of live mulch-based conservation tillage on soil properties and productivity of summer maize in Indian Himalayas. Sustainability 14:12078. https://doi.org/10.3390/su141912078

Prasad YG, Venkateswarlu, Ravindra Chary G, Srinivasarao C, Rao KV, Ramana DBV, Rao VUM, Subba Reddy G, Singh AK (2012) Contingency crop planning for 100 districts in Peninsular India. ICAR-CRIDA, Hyderabad, p 136. ISBN 978-93-80883-13-07

Ramamurthy V, Srinivas S, Kumar GDS, Qureshi A, Singh SK (2016) Identification of potential districts for sunflower in India. In: International conference on "Integrated Land Use Planning for smart Agriculture" from 10–12th Nov., 2016 at Nagpur, organized by ISSLUP

Ravindra Chary G, Kandpal B, Gopinath KA, Srinivasa Rao C, Bhaskar S, Rajanna GA, Ramamurthy V (2016a) Climate resilient crops and cropping systems. In: Venkateswarlu B, Ravindra Chary G, Singh G, Shivay YS (eds) Climate resilient agronomy. Indian Society of Agronomy, New Delhi, pp 88–119

Ravindra Chary G, Gopinath KA, Raju BMK, Srinivasarao C, Kandpal B, Bhaskar S, Kumar N, Narsimlu B (2016b) Natural resource management based approaches for enhancing pulses production in rainfed areas. Indian J Agron 61(special issue):S199–S213

Ravindra Chary G, Gopinath KA, Singh VK, Kandpal B, Bhaskar S, Srinivasa Rao C, Sahadeva Reddy B, Padmalatha Y, Yellamanda Reddy T, Nataraja KC, Malliswara Reddy A, Radha Kumari C, Babu MVS, Madhusudhana Reddy K, Bhargavi K, Narayanaswamy G, Muralikrishna C, Srinivasulu DV (2022) Over view of dryland agricultural research and achievements in scarce rainfall zone of Andhra Pradesh. Indian J Dryland Agri Res Dev 37(2):123–129

Shirahatti MS, Nandagavi RA, Momin UM, Kumara BH, Patil SB, Ravindra Chary G, Surakod VS, Gaddankeri MA, Kanthi SG, Patil HS (2022) Overview of dryland agriculture research and achievements in northern Dry Zone of Karnataka. Indian J Dryland Agri Res Dev 37(2):51–59

Singh R, Babu S, Avasthe RK, Meena RS, Yadav GS, Das A, Mohapatra KP, Rathore SS, Kumar A, Singh C (2021a) Conservation tillage and organic nutrients management improve soil properties, productivity, and economics of a maize-vegetable pea system in the Eastern Himalayas. Land Degrad Dev:1–18. https://doi.org/10.1002/ldr.4066

Singh R, Babu S, Avsthe RK, Yadav GS, Das A, Mohapatra KP, Kumar A, Singh VK, Chandra P (2021b) Crop productivity, soil health, and energy dynamics of Indian Himalayan intensified organic maize-based systems. Int Soil Water Conserv Res 9(2):260–270. https://doi.org/10.1016/j.iswcr.2020.11.003

Singh B, Bhagwat DV, Chaudhary SK, Bangar KS, Jadav ML, Kumawat N, Holkar S, Upadhyay A, Sharma SK, Gopinath KA, Chary GR, Shukla AK, Singh VK (2022) Overview of Dryland agriculture research and achievements in Malwa Plateau Zone of Madhya Pradesh. Indian J Dryland Agri Res Dev 37(2):130–134

Srinivasa Rao C, Ravindra Chary G, Rani N, Baviskar VS (2016) Real time implementation of agricultural contingency plans to cope with weather aberrations in Indian agriculture. Mausam 67(1):183–194

Velayutham M (1999) In: Singh HP, Ramakrishna YS, Sharma KL, Venkateswarlu B (eds) Crop and land use planning in dryland agriculture. 73–80. (In) Fifty years of dryland agricultural research in India. Central Research Institute for Dryland Agriculture, Hyderabad

Vittal KPR, Ravindra Chary G (2007) Horizontal and vertical diversifications of rainfed cropping system in India. In: Vittal KPR, Srivastava RL, Joshi NL, Kar A, Tewari VP, Kathju S (eds)

Dryland ecosystem – Indian perspective. CAZRI and AFRI/Allied Publishers, Jodhpur/New Delhi, pp 53–82

Vittal KPR, Ravindra Chary G, Rama Rao CA, Sankar GRM (2006) Crop diversification in rainfed farming of India. In: Sharma KD, Soni B (eds) Land use diversification for sustainable agriculture. Atlantic Publishers, and Distributors, New Delhi, pp 24–70. ISBN 978-81-2690-584-3

Yadav GS, Das A, Lal R, Babu S, Datta M, Meena RS, Somanagouda B, Patil RS (2019) Impact of no-till and mulching on soil carbon sequestration under rice (*Oryza sativa* L.)-rapeseed (*Brassica campestris* L. Var. Rapeseed) cropping system in hilly agro-ecosystem of the Eastern Himalayas, India. Agric Ecosyst Environ 275:81–92

Yadav GS, Das A, Subhash B, Mohapatra KP, Lal R, Rajkhowa DJ (2020) Potential of conservation tillage and altered land configuration to improve soil properties, carbon sequestration and productivity of maize-based cropping system in eastern Himalayas, India. Int Soil Water Conserv Res. https://doi.org/10.1016/j.iswcr.2020.12.003

Yadav GS, Babu S, Das A, Datta M, Mohapatra KP, Raghavendra S, Singh VK, Rathore SS, Chakraborty M (2021a) Productivity, soil health, and carbon management index of Indian Himalayan intensified maize-based cropping systems under live mulch-based conservation tillage practices. Field Crop Res 264(1):108080. https://doi.org/10.1016/j.fcr.2021.108080

Yadav GS, Das A, Kandpal BK, Babu S, Lal R, Datta M, Das B, Singh R, Singh VK, Mohapatra KP, Chakraborty M (2021b) The food-energy-water-carbon nexus in a maize-maize-mustard cropping sequence of the Indian Himalayas: an impact of tillage-cum-live mulching. Renew Sust Energ Rev 151:111602. https://doi.org/10.1016/j.rser.2021.111602

Chapter 14
Carbon Farming

Namrata Kashyap, Hejbina M. Hussain, A. K. Mohanty, P. K. Pathak, Aastika Pandey, and Khushboo Devi

Abstract Carbon farming is a new approach for land management that allows carbon sequestration by reducing greenhouse gas (GHG) emissions. Agriculture, forestry, and other land-use practices lead to 24% of global GHG emissions. Silvopastoral system and agroforestry generate more amount of biomass and are more efficient in the sequestration of large amounts of carbon. Carbon farming increases the carbon sinks in soil because of improvement in soil aeration from the addition of organic carbon, which reduces the process of denitrification and increases sink capacity for methane. Carbon stocking helps to increase water-holding capacity and nutrient availability of the soil which ultimately improves soil health. Also, in limited moisture conditions, biochar can efficiently improve soil moisture storage as it has a good amount of recalcitrant carbon. Again, in hot and harsh climate, *Jatropha curcas* is a well-adapted plant that has the ability to capture 17–25 t of carbon dioxide per hectare per year from the atmosphere (over a 20-year period). A bill to amend the Energy Conservation Act of 2001 was adopted by the Central Government of India in 2022 with the motive to establish voluntary carbon credit trading scheme and market within 2023. Carbon credit is a permit that allows the holding company to release a certain amount of carbon dioxide or any GHG. Conversely, government or any other authorized agency has issued carbon credit certificates to an organization/institution that has helped to remove GHGs from the atmosphere. Gujarat created the first carbon credit trading market in India in 2022. So, one carbon credit certifies that one metric ton of carbon dioxide has been removed from the atmosphere. Despite so many advantages of carbon farming,

N. Kashyap
Krishi Vigyan Kendra, Lakhimpur, Assam Agricultural University, Jorhat, India

H. M. Hussain (✉) · A. K. Mohanty
ICAR – Agricultural Technology Application Research Institute, Umiam, Meghalaya, India

P. K. Pathak
Krishi Vigyan Kendra, Nalbari, Assam Agricultural University, Jorhat, India

A. Pandey · K. Devi
Division of Agronomy, ICAR-Indian Agricultural Research Institute, New Delhi, India

S. Babu et al. (eds.), *Agricultural Diversification for Sustainable Food Production*, Sustainability Sciences in Asia and Africa,
https://doi.org/10.1007/978-981-97-7517-0_14

well-informed advisory services are necessary to motivate farmers to adopt carbon farming practices that will help improve production and soil health.

Keywords Carbon farming · Carbon sequestration · Carbon sink · Carbon credit

14.1 Introduction

Change in climate and collapse of biodiversity are intimately connected crises (Portner et al. 2021) that share some root causes and solutions. Intensification of agricultural land use and land use change for agriculture and urbanization have contributed ~25% of greenhouse gas (GHG) emissions (Dale 1997) and are the prime reason for the collapse of biodiversity (IPBES 2021). Agriculture in the recent past has received a lot of attention as it has become the root cause of global warming. It is now widely acknowledged by scientists that agriculture is a key component of the climate change debate, both because it causes and can help mitigate climate change. (Legg and Huang 2020). The world's population is still growing, and climate change is making hunger worse (Leah Samberg 2017). New research indicates that carbon-rich environments cause vitamin deficiencies in rice crops, which poses a serious threat to food security for the "600 million people worldwide whose diet consists mostly of rice." (Evich 2019). These climate changes will continue and intensify. If global emissions do not decrease by 70% by 2050 and almost 0% by the end of the century, catastrophic consequences will happen (Climate Change 2014: Synthesis Report Summary for Policymakers 2014).

Agricultural land is about 40% of the planet's surface, and an average of 6 million hectares of forest and grassland have been converted annually to feed the increasing population. One of the largest contributors to GHG emissions is the enteric fermentation processes inherent in livestock operations (Eric Toensmeier 2018). Methane emitted from ruminant livestock (cattle) is the largest contributor (Moskin et al. 2019). Since the Industrial Revolution, agricultural practices and livestock have released about 135 gigatons—135 billion metric tons—of carbon into the atmosphere (Manoff 2018). Agriculture contributes nitrous oxide emissions of about 327 million metric tons of CO_2 equivalent by 2020—behind only China and India (FAO 2020). Methane and nitrous oxide emissions from animal husbandry are about to double by 2055 (Popp et al. 2010).

Multiple climate pollutants like carbon dioxide, methane, and nitrous oxide caused by anthropogenic climate change are the three greatest individual contributors to global warming (Myhre et al. 2013). Agriculture and production are associated with these three gases, but direct agricultural emissions are dominated by methane and nitrous oxide (Lynch et al. 2021). This has led to negative effects on food security and social welfare. It has been reported that rainfall patterns and growing seasons have been altered in large geography by global warming (Linderholkm 2006). Overall decarbonization of energy generation is the priority to reduce most of these emission sources rather than specific agricultural mitigations (Lynch et al. 2021). If we take temperature as a climate change indicator, the anthropogenic emissions since preindustrial revolution have led to global warming of 1.0 °C. If these emissions continue at present rates, then we are to reach 1.5 °C of warming between 2030 and 2052—which is 0.5 °C additional warming from today's level. Agriculture has a huge role in the carbon offset market. This is because agriculture has a great potential to absorb carbon dioxide from the atmosphere, and it gets stored in the soil for a long period of time.

Reducing emissions will not adequately limit global warming. Therefore, it is critical to examine certain approaches to store existing carbon dioxide from the atmosphere. A cost-feasible approach to pursue is carbon farming. Carbon farming is defined as the explicit use of agriculture to sequester and store carbon in soil (Ries 2019). It is a holistic model to implement an array of carbon-beneficial practices and create agroecosystems that optimize economic, ecological, and carbon abundance. Carbon farming uses regenerative agricultural practices to sequester more carbon from the atmosphere than it produces (a carbon-negative practice). Research shows that carbon farming has the potential to remove substantial quantities of CO_2 from the atmosphere and is more feasible to scale, in part, because the practice itself does not require additional energy (Cornelia Rumpel et al. 2018).

Climate-smart agriculture (CSA) is another approach that transforms agricultural systems to effectively support the development and ensure food security (FAO 2020). The main objectives of CSA are to sustainably increase agricultural production and farmers' income, to build resilience to climate change, and to reduce GHG emissions. Conservation agriculture (CA) is one of the practices for achieving CSA. CA is considered as an important crop and soil management practice for sustainable agriculture. CA is beneficial for major soil and climatic constraints, which lower crop productivity in smallholder farming system of South Asia (SA) as well as sub-Saharan Africa (SSA) (Dutta et al. 2020). Minimum tillage, permanent organic soil cover, and diversified crop rotations are three important components of CA. Optimal nutrient management and water management are two key factors behind the success of conservation-agriculture practices (Vanlauwe et al. 2014; Sidhu et al. 2019; Yadav et al. 2023). Reduction of more than 1.5 Mg CO_2 equivalent/ha over business as usual (BAU) in the maize–wheat system at Indo-Gangetic Plains of India was reported by Sapkota et al. (2014). CA can reduce 0.0–2.5 Mg CO_2 equivalent/ha over BAU in different cropping systems of Bangladesh over a period of 10–30 years (Sapkota et al. 2021). This indicates that CA system creates a huge potential for developing measurable CO_2 equivalent/ha that can be traded in the international market.

14.2 Global Carbon Pools

Arrhenius (1896) recognized the importance of atmospheric carbon dioxide concentration on global temperature by the end of the nineteenth century. Carbon cycle is the movement of carbon in various forms, and it is the exchange between atmosphere, biosphere, geosphere, and oceans. The interactions that happen between the pedologic, atmospheric, and biotic carbon pools comprise important components of the global carbon cycle (GCC). So, these interactions form the basis to sequester atmospheric carbon dioxide in pedologic and biotic pools.

- The largest oceanic pool is increasing at the rate of 2.3 Pg Cyr and is estimated at 38 000 Pg.
- The geological carbon pool, which comprises fossil fuels, is 4130 Pg, of which 5.5% is oil, 5% is coal, and 3.3% is gas.
- The pedologic pool to 1 m depth is estimated at 2500 Pg. It consists of soil organic carbon (SOC) pool and soil inorganic carbon pool estimated at 1550 Pg and 950 Pg, respectively (Batjes 1996).
- The atmospheric pool is increasing at the rate of 3.5 Pg Cyr or 0.46% per yr and is of 760 Pg of carbon dioxide–carbon.
- The biotic pool is the smallest of all the pools and is estimated at 560 Pg.

14.3 Mitigation of GHG

Several agricultural practices and technologies reduce GHG emissions by improving carbon storage in soils (Table 14.1), by preserving existing soil carbon, and by reducing carbon dioxide, methane, and nitrous oxide emissions (Wani et al. 2016). Soil carbon helps to improve soil structure, soil water retention, soil fertility, and enhance soil microbial communities (Wilson et al. 2009). Improved management practices have led to carbon sink capacity of 21–51 Gt carbon, which is equivalent to all anthropogenic GHG emissions over 2–3 years (Lal 2010). The soybean–wheat system in subtropical central India, in which the annual contribution of carbon was about 22% for soybean and 32% for wheat from aboveground biomass (Kundu et al. 2001). Likewise, no-tillage increased soil organic carbon significantly by 48.3 ± 9.9 g/cm^2/yr over the conventional method of farming (Hooker et al. 2005). Organic agriculture also has great potential to reduce GHG emissions (Rodale Institute 2008).

Table 14.1 Agricultural practices and benefits

Conservation practice	GHG objectives	Additional benefit
Crops		
Conservation tillage and reduced field pass intensity	Sequestration, emission reduction	Improves soil, water, and air quality
Efficient nutrient management	Sequestration emission reduction	Improves water quality. Saves efficient nutrient management Sequestration
Crop diversity through rotations and cover crops	Sequestration	Reduces erosion and water requirements.
Animals		
Manure management	Emission reduction	On-farm sources of biogas fuel and possibly electricity for large operations provide nutrients for crops
Rotational grazing and improved forage	Sequestration, emission reduction	Reduces water requirements
Feed management	Emission reduction	Reduces quantity of nutrients. Improves water quality. More efficient use of feed

Source: NRCS. http://soils.usda.gov/survey/global_climate_change.html

14.4 Measuring Carbon Emissions

There are lot many tools to track GHG emissions and to track carbon sequestration (Fara 2016). The carbon trust label that appears on the packaging or website of a product includes a symbol, a black footprint, number of carbon dioxide grams embodied by the product, and to reduce its emissions (Carbon Trust 2023). The International Standards Organization developed ISO 14064 and ISO 14067. The development of ISO 14064 was to quantify the GHG emissions at the company level, and ISO 14067 was to quantify the carbon footprint of products (Safdie 2022). Most of the tools allow farmers to measure the emissions of GHG, but there is only one tool that measures emissions of GHG and sequestration of carbon (US Department of Agriculture 2017). The Cool Farm Tool is measuring the emissions of the GHG in the field itself, and it evaluates how GHG emissions are affected by different crops and farming practices (Cool Farm Tool 2020). The Digital ERP Tool is another farm tool that was developed by farmer-cooperative Land O′ Lakes that intends to help farmers adopt sustainable farm management practices and increase their income. It also helps food industries to monitor the sustainability credentials of supply chains (Taylor 2018).

14.5 Greenhouse Gas (GHG) Emission Reduction by Carbon Farming

Carbon farming is a new technology that describes land management practices that capture carbon dioxide from the atmosphere and store it in soil and living biomass which acts as natural sinks. Carbon farming is known to mitigate greenhouse gas (GHG) emissions across many countries (Parikh and Winfield 2020). Carbon farming also refers to the business model that aims to upscale climate mitigation by incentivizing farmers to implement climate-friendly farm management practices (European Commission 2021). Environmental and socioeconomic advantages of carbon farming are given in Fig. 14.1. Agricultural activities that add organic matter to the soil are necessary to reduce the emission of environmental carbon dioxide (Hartmann et al. 2013a, b; Mcleod et al. 2011; Babu et al. 2020a; Yadav et al. 2021a). Soil management approaches are required to restore soil carbon. Carbon deficiency is seen to occur because of soil erosion in many farming fields (Kirkels et al. 2014; Xiao et al. 2018; Yadav et al. 2023). Government has employed various measures to incentivize ecologically sustainable farming for conservation of soil carbon. In low-input areas, agroforestry increases crop productivity and helps farmers maintain soil quality (Yadav et al. 2021b). Forests reduce the GHG emission and reduce global warming by sequestration of carbon dioxide (Powlson et al. 2011; Kumar et al. 2022; Wani et al. 2023). Microalgae exhibit productive photosynthesis, which results in large amounts of carbon dioxide that are organically bound carbon in their cells (Bhola et al. 2014; Singh et al. 2016). Carbon dioxide pollution per unit may be lowered due to the carbon dioxide that is recycled and reused by algae for biomass sourced from fossil fuels (Moreira and Pires 2016).

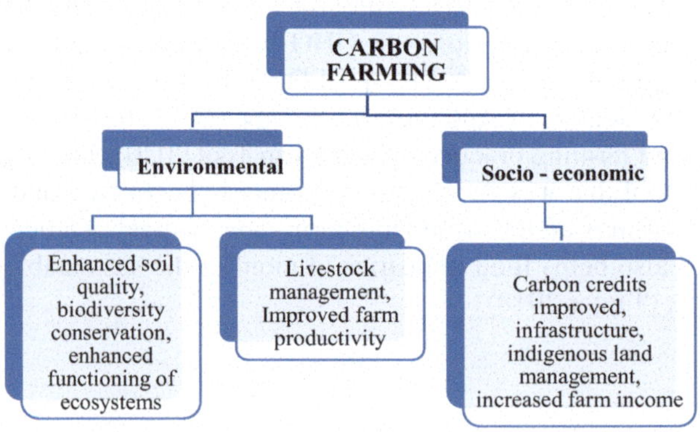

Fig. 14.1 Environmental and socio-economic advantages of carbon farming

14.5.1 Soil Carbon as a Catalyst for Agroecological Transition

Soil organic carbon (SOM) building management practices are essentially the basics of agroecology, and farmers are interested to know about the potential of soil carbon sequestration that will improve their farm income and production, and carbon farming is a major catalyst for the agroecological transition of European Union agriculture. Strong knowledge of soil biology and ecology is required to rebuild SOM. A new understanding and appreciation of soils as complex ecosystems could become truly transformational only when the farmers start on the carbon sequestration journey.

14.6 Carbon Sequestration

According to Gaikwad (2021), soil biodiversity, ecosystem productivity, soil biological activity, and environmental quality all depend on carbon. Through the natural metabolic process of photosynthesis, atmospheric carbon enters terrestrial ecosystems and is subsequently released back into the atmosphere through a number of processes collectively referred to as respiration (Gaikwad 2021). The quality of the air, water, land, and human life have all suffered in recent decades due to the overabundance of carbon released into the atmosphere (Gaikwad 2021). Since 1975, the rate of increase in global temperature has been 0.158 °C each decade. Significant environmental changes have also occurred (Greene and Pershing 2007), and wildfire frequency and intensity have risen (Running 2006; Westerling et al. 2006). The release of greenhouse gases (GHGs) brought on by human activities such as land-use change, deforestation, biomass burning, wetland drainage, soil cultivation, and fossil fuel combustion is primarily responsible for the observed changes in climate.

Increased interest in carbon sequestration techniques for reducing CO_2 emissions has been seen as a matter of environmental concern. Carbon sequestration in soil is a key strategy to reduce atmospheric carbon dioxide. Increase in soil organic carbon is usually coined as "accumulation," and the term "sequestration" can be used for climate change mitigation where there is an additional transfer of carbon from the atmosphere (Powlson et al. 2011). The term "carbon sequestration" is basically the process where carbon dioxide is either removed from the atmosphere or sidetracked from the emission sources, and it is stored in the ocean, vegetation, soils, and geological formations. Carbon sequestration is very pertinent to lowering the atmospheric CO_2; by 2030, soil carbon sequestration will account for about 90% of the total global mitigation potential in agriculture (Smith 2008). On the other hand, carbon sequestration decreases the costs of global mitigation that will promote increased food security. During the biofuel production, carbon dioxide is recycled, and biomass can be used instead of fossil fuels (Hanaki and Pereira 2018). Huge amounts of value-added biomass can be further used to make bioethanol if the process of biofuel production is carried out with proper care. So appropriate carbon management is mandatory to make sure that biomass can be used for various commercial purposes and also sequester carbon concurrently in biological media to maintain the quality of air. Climate plays a major role in carbon sequestration potential; the higher carbon sequestration was found in the semiarid region, which is followed by the semiarid subtropical region and humid subtropical region (Kumara et al. 2023). Increase in additional carbon sequestration by 18.01%, 13.02%, and 11.13% was observed through the implementation of crop rotation, no-tillage, and integrated nutrient management in humid subtropical climate (Rathore et al. 2022a, b; Yadav et al. 2021c). Under semiarid subtropical climate, higher carbon sequestration was seen through the implementation of residue retention (15.84%) and intercropping (8%) practices than in humid subtropical climate.

14.6.1 Effect of Plant Residue Quality on Carbon Dynamics

The plant residue, microbial biomass, mineralization process, and production of organic matter are important in various ecosystems. The biochemical composition of residue decides the rate of the decomposition process (Babu et al. 2023a). SOC is affected by the amount, location, and biodegradation of plant residues (Saljnikov et al. 2013). To increase the processing of biological nutrients and minerals in the soil, adequate management of plant residues is mandatory.

Conversion of the forest land to agricultural land has decreased the soil carbon levels (Babu et al. 2020b; Yadav et al. 2021a). The carbon isotope composition in the soil is a valuable technique for quantifying the rate of organic and/or nonbiological processes. C-isotopic techniques for soil organic matter concentrations investigated a number of ecosystems. It has been reported that the 13C (isotopic) properties of soil organic matter have evolved through vegetation changes between the C_3 and C_4 plants (Jou et al. 2021). Changes in the 13C concentration of the fractions can tell us about the consumption rates of soil organic carbon and provide useful information about the various control systems of management.

14.6.2 Agronomic Practices Improve Soil Carbon Sequestration

14.6.2.1 Conservation Tillage

No-till practices decreased carbon dioxide emissions; they tended to increase nitrous oxide emissions, and the carbon benefits of no-till were offset by nitrous oxide emissions (Yadav et al. 2020 Guenet et al. 2021). A meta-analysis showed that the effect of no-till differed by crop and soil type that increased nitrous oxide emissions in paddy and acid soils by 82.1% and 25.5%, respectively (Zhao et al. 2016). Soil aeration is an important factor in nitrous oxide production, as denitrification is an anaerobic process. Denitrification rates were greater under no-till as compared with conventionally tilled soils when it was determined (Rochette 2008). No-till increased nitrous oxide emissions in poorly drained soils but the effect was found to be reduced in good to medium-aerated soils. No-till practice accelerates the denitrification process by maintaining anaerobic microsites in soil structures so no-till practices are found to be more beneficial in aerated soils with dry type of climates and poorly drained soils are a hot spot of nitrous oxide emissions (Rochette 2008). Some studies suggest that no-till practices increase the volatilization of ammonia (Rochette et al. 2009) while decreasing nitrogen oxide emissions (Liu et al. 2005).

14.6.2.2 Cover Crops

Cover crops are a valuable management practice that controls erosion, improves the quality of soil, increases the rate of biological nitrogen fixation, and decreases nitrogen losses (Yadav et al. 2021b, 2023 Babu et al. 2023b). The effect of cover crops on soil nitrogen mainly depends on whether the cover crop is a nitrogen-fixing species that provides nutrients to the soil or not (Marcillo and Miguez 2017; Ansari et al. 2023; Das et al. 2022). A meta-analysis showed that the use of cover crop for suppression of early season weeds increased the yield of vegetable crops (Osipitan et al. 2018). Nitrate leaching was reduced significantly in cover crop mixtures of non-legumes and legumes more than in legume. Nonlegume cover crop management systems exhibited 70% reductions in leaching, while nitrate leaching was found to be reduced by 40% in legume-based systems relative to conventional fertilizer-based systems (Tonitto et al. 2006). The higher potential for reduced N leaching from legume fertilized systems is due to greater retention of organic nitrogen (Tonitto et al. 2006). Nitrous oxide emission reductions were found to be more extensive when cover crop residues were incorporated into the soil than non-incorporation of the crop residues. The residues of cover crops left on the field and their C:N ratio depend on the effect of cover cropping on nitrous oxide emissions and the type of cover crop that is planted, whether it is legume or nonlegume. The impacts of cover crops on nitrous oxide seem to be variable; their carbon benefits have been found to outweigh GHG emission (Yadav et al. 2021a).

14.6.2.3 Integration of Trees on Farmland

The mitigation potential of agroforestry hinges on system type, climate, and previous land use. Silvoarable and silvopastoral systems with densely planted, fast-growing trees offer high-mitigation potential (Feliciano et al. 2018). However, emissions during tree planting, due to soil disturbance, need consideration. Agroforestry's permanence in carbon removal depends on tree types and their end use. Agroforestry, delivering multiple ecosystem services, enhances soil health, prevents erosion, nitrate leaching, and flooding, and fosters biodiversity. While high-mitigation agroforestry systems may impact individual crop yields, the long-term benefits, including nutrient and light efficiency, can outweigh these concerns. Preserving and restoring established agroforestry systems prioritizes biodiversity, emphasizing the need for locally appropriate practices. Agroforestry is not recommended for peat soils due to potential greenhouse gas emissions during tree planting (COWI EI and IEEP 2021). Preserving and restoring established agroforestry systems prioritizes biodiversity, emphasizing the need for locally appropriate practices. The incorporation of trees in agricultural settings enhances carbon sequestration per hectare through the sequestration of carbon in tree biomass, the contribution of leaves and branches to the soil, and the integration of roots into deeper soil layers. The unique advantage of agroforestry lies in its ability to pump carbon into deeper soil layers, providing better assurances of permanence compared to other soil carbon sequestration methods (Babu et al. 2020b). Additionally, agroforestry introduces new income streams for farmers through the production of fruits, nuts, or timber. It can increase the overall productivity of a plot by up to 100% (Lehmann et al. 2020).

14.6.2.4 Nature-Inclusive Management of Grasslands and Croplands

Grasslands play a crucial role in both the climate and biodiversity puzzle. When managed sustainably, they serve as significant carbon sinks and vital habitats for diverse species. The conversion of even a small portion (5%) of grasslands to arable land could result in the loss of over 300 Mt CO_2eq in the next 50 years (Lugato et al. 2014). On the contrary, restoring diversity to intensive grasslands has the potential for substantial carbon removal. However, these seminatural grasslands face threats from abandonment and intensified management. Protecting biodiversity-rich grasslands, preventing their conversion to arable land, promoting additional conversion from arable to grassland, and adopting nature and climate-friendly management practices are critical priorities for safeguarding and enhancing carbon sinks. Converting grasslands to forests may not be a wise strategy from both climate and biodiversity perspectives. Afforestation, while increasing carbon stocks in living biomass, could offset this gain by decreasing soil carbon, especially in moist regions and when conifer trees are planted (Burrascana et al. 2016). Moreover, as seminatural grasslands are vital habitats for biodiversity, further afforestation could have negative impacts on diverse grassland species.

14.6.2.5 Increase Root and Rhizosphere Sink Strength

Optimizing plant root systems, along with the associated microbial communities, represents a promising avenue for increasing soil carbon sequestration and addressing climate change challenges. Increasing soil carbon stocks is crucial for mitigating climate change, and one such approach involves manipulating plant root systems. This can be achieved by developing or employing crop cultivars with enhanced root systems that contribute more carbon to the soil. Additionally, promoting the growth of deeper root systems provides a pathway to enhance carbon input and retention, as carbon deposited at depth has a longer residence time. Deeper roots not only contribute to long-term soil carbon storage but also offer resilience against droughts. Manipulating the quantity and composition of root exudates emerges as a promising strategy. Altering the composition of root exudates influences rhizosphere microbial community succession, offering a way to steer the composition of crop rhizobium favorably. Tailoring the rhizosphere microbiome to combine increased sink strength with enhanced plant growth-promoting properties presents a dual benefit of improving the soil carbon budget and boosting productivity. This approach may involve designing synthetic microbial communities that act as obligate symbionts to the plant host, providing a biocontainment strategy (Jansson et al. 2021).

14.6.2.6 Biochar

According to the International Biochar Initiative (IBI), biochar is a charcoal whose primary application is as a soil amendment for increasing soil fertility while ensuring long-term carbon storage (IBI 2015). Biochar is a carbon-rich by-product from pyrolysis of organic material whose potential to improve soil properties and functions has been extensively studied (Schmidt et al. 2021). The application of biochar to soil has generally demonstrated beneficial effects on the cycling of nitrogen. Biochar can improve plant N absorption, boost symbiotic biological nitrogen fixation in crops, and increase soil N availability (Almaraz et al. 2021). Determining the fate of carbon in the soil and its consequences for the global carbon cycle requires an understanding of the interactions between different soil particles and biochar (Idapuganti et al. 2023). Understanding the mechanisms influencing carbon stability and cycling through the interaction of soil microorganisms with biochar is important because these microorganisms play a crucial role in mediating the transformation of organic matter.

14.6.2.7 Soil Organic Matter, Mineral, and Biochar Interaction

Biochar serves as a core component of soil aggregates, enhancing microbial activity and influencing soil aggregation indirectly through effects on microbial activity, plant root architecture, root secretion, and symbiotic relationships with fungi (Cross and Sohi 2011). The degradation of biochar in soil follows a double exponential model, involving a rapid degradation phase of labile fractions and a slower oxidation phase of recalcitrant portions. The addition of easily oxidizable organic matter through biochar induces a "positive priming effect," accelerating the mineralization rate of original organic matter, while a "negative priming effect" may occur, depending on factors such as biochar production materials, soil texture, and soil organic matter composition. For instance, the priming effect can vary with pyrolysis temperature, as grass wood biochar pyrolyzed at 250–450 °C exhibits a strong positive priming effect, whereas woody plant biochar pyrolyzed at 450–700 °C shows no strong priming effect due to different degrees of carbonization (Zimmerman 2010). While "negative priming effect" is attributed to the absorption of soil organic matter within biochar pores or the utilization of biochar-released organic carbon as the primary carbon source for soil microorganisms instead of the mineralization of original organic matter (Li et al. 2021). The pyrogenic organic matter positively primes native soil organic matter in the initial 20 days, followed by a negative priming effect in later stages (Maestrini et al. 2015). Furthermore, biochar with lower carbon content appears to induce a higher positive priming effect on native soil organic carbon.

14.6.2.8 Organic Amendments

Organic additions in agricultural soils are a sustainable approach that can be used to replace or supplement inorganic fertilizers. This strategy not only promotes soil C enrichment, but also decreases dependency on fossil fuels and diverts waste from landfills, helping to cut emissions (Chen et al. 2018). Organic additions include a wide range of organic resources such as compost, agricultural residues, manure, biosolids, sawdust, and wood ash. This debate primarily focuses on techniques known to reduce emissions, such as compost additions, while separating them from others that may raise greenhouse gas (GHG) emissions. Organic amendments are critical for nutrient recycling, improving crop nutrition, and increasing yields. Notably, a one-time application of municipal compost boosted soil N in grasslands considerably over a three-year period. Furthermore, in relevant meta-analyses, C-based additions in rangelands were observed to boost plant tissue N and aboveground net primary output. However, research suggests that organic additions lower nitrogen emissions, presumably because they improve soil C, biomass, and water-holding capacity. Organic manure has been found to lower ammonia (NH_3) emissions (Huang et al. 2016), while other organic amendments reduce N_2O emissions when compared to conventional agriculture. The inhibition of oxygen consumption in soil pores, which creates ideal anaerobic conditions for the conversion of N_2O to N_2, may be the cause of the decrease in N_2O emissions (Chen et al. 2013).

14.6.2.9 Silicate Rock Amendments

Recent discussions in carbon farming literature have shed light on the significant potential of pulverized rock amendments, primarily composed of calcium and magnesium silicates. These finely ground silicate rocks, including basalt, wollastonite, and olivine, are by-products of mining and can be strategically applied to soil, accelerating chemical weathering processes and effectively consuming atmospheric CO_2 (David et al. 2019). Unlike practices aimed at increasing SOC, such as cover cropping, agroforestry, and biochar, silicate rock amendments facilitate dissolved and precipitated carbonate formation through a natural weathering process. Silicate rock amendments have shown promise in improving soil health indicators like water-holding capacity, nutrient content, cation exchange, and pH buffering, all of which contribute to enhanced crop yields. Studies suggest that silicate rock amendments indirectly increase soil nitrogen through higher nitrification rates and reduced nitrogen losses mediated by pH changes. Silicate rock additions have also demonstrated the ability to decrease N_2O emissions from cropping systems, showcasing their potential in reducing nitrogen losses (Blanc-Betes et al. 2021). While the positive impact of silicate rock amendments on soil fertility is evident, there is a need for further research, especially at the field scale.

14.7 Carbon Farming

Carbon farming involves systemic shifts, including land-use changes. Carbon farming can deliver societal co-benefits like biodiversity, soil health, and water quality. One key aspect of carbon farming involves the adoption of agroforestry, afforestation, and reforestation initiatives. Trees act as powerful carbon sinks, absorbing and storing large amounts of carbon dioxide. Integrating these practices into conventional farming systems can not only offset emissions but also diversify agricultural landscapes, enhance biodiversity, and improve water management. Furthermore, regenerative agricultural practices, such as cover cropping and no-till farming, can enhance soil carbon sequestration. Healthy soils with increased organic matter not only sequester carbon but also improve water retention, nutrient cycling, and overall soil fertility. The potential of carbon farming on a global level is vast, offering a multifaceted approach to address climate change, promote sustainable agriculture, and enhance overall environmental well-being. Implementation of these practices requires collaborative efforts among governments, farmers, and communities to realize a future where agriculture serves as a powerful ally in the fight against climate change.

14.8 Carbon Economy

Carbon economy refers to an economy in which carbon plays an important role in energy generation and thus functioning of the entire ecosystems. With its growth, carbon has brought about higher pollution and global warming. As a result, the world today is trying to reach a future in which the amount of carbon used is less, a low-carbon economy. Retailers are changing their approach to governance spaces created by calculating a product's carbon footprint. Instead of taking control, they are establishing new corporate responsibility regimes to measure and manage C-emissions. In this shift, specific behaviors are promoted, legitimized, or eliminated to make individual suppliers active participants in their own governance (Green 2010; Ormond 2015). This also applies to C reduction in agricultural products, aligning with the values and ambitions of supply chain partners. Retailers are increasingly playing a more significant role in measuring and reducing the carbon footprint of products to solidify their position in society, ensuring long-term viability and their license to operate (Boot-Handford et al. 2014). Nevertheless, retailers, given their size and ubiquity at the intersection of daily consumption and production practices, have sparked a significant reimagining of the carbon economy (Boot-Handford et al. 2014). They are reshaping supply chain models by considering the entire life cycle of products through carbon foot printing, introducing new norms and priorities into existing relationships (Haslam et al. 2014; Higgins et al. 2015). The concept of product's carbon footprint places responsibility on the suppliers for environmental and social impacts of their products as they move through the supply chain (Derqui 2020). Suppliers at all levels of a product's supply chain are encouraged to collect and share their most accurate carbon footprint data (Rissman et al. 2020). Multinational corporations play a significant role in influencing sustainability practices. Their priorities, ambitions, and assumed responsibilities, such as halting deforestation and reducing carbon emissions, have far-reaching effects on consumers, suppliers, society, and global ecosystems (Horton and Horton 2019). In the absence of well-defined and well-enforced international environmental standards, these corporations are increasingly shaping the definition of sustainability based on their corporate interests (Dauvergne and Lister 2012). Hence, there is a need to examine the role of consumption patterns and corporate power in creating a market for low-carbon products.

14.8.1 Co-benefits of Carbon Farming for Landholders

Carbon farming initiatives offer a range of benefits. These benefits can be categorized into private and public benefits, with private benefits directly impacting the landholder and public benefits affecting everyone (Pannell et al. 2006). Through revegetation efforts, carbon sequestration occurs, serving as a public benefit. Landholders gain financially through carbon offset payments, representing a private benefit. Carbon farming contributes to nonfinancial public environmental benefits, including enhanced air and water quality resulting from reduced chemical inputs. Initiatives like forest restoration have shown the recovery of ecosystem functions and biodiversity over time, providing public benefits. Increased tree plantings and revegetation can boost pollination rates and crop yields, benefiting around 75% of globally grown crops. Carbon farming contributes to ecological infrastructure, supporting natural enemies that aid in pest suppression, and offering economic benefits to landholders. Tree cover provides windbreaks, protecting crops from extreme weather events and offering additional benefits like reduced soil erosion and improved microclimate (Pannell et al. 2006).

14.8.2 Opportunities in Carbon Farming

Farmers today have opportunities to receive payments for their role in sequestering carbon in the soil. Two primary mechanisms that drive these payments are offset markets and inset markets. Voluntary offset programs and regulated carbon cap-and-trade programs (such as the United Nations Framework on Climate Change's clean development mechanism markets) facilitate these transactions. Under cap-and-trade programs, regulated industries are allotted permits, each permitting the emission of one unit of pollution. These permits can be traded among regulated polluters, and offset purchases from unregulated sources are often allowed. Although major cap-and-trade markets in the United States permit such offset purchases, row crop agriculture is currently excluded. Inset markets signify initiatives taken by companies to combat emissions within their supply chains. This approach involves internal efforts by companies to actively engage and collaborate with their supply chain components, providing education, technical assistance, and financial support. By implementing regenerative practices, these companies aim to sequester carbon directly, contributing to emission reduction efforts. These collaborations involve implementing carbon-sequestering practices, verifying offsets, and connecting with buyers interested in supporting carbon emission reduction initiatives (Feliciano et al. 2018). India too presents distinct business opportunities in the realm of carbon farming. Carbon sink of about 28% of India's total estimated carbon stock of 7,124 million tons is found to be in northeastern India. Bamboos, which are mainly grown in northeastern India are highly popular in carbon stocking, and carbon sequestration rate ranges between 1–2.3 Mg/ha/yr in aboveground biomass and 0.14–0.39 Mg/ha/yr in soil. Technological advancements for credit certification, including measuring and calculating soil carbon, offer a significant avenue. To address concerns about potential yield reduction during the transition to carbon farming, there's a growing need for technologies sustaining yields with reduced fertilizer usage. Financial incentives beyond carbon credits, like connecting farmers with consumers willing to pay a premium for sustainable efforts, open up new business prospects. Exploring sustainable finance schemes with preferential interest rates for participating farmers is another avenue. Notably, dairy farming contributes significantly to methane emissions. Innovations like methane emission-reducing feeds could be explored in the Indian context. With rice cultivation also contributing to methane emissions, adopting methods like alternate wetting and drying (AWD) irrigation becomes crucial. Tackling field burning, a source of methane and nitrous oxide emissions, has become a focus, offering avenues for new businesses (Nozaki and Div 2023). While promising, accessing India's vast agricultural community, comprising 146 million farmers, mostly smallholders with less than one hectare of farmland, remains a challenge. However, in this regard, the use of local human resources in rural areas, FPOs, which have proliferated in recent years, and agri-tech companies that have contacts with farmers may serve as channels for reaching these farmers. The affordability and financial incentives of carbon farming are likely to be well-received by low-income, smallholder farmers, especially if policies shift away from subsidy-centric approaches.

14.8.3 Carbon Credit and Carbon Markets

Carbon credit was originated in 1997 during the United Nations Framework Convention on Climate Change (UNFCCC) and was part of the Kyoto Protocol. Carbon credit, also known as offsets, permits individuals or organizations to emit a specified amount of greenhouse gases. Carbon market was established in 2001 after the ratification of the Kyoto Protocol and is of two types—compliance market and voluntary markets (Kenton et al. 2022). Compliance markets also known as cap-and-trade are regulated by regional or international carbon reduction regimes. While voluntary markets, set up by private entities, offer flexibility without geographical constraints. Voluntary markets function outside of the compliance markets and enable companies and individuals to purchase carbon offsets on voluntary basis. Such carbon credits cannot be used to meet legally binding emissions reduction targets (Favasuli and Sebastian 2021). The voluntary carbon markets come with challenges related to credit project design. Concerns include accurate monitoring, reporting, and verification (MRV), uncertainty in soil carbon models, and the risk of fake offsets.

14.8.4 Creation of Regulated Carbon Market

Establishing a regulated carbon market ensures increased farmer participation. Regulated carbon markets offer transparency, and it prompts the organizations to ensure that credits are from verifiable sources. Increased due diligence procedures contribute to farmers gaining a positive reputation as preferred suppliers and meeting compliance requirements. This aligns with the goals of organizations looking for sustainable practices and strengthens the reputation of cooperatives engaged in carbon farming. Large corporations may invest significantly in creating reserves of carbon credits. Safeguards are essential to prevent undue advantage, such as setting a ceiling on the number of carbon credits purchasable by a single entity based on annual income or market capitalization.

14.8.5 Farming Credit Schemes

Private financing for carbon farming can thrive through collaborative models. Policymakers should incentivize private companies, including those beyond agriculture like insurance firms, to establish schemes supporting farmers through upskilling and financial aid. To ensure effectiveness, result-based carbon farming schemes are vital. Key provisions include enrolling entire farms, accounting for all greenhouse gas fluxes, defining the scope of carbon farming as nature-based solutions, ensuring permanence through long-term commitments, and maintaining a publicly accessible database of projects, including emitter and buyer details. A comprehensive approach involves promoting collaboration, setting clear rules, and ensuring inclusive support. Policymakers also play a pivotal role in incentivizing private initiatives, establishing standardized practices, and addressing concerns to foster a sustainable and equitable carbon farming ecosystem.

14.8.6 Carbon Farming Certification

The Carbon Farming Certification emerges as a crucial component in the network-of-resources approach, presenting a feasible and expansive solution. By placing the certification on farm products, a carbon label is introduced, which allows consumers to identify low-carbon footprint products. The certification plays a pivotal role in scaling up carbon farming by integrating various stakeholders, expanding financial resources for farmers, and offering consumers a tool for informed decisions. It introduces a price premium for certified farm products, providing a significant incentive for farmers. Additionally, it becomes a mechanism for government and producers to meet sustainability goals and engage in carbon trading (Marks and Brunet 2020). Certifications, such as organic labels, offer farmers a price premium and increased collaboration opportunities. Certifications also contribute significantly to environmental benefits that include improved carbon stock, biodiversity, soil health, and water conservation. Consumer recognition of carbon certifications remains lower compared to other eco-labels, requiring comprehensive awareness campaigns. The Carbon Farming Certification is a linchpin in fostering sustainable agriculture, promoting informed consumer choices, and incentivizing farmers. As a comprehensive tool, it addresses challenges, enhances environmental impact, and sets the stage for a more sustainable and equitable farming ecosystem.

14.8.7 Costs, Funding, and Incentives

14.8.7.1 Financial Costs

From a business standpoint, carbon farming is viable for farmers if the rewards including payments and co-benefits, outweigh the costs. Societally, decisionmakers must ensure that the benefits, such as climate mitigation and biodiversity, surpass both farmer and administrator costs. Costs for setting up and running carbon farming mechanisms vary based on type, subcategory, specific measures, and local context. Private and corporate finance can play a significant role in carbon farming, particularly in voluntary carbon markets. Voluntary carbon markets involve result-based payments, where private actors purchase offset credits. Private sector payments face challenges, including high upfront development costs. Corporate supply chain finance, where companies incentivize farmers, also contributes to carbon farming.

14.8.7.2 Public Source of Funding for Carbon Farming

The primary source of public funding for carbon farming is expected to come from the government side. Private and corporate financing plays a predominant role in carbon farming mechanisms associated with voluntary carbon markets or companies within the agricultural supply chain. These private sector payments offer the potential to enhance funding for carbon farming, but they encounter certain challenges. Corporate supply chain finance for carbon farming comes from companies seeking to reduce their product carbon footprint. They provide farmers in their supply chain with financial incentives for implementing action-based carbon farming, often with lower monitoring, reporting, and verification (MRV) costs compared to carbon markets. Several factors influence the amount of private finance available, with a key limitation being the perceived environmental robustness of carbon farming offset credits. Concerns about permanence, non-additionality, and uncertainty in measuring mitigation impact pose risks and may limit demand within voluntary carbon markets.

14.9 Challenges in Carbon Farming

Carbon Farming Initiatives (CFI) require agroenvironmental policies to incentivize farmers to adopt best farm management practices. However, engaging farmers in such programs is often challenging due to the complex scheme design, implementation issues, and conflicting targets between policymakers and farmers (Castelo and Marcelo 2017). Several factors, including personal interests of landholders, farm characteristics, inadequate skills, and management abilities, contribute to the barriers in adopting new farm management practices (Valdivia et al. 2012; Liu et al. 2018). Political instability is another factor affecting the acceptance and implementation of such practices. Landholders' interests and lack of skills or management abilities are direct barriers to carbon farming. Political instability and uncertainties about environmental impacts, coupled with lack of awareness of schemes and policies further hinder its adoption (Funk et al. 2014; Francis 2016). Farmers lack sufficient information about available options for carbon farming, many of them not fully understanding its meaning and the pros and cons associated with it. High input costs and concerns about the impact of carbon farming on yield and farm productivity exacerbate the situation. CFI faces additional barriers, including lack of approved methods and procedures, high administrative expenses, and challenges in obtaining certification as a qualified carbon offset provider (Hartman and Dhanda 2011; Macintosh and Waugh 2012). Capital investment requirements, incompatibility of carbon farming with existing farm practices, and potential impacts on farmers' ability to secure financial assistance from banks are identified as significant challenges (Francis 2016; Evans et al. 2015). Specific barriers in carbon farming include instability in carbon prices, uncertainty about benefits, difficulty in monitoring progress, and uncertainty about carbon market selling practices (Narassimhan et al. 2018; Lo 2016). Some farmers believe that carbon farming policies reward improper land management practices, further hindering their involvement (Tesfahunegn 2019). This suggests that farmers' interests or sentiments act as barriers to CFI, and financial incentives alone may not be sufficient to address the challenges they face.

14.10 Conclusion

Carbon farming holds significant potential on a global scale as a sustainable solution to mitigate climate change and promote environmental resilience. This agricultural approach focuses on sequestering carbon dioxide from the atmosphere and enhancing soil health, thereby contributing to carbon neutrality and fostering climate-smart practices. This promotes resilient and sustainable food production systems while reducing the reliance on synthetic inputs. Carbon farming also aligns with international efforts to achieve carbon neutrality. By sequestering carbon in agricultural soils and vegetation, countries can contribute to meeting their greenhouse gas reduction targets outlined in global agreements such as the Paris Agreement. Policies must incorporate safeguards to prevent negative impacts, monitor carbon farming's effects, and incentivize actions maximizing multiple societal benefits. Monitoring and certification should capture and reward non-mitigation impacts. Developing low-cost monitoring methods for carbon farming's mitigation impact is crucial. Carbon farming requires new skills; farmers and administrators should be supported in developing and exchanging these skills. Developing a robust, transparent, and science-based certification system for carbon removals is essential for the integrity of carbon farming initiatives globally.

References

Almaraz M, Wong MY, Geoghegan EK, Houlton BZ (2021) A review of carbon farming impacts on nitrogen cycling, retention, and loss. In Annals of the New York Academy of Sciences 1505(1): 102–117. Wiley. https://doi.org/10.1111/nyas.14690

Ansari MA, Ravisankar N, Ansari MH, Babu S, Layek J, Panwar AS (2023) Integrating conservation agriculture with intensive crop diversification in the maize-based organic system: impact on sustaining food and nutritional security. Front Nutr 10:1137247. https://doi.org/10.3389/fnut.2023.1137247

Arrhenius S (1896) On the influence of carbonic acid in the air upon the temperature of the ground. Philos Mag J Sci 41:237. https://doi.org/10.1080/14786449608620846

Babu S, Mohapatra KP, Das A, Yadav GS, Tahasildar M, Singh R, Panwar AS, Yadav V, Chandra P (2020a) Designing energy-efficient, economically sustainable and environmentally safe cropping system for the rainfed maize–fallow land of the eastern Himalayas. Sci Total Environ. https://doi.org/10.1016/j.scitotenv.2020.137874

Babu S, Mohapatra KP, Yadav GS, Lal R, Singh R, Avasthe RK, Das A, Chandra P, Gudade BA, Kumar A (2020b) Soil carbon dynamics in diverse organic land use systems in north eastern Himalayan ecosystem of India. Catena. https://doi.org/10.1016/j.catena.2020.104785

Babu S, Singh R, Avasthe RK, Yadav GS, Mohapatra KP, Selvan T, Das A, Singh VK, Valente D, Petrosillo I (2020c) Soil carbon dynamics in Indian Himalayan intensified organic rice-based cropping sequences. Ecol Indic. https://doi.org/10.1016/j.ecolind.2020.106292

Babu S, Anup D, Raghavendra S, Mohapatra KP, Kumar S, Rathore SS, Yadav SK, Poonam Y, Ansari MA, Panwar AS, Wani OA, Singh M, Ravishankar N, Layek J, Chandra P, Singh VK (2023a) Designing an energy efficient, economically feasible, and environmentally robust integrated farming system model for sustainable food production in the Indian Himalayas. Sustain Food Technol. https://doi.org/10.1039/D2FB00016D

Babu S, Singh R, Avasthe R, Rathore SS, Kumar S, Das A, Layek J, Sharma V, Wani OA, Singh VK (2023b) Conservation tillage and diversified cropping enhance system productivity and eco-efficiency and reduce greenhouse gas intensity in organic farming. Front Sustain Food Syst 7:1114617. https://doi.org/10.3389/fsufs.2023.1114617

Babu S, Singh R, Avasthe RK, Kumar S, Rathore SS, Singh VK, Ansari MA, Valente D, Petrosillo I (2023c) Soil carbon dynamics under organic farming: impact of tillage and cropping diversity. Ecol Indic. https://doi.org/10.1016/j.ecolind.2023.109940

Batjes NH (1996) Total C and N in soils of the world. Eur J Soil Sci 47:151–163. https://doi.org/10.1111/j.1365-2389.1996.tb01386.x

Bhola V, Swalaha F, Kumar RR, Singh M, Bux F (2014) Overview of the potential of microalgae for CO_2 sequestration. Int J Environ Sci Technol 11:2103–2118. https://doi.org/10.1007/s13762-013-0487-6

Blanc-Betes E, Kantola IB, Gomez-Casanovas N, Hartman MD, Parton WJ, Lewis AL, Beerling DJ, DeLucia EH (2021) In silico assessment of the potential of basalt amendments to reduce N_2O emissions from bioenergy crops. GCB Bioenergy 13(1):224–241. https://doi.org/10.1111/gcbb.12757

Boot-Handford ME, Abanades JC, Anthony EJ, Blunt MJ, Brandani S, Mac Dowell N, Fernández JR, Ferrari MC, Gross R, Hallett JP, Haszeldine RS, Heptonstall P, Lyngfelt A, Makuch Z, Mangano E, Porter RTJ, Pourkashanian M, Rochelle GT, Shah N, Fennell PS (2014) Carbon capture and storage update. Energy Environ Sci 7(1):130–189. https://doi.org/10.1039/c3ee42350f

Burrascana S, Chytry M, Kuemmerle T, Giarrizzo E, Luyssaert S, Sabatini FM, Blasi C (2016) Current European policies are unlikely to jointly foster carbon sequestration and protect biodiversity. Biol Conserv 201:370–376. https://doi.org/10.1016/j.biocon.2016.08.005

Carbon Trust (2023) Guides, reports and tools. https://www.carbontrust.com

Castelo S, Marcelo E (2017) The role of factors that influence the adoption of the Australian carbon farming initiative-emissions reduction fund: a mixed methods study

Chen H, Li X, Hu F, Shi W (2013) Soil nitrous oxide emissions following crop residue addition: a meta-analysis. Glob Chang Biol 19(10):2956–2964. https://doi.org/10.1111/gcb.12274

Chen Y, Camps-Arbestain M, Shen Q, Singh B, Cayuela ML (2018) The long-term role of organic amendments in building soil nutrient fertility: a meta-analysis and review. Nutr Cycl Agroecosyst 111(2–3):103–125. https://doi.org/10.1007/s10705-017-9903-5

Cool Farm Tool (2020) FCRN. https://www.fcrn.org.uk/research-library/cool-farm-tool

COWI Ecological Institute, IEEEP (2021) Annexes to technical guidance handbook – setting up and implementing result-based carbon farming mechanisms in the EU. Report to the European commission, DG climate action on contract no. CLIMA/C.3/ETU/2018/007. COWI, Kongens Lyngby, pp 1–297

Cross A, Sohi SP (2011) The priming potential of biochar products in relation to labile carbon contents and soil organic matter status. Soil Biol Biochem 43(10):2127–2134. https://doi.org/10.1016/j.soilbio.2011.06.016

Curt Ries (2019) A green new deal must prioritize regenerative agriculture, truthout. https://truthout.org/articles/a-green-new-deal-must-prioritize-regenerative-agriculture

Dale VH (1997) The relationship between land-use change and climate change. Ecol Appl 7(3):753–769. https://doi.org/10.1890/10510761(1997)007[0753,TRBLUC]2.0.CO;2

Das A, Babu S, Datta M, Kumar S, Singh R, Avasthe RK, Rathore SS, Yadav SK, Singh VK (2022) Restoring soil carbon in marginal land of Indian Himalayas: impact of crop intensification and conservation tillage. J Environ Manag. https://doi.org/10.1016/j.jenvman.2022.115603

Dauvergne P, Lister J (2012) Big brand sustainability: Governance prospects and environmental. Glob Environ Chang 22(1):36–45. https://doi.org/10.1016/j.gloenvcha.2011.10.007

David J, Beerling DJ, Leake JR, Long SP, Scholes JD, Ton J, Delucia E, Kantola I, Müller C, Rau GH, Hansen J (2019) Farming with crops and rocks to address global climate, food and soil security. Nat Plants 4(3):138–147. https://doi.org/10.1038/s41477-018-0108-y

Derqui B (2020) Towards sustainable development: evolution of corporate sustainability in multinational firms. Corp Soc Responsib Environ Manag 27(6):2712–2723

Dutta SK, Njoroge S, Tangi A, Moulay H, Zingore S, El Gharous M, Majumdar K (2020) Nutrient management at conservation agriculture – a special focus on smallholder farmers of South Asia and Sub–Saharan Africa. SATSA Mukhapatra – Annual Tech Issue 24:46–68

European Commission (2021) Commission implementing regulation (EU) 2021/1165 of 15 July 2021 authorising certain products and substances for use in organic production and establishing their lists

Evans MC, Carwardine J, Fensham RJ, Butler DW, Wilson KA, Possingham HP, Martin TG (2015) Carbon farming via assisted natural regeneration as a cost-effective mechanism for restoring biodiversity in agricultural landscapes. Environ Sci Pol 50:114–129. https://doi.org/10.1016/j.envsci.2015.02.003

Helena Bottemiller Evich (2019) Agriculture Department Buries Studies Showing Dangers of Climate Change, POLITICO, June 23, 2019, 5:04 PM, https://www.politico.com/story/2019/06/23/agriculture-department-climate-change-1376413

Fara M (2016) Greenhouse gas emission metrics, FM LINK

Favasuli S, Sebastian V (2021) Voluntary carbon markets: how they work, how they're priced and who's involved. https://www.spglobal.com/commodityinsights/en/market-insights/blogs/energy-transition/061021-voluntary-carbon-markets-pricing-participants-trading-corsia-credits

Feliciano D, Ledo A, Hillier J, Nayak DR (2018) Which agroforestry options give the greatest soil and above ground carbon benefits in different world regions? Agric Ecosyst Environ 254:117–129. https://doi.org/10.1016/j.agee.2017.11.032

Food and Agriculture Organization (2020) Emissions due to agriculture. Global, regional and country trends 2000–2018. FAOSTAT analytical brief series no 18. Rome

Francis C (2016) The carbon farming solution: a global toolkit of perennial crops and regenerative agriculture practices for climate change mitigation and food security, by Eric Toensmeier. Agroecol Sustain Food Syst 40(9):1039–1040. https://doi.org/10.1080/21683565.2016.1214861

Funk JM, Field CB, Kerr S, Daigneault A (2014) Modeling the impact of carbon farming on land use in a New Zealand landscape. Environ Sci Pol 37:1–10. https://doi.org/10.1016/j.envsci.2013.08.008

Gaikwad AS (2021) Quantification of soil carbon pools and carbon sequestration rate as influenced by bamboo plantation grown on Entisol of semiarid climate of Maharashtra. Master's thesis, Post Graduate Institute, Mahatma Phule Krishi Vidyapeeth, Rahuri, Ahmednagar, Maharashtra State, India. https://krishikosh.egranth.ac.in/handle/1/5810183402

Green JF (2010) Private standards in the climate regime: the greenhouse gas protocol. Business and Politics De Gruyter vol. 12(3):1–39, October

Greene CH, Pershing AJ (2007) Climate drives sea change. Science 315(5815):1084–1085. https://doi.org/10.1126/science.1136495

Guenet B, Gabrielle B, Chenu C (2021) Can N_2O emissions offset the benefits from soil organic carbon storage? Glob Chang Biol 27:237–256

Hanaki K, Pereira JP (2018) The effect of biofuel production on greenhouse gas emission reductions. In: Biofuels and sustainability. ISBN: 978-4-431-54894-2

Hartman LP, Dhanda KK (2011) The ethics of carbon neutrality – a critical examination of voluntary carbon offset providers. J Bus Ethics 100(1):119–149. https://doi.org/10.1007/s10551-011-0766-4

Hartmann F, Perego P, Young A (2013a) Carbon accounting: challenges for research in management control and performance measurement. Abacus 49(4):539–563. https://doi.org/10.1111/abac.12018

Hartmann J, West AJ, Renforth P, Kohler P, De La Rocha CL, Wolf-Gladrow DA, Durr HH, Scheffran J (2013b) Enhanced chemical weathering as a geoengineering strategy to reduce atmospheric carbon dioxide, supply nutrients, and Mitigate Ocean acidification. Rev Geophys 51:113–149. https://doi.org/10.1002/rog.20004

Haslam C, Butlin J, Andersson T, Malamatenios J, Lehman G (2014) Accounting for carbon and reframing disclosure: a business model approach. Account Forum 38(3):200–211. https://doi.org/10.1016/j.accfor.2014.04.002

Higgins V, Dibden J, Cocklin C (2015) Private agri-food governance and greenhouse gas abatement: constructing a corporate carbon economy. Geoforum 66:75–84. https://doi.org/10.1016/j.geoforum.2015.09.012

Hooker BA, Morris TF, Peters R, Cardon ZG (2005) Long – term effects of tillage and corn stalk on soil carbon dynamics. Soil Sci Soc Am J 69:188–196. https://doi.org/10.2136/sssaj2005.0188

Horton P, Horton BP (2019) Re-defining sustainability: living in harmony with life on earth. One Earth 1(1):86–94. https://doi.org/10.1016/j.oneear.2019.08.019

Huang SL, Lv W, Bloszies S, Shi Q, Pan X, Zeng Y (2016) Effects of fertilizer management practices on yield-scaled ammonia emissions from croplands in China: a meta-analysis. Field Crop Res 192:118–125. https://doi.org/10.1016/j.fcr.2016.04.023

IBI (2015) Standardized product definition and product testing guidelines for biochar that is used in soil. International Biochar Initiative, November, 23

Idapuganti RG, Das A, Sungoh H, Layek J, Mandal S, Verma BC, Lal R, Rangappa KB, Hazarika S (2023) Can biochar conserve soil moisture and improve soil properties for sustainable intensification of acid soils in the eastern Indian Himalayas? Land Degrad Dev:1–17. https://doi.org/10.1002/ldr.4981

Intergovernmental Panel on Climate Change, Climate Change 2014: Synthesis Report Summary for Policymakers, 10 (2014)

Jansson C, Faiola C, Wingler A, Zhu XG, Kravchenko A, de Graaff MA, Ogden AJ, Handakumbura PP, Werner C, Beckles DM (2021) Crops for carbon farming. Front Plant Sci 12. https://doi.org/10.3389/fpls.2021.636709

Kenton W, Boyle MJ, Costagliola D (2022) Carbon credits and how they can offset your carbon footprint. Economy: government and policy. https://www.investopedia.com/terms/c/carbon_credit.asp

Kirkels F, Cammeraat LH, Kuhn NJ (2014) The fate of soil organic carbon upon Erosion, transport and deposition in agricultural landscapes – a review of different concepts. Geomorphology 226:94–105. https://doi.org/10.1016/j.geomorph.2014.07.023

Kumar SS, Mir SA, Wani OA, Babu S, Yeasin M, Bhat MA, Hussain N, Wani AA, Kumar R, Devideen Y, Dar SR (2022) Land-use systems regulate carbon geochemistry in the temperate Himalayas, India. J Environ Manag. https://doi.org/10.1016/j.jenvman.2022.115811

Kumara KTM, Pal S, Chand P, Kandpal A (2023) Carbon sequestration potential of sustainable agricultural practices to mitigate climate change in Indian agriculture: a meta-analysis. Sustain Prod Consump 35:697–708. https://doi.org/10.1016/j.spc.2022.12.015

Kundu S, Singh M, Saha JK, Biswas A, Tripathi AK, Acharya CL (2001) Relationship between C addition and storage in a vertisol under soybean-wheat cropping system in subtropical central. J Plant Nutr Soil Sci 164:483–486. https://doi.org/10.1002/1522-2624(200110)164:5<483::AID-JPLN483>3.0.CO;2-Y

Lal R (2010) Soil carbon sequestration impacts on global climate change and food security. Science 304:1623–1627. https://doi.org/10.1126/science.1097396

Legg W, Huang H (2020) Climate change and agriculture, OECD observer. https://oecdobserver.org/news/archivestory.php/aid/3213/Climate_change_and_agriculture.html

Lehmann LM, Smith J, Westaway S, Pisanelli A, Russo G, Borek R, Sandor M, Gliga A, Smith L, Ghaley BB (2020) Productivity and economic evaluation of agroforestry systems for sustainable production of food and non-food products. Sustainability (Switzerland) 12(13):5429. https://doi.org/10.3390/su12135429

Li M, Xiong Y, Cai L (2021) Effects of biochar on the soil carbon cycle in agroecosystems: an promising way to increase the carbon pool in dryland. IOP Conference Series: Earth and Environmental Science 693(1):012082. https://doi.org/10.1088/1755-1315/693/1/012082

Linderholkm HW (2006) Growing season changes in the last century. Agric For Meteorol 137:1–14. https://doi.org/10.1016/j.agrformet.2006.03.006

Liu XJ, Mosier AR, Halvorson AD, Zhang FS (2005) Tillage and nitrogen application effects on nitrous and nitric oxide emissions from irrigated corn fields. Plant Soil 276:235–249. https://doi.org/10.1007/s11104-005-4894-4

Liu T, Bruins RJF, Heberling MT (2018) Factors influencing farmers' adoption of best management practices: a review and synthesis. Sustainability (Switzerland) 10(2):432. https://doi.org/10.3390/su10020432

Lo AY (2016) Challenges to the development of carbon markets in China. Clim Pol 16(1):109–124. https://doi.org/10.1080/14693062.2014.991907

Lugato E, Bampa F, Panagos P, Montanarella L, Jones A (2014) Potential carbon sequestration of European arable soils estimated by modelling a comprehensive set of management practices. Glob Chang Biol 20(11):3557–3567. https://doi.org/10.1111/gcb.12551

Lynch J, Cain M, Frame D, Pierrehumbert D (2021) Agriculture's contribution to climate change and role in mitigation is distinct from predominantly fossil CO_2-emitting sectors. Front Sustain Food Syst 4:518039. https://doi.org/10.3389/fsufs.2020.518039

Macintosh A, Waugh L (2012) An introduction to the carbon farming initiative: key principles and concepts. Environ Plan Law J 29(6):439–461

Maestrini B, Nannipieri P, Abiven S (2015) A meta-analysis on pyrogenic organic matter induced priming effect. GCB Bioenergy 7(4):577–590. https://doi.org/10.1111/gcbb.12194

Manoff MV (2018) Can dirt save the earth? New York Times Magazine

Marcillo GS, Miguez FE (2017) Corn yield response to winter cover crops: an updated meta-analysis. J Soil Water Conserv 72:226–239. https://doi.org/10.2489/jswc.72.3.226

Marks AB, Brunet A (2020) Carbon farming our way out of climate change. 97 Denv L Rev 497:503

Mcleod E, Chmura GL, Bouillon S, Salm R, Bjork M, Duarte CM, Lovelock CE, Schlesinger WH, Silliman BR (2011) A blueprint for blue carbon: toward an improved understanding of the role of vegetated coastal habitats in sequestering CO_2. Front Ecol Environ 9:552–560. https://doi.org/10.1890/110004

Moreira D, Pires JC (2016) Atmospheric CO_2 capture by algae: negative carbon dioxide emission path. Bioresour Technol 215:371–379. https://doi.org/10.1016/j.biortech.2016.03.060
Moskin J, Plumer B, Lieberman R, Weingart E (2019) Your questions about food and climate change, answered. New York Times
Myhre G, Shindell D, Bréon FM, Collins W, Fuglestvedt J, Huang J, Koch D, Lamarque JF, Lee D, Mendoza B, Nakajima T, Robock A, Stephens G, Takemura T, Zhang H (2013) Anthropogenic and natural radiative forcing. In: Climate change 2013: the physical science basis. contribution of working group I to the fifth assessment report of the intergovernmental panel on climate change. Cambridge University Press, Cambridge, UK/New York
Narassimhan E, Gallagher KS, Koester S, Alejo JR (2018) Carbon pricing in practice: a review of existing emissions trading systems. Clim Pol 18(8):967–991. https://doi.org/10.1080/14693062.2018.1467827
Nozaki Y, Div IS (2023) Rise of carbon farming in India – world's largest agrarian country expected to become the leading market for carbon farming credits –. February, 1–8
Ormond J (2015) New regimes of responsibilization: practicing product carbon footprinting in the new carbon economy. Econ Geogr 91(4):425–448. https://doi.org/10.1111/ecge.12095
Osipitan OA, Dille JA, Assefa Y, Knezevic SJ (2018) Cover crop for early season weed suppression in crops: systematic review and meta-analysis. Agron J 110:2211–2221. https://doi.org/10.2134/agronj2017.12.0752
Pannell DJ, Marshall GR, Barr N, Curtis A, Vanclay F, Wilkinson R (2006) Understanding and promoting adoption of conservation practices by rural landholders. Aust J Exp Agric 46(11):1407–1424. https://doi.org/10.1071/EA05037
Parikh SJ, Winfield E (2020) Climate-smart agriculture: biochar amendments. USDA California Climate Hub, Web, Davis
Popp A, Campen HL, Bodirsky B (2010) Food consumption, diet shifts and associated non-CO_2 greenhouse gases from agricultural production. Glob Environ Chang 451:455–456. https://doi.org/10.1016/j.gloenvcha.2010.02.001
Portner HO, Scholes RJ, Agard J, Archer E, Arneth A, Bai X, Barnes D, Burrows M, Chan L, Cheung W, Diamond S, Donatti CI, Duarte CM, Eisenhauer N, Foden W, Gasalla MA, Handa C, Hickler T, Hoegh-Guldberg O, Ngo H (2021) Scientific outcome of the IPBES-IPCC co-sponsored workshop on biodiversity and climate change. https://doi.org/10.5281/zenodo.5101125
Powlson DS, Glendining MJ, Coleman K, Whitmore AP (2011) Implications for soil properties of removing cereal straw – results from long-term studies. Agron J 103:279–287
Rathore SS, Babu S, El-Sappah AH, Shekhawat K, Singh VK, Singh RK, Upadhyay PK, Singh R (2022a) Integrated agroforestry systems improve soil carbon storage, water productivity, and economic returns in the marginal land of the semi-arid region. Saudi J Biol Sci. https://doi.org/10.1016/j.sjbs.2022.103427
Rathore SS, Babu S, Shekhawat K, Singh R, Yadav SK, Singh VK, Singh C (2022b) Designing energy cum carbon-efficient environmentally clean production system for achieving green economy in agriculture. Sustain Energy Technol Assess. https://doi.org/10.1016/j.seta.2022.102190
Rissman J, Bataille C, Masanet E, Aden N, Morrow WR, Zhou N, Elliott N, Dell R, Heeren N, Huckestein B, Cresko J, Miller SA, Roy J, Fennell P, Cremmins B, Koch Blank T, Hone D, Williams ED, de la Rue du Can S, Helseth J (2020) Technologies and policies to decarbonize global industry: review and assessment of mitigation drivers through 2070. Appl Energy 266:114848. https://doi.org/10.1016/j.apenergy.2020.114848
Rochette P (2008) No-till only increases N_2O emissions in poorly-aerated soils. Soil Tillage Res 101:97–100. https://doi.org/10.1016/j.still.2008.07.011
Rochette P, Angers DA, Chantigny MH (2009) Ammonia volatilization following surface application of urea to tilled and no-till soils: a laboratory comparison. Soil Tillage Res 103:310–315. https://doi.org/10.1016/j.still.2008.10.028
Rodale Institute (2008) Regenerative organic farming: a solution to global warming

Rumpel C et al (2018) Put more carbon in soils to meet Paris climate pledges. Nature. https://www.nature.com/articles/d41586-018-07587-4; Velasquez-Manoff, supra note 27

Running SM (2006) Is global warming causing more, large wildfires? Science 313(5789):927–928. https://doi.org/10.1126/science.1130370

Safdie S (2022) ISO 14067: meaning, standard and requirements. US Copywriter, Greenly

Samberg L (2017) World hunger is increasing thanks to wars and climate change. Conversation (Oct. 17, 2017, 7:51 PM). https://theconversation.com/world-hunger-is-increasingthanks-to-wars-and-climate-change-84506

Sapkota TB, Majumdar K, Jat ML, Kumar A, Bishnoi DK, McDonald AJ, Pampolino M (2014) Precision nutrient management in conservation agriculture based wheat production of Northwest India: profitability, nutrient–use efficiency and environmental footprint. Field Crop Res 155:233–244. https://doi.org/10.1016/j.fcr.2013.09.001

Sapkota TB, Khanam F, Mathivanan GP, Vetter S, Hussain GK, Pilat AL, Shahrin S, Hossain MK, Sarker NR, Krupnik TJ (2021) Quantifying opportunities for greenhouse gas emissions mitigation using big data from smallholder crop and livestock farmers across Bangladesh. Sci Total Environ 786:147344. https://doi.org/10.1016/j.scitotenv.2021.147344

Schmidt HP, Kammann C, Hagemann N, Leifeld J, Bucheli TD, Sánchez Monedero MA, Cayuela ML (2021) Biochar in agriculture – a systematic review of 26 global meta-analyses. GCB Bioenergy 13(11):1708–1730. https://doi.org/10.1111/gcbb.12889

Sidhu HS, Jat ML, Singh Y, Sidhu RK, Gupta N, Singh P, Singh P, Jat HS, Gerard B (2019) Sub–surface drip fertigation with conservation agriculture in a rice–wheat system: a breakthrough for addressing water and nitrogen use efficiency. Agric Water Manag 216:273–283. https://doi.org/10.1016/j.agwat.2019.02.019

Singh SK, Sundaram S, Sinha S, Rahman MA, Kapur S (2016) Recent advances in CO_2 uptake and fixation mechanism of cyanobacteria and microalgae. Crit Rev Environ Sci Technol 46:1297–1323. https://doi.org/10.1080/10643389.2016.1217911

Smith P (2008) Land use change and soil organic carbon dynamics. Nutr Cycl Agroecosyst 81:169–178. https://doi.org/10.1007/s10705-007-9138-y

Taylor LB (2018) Land O' Lakes launches digital erp tool to measure profitability of sustainable farming practices. Agfunder Network Partners

Tesfahunegn GB (2019) Farmers' perception on land degradation in northern Ethiopia: implication for developing sustainable land management. Soc Sci J 56(2):268–287. https://doi.org/10.1016/j.soscij.2018.07.004

Toensmeier E (2018) How we can make beef less terrible for the environment. The Washington Post

Tonitto C, David MB, Drinkwater LE (2006) Replacing bare fallows with cover crops in fertilizer-intensive cropping systems: a meta-analysis of crop yield and N dynamics. Agric Ecosyst Environ 112:58–72

US Department of Agriculture (2017) Comet-planner tool. http://comet-planner.com/

Valdivia C, Barbieri C, Gold MA (2012) Between forestry and farming: policy and environmental implications of the barriers to agroforestry adoption. Can J Agric Econ 60(2):155–175. https://doi.org/10.1111/j.1744-7976.2012.01248.x

Vanlauwe B, Wendt J, Giller KE, Corbeels M, Gerard B, Nolte C (2014) A fourth principle is required to define conservation agriculture in sub–Saharan Africa: the appropriate use of fertilizer to enhance crop productivity. Field Crop Res 155:1013. https://doi.org/10.1016/j.fcr.2013.10.002

Wani SA, Sofi MN, Chand S, Hakeem KR (2016) Soil carbon sequestration: as a climate change adaptation and mitigation strategy – an overview. Int J Plant Anim Environ Sci 6(1):227–232

Wani OA, Sharma V, Kumar SS, Babu S, Sharma KR, Rathore SS, Marwaha S, Ganai NA, Dar SR, Yeasin M, Raghavendra S, Tomar J (2023) Climate plays a dominant role over land management in governing soil carbon dynamics in North Western Himalayas. J Environ Manag. https://doi.org/10.1016/j.jenvman.2023.117740

Westerling AL, Hidalgo HG, Cayan DR, Swetnam TW (2006) Warming and earlier spring increase Western U.S. Forest wildfire activity. Science 313(5789):940–943. https://doi.org/10.1126/science.1128834

Wilson GW, Rice CW, Rillig MC, Springer A, Hartnett DC (2009) Soil aggregation and carbon sequestration are tightly correlated with the abundance of arbuscular mycorrhizal fungi: results from long-term field experiments. Ecol Lett 12(5):452–461. https://doi.org/10.1111/j.1461-0248.2009.01303.x

Xiao H, Li Z, Chang X, Huang B, Nie X, Liu C, Liu L, Wang D, Jiang J (2018) The mineralization and sequestration of organic carbon in relation to agricultural soil Erosion. Geoderma 329:73–81

Yadav GS, Das A, Babu S, Mohapatra KP, Lal R, Rajkhowa DJ (2020) Potential of conservation tillage and altered land configuration to improve soil properties, carbon sequestration and productivity of maize based cropping system in eastern Himalayas, India. Int Soil Water Conserv Res. https://doi.org/10.1016/j.iswcr.2020.12.003

Yadav GS, Babu S, Das A, Datta M, Mohapatra KP, Singh R, Singh VK, Rathore SS, Chakraborty M (2021a) Productivity, soil health, and carbon management index of Indian Himalayan intensified maize-based cropping systems under live mulch-based conservation tillage practices. Field Crop Res. https://doi.org/10.1016/j.fcr.2021.108080

Yadav GS, Das A, Kandpal BK, Babu S, Rattan L, Datta M, Das B, Singh R, Singh VK, Mohapatra KP, Chakraborty M (2021b) The food-energy-water-carbon nexus in a maize-maize-mustard cropping sequence of the Indian Himalayas: an impact of tillage-cum-live mulching. Renew Sust Energ Rev. https://doi.org/10.1016/j.rser.2021.111602

Yadav GS, Kandpal BK, Das A, Babu S, Mohapatra KP, Devi AG, Devi HL, Chandra P, Singh R, Barman KK (2021c) Impact of 28-year-old agroforestry systems on soil carbon dynamics in eastern Himalayas. J Environ Manag. https://doi.org/10.1016/j.jenvman.2021.111978

Yadav D, Singh D, Babu S, Madegowda M, Singh DV, Mandal D, Rathore AC, Sharma VK, Singhal V, Kumawat A, Yadav DK, Yadav RK, Kumar S (2023) Intensified cropping reduces soil erosion and improves rainfall partitioning and soil properties in the marginal land of the Indian Himalayas. International Soil and Water Conservation Research. https://doi.org/10.1016/j.iswcr.2023.10.002

Zhao X, Liu SL, Pu C (2016) Methane and nitrous oxide emissions under no-till farming in China: a meta-analysis. Glob Chang Biol 22:1372–1384. https://doi.org/10.1111/gcb.13185

Zimmerman AR (2010) Abiotic and microbial oxidation of laboratory-produced black carbon (biochar). Environ Sci Technol 44(4):1295–1301. https://doi.org/10.1021/es903140c

Chapter 15
Agroecological Approaches for Sustainable Intensification

Nitin N. Gudadhe, Subhash Babu, Raghavendra Singh, G. I. Ramkrushna, Aastika Pandey, Khushboo Devi, Dileswar Nayak, and Sanjeev Kumar

Abstract Today, the term "sustainable intensification" is frequently used to characterize how agricultural and food production will develop in the future in order to solve the issues of resource conservation, climate change, food security, and the growing global population. While some people understand sustainable intensification to mean boosting output through more effective but possibly more intensive use of technology and inputs, environmental protection, which includes the preservation and renewal of natural capital and the output of ecosystem services, must also be considered. There is growing agreement that sustainable intensification ought to actively promote environmental advantages in addition to preventing further harm to the environment. This includes addressing issues of consumption, waste, biodiversity conservation, and resource use while ensuring sufficient overall levels of production to meet human needs. "Agroecology" is also now receiving increasing attention as an approach to agriculture that attempts to reconcile environmental, sustainability, and production goals by emphasizing the application of ecological concepts and principles to the design and management of agricultural systems.

N. N. Gudadhe (✉)
Navsari Agricultural University, Navsari, Gujarat, India

S. Babu · A. Pandey · K. Devi
Division of Agronomy, ICAR-Indian Agricultural Research Institute, New Delhi, India

R. Singh
ICAR-Indian Institute of Farming Systems Research, Modipuram, Meerut, Uttar Pradesh, India

G. I. Ramkrushna
ICAR-Central Institute for Cotton Research, Nagpur, Maharashtra, India

D. Nayak
Navsari Agricultural University, Navsari, Gujarat, India

School of Environmental Sciences, Jawaharlal Nehru University, New Delhi, India

S. Kumar
Sardar Vallabhbhai Patel University of Agriculture and Technology, Meerut, Uttar Pradesh, India

S. Babu et al. (eds.), *Agricultural Diversification for Sustainable Food Production*, Sustainability Sciences in Asia and Africa,
https://doi.org/10.1007/978-981-97-7517-0_15

Agroecology can be seen as part of a broader approach to sustainable intensification focusing on ecological and knowledge intensification alongside technological intensification. Agroecological-based sustainable intensification interventions such as conservation tillage, legumes, cover crops, and catch crops in rotations, integrated pest management, soil conservation, on-farm mechanization, smart, precision technologies for irrigation and nutrient use efficiency, high-yielding varieties, and animal–crop integration can be used for maintaining sustainability.

Keywords Economic diversification · Mechanization · Soil health · Synergy

15.1 Introduction

The anticipated growth in world population of up to 9.1 billion in 2050 (United Nations 2009) will require a huge effort to increase food production for an additional two billion people compared to now (Wezel et al. 2014). Everyone agrees that the global food system is falling short on a number of critical indicators, such as rates of hunger and malnutrition, respectable agricultural livelihoods, and the impact of agriculture on the environment (Wezel et al. 2020). A profound transformation is needed at multiple scales to meet the interacting challenges of increased pressure and competition over renewable resources, persistent malnutrition, rural poverty, increased power and concentration of agricultural and food industries, growing consequences of climatic change, and alarming losses of biodiversity (IPCC 2019).

Agricultural food production techniques must be developed for all forms of agriculture, including organic, integrated, and conventional agriculture. The best agricultural production techniques to achieve the objective of increased and sustainable food production are the subject of a contentious, ongoing discussion. Options for agriculture range from highly technological methods to ecological methods. On the one hand, precision farming (Srinivasan 2006) or the adoption of genetically modified crops (Phipps and Park 2002) could assist match the future food need. However, other options include no or reduced tillage, which increases soil biota activity and improves soil fertility, or natural biological control of pests, which involves integrating natural landscape elements into agricultural landscapes to decrease pesticide use (Gurr et al. 2004). Biological pest control, crop rotation, and organic crop fertilization are examples of well-known agricultural techniques that have been in use for a very long time. But during the past 20 years, they have come to be known more and more as agroecological techniques (Altieri 1995).

Today, the term "sustainable intensification" is frequently used to characterize the course that agriculture and food production will take in the future in order to meet the issues of resource conservation, climate change, food security, and the growing global population. While some people understand sustainable intensification to mean boosting output through more effective but possibly more intensive use of technology and inputs, environmental protection, which includes the preservation and renewal of natural capital and the output of ecosystem services, must also be considered. There is growing agreement that sustainable intensification ought to actively promote environmental advantages in addition to preventing more environmental harm. These entail managing concerns related to waste, consumption (including diets), biodiversity preservation, and resource utilization, all the while guaranteeing adequate production levels to fulfill human requirements.

As agroecology developed in the 1980s, the phrase "agroecological practices" came into being (Wezel et al. 2009). Currently, there are three main currents or interpretations of agroecology: a scientific field, a movement, and agroecology as a practice. The goal is to reshape agriculture to create locally relevant food systems that support rural communities' economic sustainability through short marketing chains and equitable and secure food production. Furthermore, whatever more recent developments might be regarded as cutting-edge agroecological techniques and show promise for creating a more sustainable agricultural output? A practice that is novel for improvement can be entirely new, but it can also be founded on time-tested ideas or methods that have received little attention (Uphoff 2002) and are modified in a novel way. "Agroecology," an approach to agriculture that emphasizes the application of ecological concepts and principles to the design and management of agricultural systems, is also gaining popularity as a way to try and harmonize environmental, sustainability, and productivity goals. Agroecology can be viewed as a component of a more comprehensive strategy for sustainable intensification that emphasizes knowledge and ecological (or eco-functional) intensification in addition to technological intensification. The agroecological approaches to sustainable farming techniques are given additional emphasis in this chapter.

15.2 What Is Agroecology?

The use of ecological concepts and methods in agricultural sciences was first described by the term "agroecology" in the early 1900s. The word "agroecosystem," first used by Odum in 1969, refers to an ecosystem that has been tamed and is under human management. Twenty years later, agroecology began to move from the farm and field sizes to the whole agroecosystem, with important contributions highlighting the processes of intercultural creation of local knowledge incorporated into agroecological knowledge (Xolocotzi 1977). According to the Food and Agriculture Organization (FAO), "Agroecological innovations apply ecological principles-such as recycling, resource use efficiency, reducing external inputs, diversification, integration, soil health and synergies, for the design of farming systems" (FAO 2016) (Table 15.1). Based on these advancements, Altieri defined agroecology as "the application of ecological concepts and principles to the design and management of sustainable agroecosystems" (Altieri 1995).

Agroecology's focus was broadened a few years later to include a food system focus, which covered consumer-producer relationships, agri-food systems, markets, agronomy, ecology, sociology, and economics, as well as local and alternative food networks. This resulted in the creation of "an integrated discipline that includes

Table 15.1 Principles of agroecology

1. Recycling	Use renewable local resources whenever possible. Fertilizer and biomass resource cycles as closely as possible.
2. Input reduction	Enhance self-sufficiency by reducing or eliminating reliance on input purchases.
3. Soil health	Specifically, by regulating organic matter and improving soil biological activity, secure and improve soil health and functioning for better plant growth.
4. Animal health	Ensure animal health and welfare.
5. Biodiversity	Maintain and improve the diversity of species, functions, and genetic resources, preserving the overall spatial biodiversity of agroecosystems at the field, farm, and landscape scales.
6. Synergy	Enhance the harmony, integration, complementarity, and beneficial ecological interaction between the various agroecological components (plants, animals, trees, soil, and water).
7. Economic diversification	Increase small-scale farmers' financial independence and chances for value addition so they can better respond to consumer demand, thereby diversifying their on-farm incomes.
8. Co-creation of knowledge	Improve knowledge co-creation and horizontal sharing, particularly through farmer-to-farmer exchange, to include regional and scientific innovation.
9. Social values and diets	Develop local food systems that offer wholesome, varied, seasonally appropriate diets based on the culture, identity, tradition, social justice, and gender equity of the community.
10. Fairness	Based on fair trade, fair employment, and fair use of intellectual property rights, support robust and dignified livelihoods for all participants in the food systems, particularly small-scale food producers.
11. Connectivity	Ensure closeness and trust between farmers and customers by fostering fair and condensed distribution networks and reincorporating food systems into regional economies.
12. Land and natural resource governance	Enhance institutional frameworks, particularly those that recognize and assist smallholders, peasant food producers, and family farmers as responsible stewards of genetic resources and natural resources.
13. Participation	To encourage decentralized governance and local adaptive management of agricultural and food systems, promote social organization, and increase participation in decision-making by food producers and consumers.

elements from agronomy, ecology, sociology, and economics" (Dalgaard et al. 2003). Since then, agroecology has adopted a transdisciplinary focus in place of its previous interdisciplinary one. It now seeks revolutionary answers to contemporary difficulties in the establishment of sustainable food systems while aiming for inclusive participation of all stakeholders in the generation of knowledge (Méndez et al. 2013; Sinclair et al. 2019).

15.3 Sustainable Intensification

An innovative strategy called "sustainable intensification" aims to boost output on already cultivated land while having a good social and environmental impact. Sustainable intensification is emerging as the most commonly discussed new paradigm of agricultural production in scientific and development literature (Pretty et al. 2011). A more recent and often cited definition of sustainable intensification is found in the FAO's Save and Grow publication from 2011, which defines it as "producing more from the same area of land while conserving resources, reducing negative impacts on the environment, and enhancing natural capital and the flow of ecosystem services." In their publications, conference speeches, and reports, a range of companies, environmentalists, and farmers explore the future of food production under the term "sustainable intensification." Though some environmental movements have expressed mistrust and criticism over its unclear definition—which is more of an aim to increase yields without negatively impacting the environment—it appears that environmental movements are becoming more aware of this concept. One common example of this kind of criticism is when Friends of the Earth International refers to sustainable intensification in conventional agriculture as "a wolf in sheep's clothing" (Collins and Chandrasekaran 2012).

In order to address the concerns of resource conservation, climate change, food security, and an increasing global population, agriculture and food production will likely follow a path known as "sustainable intensification" in the future. While some define sustainable intensification as increasing output while utilizing technology and inputs more extensively—or more effectively—there is also a need to consider environmental protection, which includes the output of ecosystem services and the preservation and renewal of natural capital. There is growing consensus that, in addition to halting more environmental harm, sustainable intensification should actively promote environmental benefits. Many industry players are pushing for the intensification of agriculture. However, it is neither clear nor simple to agree upon what intensification actually means. The three primary intensification concepts utilized today are "ecological intensification," "sustainable intensification," and, most recently, "agroecological intensification" (Wezel et al. 2015) (Table 15.2). "Agroecology," an approach to agriculture that emphasizes the application of ecological concepts and principles to the design and management of agricultural systems, is also gaining popularity as a way to try and harmonize environmental, sustainability, and productivity goals. Agroecology fits into a larger paradigm of sustainable intensification that emphasizes knowledge and ecological (or eco-functional) intensification in addition to technological intensification (Lampkin et al. 2015).

Table 15.2 Ecological, agroecological, and sustainable intensification farming practices

Ecological intensification	Agroecological intensification	Sustainable intensification
Conservation tillage	Mulching	Conservation effective tillage
Mixed cropping systems	Intercropping	Legumes, cover crops, and catch crops in sequence
Diversified crop rotation/sequence	Crop rotations/sequence	Integrated pest management (IPM)
Cover crops	Integrated pest management and biological control	Soil and water conservation
Minimization of soil Compaction	Integrated soil and nutrient management	On-farm mechanization
Integrated pest management	Balanced and efficient use of fertilizers	Smart, precision technologies for irrigation and nutrient use efficiency
Multi-nutrients fertilizer	Use of quality seeds of well-suited hybrids/varieties	High-yielding varieties
Optimum nutrient management	Water conservation	Animal–crop-tree integration
Direct seeding	Organic inputs	
Biodiversity preservation	"Right-time" planting/sowing	
	Crop–tree–livestock interactions	
	Recycling of biomass and agricultural by-products	

15.4 Agroecological Approaches for Sustainable Intensification

15.4.1 Conservation Tillage

In the current semiarid and arid conditions brought on by changing climatic conditions, the application of conservation tillage, which is customized to the type and properties of soil and agroecological conditions in the production area, can significantly contribute to water conservation, prevent soil degradation, and achieve high and stable yields (Jug et al. 2019). Conservation tillage is one of the main principles of conservation agriculture. According to the FAO (FAO 2016), conservation agriculture is a method of managing agroecosystems for improved and sustained production, higher profits, and food security, all the while preserving and improving the environment and resource base (Yadav et al. 2018, 2021b).

Several studies have shown that conservation tillage improves soil properties and raises crop yields (Hirzel et al. 2020; Yadav et al. 2019a, 2020b). It has been established that conservation tillage practices minimize the degradation of soil organic matter (Lal 2007) and support crop production systems that are sustainable (Zhang et al. 2012; Yadav et al. 2020a). Conservation tillage can lead to less soil production even if it is a sustainable management approach. Conservation tillage should be used in conjunction with other management strategies (such as crop rotation and cover crops) to increase soil organic matter, carbon sequestration (Lu et al. 2018; Yadav et al. 2019b), and nitrogen storage (Jug et al. 2021).

Conservation tillage, commonly referred to as minimum tillage, is a widely defined practice that includes no-till, strip-till, ridge-till, and mulch-till systems. These techniques maintain plant residues on at least 30% of the soil's surface after tillage operations. When compared to conventional approaches, minimum tillage techniques can reduce tillage passes by 40% or more (Babu et al. 2023a). Over the last few decades, conservation tillage has gained popularity worldwide due to its capacity to reduce production costs, maintain soil moisture, and protect soils from erosion and compaction. Crop residue that is left on the soil's surface all year round can serve as a cover, reducing water and wind erosion as well as runoff, particulate matter, and nutrient loss (Yadav et al. 2018; Babu et al. 2023b). The air and water quality are enhanced as a result. Lowering tillage can increase biological activity, enhance soil aggregation, and increase infiltration and water-holding rates (Yadav et al. 2021a). Because conservation tillage eliminates the need for as many field visits for seedbed preparation, it saves a substantial amount of money on fuel, labor, and machinery (Lithourgidis et al. 2009; Singh et al. 2021a).

15.4.2 Integrated Pest Management

The concept of integrated plant protection, which emphasizes ecological considerations in chemical pest management, emerged as a ground-breaking innovation in the middle of the twentieth century. The eight guiding principles of integrated pest management (IPM)—prevention and suppression, monitoring, decision-making based on thresholds and monitoring, nonchemical methods of control, pesticide selection, decreased use, anti-resistance strategies, and evaluation—were organized along with the key features of IPM as the toolkit grew over time (Bažok 2022). IPM, or integrated pest management, provides an illustration of how crop security has evolved during the preceding 60 years.

Three categories are commonly used to describe modern agroecology: science, practice, and movement. The first is the essential feature, the scientific approach that made the other two possible to begin with, even though the latter two now play important and varied roles in different parts of the world. Over this time, IPM has addressed many environmental, health, and socioeconomic concerns as well as advanced sustainable agriculture practices and drastically decreased the usage of synthetic pesticides. However, the global usage of pesticides has mostly continued unabated, having a negative impact on biodiversity protection, farmer livelihoods, and everyone's right to food (Deguine et al. 2021).

Ecological plant protection and IPM set preferences of sustainability, e.g., the use of mechanical or biological protection methods and lowering the rate of agrochemical protection, but they have failed to establish transparent sustainability requirements that are easy to comprehend by general consumers. Contrarily, ecological (organic) agriculture was able to create very explicit regulations (a total prohibition on synthetic pesticides), which is reflected in consumers' growing preference for organic produce. However, this approach was unable to demonstrate that the health benefits of organic produce are attributable to the absence of those pesticide residues. Conversely, the established agroecological objectives represent a significant current ecological approach presence. Looking back, it is especially striking to see how integrated pest management (IPM), sustainable agriculture, and the three facets of agroecology have all developed since the ecological projects were founded in the late 1950s (Székács and Darvas 2022).

15.4.3 On-farm Mechanization

According to Mrema et al. (2014), a sustainable agricultural mechanization strategy is a planning approach that promotes social benefits, inclusive economic growth, and food self-sufficiency while also advancing the objectives of sustainable agriculture. It is difficult to maintain natural resources while simultaneously increasing food production. The Green Revolution, which produced more than twice as much food in the latter half of the twentieth century, is not doing well right now. The principal crop yields (corn, rice, and wheat) are growing at a slower rate. The development of new mechanization technology will be necessary to increase food production through resource-friendly approaches (Sims and Kienzle 2015). Rosegrant et al. (2014) found that although there are several technological choices, there is comparatively little evidence to support using technology to support decision-making. The low degree of engineering technology in agriculture, however, is one of the main obstacles to developing and modernizing production in developing nations (Fonteh 2010).

Mechanization is still viewed as an input, much like other inputs like fertilizers, seeds, and chemical protection products, despite its many benefits and undeniable values, which are briefly mentioned above. In the best-case scenario, however, it is one of the combined management tools aimed at maximizing farmers' profits and productivity (Clarke 2000). Even though agricultural mechanization has a positive impact on the process of agricultural production, there have always been disagreements and skepticism regarding the role of mechanization in the agricultural industry because there has not been enough thorough research done on the full impact of mechanization components on production and its fields of application.

This has stopped agricultural mechanization in the agricultural sector—as well as in the industrial and service sectors in developing nations—from stabilizing and improving. Research indicates that the benefits of crop production are greater in regions with access to agricultural mechanization (Haruna and Junior 2013). Research shows that yields in mechanized versus non-mechanized agricultural areas were much higher. Additionally, fertilizers and land were used as efficiently as possible, and pesticide application was more successful. This is the necessity of the globe today, establishing food security by safeguarding natural resources for future generations (Emami et al. 2018).

15.4.4 Maintaining Soil Health

In order to turn current agriculture into sustainable agriculture, agroecological approaches for sustainable soil management explain how to stop the degradation of soil quality, irrigation water quality, reuse of wastewater in agriculture after treatment, organic fertigation, and corporate fertigation. It also presents affordable technologies for the sustainable development of site-specific ecosystems (Prasad and Kumar 2023).

Agroecological approaches for sustainable soil management can address the following topics: (1) leading advances in precision agriculture, organic farming, climate-resilient forage and fodder cultivation, and animal management for the regeneration and utilization of degraded areas; (2) the development of a variety of products, such as biofuel, fiber, fodder, timber, and herbal products resulted in the creation of social capitals; (3) the promotion of corporate fertigation, agro-forestry-based apiculture, silviculture, and sericulture; and (4) the reclamation of urban brownfields and industrial areas; and (5) ecology of intercropping systems, tree-cover dynamics of grazing pastures, and cover crops for soil management.

Soil health is one of the 13 HLPE principles of agroecology, and it is reflected in three of the ten FAO agroecology elements: variety, synergy, and resilience (Wezel et al. 2020). According to estimates, 25% of croplands globally, regardless of environmental conditions, have degraded soils, or soils with poor quality chemical and physical features and poor health (Le et al. 2016). Two core pillars of agroecology recycling and biodiversity, as well as related practices using organic amendments and agroforestry, are crucial to sustaining and enhancing soil health (Yadav et al. 2021c; Singh et al. 2021b). On deteriorated soils, mineral fertilizers usually do not perform well (Nezomba et al. 2015). To protect or promote soil health, organic resources must be applied or retained in place (Laub et al. 2023).

Crop residue is the most abundant and easily obtainable organic resource, and preserving it in croplands can have a significant influence. However, due to competing needs for fuel, animal feed, and construction, the amounts of agricultural waste returned to cropland are usually negligible (Valbuena et al. 2012). In this case, manure is typically the most critical resource to protect the health of the soil; nonetheless, smallholders may find the needed application rates to be prohibitive. However, by combining better herd management practices (such as cow housing) and improved manure storage techniques (such as heap covering), there is a large possibility to increase the biomass and nutrients being recycled through dung (Rufino et al. 2007).

Therefore, to maintain soil organic matter, better recycling is required. In soils with insufficient cation exchange and pH buffering capabilities, the application of manure helps mitigate the acidity of the soil that results from frequent N fertilizer applications. Perennial agroforestry plants fix more carbon in the soil than annuals do, they can draw nutrients from deeper soil layers, and they can produce biomass and litter for longer periods, all of which contribute to an increase in the quantity of organic carbon supplied to the soil when integrated into cropping systems. But because of stoichiometric relationships in plant biomass, organic matter, and microbial biomass, this can also lead to nutrient cycles (e.g., N, P) that are more closely related to the C cycle and less leaky (Lemaire et al. 2014; Falconnier et al. 2023).

15.4.5 Legumes, Cover Crops, and Catch Crops in Rotations

Several management strategies have been put into practice and researched to increase the sustainability of cropping systems. Growing cover crops is one of the most promising methods recommended to stop soil degradation (Alaoui et al. 2020). A cover crop can be defined as a crop produced primarily to cover the soil to protect it from soil erosion and nutrient losses between times of crop production (Sarrantonio and Gallandt 2003). Depending on the growing conditions, legume cover crops such as red clover, crimson clover, vetch, peas, and beans can fix a significant amount of nitrogen (N) for succeeding crops, often between 50 and 150 pounds per acre.

After a legume, you can normally cut back on your nitrogen fertilizer inputs; nevertheless, legumes are not particularly excellent at scavenging nitrogen that remains after your cash crops. Nonlegume cover crops include cereals (rye, wheat, barley, oats, triticale), forage grasses (annual ryegrass), and broadleaf species (buckwheat, mustards, and brassicas, including the forage radish). Benefits and uses of cover crops, including reduction of water runoff and soil erosion, the addition of organic matter, improved soil structure and tilth, addition, and recycling of nitrogen, greater soil productivity, and weed, pest, and disease control, have been previously documented and reviewed (Blackshaw et al. 2005).

By absorbing and immobilizing water-soluble nutrients in their roots and aerial portions, they help prevent groundwater pollution and lessen the chance of nutrient leaking during the rainy season (Gabriel et al. 2012). Green manure-growing crops can be categorized as a specific kind of cover crop. To nourish the soil, fresh plant material is explicitly referred to as "green manures." Therefore, the main purpose of growing green manure crops is to assimilate them as organic matter into the soil while they are still fresh and green. Compared to conventional crop rotations cover crops, and green manures typically provide larger organic matter inputs, which improve soil fertility and structure (Fageria 2007; Yadav et al. 2023).

Additionally, they can directly compete with weeds, suppressing their growth throughout the fallow season and, consequently, lowering seed bank deposits (Büchi et al. 2020). Incorporating flowering cover crops into the cropping system has the potential to decrease plant diseases and draw in more beneficial insects. In addition, cover crops can promote water retention and have a positive impact on the microbial community relevant for nutrient cycling (Brennan and Acosta-Martinez 2017). In addition to causing notable alterations in the characteristics of the soil microbial community and a rise in microbial biomass and activity (Liu et al. 2010), green manures also alter microbial communities in ways that set them apart from other forms of organic matter amendments (Hollister et al. 2013). The degree of development and plant density of cover crops determines their ability to deliver agroecological services. To efficiently provide agroecological services and enable sustainable intensification of cropping systems, cover crop species adaptation to local conditions is a crucial and essential component (Boulet et al. 2021).

Smother crops are produced between cash crops, much like catch crops and green manures, but their primary function is to suffocate weeds via intense competition from both roots and foliage. They are mostly used to control weeds that are extremely difficult. To guarantee that they can properly compete with the weed, the smother crop species must be matched to certain weeds. Red clover, *Trifolium pratense* L., grows higher than *E. repens* and has dense, thick foliage that inhibits the light from reaching the *E. repens* leaves, making it an excellent smother crop. As a legume with the ability to fix nitrogen on its own, red clover has an advantage over couch grass in situations when the soil lacks nitrogen. Therefore, contrasted to biological/ecological measures that can improve soil health, smothering crops provide a particularly clear example of how physical and chemical approaches can harm the soil (Bandyopadhyay et al. 2023).

15.4.6 Precision Technologies for Irrigation and Nutrient Use Efficiency

Rather than distributing inputs such as insecticides, fertilizers, and herbicides evenly throughout the field, precision farming involves managing any agricultural issue for a particular area on a site-specific basis. By recognizing the geographical and temporal variability within and across fields, precision farming aids in the optimization of inputs such as water and fertilizers by crop requirements. Higher nutrient use efficiency can be achieved than with conventional practices thanks to precision nutrient management technologies such as site-specific nutrient management options, sensors for estimating soil and crop nutrient content and estimating the spatial variability of nutrients, and mapping for variable rate fertilizer application. Similarly, precision water management such as sensors for moisture content determination and irrigation scheduling increases water use efficiency (Tripathi et al. 2021).

Fertilizer use is essential for guaranteed crop production success. However, farmers' fields with indiscriminate fertilizer use yield higher agricultural yields at the expense of environmental sustainability. The notion of sustainability will mean nothing if we do not increase the efficiency with which we use our resources. For obtaining sustained system productivity, management modifications have to be implemented in individual farmers' fields. Given the current situation, when resources are becoming scarcer and fertilizer prices are rising, we must turn to more affordable alternatives.

Nano fertilizer is a potential emerging area, but its use is limited by factors like cost, lack of legal regulations, and accessibility. Modern techniques for improving efficient nutrient management include the use of bio-stimulants, models like

QUEFTS for yield monitoring, precision farming for accurate and need-based application of nutrients, and precision farming. It is necessary to encourage the integration of traditional and contemporary tools and approaches. In an integrated nutrient management approach, the use of mineral fertilizers and organic sources is equally important. Utilizing the potential of native microorganisms, soil fertility is increased, and reliance on imported mineral fertilizers is decreased. Nutrient utilization efficiency is further enhanced by the application of fertilizers in conjunction with irrigation water.

Another strategy that is becoming more and more popular is conservation agriculture, which sustainably improves soil health and fertility. In addition, the broad use of precision nutrient management approaches utilizing aerial imagery, chlorophyll meters, optical sensors, and leaf color charts is being encouraged in the contemporary agricultural environment. A deeper understanding of cutting-edge technologies is required for improved nutrition management. Although enhanced knowledge is crucial for the enhancement of the current management scenario, a fundamental problem is to make such technologies user accessible, inexpensive, and attractive for adoption by all levels of farmers (Srinivasarao et al. 2021).

15.4.7 Selection of High-Yielding Varieties

India's agricultural sector needs to boost crop production per unit if it wants to compete with the rest of the world. This necessitates the creation and manufacturing of seeds with higher yields, disease resistance, susceptibility to insect pest attack, and ability to tolerate harsh weather conditions. The most fundamental ingredient for achieving steady growth in agricultural output is seed. The quality of high-yielding variety seeds (HYV seeds) is higher than that of regular-quality seeds. Some people refer to them as "miracle seeds." Compared to regular seeds, these yield a little bit more product. Conventional seed has a low yield. High-yielding varieties of paddy and wheat: Without these, the Sixties Green Revolution would never have happened. Regretfully, our current crop yields pale in comparison to those of other nations. This has significantly decreased our overall output.

To ensure a robust and abundant harvest, HYV seeds are a superior choice. Because of their strong immune systems, these seeds can fend off illnesses and insects. These seeds also require relatively little maintenance in the form of irrigation. A particular focus of the new agricultural policy is the creation and broad implementation of high-yielding seed types. The government of India maintained a target of 70 million hectares for the area covered by HYV in its Seventh Plan. In contrast, after the Seventh Plan, the actual area covered by HYV was only 63.1 million hectares. In 1998–99, the coverage rose to 78.4 million hectares. The Centre and State governments, as well as registered seed farmers, fostered the production of improved seeds, particularly high-yielding kinds. Even though the private sector has begun to play a major role in the production and distribution of seeds, especially after the New Seed Policy of 1988 was introduced, the public sector still holds a dominant position in the organized seed sector, especially for food crops and cereals. As far as the distribution of certified/quality seeds is concerned, it grew from 25 lakh quintals in 1980–81 to 277.3 lakh quintals in 2010–11 (UCBMSH 2024).

When 75–90% of staple foods are produced and consumed locally, sustainable intensification of agriculture is a good alternative to prevent localized chronic food and nutrition insecurity. The System of Rice Intensification (SRI) has become a viable substitute for traditional rice cultivation techniques. SRI produces noticeably better yields while utilizing less water and more effectively the water that is available. Thus, the creation of decent work, sustainable agriculture, and global food and nutrition security all depend on the full potential of smallholders, including women farmers. High-yielding varieties are indeed innovative in terms of replacing lost land, conserving water, requiring less labor, and creating jobs; yet, they are also extremely fragile and sensitive and need careful handling to reap a good crop (Jagranjosh 2024).

15.4.8 Animal–Crop Integration

Livestock systems are a big worldwide asset contributing to food security and poverty alleviation, yet livestock supply chains have major environmental implications at a global scale (Godde et al. 2021). Large and medium-sized operations are showing a revived interest in integrated crop–livestock systems (ICLS), which are diverse cropping systems that reintegrate animal and crop industries (Ryschawy et al. 2017). They are essential to ecological intensification techniques, which seek to substitute services from improved ecosystem functioning for some of the anthropogenic inputs (Bommarco et al. 2013; Peterson et al. 2020).

Integration of livestock in a crop production system intensifies crop yield and soil quality by leveraging synergies between multiple agroecosystem components and the added complexity due to trophic level (i.e., grazing animals) rather than relying on the increased conventional input levels such as fertilizers (Peterson et al. 2019). ICLS may increase overall farm profitability (Hanson and Franzluebbers 2008), reduce weed and pest, improve nutrient cycling as a result of rotating livestock, and prevent the accumulation of harmful nutrient levels from animal manure, in addition to facilitating diversification that benefits the environment and soil health (Berry et al. 2003).

The biological, physical, and chemical characteristics of soil can be influenced by integrated crop–livestock systems (Deiss et al. 2019). Improved stable aggregate fraction is linked to higher total nitrogen (N), carbon (C) sequestration, soil microbial biomass C, and a decrease in penetration resistance, as demonstrated by ICLS experiments (Tobin et al. 2020). Crop–livestock integration seeks three aims: minimizing the openness of nutrient cycles, following the rationale of industrial ecology, organizing land use and agricultural techniques to promote ecosystem services, and boosting farm resilience to adverse climatic and economic shocks. Two types of integration—co-location vs. segregation and synchronization vs. rotation—between crops, grasslands, and animals can be controlled throughout time and space to accomplish these goals (Bell and Moore 2012). Crop–livestock integration extends beyond the farm level and encourages the adoption of cover crops, legumes, and short-duration grasslands. It begins with the cropping systems of specialized cash crop farms. These modifications support the C, N, and P cycles' recoupling. By incorporating grassland areas into cropping landscapes, preserving semi-natural grasslands in livestock systems, and integrating these areas (in time and space) within the current simplified arrangement of crops (configurational heterogeneity), integrating crops and livestock beyond the farm level allows increasing heterogeneity of both dimensions (Martin et al. 2016).

Increased nutrient cycling (Nunes de Albuquerque et al. 2021), higher microbial functional diversity (Sekaran et al. 2021), and improved soil structure and organic matter indicate increased biophysical protection to less favorable ecological conditions in cattle grazed systems. These and other ecosystem processes need to be further evaluated and compared with native grassland systems (Peterson et al. 2018). On the other hand, if the goal is to use ICLS as a tool for sustainable growth, it is critical to assess how effectively it contributes to both long-term stability and a productive system (Bansal et al. 2022). According to Martin et al. (2016), three types of crop–livestock integration occur outside of farms: complementarity, synergy, and local cohabitation. These types of integration have varying degrees of organizational, temporal, and spatial coordination among farms. Crop–farm diversity may encourage the establishment of aesthetically pleasing outdoor environments and more palatable agricultural practices.

15.5 Conclusion

To feed a growing world population, we need practices that provide sufficient food that are not detrimental or risky to the environment and that assure economic viability for farmers. Here, agroecological cropping practices can and should play a central role. Agroecological practices such as fertilizers, organic fertilization, natural pesticides, crop choice and rotations, intercropping and relay intercropping, agroforestry, and allelopathic plants have, so far, a low integration in today's agriculture. They have only low or medium potential to be more broadly implemented in the next decade for sustainable intensification. In contrast agroecological practices, viz., conservation tillage; legumes, cover crops, and catch crops in rotations; integrated pest management; soil conservation; on-farm mechanization, smart, precision technologies for irrigation and nutrient use efficiency; and high-yielding cultivar choice, animal–crop integration have already medium or high integration levels in today's agriculture and medium or high potential for the future sustainable intensification.

References

Alaoui A, Barão L, Ferreira C, Schwilch G, Basch G, Garcia-Orenes F et al (2020) Visual assessment of the impact of agricultural management practices on soil quality. Agron Soils Environ Qual 112(4):2608–2623. https://doi.org/10.1002/agj2.20216

Altieri MA (1995) Agroecology: the science of sustainable agriculture. Westview Press, Boulder

Babu S, Singh R, Avasthe RK, Kumar S, Rathore SS, Singh VK, Ansari MA, Valente D, Petrosillo I (2023a) Soil carbon dynamics under organic farming: impact of tillage and cropping diversity. Ecol Indic 147:109940. https://doi.org/10.1016/j.ecolind.2023.109940

Babu S, Singh R, Avasthe RK, Rathore SS, Kumar S, Das A, Layek J, Sharma V, Wani OA, Singh VK (2023b) Conservation tillage and diversified cropping enhance system productivity and ecoefficiency and reduce greenhouse gas intensity in organic farming. Front Sustain Food Syst 7:1114617. https://doi.org/10.3389/fsufs.2023.1114617

Bandyopadhyay KK, Acharya CL, Hati KM (2023) Mulches and cover crops part I: types. In: Goss MJ, Oliver M (eds) Encyclopedia of soils in the environment, 2nd edn. Academic, Amsterdam. ISBN 9780323951333

Bansal S, Chakraborty P, Kumar S (2022) Crop–livestock integration enhanced soil aggregate-associated carbon and nitrogen, and phospholipid fatty acid. Sci Rep 12:2781. https://doi.org/10.1038/s41598-022-06560-6

Bažok R (2022) Integrated pest management of field crops. Agriculture 12:425. https://doi.org/10.3390/agriculture12030425

Bell LW, Moore AD (2012) Integrated crop-livestock systems in Australian agriculture: trends, drivers and implications. Agric Syst 111:1–12. https://doi.org/10.1016/j.agsy.2012.04.003

Berry PM, Stockdale EA, Sylvester-Bradley, R, Philipps L, Smith KA, Lord EI, Watson CA, Fortune S (2003) N, P and K budgets for crop rotations on nine organic farms in the UK. Soil Use Management 19(2):112–118

Blackshaw RE, Moyer JR, Huang HC, Pandalai SG (2005) Beneficial effects of cover crops on soil health and crop improvement. Recent Res Dev Soil Sci 1:15–35

Bommarco R, Kleijn D, Potts SG (2013) Ecological intensification: harnessing ecosystem services for food security. Trends Ecol Evol 28(4):230–238. https://doi.org/10.1016/j.tree.2012.10.012. PMID: 23153724

Boulet AK, Carlos A, Carla F, Zahra K, Adelcia V, Lara C, Antonio F, Rudi H (2021) Agro-ecological services delivered by legume cover crops grown in succession with grain corn crops in the Mediterranean region open. Agriculture 6(1):609–626. https://doi.org/10.1515/opag-2021-0041

Brennan EB, Acosta-Martinez V (2017) Cover cropping frequency is the main driver of soil microbial changes during six years of organic vegetable production. Soil Biol Biochem 109:188–204. https://doi.org/10.1016/j.soilbio.2017.01.014

Büchi L, Wendling M, Amossé C, Jeangros B, Charles R (2020) Cover crops to secure weed control strategies in a maize crop with reduced tillage. Field Crop Res 247:107583. https://doi.org/10.1016/j.fcr.2019.107583

Clarke L (2000) Strategise for agricultural mechanization development. Agricultural Support System Division, FAO, Rome

Collins ED, Chandrasekaran K (2012) A wolf in Sheep's clothing? An analysis of the 'sustainable intensification' of agriculture. Friends of the Earth International, Amsterdam

Dalgaard T, Hutchings NJ, Porter JR (2003) Agroecology, scaling and interdisciplinarity. Agric Ecosyst Environ 100:39–51

Deguine JP, Aubertot JN, Flor RJ et al (2021) Integrated pest management: good intentions, hard realities. A review. Agron Sustain Dev 41:38. https://doi.org/10.1007/s13593-021-00689-w

Deiss L et al (2019) Soil chemical properties under no-tillage as affected by agricultural trophic complexity. Eur J Soil Sci 71:1090–1105

Emami M, Almassi M, Bakhoda H et al (2018) Agricultural mechanization, a key to food security in developing countries: strategy formulating for Iran. Agric & Food Secur 7:24. https://doi.org/10.1186/s40066-018-0176-2

Fageria NK (2007) Green manuring in crop production. J Plant Nutr 30:691–719. https://doi.org/10.1080/00103620500303939

Falconnier GN, Cardinael R, Corbeels M, Baudron F, Chivenge P, Couëdel A, Ripoche A, Affholder F, Naudin K, Benaillon E, Rusinamhodzi L, Leroux L, Vanlauwe B, Giller KE (2023) The input reduction principle of agroecology is wrong when it comes to mineral fertilizer use in sub-Saharan Africa. Outlook Agric 52(3):311–326. https://doi.org/10.1177/00307270231199795

FAO (2016) Achieving sustainable rural development through agricultural innovation: COAG/2016/6. FAO, Rome. http://www.fao.org/3/amr236e.pdf

Fonteh MF (2010) Agricultural mechanization in Mali and Ghana: strategies, experiences, and lessons for sustained impacts. FAO, Rome

Gabriel JL, Muñoz-Carpena R, Quemada M (2012) The role of cover crops in irrigated systems: water balance, nitrate leaching and soil mineral nitrogen accumulation. Agric Ecosyst Environ 155:50–61. https://doi.org/10.1016/j.agee.2012.03.021

Godde CM, Mason-D'Croz D, Mayberry DE, Thornton PK, Herrero M (2021) Impacts of climate change on the livestock food supply chain; a review of the evidence. Glob Food Sec 28:100488. https://doi.org/10.1016/j.gfs.2020.100488. ISSN 2211-9124

Gurr GM, Wratten SD, Altieri MA (2004) Ecological engineering for pest management. Advances in habitat manipulation for arthropods. CSIRO, Collingwood

Hanson JD, Franzluebbers A (2008) Principles of integrated agricultural systems. Renew Agric Food Syst 23(04):263–264

Haruna IM, Junior JA (2013) Mechanization practice: a tool for agricultural development in Nigeria: a case study of Ifelodun local government area of Kwara state. Int J Basic Appl Sci 2(3):98–106

Hirzel J, Undurraga P, Panichini LLM, Carrasco J, González J, Matus I (2020) Different residues affect wheat nutritional composition. J Soil Sci Plant Nutr 20:75–82

Hollister EB, Hu P, Wang AS, Hons FM, Gentry TJ (2013) Differential impacts of brassicaceous and non brassicaceous oilseed meals on soil bacterial and fungal communities. FEMS Microbiol Ecol 83:632–641. https://doi.org/10.1111/1574-6941.12020

IPCC (2019) Climate change and land. IPCC, Intergovernmental Panel on Climate Change, Cambridge

Jagranjosh (2024) Jagran Josh: education news, complete career guide for board exams, colleges and jobs. https://www.jagranjosh.com/general-knowledge/. Accessed 16 Jan 2024

Jug D, Đurdevi'c B, Birkás M, Brozovi'c B, Lipiec J, Vukadinovi'c V, Jug I (2019) Effect of conservation tillage on crop productivity and nitrogen use efficiency. Soil Tillage Res 194:1–15

Jug I, Brozović B, Đurđević B, Wilczewski E, Vukadinović V, Stipešević B, Jug D (2021) Response of crops to conservation tillage and nitrogen fertilization under different agroecological conditions. Agronomy 11(11):2156. https://doi.org/10.3390/agronomy11112156

Lal R (2007) Carbon management in agricultural soils. Mitig Adapt Strateg Glob Chang 12:303–322

Lampkin NH, Pearce BD, Leake AR, Creissen H, Gerrard CL, Girling R, Lloyd S, Padel S, Smith J, Smith LG, Vieweger A, Wolfe MS (2015) The role of agroecology in sustainable intensification. Report for the Land Use Policy Group. Organic Research Centre, Elm Farm and Game & Wildlife Conservation Trust

Laub M, Corbeels M, Couëdel A et al (2023) Managing soil organic carbon in tropical agroecosystems: evidence from four long-term experiments in Kenya. Soil 9(1):301–323

Le QB, Nkonya E, Mirzabaev A (2016) Biomass productivity-based mapping of global land degradation hotspots. In: Nkonya E, Mirzabaev A, von Braun J (eds) Economics of land degradation and improvement – a global assessment for sustainable development. Springer International Publishing, Cham, pp 55–84

Lemaire G, Franzluebbers A, Carvalho PCF et al (2014) Integrated crop–livestock systems: strategies to achieve synergy between agricultural production and environmental quality. Agric Ecosyst Environ 190:4–8
Lithourgidis AS, Damalas CA, Eleftherohorinos IG (2009) Conservation tillage: a promising perspective for sustainable agriculture in Greece. J Sustain Agric 33(1):85–95. https://doi.org/10.1080/10440040802587280
Liu G, Li Z, Jing H, Ye X, Shi H, Wang Y et al (2010) Effects of consecutive turnover of green manures on soil microbial biomass and enzyme activity. Plant Nutr Fert Sci 16(6):1472–1478
Lu X, Lu X, Liao Y (2018) Conservation tillage increases carbon sequestration of winter wheat-summer maize farmland on Loess Plateau in China. PLoS One 13:1–16
Martin G, Moraine M, Ryschawy J et al (2016) Crop–livestock integration beyond the farm level: a review. Agron Sustain Dev 36:53. https://doi.org/10.1007/s13593-016-0390-x
Méndez VE, Bacon CM, Cohen R (2013) Agroecology as a transdisciplinary, participatory, and action-oriented approach. Agroecol Sustain Food Syst 37(1):3–18
Mrema G, Soni P, Rolle RS (2014) A regional strategy for sustainable agricultural mechanization: sustainable mechanization across agri-food chains in Asia and the Pacifc region. FAO, Bangkok
Nezomba H, Mtambanengwe F, Tittonell P et al (2015) Point of no return? Rehabilitating degraded soils for increased crop productivity on smallholder farms in eastern Zimbabwe. Geoderma 239–240:143–155
Nunes de Albuquerque PA et al (2021) Livestock integration into soybean systems improves long-term system stability and profits without compromising crop yields. Sci Rep 11(1):1649
Peterson CA, Eviner VT, Gaudin ACM (2018) Ways forward for resilience research in agroecosystems. Agric Syst 162:19–27
Peterson CA et al (2019) Winter grazing does not affect soybean yield despite lower soil water content in a subtropical crop–livestock system. Agron Sustain Dev 39(2):1–10
Peterson CA, Deiss L, Gaudin ACM (2020) Commercial integrated crop-livestock systems achieve comparable crop yields to specialized production systems: a meta-analysis. PLoS One 15(5):e0231840. https://doi.org/10.1371/journal.pone.0231840
Phipps RH, Park JR (2002) Environmental benefits of genetically modified crops: global and European perspectives on their ability to reduce pesticide use. J Anim Feed Sci 11:1–18
Prasad MNV, Kumar C (eds) (2023) Agroecological approaches for sustainable soil management. Wiley, New York. ISBN 978-1-119-91198-2
Pretty JN, Toulmin C, Williams S (2011) Sustainable intensification in African agriculture. Int J Agric Sustain 9(1):5–24. https://doi.org/10.3763/ijas.2010.0583
Rosegrant MW, Koo J, Cenacchi N, Ringler C, Robertson R, Fisher M, Sabbagh P (2014) Food security in a world of natural resource scarcity, the role of agricultural technologies. IFPRI, Washington, DC
Rufino MC, Tittonell P, van Wijk MT et al (2007) Manure as a key resource within smallholder farming systems: analysing farm-scale nutrient cycling efficiencies with the NUANCES framework. Livest Sci 112(3):273–287
Ryschawy J, Martin G, Moraine M, Duru M, Therond O (2017) Designing crop–livestock integration at different levels: toward new agroecological models? Nutr Cycl Agroecosyst 108(1):5–20
Sarrantonio M, Gallandt E (2003) The role of cover crops in North American cropping systems. J Crop Prod 8(1/2):53–74. https://doi.org/10.1300/J144v08n01_04
Sekaran U, Kumar S, Luis GHJ (2021) Integration of crop and livestock enhanced soil biochemical properties and microbial community structure. Geoderma 381:114686
Sims B, Kienzle J (2015) Mechanization of conservation agriculture for smallholders: issues and options for sustainable intensification. Environments 2:139–166
Sinclair F, Wezel A, Mbow C, Chomba S, Robiglio V, Harrison R (2019) The contribution of agroecological approaches to realizing climate-resilient agriculture. Global Commission on Adaptation, Rotterdam/Washington, DC

Singh R, Babu S, Avasthe RK, Meena RS, Yadav GS, Das A, Mohapatra KP, Rathore SS, Kumar A, Singh C (2021a) Conservation tillage and organic nutrients management improve soil properties, productivity, and economics of a maize-vegetable pea system in the Eastern Himalayas. Land Degrad Dev:1–18. https://doi.org/10.1002/ldr.4066

Singh R, Babu S, Avsthe RK, Yadav GS, Das A, Mohapatra KP, Kumar A, Singh VK, Chandra P (2021b) Crop productivity, soil health, and energy dynamics of Indian Himalayan intensified organic maize-based systems. Int Soil Water Conserv Res 9(2):260–270. https://doi.org/10.1016/j.iswcr.2020.11.003

Srinivasan A (2006) Handbook of precision agriculture: principles and applications. Haworth Press, New York

Srinivasarao C, Balakrishnan M, Krishnan P, Sumanth Kumar VV (2021) Agricultural research, technology and policy: innovations and advances. ICAR – National Academy of Agricultural Research Management (NAARM), Hyderabad, p442

Székács A, Darvas B (2022) Attempts for undoing the ecological incompatibility of agricultural technologies: from ecological pest management to agroecology. Ecocycles 8(2):12–22. https://doi.org/10.19040/ecocycles.v8i2.222

Tobin C et al (2020) Demonstrating short-term impacts of grazing and cover crops on soil health and economic benefits in an integrated crop-livestock system in South Dakota. Open J Soil Sci 10(03):109–136

Tripathi R, Kumar A, Guru P, Debnath M, Mohapatra SD, Mohanty S, Khanam R, Shahid M, Nayak AK (2021) Precision farming technologies for water and nutrient management in rice: challenges and opportunities. Oryza 58:126–142

UCBMSH (2024) Uttaranchal College of Biomedical Sciences & Hospital. https://www.ucbmsh.org/colleges-in-dehradun/. Accessed 11 Jan 2024

United Nations (2009) World population prospects: the 2008 revision. Population division of the department of Economic and social affairs of the United Nations Secretariat. http://esa.un.org/unpp. Accessed Mar 2011

Uphoff N (2002) Agroecological innovations. Increasing food production with participatory development. Earthsan, London

Valbuena D, Erenstein O, Homann-Kee Tui S et al (2012) Conservation agriculture in mixed crop-livestock systems: scoping crop residue trade-offs in Sub-Saharan Africa and South Asia. Field Crop Res 132:175–184

Wezel A, Bellon S, Doré T, Francis C, Vallod D, David C (2009) Agroecology as a science, a movement or a practice. A review. Agron Sustain Dev 29:503–515. https://doi.org/10.1051/agro/2009004

Wezel A, Casagrande M, Celette F, Vian JF, Ferrer A et al (2014) Agroecological practices for sustainable agriculture-A review. Agron Sustain Dev 34(1):1–20. https://doi.org/10.1007/s13593-013-0180-7ff. final-01234800

Wezel A, Soboksa G, McClelland S et al (2015) The blurred boundaries of ecological, sustainable, and agroecological intensification: a review. Agron Sustain Dev 35:1283–1295. https://doi.org/10.1007/s13593-015-0333-y

Wezel A, Herren BG, Kerr RB et al (2020) Agroecological principles and elements and their implications for transitioning to sustainable food systems-a review. Agron Sustain Dev 40:40. https://doi.org/10.1007/s13593-020-00646-z

Xolocotzi HE (1977) Agroecosistemas de México: Contribuciones a La Enseñanza, Investigación y Divulgación Agrícola. Colegio de Postgraduados, Chapingo

Yadav GS, Das A, Lal R, Subhash B, Meena RS, Saha P, Singh R, Datta M (2018) Energy budget and carbon footprint in a no-till and mulch-based rice–mustard cropping system. J Clean Prod 191:144–157

Yadav GS, Subhash B, Das A, Bhowmik SN, Datta M, Singh R (2019a) Soil carbon dynamics and productivity of rice–rice system under conservation tillage in submerged and unsubmerged ecologies of eastern Indian Himalaya. Carbon Manag 10(1):51–62. https://doi.org/10.1080/17583004.2018.1545518

Yadav GS, Das A, Lal R, Babu S, Datta M, Meena RS, Patil SB, Singh R (2019b) Impact of no-till and mulching on soil carbon sequestration under rice (*Oryza sativa* L.)-rapeseed (*Brassica campestris* L. Var. Rapeseed) cropping system in hilly agro-ecosystem of the Eastern Himalayas, India. Agric Ecosyst Environ 275:81–92

Yadav GS, Subhash B, Das A, Mohapatra KP, Singh R, Avasthe RK, Roy S (2020a) No till and mulching enhance energy use efficiency and reduce carbon footprint of a direct-seeded upland rice production system. J Clean Prod. https://doi.org/10.1016/j.jclepro.122700

Yadav GS, Das A, Subhash B, Mohapatra KP, Lal R, Rajkhowa DJ (2020b) Potential of conservation tillage and altered land configuration to improve soil properties, carbon sequestration and productivity of maize-based cropping system in eastern Himalayas, India. Int Soil Water Conserv Res. https://doi.org/10.1016/j.iswcr.2020.12.003

Yadav GS, Babu S, Das A, Datta M, Mohapatra KP, Singh R, Singh VK, Rathore SS, Chakraborty M (2021a) Productivity, soil health, and carbon management index of Indian Himalayan intensified maize-based cropping systems under live mulch-based conservation tillage practices. Field Crop Res 264:108080. https://doi.org/10.1016/j.fcr.2021.108080

Yadav GS, Das A, Kandpal BK, Babu S, Lal R, Datta M, Das B, Singh R, Singh VK, Mohapatra KP, Chakraborty M (2021b) The food-energy-water-carbon nexus in a maize-maize-mustard cropping sequence of the Indian Himalayas: an impact of tillage-cum-live mulching. Renew Sust Energ Rev 151:111602. https://doi.org/10.1016/j.rser.2021.111602

Yadav GS, Kandpal BK, Das A, Babu S, Mohapatra KP, Devi G, Chandra P, Singh R, Barman KK (2021c) Impact of 28-year-old agroforestry systems on soil carbon dynamics in Eastern Himalayas. J Environ Manag 283:111978. https://doi.org/10.1016/j.jenvman.2021.111978

Yadav D, Singh D, Subhash B, Madegowda M, Singh DV, Mandal D, Rathore AC, Sharma VK, Singhal V, Kumawat A, Yadav DK, Yadav RK, Kumar S (2023) Intensified cropping reduces soil erosion and improves rainfall partitioning and soil properties in the marginal land of the Indian Himalayas. Int Soil Water Conserv Res. https://doi.org/10.1016/j.iswcr.2023.10.002

Zhang SX, Li Q, Zhang XP, Wei K, Chen LJ, Liang WJ (2012) Effects of conservation tillage on soil aggregation and aggregate binding agents in black soil of Northeast China. Soil Tillage Res 124:196–202

Chapter 16
Solar Energy for Farm Diversification: Implications in Sustainable Food Production

Santosh S. Mali, Akram Ahmed, Ved Prakash, and Anup Das

Abstract The energy sector is facing sustainability challenges along with increasing environmental concerns. Globally, the energy sector is transitioning from fossil fuel and electricity-based sources to cleaner and greener alternatives, such as solar energy. Implementing solar energy in agriculture not only reduces the carbon footprint but also provides a reliable power source, especially in remote or off-grid locations. Farm diversification is considered a key strategy for building resilience of the agricultural systems to changing climatic conditions and enhancing the income of farming communities worldwide. However, the diversification of farms through the inclusion of a variety of enterprises is energy-intensive. Solar energy, in both direct and indirect forms, can be harnessed to meet the specific energy requirements of these enterprises, leading to reduced dependence on conventional energy sources. Solar-powered applications like cooling, heating, and electricity production have shown enormous potential in diversifying Indian farms, making it feasible to include poultry, piggery, fishery, and dairy as part of mainstream agriculture. Adoption of greener solar energy will greatly facilitate the sprouting of small-scale agro-industries, fostering employment generation and poverty alleviation. Renewable sources can not only reduce the dependence on conventional fuels like coal, oil, and natural gas, resources, which are depleting at an alarming rate, but also decrease global greenhouse gas emissions. To fulfill the future requirements of modern sustainable agriculture, researchers, governments, and other stakeholders should focus on the development of innovative, affordable, and effective solar-powered technologies for the development of agriculture.

Keywords Agri-photovoltaic · Renewable energy · Solar energy · Water pump

---

S. S. Mali (✉)
Farming Systems Research Center for Hill and Plateau Region, ICAR-Research Complex for Eastern Region, Ranchi, Jharkhand, India

A. Ahmed · V. Prakash · A. Das
ICAR-Research Complex for Eastern Region, Patna, Bihar, India

S. Babu et al. (eds.), *Agricultural Diversification for Sustainable Food Production*, Sustainability Sciences in Asia and Africa,
https://doi.org/10.1007/978-981-97-7517-0_16

16.1 Introduction

Water, energy, and food are fundamental to sustaining life on Earth. There are global concerns over the sustainability of these resources in the context of increasing demands and limited accessibility due to rapid population growth (FAO 2019). UN projections indicate that the global population will rise from 8.5 billion in 2030 to 9.7 billion by 2050, reaching approximately 10.4 billion in the 2080s (Nations Unis 2022). As the population expands and industry grows, the need for alternative and innovative methods of food and energy production becomes increasingly urgent (Amaducci et al. 2018). Due to population and economic growth, global energy demand is anticipated to double by mid-century and renewable, environment-friendly energy will be vital in meeting the rising demands (Rodman and Meentemeyer 2006). Among all renewable energies, solar energy is the fastest-growing, most abundant, and most accessible source (Rogner 2012).

Farm diversification involves expanding agricultural activities into innovations or branching out into non-agricultural business ventures. Diversification plays a crucial role in the structural transformation of an agrarian economy. Certain subsectors within the agriculture sector are increasingly assuming greater importance than traditional crop production. Factors promoting diversification and the pace at which it occurs depend on a variety of technical, social, economic, and environmental determinants. Among others, energy is one of the critical factors that drive the diversification of farms. Currently, farming businesses are encountering growing energy challenges, alongside concerns about agriculture's environmental impact in terms of resources and emissions (Barnes et al. 2018; Loizou et al. 2019; Morris and Bowen 2020).

Given the context that many of the farm enterprises are energy intensive and that the production, productivity, and efficiency of farms are directly linked to the on-farm energy availability, energy remains a critical input in farm diversification. The agriculture sector's energy demand has risen sharply to support energy-intensive technologies in agricultural production systems. This underscores the need for alternative energy sources to reduce reliance on fossil fuels and conventional energy. Advancements in solar technology with simultaneous reductions in the prices, particularly of solar panels, have opened enormous opportunities for not only shifting the nexus into a virtuous cycle but also transforming rural livelihoods. By neglecting farm diversification and relying solely on conventional nonrenewable energy sources, we risk missing a significant opportunity to boost farm income, enhance climate resilience in agriculture, and reduce rural India's dependence on agriculture.

Adoption of solar energy-based enterprises and solutions can revolutionize the agricultural sector through water conservation, bringing energy self-sufficiency, and saving power costs in the long run. Since, farm diversification is one strategy that emphasizes the entrepreneurial side of farmers (Vik and McElwee 2011), solar-based interventions enable the farmers to set up new solar energy-based start-ups in farming. Solar energy suitably provides valuable opportunities for diversification of production activities in response to current farming challenges and uncertainties.

This chapter identifies the solar energy-based alternatives to diversify the farms venturing into the adoption of sustainable energy.

16.2 Solar Energy: The Indian Scenario

Solar energy is increasingly recognized as a dependable and promising power source for various agricultural operations. Although large-scale adoption of solar in Indian agricultural systems started a few years back, there have been many promising advancements over time, both in terms of solar-based technologies and their adoption. Over the years, India has been a hub for numerous solar-based technologies and has emerged as a key player in the worldwide solar transformation. In the past 5 years, India installed 48,445 MW of additional solar capacity (Fig. 16.1) (MNRE 2023). Setting ambitious targets, India is positioning itself to establish 500 GW of renewable energy capacity by 2030. The focus is on the declining costs of solar technologies alongside advancements in energy storage solutions and grid infrastructure.

The government of India is implementing several solar-based schemes and policies to progress towards a sustainable energy future. The Production Linked Incentive Scheme (PLI), with an allocation of INR 24,000 crore, aims to stimulate domestic manufacturing and strengthen India's position as a leader in solar energy. Among the Indian states, the state of Gujarat has made more progress in the development of solar energy. At Charaka Village, the state has officially opened the largest solar park in Asia with a capacity of 500 MW. Gujarat is second to Rajasthan in terms of solar energy production and is moving in the direction of development. The biggest solar farm in the world with a capacity of 2.245 GW, Bhadla Solar Park, is

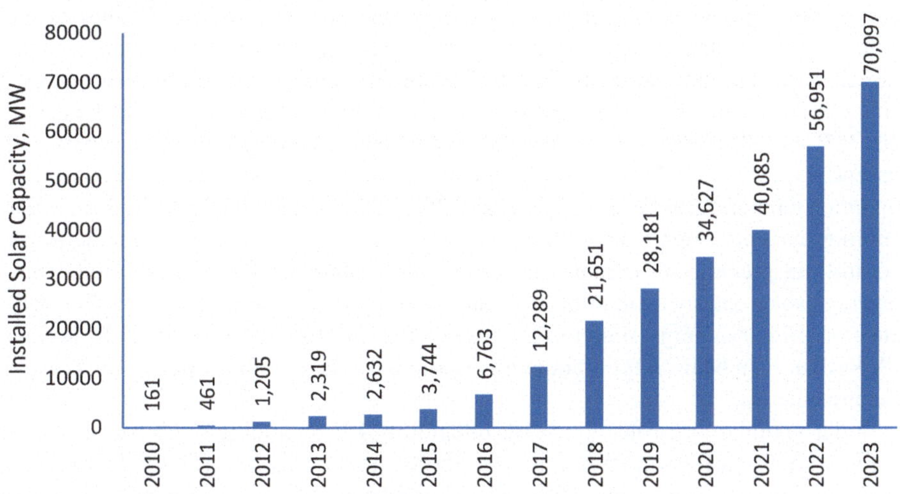

Fig. 16.1 Timeline of installed solar capacity in India

in Rajasthan's Jodhpur and has a spatial coverage of 14,000 acres. The Rajasthan state is leading in total solar energy production. This state is producing the amount of 2245.0 MWp. The park has currently achieved a solar power generation capacity of 1614 MW, which is 71.89% of its planned capacity.

16.3 Solar-Powered Smart Greenhouses

Solarization of agriculture can mitigate the effects of climate change to some extent. Among other solar-based agricultural interventions, solar-powered greenhouses offer a sustainable solution to diversify farm enterprises. Protected cultivation structures like greenhouses offer great potential for off-season cultivation, wherein farmers get the advantage of higher prices through off-season production of good-quality agricultural produce. Simply put, a solar energy greenhouse is a structure that has solar panels installed to harvest the sunlight and transform it into heat. The trapped heat is then used to provide the right environmental conditions required for optimal plant growth, even in unfavorable outside climatic conditions. The addition of solar panels on the roof of a solar energy greenhouse can assist plants in regulating the process of photosynthesis.

Solar-powered greenhouses offer various advantages in terms of greater energy use efficiency and lower production expenses. Solar-powered greenhouses can be designed to harvest and store solar energy in the form of heat for use during nights or cloudy days. Small growers prefer passive solar greenhouses because they are economical and effective in extending the crop-growing season (Acosta-Silva et al. 2019). Active solar greenhouses require an additional energy source to transfer solar-heated air or water to different parts of the greenhouse. (Ron and Shapiro 1981).

In greenhouse-based production, heating and cooling systems are needed. In greenhouse operations, winter heating accounts for 70% of production costs. Adopting renewable energy-based systems effectively reduces reliance on conventional energy (Mahmoudi et al. 2008). Customizing greenhouse designs to suit specific climates can enhance production efficiency while lowering costs (Beyhan et al. 2013; Bot et al. 2005). Researchers have innovated a side vent controller powered independently by a 0.78 m^2 a-Si PV module, ensuring precise ventilation control to maintain temperatures between 20 °C and 25 °C (Yano et al. 2007). Moreover, Solar cooling systems offer a viable alternative to traditional evaporative methods for controlling greenhouse climates in regions with abundant solar irradiation. They leverage solar energy availability to match cooling demands, thereby reducing reliance on primary energy sources (Campiotti et al. 2016). These results indicate that PV systems can be used for maintaining optimal levels of climatic parameters inside the greenhouse.

Solar greenhouse projects involve integrating PV technologies above or within the greenhouse infrastructure. This type of setup serves to provide partial shade and regulate light exposure, enhancing agricultural productivity while generating solar energy (Cossu et al. 2020; Marucci and Cappuccini 2016). PV modules on greenhouse rooftops and walls create a shade that can hamper the growing pattern of plants, which can be overcome using bifacial PV modules (Mohsenipour et al. 2020; Yusaf et al. 2011). Furthermore, PVT modules increase efficacies in both heat and electricity, and the use of concentration modules reduces shading.

Solar panel options can range from transparent to semitransparent or opaque modules installed on or integrated into building structures. These installations can modulate light availability and quality throughout the day. Solar-powered greenhouses are highly energy-efficient and less reliant on fossil fuels, employing technologies like automated climate controls and LED lighting to reduce energy usage and greenhouse gas emissions, thus aiding in mitigating climate change. The running cost of solar-powered greenhouses is much less as compared to conventional greenhouses. Considering the carbon emissions and environmental impacts, solar greenhouses offer tremendous potential in preventing climate change. More technological advances and cost reductions are necessary for solar greenhouse technology to become an alternative to sustainable growth, even though it has advanced (Hassanien et al. 2016). Technical advancements in sensors and wireless communications systems have made it possible to use greenhouse parameters in precision agriculture in tracking and monitoring (Henten 2004; Arif and Abbas 2015).

16.4 Agri-photovoltaic Systems

Photovoltaics (PV) are crucial in combating climate change and achieving carbon neutrality and are the most widely adopted technology in renewable energy sectors (Hayat et al. 2019; Olabi and Abdelkareem 2022). However, a large land area is required for the installation of solar panels, which leads to a reduction in the croplands. To overcome this challenge, agrivoltaic systems (Agri-PV) are rapidly garnering attention that use solar energy for crop growth and electricity production through the installation of PV modules above the cultivation area. The substantial requirement of land area for solar panel installation can diminish available croplands. In this context, agrivoltaic systems (Agri-PV) are a potential alternative to harness solar energy for crop growth and electricity production by positioning PV modules above the cultivation area (Yoon et al. 2019; Katsikogiannis et al. 2022). Agri-PV is a cutting-edge smart-farm technology that boosts land use efficiency and sustainability through the integration of agriculture with energy production (Elamri et al. 2018).

16.4.1 Concept of Agri-PV

The Agri-PV concept was first proposed by Goetzberger and Zastrow in 1982. The proposed integration of energy production using solar panels with crops in open fields (Fig. 16.2). The concept of Agri-PV, or combining agriculture with solar

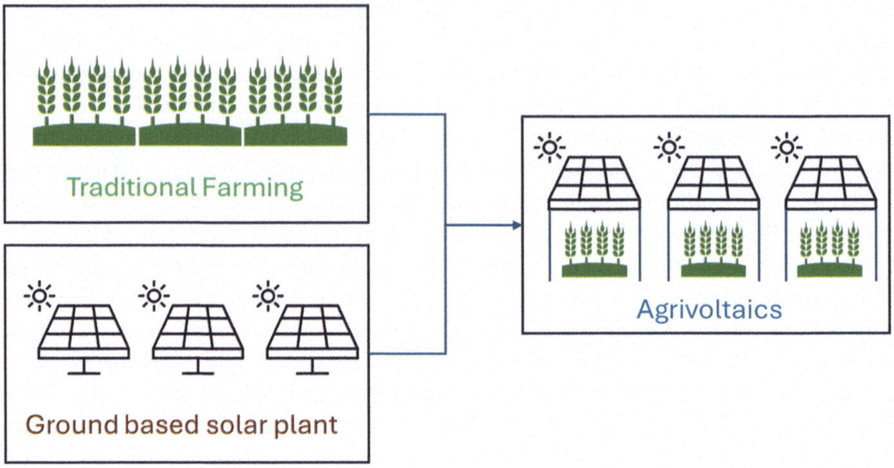

Fig. 16.2 Illustration of typical Agri-Photovoltaic system

power generation, is a relatively new field of research and practice. It attempts to solve multiple problems at once by increasing renewable energy production, increasing sustainable food production, and preserving land and water resources. Agri-PV aims to maximize land productivity and offers potential benefits including reduced irrigation costs, increased crop yields, preservation of agricultural land, and enhanced socio-economic welfare for farmers (Weselek et al. 2019). In an Agri-PV system, the solar panels are mounted on structures or racks elevated above the ground, allowing sunlight to pass through and reach the crops grown below (Goetzberger and Zastrow 1982). This arrangement provides dual benefits: harvesting renewable energy from the sun while utilizing the land for agriculture.

Agri-PV systems are gaining global attention as solar photovoltaic energy expands, necessitating effective solutions for integrating landscapes that minimize ecological impact and benefit local communities (Barron-Gafford et al. 2019). The key idea behind Agri-PV is to optimize land use efficiency and resource utilization. The solar panels provide shade to the crops, reducing evaporation and creating a microclimate that can be advantageous for certain crops, particularly in hot and arid regions. Additionally, the panels help protect crops from extreme weather conditions such as hail or heavy rainfall. Agrivoltaics involves sharing sunlight between photovoltaic and photosynthesis systems. Agricultural activities encompass meeting human needs for food, fiber, and fuel, alongside efforts to improve environmental quality and sustain natural resources crucial for agricultural economies (USDA 2007).

16.4.2 Categories of AVS

Agri-PV systems can be configured in various ways based on specific farming activities (Fig. 16.3). These systems allow for simultaneous multiple on the same site. Projects can be designed to have specific zones where activities like crop production or ecosystem service production occur. An alternative configuration for these systems involves using shared zones for activities such as targeted grazing and pollinator rearing.

Agri-PV-based production systems integrate annual or perennial crops. The partial shade from PV panels influences crop growth, allowing crops to be cultivated directly beneath and between rows of PV arrays (Mamun et al. 2022; Weselek et al. 2021). These systems can be adapted to ground-mounted installations by adjusting PV panel height, spacing, row arrangements, and other design elements to optimize sunlight and compatibility with farming practices (Riaz et al. 2021; Tahir and Butt 2022). Ongoing research explores innovative technologies and configurations to support diverse crops such as orchards, vineyards, field crops, and fruits (Gorjian et al. 2022). Research so far has shown that all types of crops can be cultivated under an Agri-PV system, although yields may vary due to the shading effect. Shade-tolerant crops like leafy vegetables, forage species, and fruit crops are particularly suitable for such systems.

Agri-PV systems having the component of animal grazing directly below, around, and in the vicinity of solar panels are known as animal-based Agri-PV

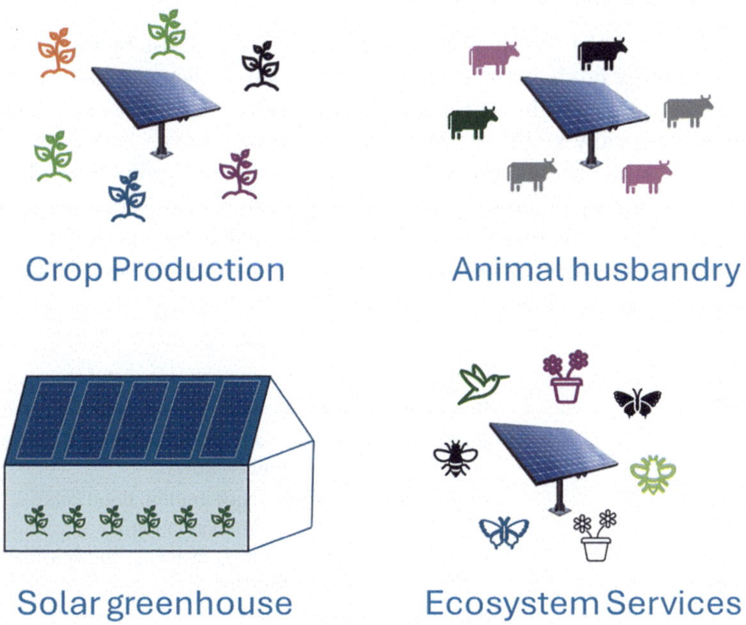

Fig. 16.3 The different types of Agrivoltaic systems

systems. Previous research has successfully integrated sheep, cattle, poultry, honeybees, and rabbits into the Agri-PV systems (Sharpe et al. 2021; Maia et al. 2020). Animals may need additional nutrition from off-site sources and can be shifted to new locations periodically. There's a growing trend of integrating solar infrastructure with aquaculture, though not all countries recognize this as agrivoltaics (Vo et al. 2021).

Agri-PV systems are valued for creating or restoring habitats, improving soil, and providing essential ecosystem services, including supporting, provisioning, regulating, and sociocultural services (Walston et al. 2021). Ecosystem services offered by Agri-PV systems are summarized in Table 16.1. The Agri-PV sites reserved for ecosystem services typically need minimal modification of traditional solar designs, but raising panel heights to allow beneficial vegetation growth is crucial for maximizing their potential to provide ecosystem services.

16.4.3 Agri-PV in Sustainable Agriculture and Energy Production

An Agri-PV system integrates photovoltaics with agriculture, offering a promising strategy to meet global renewable energy and sustainability demands. Agri-PV represents a sustainable land use practice that combines renewable energy generation with ecologically responsible agriculture. This aligns with broader goals of sustainability, climate mitigation, and conservation.

Simultaneous Land Use Agrivoltaic systems optimize land use by integrating agriculture with energy production. This means that a single piece of land can generate both food and energy. By diversifying income streams, agrivoltaics can make farming more economically viable. Farmers can earn revenue from both crop sales and the sale of renewable energy, potentially increasing overall profitability. In regions where land availability is limited, combining agriculture with solar energy production mitigates the competition between land designated for agriculture and land designated for solar farms. This alleviates the need to choose one land use over the other.

Table 16.1 Examples of ecosystem services

Category	Ecosystem services examples
Supporting	Soil formation, photosynthesis, biodiversity
Provisioning	Clean water, food, and other useful plants (e.g., timber, fiber)
Regulating	Carbon sequestration, flood control, temperature regulation
Cultural	Recreational, aesthetic, and spiritual benefits (e.g., tourism)

Source: USDA (2007)

Enhanced Land Productivity By strategically placing solar panels, agrivoltaic systems can enhance land productivity through several mechanisms. Solar panels are raised above the ground, allowing sunlight to filter through to the crops below. This controlled shading can create an optimal microclimate for plant growth. Different crops have varying light requirements, and agrivoltaics can be designed to meet these specific needs. Solar panels provide a degree of shading, which helps regulate temperature extremes. This can be particularly beneficial in regions with harsh climates. By mitigating the effects of extreme heat or cold, agrivoltaic systems can extend growing seasons and improve overall crop yields. The shading provided by solar panels can reduce water evaporation from the soil. This means that less irrigation water is required to maintain soil moisture levels. This is especially crucial in areas where water availability is a limiting factor for agriculture. Some crops are better suited for agrivoltaic systems due to their shade tolerance. By selecting crops that thrive in the partial shade provided by solar panels, farmers can diversify their agricultural output and potentially increase overall productivity.

Economic Viability The economic viability of AVS depends on various factors, including location, crop selection, system design, and local energy markets. Here are some key considerations regarding the economic viability of agrivoltaic systems. Agrivoltaic systems offer multiple revenue streams. In addition to income from crop sales, farmers can generate revenue from the sale of renewable energy. This diversification of income can improve the overall economic viability of the land. If the farm consumes electricity, using on-site solar energy from the agrivoltaic system can reduce or even eliminate the need to purchase electricity from the grid. This can lead to significant cost savings over time. In regions with favorable policies, agrivoltaic system owners may be eligible for net metering or feed-in tariffs. Net metering allows excess electricity to be sold back to the grid, while feed-in tariffs offer fixed payments for renewable energy production. The upfront capital costs of installing an agrivoltaic system can be a significant investment. However, government incentives, grants, and subsidies may be available to offset some of these costs.

Environmental Benefits Agrivoltaic systems offer a range of environmental benefits that contribute to a more sustainable and resilient future. By generating renewable energy from sunlight, agrivoltaic systems help reduce reliance on fossil fuels for electricity generation. This leads to a decrease in greenhouse gas emissions, which contribute to global climate change. Agrivoltaic systems do not produce air pollutants or emissions during operation. This contrasts with traditional fossil fuel-based power generation, which can release harmful pollutants into the atmosphere, such as sulfur dioxide, nitrogen oxides, and particulate matter. Agrivoltaics allow for the dual use of land, meaning that a single plot can simultaneously support agriculture and renewable energy production. This can reduce the need for converting additional land into solar farms or agricultural fields, helping to conserve natural habitats and ecosystems. Agrivoltaic systems can enhance the resilience of agricultural operations to the impacts of climate change. By providing a degree of protection from extreme weather events, such as hail or heavy rainfall, these systems can help maintain crop productivity in the face of increasingly unpredictable weather patterns. Farm diversification opportunities offered by Agri-PV systems depend on local biophysical and climatic conditions. Studies have shown that Agri-PV systems are financially beneficial and offer low-stress environments for farmers. Agri-PV installations can enhance productivity and income alongside traditional farming, unaffected by uncertainties in unfavorable economic and climatic conditions. This is especially relevant in densely populated industrial countries, where expanding renewable energy is crucial, yet preserving productive farmlands is equally important.

Overall, agrivoltaic systems offer a promising solution for addressing land-use conflicts and advancing sustainable agriculture practices. Through thoughtful planning and ongoing research, agrivoltaics have the potential to transition towards more sustainable and integrated approaches to energy and food production. In conclusion, agrivoltaic systems represent an innovative and sustainable approach to land use, combining agricultural production with solar energy generation. This integrated system offers a range of potential benefits, but it also requires careful planning, management, and monitoring to ensure its success.

16.4.4 Photovoltaic Water Pumping Systems for Irrigation

Assured irrigation supplies play a pivotal role in the diversification of farming. Optimal availability of water during critical times of the year paves the way for adopting newer farm enterprises, enhancing the income of farming communities. The increasing global demand for food production, coupled with the challenges posed by climate change and depleting fossil fuel resources, has prompted a shift towards sustainable and renewable energy-based irrigation pumping solutions in agriculture. Among these solutions, photovoltaic (PV) irrigation systems stand out as a promising avenue for the efficient use of solar energy in agriculture. Here we explore the implications of integrating solar energy into farming practices, specifically through the deployment of photovoltaic water pumping systems. Traditional irrigation methods often depend on nonrenewable energy sources, contributing to environmental degradation and escalating operational costs for farmers. The inclusion of solar energy in agriculture addresses these challenges by providing a sustainable alternative. Photovoltaic water pumping systems convert sunlight into electrical energy, offering an eco-friendly solution for powering irrigation systems while reducing dependence on conventional energy sources.

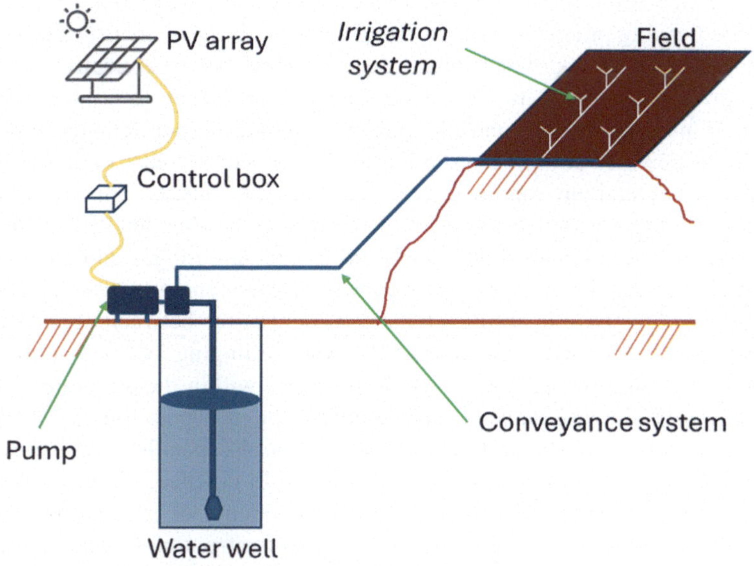

Fig. 16.4 Components of solar-powered water pumping system

16.4.4.1 Components of Photovoltaic Water Pumping Systems

PV water pumping systems consist of solar panels, a pump, and a water storage facility. A generalized schematic diagram of the components of the solar-powered water pumping system is shown in Fig. 16.4. Solar panels harness sunlight and transform it into electricity, which powers the pump to lift water from a source (such as a well or a reservoir) to irrigate crops. The system operates efficiently with minimal maintenance requirements, making it a viable option for rural and remote farming communities.

16.4.4.2 Advantages of PV Water Pumping Systems

Energy Independence Harnessing solar power in agriculture offers farmers a pathway to energy independence, marking a departure from traditional dependence on grid electricity or diesel generators. This transition holds multifaceted advantages for farmers and their operations. Solar energy serves as a dependable and consistent power source, especially in regions where access to a stable power grid is challenging. The intermittent nature of power supply in such areas can disrupt critical farming activities, affecting irrigation schedules, automated machinery, and other essential processes. Solar power systems, however, provide a reliable and continuous source of energy, mitigating the challenges posed by inconsistent power supply and contributing to the overall resilience of farming operations (Pascaris et al. 2021).

Environmental Sustainability PV water pumping systems are pivotal for advancing sustainable farming practices through the generation of clean energy. Unlike traditional energy sources like fossil fuels, these systems produce electricity without emitting greenhouse gases, making a significant contribution to reducing agriculture's environmental impact and aligning with global climate change mitigation efforts. Solar-powered pumping systems allow farmers to decrease their dependence on nonrenewable and environmentally harmful energy sources. This transition not only improves the ecological footprint of farming operations but also addresses a major contributor to climate change by avoiding greenhouse gas emissions associated with conventional energy. This move toward cleaner energy aligns with broader sustainability objectives, fostering a balanced coexistence between agriculture and the environment. Moreover, integrating PV water pumping systems showcases a dedication to environmentally responsible practices within the agricultural sector. As awareness of climate change expands globally, the incorporation of clean energy solutions not only boosts operational efficiency but also positions agriculture as a positive contributor to environmental stewardship. In essence, PV water pumping systems emerge as tangible and impactful components of sustainable farming, actively participating in the collective endeavor to confront and alleviate the challenges posed by global climate change. Kumar and Kandpal's (2007) focus on India emphasized the environmental benefits of CO_2 emissions mitigation, especially when substituting diesel or electricity. The study establishes the potential of PV pumps to contribute to reduced carbon emissions in the context of sustainable practices. Although these systems come with high installation costs, there are considerable savings, showcasing the environmental efficiency of solar-based pumping systems (Rezae et al. 2013). Technological and policy issues related to PV water pumping systems need to be addressed to ensure the environmental sustainability of photovoltaic pumping systems (Fedrizzi et al. 2009). Meah et al. (2008) underscored the environmental benefits of using PV pumping in drought-prone states of the USA. The analysis of 75 operational systems in Wyoming showcased excellent environmental performance, particularly in terms of reducing carbon emissions, highlighting the potential for sustainable water supply to remote livestock locations. Overall, these studies collectively emphasize the environmental advantages of employing PV water pumping systems across different geographical contexts.

Cost Savings Although the capital costs of solar irrigation systems appear significant, the enduring advantages encompass decreased energy costs and diminished maintenance expenses. Governments and institutions frequently offer incentives and subsidies to promote the embrace of solar technologies, alleviating the financial strain on farmers. Assessments on the economic viability of solar and diesel pumping systems highlighted the need for critical analysis to match the water demand and supply patterns (Odeh et al. 2006). Techno-economic analysis of solar water pumping systems for an academic institution in New Delhi showed a four-year payback period and an estimated 20-year lifespan, resulting in sixteen years of free water supply (Jamil et al. 2012). This highlights substantial cost savings and reduced grid dependency compared to conventional systems. Analysis of an experimental PV water pumping system in Greece, serving over 200 consumers daily, showed a reliable system operation with low electrical losses, up to 10%, emphasizing the environmental friendliness of PV water pumping applications (Kaldellis et al. 2011). Performance evaluation of PV water pumping systems in India indicated that, despite financial incentives, higher costs might make PV pumps financially unviable for potential users in India.

16.4.4.3 Implications of Solar Pumping Systems

Increased Crop Yield Reliable and efficient irrigation facilitated by photovoltaic water pumping systems enhances crop yield and quality. This is particularly crucial in regions prone to water scarcity, as solar-powered irrigation ensures a consistent and sufficient water supply for crops. Adoption of Solar-Powered Pumping Systems (SPPS) in tomato cultivation in Saudi Arabia showed a significant (25%) increase in tomato yield over conventional irrigation methods. Similarly, SPPS implementation in Pakistan led to a 13.3 % increase in cucumber yield compared to diesel powered drip irrigation system (Akram and Asif, 2020).

Diversification of Agricultural Activities Solar energy allows farmers to diversify their agricultural activities. Beyond crop cultivation, solar-powered irrigation supports aquaculture, livestock farming, and other income-generating ventures, contributing to the overall resilience and sustainability of farming communities.

Community Empowerment The adoption of photovoltaic water pumping systems empowers rural communities by providing them with the means to control their energy production. This not only fosters self-sufficiency but also strengthens the resilience of farming communities in the face of external challenges. Experiences in the implementation of community-based solar irrigation pumping systems showed that the collective approach was successful not only in empowering the farming communities but was also effective in the management and maintenance of the irrigation infrastructure created (Fig. 16.5).

Fig. 16.5 A community-based solar pumping system in Madhubani, Bihar

Photovoltaic water pumping systems for irrigation represent a transformative approach to sustainable agriculture, aligning with the broader goals of environmental conservation and resource efficiency. As the world grapples with the need to enhance food production while minimizing environmental impact, solar energy emerges as a key enabler for farm diversification and the promotion of sustainable food production practices. This section highlights the potential of photovoltaic water pumping systems as a cornerstone in the evolution toward a more resilient and sustainable agricultural future.

16.5 Diversification to Livestock Production

In the pursuit of sustainable and environmentally friendly farming practices, the integration of solar power in agriculture extends beyond crop cultivation to include livestock production. This section explores the implications of employing solar energy in various facets of livestock farming, highlighting the benefits and challenges associated with harnessing the sun's power for sustainable food production.

16.5.1 Solar Energy for Livestock Facilities

Off-Grid Power Supply Livestock facilities often operate in remote or rural areas where access to a reliable power grid may be limited. Solar power provides an off-grid solution, allowing farmers to electrify barns, sheds, and other facilities essential for the well-being of the animals. A case study in a dairy cattle barn in Bursa province in Turkey is presented, where the orientation of photovoltaic panels (North-South, East-West, Northeast-Southwest, Northwest-Southeast) is analyzed using PVSOL software. The results indicate that energy production exceeds the barn's annual requirement by 5–7 times, demonstrating the viability of solar energy for meeting the energy needs of agricultural facilities and reducing dependence on conventional sources (Delice and Yaslioğlu 2021).

Feeding and Watering Systems Solar energy can be harnessed to power automated feeding and watering systems, ensuring a consistent and efficient supply of food and water for livestock. This not only reduces manual labor but also enhances the overall management of the farm. Morris et al. (2002) discussed in detail solar-powered livestock-watering systems for Montana, America, covering aspects such as cost, components, and terminology. It also offers insights into the design and installation of these systems, comparing the strengths and weaknesses of solar pumping with alternative options like mechanical windmills, wind turbines, and portable generators.

16.5.2 Solar-Powered Fencing and Security

Electric Fencing Solar-powered electric fencing offers an effective and eco-friendly solution for securing pasture areas. This technology not only provides containment for livestock but also deters predators, enhancing the safety and well-being of the animals. Kadam et al. (2011) developed a solar power fence system that operates on solar energy. The average input-output energy from the solar panel was measured at 172 and 23 watts, respectively. The fence live wire maintains a voltage range of 2–11 kV, with pulsating current ranging from 0.005 to 0.008 Amp, featuring pulses lasting 0.0003 s spaced approximately 1.0 s apart. A 35 kW panel and 12 V battery effectively powered a 3.5 km fence line, but performance declines as sunlight hours decrease and the battery discharges, resulting in decreased fence wire voltage and reduced effectiveness in controlling animals.

Surveillance and Lighting Solar-powered surveillance cameras and lighting systems contribute to the security of livestock facilities. These systems not only discourage theft and unauthorized access but also aid in monitoring the health and behavior of the animals, enabling timely intervention when needed (Jabbar et al. 2023).

16.5.3 Impacts on Livestock Welfare

Temperature Control Livestock and dairy operations require efficient air and water heating to maintain optimal conditions for animal health and growth. Solar heaters integrated into farm building design can help mitigate the energy-intensive process of heating air by preheating incoming fresh air and improving temperature and air quality control. These solar systems can supplement natural ventilation during the summer, offering a sustainable and energy-efficient solution for agricultural facilities (Chikaire et al. 2010).

Water Heating Solar water heating systems provide warm water for cleaning and sterilization purposes, promoting hygiene in livestock farming. This contributes to disease prevention and overall improvement in animal health. Hosouli et al. (2023) developed a cost-effective renewable energy source solution tailored for the LVAT-ATB dairy farm in Germany. A solar photovoltaic thermal (PVT) system, utilizing heat recovery from milk coolers and enhancing an electric boiler's efficiency, was designed and evaluated using Solarus and DualSun collectors. The DualSun collector exhibited higher electricity output, while Solarus reached higher temperatures. The integration of an existing heat recovery system, coupled with a semiautomated approach, maximizes thermal energy storage, minimizing electricity consumption.

16.5.4 Economic and Environmental Implications

Cost Savings The adoption of solar power in livestock production can prove to be cost-effective over time. Reduced energy bills, government incentives, and the potential for additional income generation through excess energy production contribute to the economic viability of solar-powered livestock farms.

Environmental Stewardship Solar energy in livestock production aligns with the broader goals of environmental conservation. By reducing reliance on conventional energy sources, farmers can minimize their carbon footprint and contribute to sustainable agricultural practices.

The integration of solar power into livestock production represents a pivotal step in the diversification of farming practices toward sustainability. Beyond crop cultivation, solar energy plays a crucial role in enhancing the efficiency, cost-effectiveness, and environmental friendliness of livestock farming. This section underscores the potential of solar power to revolutionize the way we approach livestock production, promoting a more resilient and sustainable future for agriculture as a whole. As the agricultural sector continues to evolve, solar energy emerges as a cornerstone in the broader movement toward farm diversification and the promotion of sustainable food production practices.

16.6 Solar Applications in the Fisheries Sector

The fisheries sector, critical for global food security, is undergoing transformative changes driven by the need for sustainability and reduced environmental impact. This chapter explores the diverse applications of solar energy in the fisheries sector, highlighting its implications for enhancing operational efficiency, reducing costs, and contributing to sustainable food production.

16.6.1 Off-Grid Power Solutions for Aquaculture

Pond Aeration Systems Water aeration systems play a pivotal role in aquaculture by maintaining optimal oxygen levels for fish. Solar aeration systems are advantageous in remote areas where access to the power grid is limited, providing an off-grid solution for small-scale fish farmers. Sarma et al. (2018) investigated the impact of a solar aerator on pond water's dissolved oxygen (DO) concentration and the growth and survival of rohu (Labeo rohita). The aerator, powered by a 750 W DC induction motor and a 900 Wp solar array, increased the pond's overall oxygen reserve by 26.1% compared to the control. Significant improvements were observed in rohu's growth performance, including weight gain, percentage weight gain, percentage specific growth rate, and FCR, in the aerated pond compared to the control. This study concludes that solar-powered aerators efficiently enhance dissolved oxygen levels, leading to improved fish growth and survival.

Water Circulation and Filtration Solar pumps can be employed for water circulation and filtration in aquaculture facilities. By harnessing the sun's energy, farmers can ensure a consistent and clean water supply, creating favorable conditions for fish growth and minimizing the risk of diseases. A report was published by the Food and Agriculture Organization (FAO 2012), which discussed the use of solar pumps for water circulation and filtration in aquaculture facilities in developing countries. It provides case studies of successful solar-powered aquaculture systems in Africa, Asia, and Latin America.

Fish Vending Cart The solar-powered fish vending cart is a sustainable solution for mobile fish vendors. Using solar panels ensures fresh and high-quality fish by powering a refrigeration unit. This eco-friendly approach reduces operational costs and addresses power reliability issues, making it an innovative and economically viable solution for small-scale fish businesses. ICAR Research Complex for the Eastern Region has designed and developed a portable fish vending cart prototype. The cart is provided with two rear wheels with tires to handle additional load. The cart's layout included separate zones for storage of live and frozen fish, cutting and processing, water storage, and utility space. Aeration was facilitated by two solar panels having a capacity of 40 Wp used to charge the battery, while a simple chain drive served as the driving mechanism (ICAR-RCER Annual Report 2021).

16.6.2 Solar-Powered Fishing Boats

Electric Propulsion Solar-powered electric propulsion systems offer an eco-friendly alternative to traditional fuel-powered engines for fishing boats. This not only reduces the carbon footprint but also lowers operational costs, making it an economically viable solution for small-scale fisheries. Minak (2023) explore the feasibility of integrating a solar-powered propulsion system into multiday fishing boats. Since boat designs are flexible to adopt any shape and size of PV panels and boats require lower aerodynamic requirements, they can be adopted on wider scales (Minak 2023).

On-Board Energy Storage Solar panels on fishing boats can charge on-board energy storage systems, providing power for navigation, communication, and refrigeration. This promotes sustainable and efficient fishing practices by enabling longer journeys and the preservation of catch quality. Sultoni et al. (2020) introduced a PV/gasoline hybrid power system for fishing hulls in Indonesia, aiming to replace traditional gasoline propulsion with solar energy. The design incorporates eight 200 W PV modules on the ship's roof, combined with a 3 kW generator. An automatic switch selects PV or gasoline based on battery power levels. Tested on a 7 m × 2 m fishing hull, the hybrid system reaches a maximum speed of 8.1 km/h, consuming a maximum of 1425 Watts. Economic analysis reveals the hybrid system costs only one-tenth of a conventional gasoline engine, showcasing its potential as a cost-effective and sustainable alternative.

16.6.3 Solar Cold Storage Facilities

Off-Grid Refrigeration Solar energy can be harnessed to power off-grid cold storage facilities for preserving fish and seafood. This is particularly beneficial in remote coastal areas where reliable electricity supply is a challenge. Solar-powered cold storage helps reduce postharvest losses and ensures the freshness of the catch. Nagori et al. (2016) developed a small refrigeration prototype for chilling fresh fish using solar photovoltaic (PV) technology at the Central Institute of Fisheries Technology in Cochin. The system includes a cooling chamber made of an 18-gauge GI sheet, a condensing unit with a DC compressor and DC fan, and PV panels with batteries. The 30-liter chamber, shaped like an inverted frustum of a cone, utilizes copper cooling coils and thermocol/PUF insulation. Power is sourced from 250 W PV panels connected to two 12 V, 100 Ah batteries. Preliminary trials with 3 kg of Mackerel showed the cooling chamber temperature reducing to −2 °C from an ambient temperature of 32 °C within an hour. The fish core temperature dropped below 10 °C from 22 °C within 45–50 min, exhibiting minimal variation. The cooling chamber's temperature reduction was significant, reaching 0.2 °C after 50 min, showcasing improved fish quality and longer shelf life compared to traditional methods using crushed block ice.

Ice-Making Equipment Solar-powered ice-making machines contribute to the preservation of fish quality during transportation and storage. This is crucial for maintaining the value chain and ensuring that fish reach consumers in optimal condition, especially in areas where traditional ice-making facilities are scarce.

16.6.4 Economic and Environmental Benefits

Cost Savings Solar applications in the fisheries sector can lead to substantial cost savings in energy expenses, fuel costs, and postharvest losses. This is especially relevant for small-scale fishers and aquaculture operations, contributing to the economic sustainability of the sector.

Environmental Sustainability Solar energy in fisheries aligns with global efforts to reduce the environmental impact of fishing activities. By transitioning to solar-powered solutions, the sector can contribute to lower greenhouse gas emissions, reduced reliance on fossil fuels, and overall environmental stewardship.

The integration of solar energy applications in the fisheries sector presents a promising avenue for sustainable and resilient food production. From aquaculture to fishing operations and postharvest processing, solar solutions offer a range of benefits, including cost savings, reduced environmental impact, and increased energy independence. This section highlights the transformative potential of solar energy in diversifying and enhancing the sustainability of the fisheries sector, contributing to the broader goal of sustainable food production practices. As the world grapples with the challenges of feeding a growing population, solar energy emerges as a key ally in reshaping the future of fisheries towards a more sustainable and environmentally conscious industry.

16.7 Solar Energy-Based Initiatives in India

The government of India is taking giant strides to improve the technological advancements and mesoscale adoption of solar-based solutions in agriculture. In 2019, the government of India launched an ambitious scheme "Pradhan Mantri Kisan Urja Suraksha evam Utthaan Mahabhiyaan (PM KUSUM)," which aims to achieve an additional solar power capacity of 30.8 GW by March 2026. The scheme aims to enhance energy security for Indian farmers while supporting India's commitment to increase the share of non-fossil-fuel sources in electric power to 40% by 2030. The scheme consists of three components A, B, and C. Component "A" involves developing 10,000 MW of solar capacity through small solar power plants (<2 MW), Component "B" focuses on installing 20 lakh standalone solar-powered agriculture pumps, and Component "C" involves solarizing 15 lakh grid-connected agriculture pumps.

These large-scale schemes are transforming India's agriculture by establishing decentralized solar plants, replacing liquid fuel pumps with solar-powered pumps, and converting grid electricity pumps into solar-powered pump sets. Realizing that increased water availability on spatial and temporal scales can lead to farm diversification and improved livelihoods in the farming communities, PM-KUSUMs major focus is on improving irrigation infrastructure across the country. The scheme will provide farmers with a stable income, supporting farm diversification and enabling cultivation of fallow lands. The concept of Agri-PV is also being integrated into the scheme such that in cases where cultivated fields are chosen for setting up solar power projects, the farmers could continue to grow crops. This has been achieved by developing guidelines on minimum above-ground height of solar panels to ensure agriculture below the panels.

India's vision for 2030 comprises a mix of renewable energy targets, climate change mitigation, and energy transition. India aspires to adopt innovative approaches, advanced technologies, and profitable solar-based business models that promise secure, affordable, and sustainable energy consumption patterns. With a robust policy framework and the government's unwavering support, solar energy-based farm diversification can be an achievable target.

India's vision for 2030 includes renewable energy targets, climate change mitigation, and energy transition. India aims to embrace innovative approaches, advanced technologies, and profitable solar-based business models to ensure secure, affordable, and sustainable energy consumption. Backed by a robust policy framework and steadfast government support, the diversification of farms using solar energy is within the reach of common farmers.

16.8 Conclusion

Farm diversification through solar energy integration represents a forward-thinking approach to agriculture, offering multifaceted benefits. Solar energy integrated farming system is a sustainable and innovative farming approach, offering economic, social, and environmental benefits. Such integration also comes with a reduced carbon footprint and significantly contributes to combating climate change. Transitioning from conventional, non-environmentally friendly energy sources to cleaner alternatives reduces pollution, promoting a healthier and more sustainable environment. Socially, it supports rural communities with stable incomes, offers climate resilience to farming systems, and paves the way towards a more sustainable future for generations. Solar power's environmental benefits are essential for ensuring a sustainable future for our planet.

References

Acosta-Silva YJ, Torres-Pacheco I, Matsumoto Y, Toledano-Ayala M, Soto-Zarazúa GM, Zelaya-Ángel O, Méndez-López A (2019) Applications of solar and wind renewable energy in agriculture: a review. Sci Prog 102(2):127–140. https://doi.org/10.1177/0036850419832696. Epub 2019 Mar 28. PMID: 31829840; PMCID: PMC10424537

Akram MM, Asif M (2020) Suitability study of on-form solar system as an energy source for drip irrigation. Sci Lett 8:1–6

Amaducci S, Yin X, Colauzzi M (2018) Agrivoltaic systems to optimize land use for electric energy production. Appl Energy 220:545–561

Arif KI, Abbas HF (2015) Design and implementation a smart greenhouse. Int J Comput Sci Mob Comput 4(8):335–347

Barnes A, Thomson S, Ferreira J (2018) Disadvantage and economic viability: evidence from Scottish Upland farming (No. 2038-2018-2969)

Barron-Gafford GA, Pavao-Zuckerman MA, Minor RL, Sutter LF, Barnett-Moreno I, Blackett DT, Thompson M, Dimond K, Gerlak AK, Nabhan GP et al (2019) Agrivoltaics provide mutual benefits across the food–energy–water nexus in drylands. Nat Sustainability 2:848–855

Beyhan B, Paksoy H, Daşgan Y (2013) Root zone temperature control with thermal energy storage in phase change materials for soilless greenhouse applications. Energy Convers Manag 74:446–453

Bot G, Van De Braak N, Challa H, Hemming S, Rieswijk TH, Van Straten G, Verlodt I (2005) The solar greenhouse: state of the art in energy saving and sustainable energy supply. Acta Hortic 691(2):501

Campiotti CA, Morosinotto G, Puglisi G, Schettini E, Vox G (2016) Performance evaluation of a solar cooling plant applied for greenhouse thermal control. Agric Agric Sci Procedia 8:664–669. https://doi.org/10.1016/j.aaspro.2016.02.076
Chikaire J, Nnadi FN, Nwakwasi RN, Anyoha NO, Aja OO, Onoh PA, Nwachukwu CA (2010) Solar energy applications for agriculture. J Agric Vet Sci 2:58–62
Cossu M, Yano A, Solinas S, Deligios PA, Tiloca MT, Cossu A, Ledda L (2020) Agricultural sustainability estimation of the European photovoltaic greenhouses. Eur J Agron 118:126074
Delice H, Yaslioğlu E (2021) The effect of building orientation on utilization of solar energy in dairy cattle barns. Tekirdağ Ziraat Fakültesi Dergisi 18(3):419–427
Elamri Y, Cheviron B, Lopez J-M, Dejean C, Belaud G (2018) Water budget and crop modeling for agrivoltaic systems: application to irrigated lettuces. Agric Water Manag 208:440–453. https://doi.org/10.1016/j.agwat.2018.07.001
FAO (2012) Use of solar pumps for water circulation and filtration in aquaculture facilities. Food and Agriculture Organization of the United Nations, FAO, Rome
FAO (2019) Perspectives Agricoles de l'OCDE et de la FAO 2019–2028. OECD and Food and Agriculture Organization of the United Nations, Rome
Fedrizzi MC, Ribeiro FS, Zilles R (2009) Lessons from field experiences with photovoltaic pumping systems in traditional communities. Energy Sustain Dev 13(1):64–70
Goetzberger A, Zastrow A (1982) On the coexistence of solar-energy conversion and plant cultivation. Int J Sol Energy 1:55–69
Gorjian S, Bousi E, Özdemir ÖE, Trommsdorff M, Kumar NM, Anand A, Kant K, Chopra SS (2022) Progress and challenges of crop production and electricity generation in agrivoltaic systems using semi-transparent photovoltaic technology. Renew Sust Energ Rev 158:112126. https://doi.org/10.1016/j.rser.2022.112126
Hassanien RHE, Li M, Lin WD (2016) Advanced applications of solar energy in agricultural greenhouses. Renew Sust Energ Rev 54:989–1001
Hayat MB, Ali D, Monyake KC, Alagha L, Ahmed N (2019) Solar energy – a look into power generation, challenges, and a solar-powered future. Int J Energy Res 43:1049–1067. https://doi.org/10.1002/er.4252
Hosouli S, Gomes J, Loris A, Pazmiño IA, Naidoo A, Lennermo G, Mohammadi H (2023) Evaluation of a solar photovoltaic thermal (PVT) system in a dairy farm in Germany. Sol Energy Adv 3:100035
ICAR-RCER Annual Report (2021) ICAR Research Complex for Eastern Region, Patna-179p
Jabbar MA, Asif SM, Khan MA (2023) Use of solar-powered surveillance and lighting systems to improve livestock welfare and security. Pak J Agric Res 66(4):987–994
Jamil M, Anees AS, Rizwan M (2012) SPV-based water pumping system for an academic institution. Am J Electr Power Energy Syst 1(1):1–7
Kadam DM, Dange AR, Khambalkar VP (2011) Performance of solar power fencing system for agriculture. J Agric Technol 7(5):1199–1209
Kaldellis JK, Meidanis E, Zafirakis D (2011) Experimental energy analysis of a stand-alone photovoltaic-based water pumping installation. Appl Energy 88(12):4556–4562
Katsikogiannis OA, Ziar H, Isabella O (2022) Integration of bifacial photovoltaics in agrivoltaic systems: a synergistic design approach. Appl Energy 309:118475
Kumar A, Kandpal TC (2007) Potential and cost of CO_2 emissions mitigation by using solar photovoltaic pumps in India. Int J Sustainable Energy 26(3):159–166
Loizou E, Karelakis C, Galanopoulos K, Mattas K (2019) The role of agriculture as a development tool for a regional economy. Agric Syst 173:482–490
Mahmoudi H, Abdul-Wahab SA, Goosen MFA, Sablani SS, Perret J, Ouagued A, Spahis N (2008) Weather data and analysis of hybrid photovoltaic–wind power generation systems adapted to a seawater greenhouse desalination unit designed for arid coastal countries. Desalination 222(1–3):119–127

Maia ASC, de Andrade Culhari E, de França Carvalho Fonsêca V, Milan HFM, Gebremedhin KG (2020) Photovoltaic panels as shading resources for livestock. J Clean Prod 258:120551. https://doi.org/10.1016/j.jclepro.2020.120551

Mamun MA, Al PD, Wadley D, Zulkarnain NA, Aziz AA (2022) A review of research on agrivoltaic systems. Renew Sust Energ Rev 161:112351

Marucci A, Cappuccini A (2016) Dynamic photovoltaic greenhouse: energy balance in completely clear sky condition during the hot period. Energy 102:302–312. https://doi.org/10.1016/j.energy.2016.02.053

Meah K, Fletcher S, Ula S (2008) Solar photovoltaic water pumping for remote locations. Renewable and sustainable energy reviews, 12(2):472–487. https://doi.org/10.1016/j.rser.2006.10.008

Minak G (2023) Solar energy-powered boats: state of the art and perspectives. J Mar Sci Eng 11(8):1519. https://doi.org/10.3390/jmse11081519

MNRE (2023) Ministry of New and Renewable Energy. https://pib.gov.in/PressReleaseIframePage.aspx?PRID=1947140

Mohsenipour M, Ebadollahi M, Rostamzadeh H, Amidpour M (2020) Design and evaluation of a solar-based trigeneration system for a nearly zero energy greenhouse in arid region. J Clean Prod 254:119990

Morris W, Bowen R (2020) Renewable energy diversification: considerations for farm business resilience. J Rural Stud 80:380–390. https://doi.org/10.1016/j.jrurstud.2020.10.014

Morris M, Lynne V, Engineer NE, Lent C (2002) Solar-powered livestock watering system. National Sustainable Agriculture Information Service

Nagori A, Zynudheen AA, Ravishankar CN (2016) Solar PV-based chilled storage system for preservation of fresh fish. Fish Technol 53:86–88

Nations Unis (2022) La population Mondiale Atteindra Milliards le 15 Novembre 2022. Nations Unis, New York

Odeh I, Yohanis YG, Norton B (2006) The economic viability of photovoltaic water pumping systems. Sol Energy 80(7):850–860

Olabi AG, Abdelkareem MA (2022) Renewable energy and climate change. Renew Sustain Energy Rev 158:112111. https://doi.org/10.1016/j.rser.2022.112111

Pascaris AS, Schelly C, Burnham L, Pearce JM (2021) Integrating solar energy with agriculture: industry perspectives on the market, community, and socio-political dimensions of agrivoltaics. Energy Res Soc Sci 75:102023

Rezae A, Gholamian SA, Asghar Gholamian S (2013) Technical and financial analysis of photovoltaic water pumping system for GORGAN, Iran. Int J Cybern Inform 2(2):21–31

Riaz MH, Imran H, Younas R, Butt NZ (2021) The optimization of vertical bifacial photovoltaic farms for efficient agrivoltaic systems. Sol Energy 230:1004–1012. https://doi.org/10.1016/j.solener.2021.10.051

Rodman LC, Meentemeyer RK (2006) A geographic analysis of wind turbine placement in Northern California. Energy Policy 34:2137–2149

Rogner HH (2012) Towards a sustainable future, Chapter 7. In: Global energy assessment. Cambridge University Press, Cambridge, pp 423–512

Ron A, Shapiro A (1981) Low-cost passive solar greenhouses. National Center for Appropriate Technology, Butte. 173 p

Sarma K, Rahman A, Dey A (2018) Impact of solar operated aerator on dissolved oxygen and fish growth. J Exp Zool India 21(2):1041–1046

Sharpe KT, Heins BJ, Buchanan ES, Reese MH (2021) Evaluation of solar photovoltaic systems to shade cows in a pasture-based dairy herd. J Dairy Sci 104(3):2794–2806. https://doi.org/10.3168/jds.2020-18821

Sultoni HI, Ali MM, Aji ZP (2020) Design and fabrication of PV/gasoline hybrid drive propulsion for fishing hull. Int Rev Mech Eng 14. https://doi.org/10.15866/ireme.v14i1.1746125D

Tahir Z, Butt NZ (2022) Implications of spatial-temporal shading in agrivoltaics under fixed tilt & tracking bifacial photovoltaic panels. Renew Energy 190:167–176. https://doi.org/10.1016/j.renene.2022.03.078

USDA (2007) Sustainable agriculture: definitions and terms|Alternative Farming Systems Information Center|NAL|USDA

Van Henten E (2004) Greenhouse mechanization: state of the art and future perspective. In: International symposium on greenhouses, environmental controls and in-house mechanization for crop production in the tropics, vol 710, pp 55–70

Vik J, McElwee G (2011) Diversification and the entrepreneurial motivations of farmers in Norway. J Small Bus Manag 49:390–410

Vo TT, Em HK, Huh J-H, Park N (2021) Overview of solar energy for aquaculture: the potential and future trends. Energies 14(21):6923. https://doi.org/10.3390/en14216923

Walston LJ, Li Y, Hartmann HM, Macknick J, Hanson A, Nootenboom C, Lonsdorf E, Hellmann J (2021) Modeling the ecosystem services of native vegetation management Practices at solar energy facilities in the Midwestern United States. Ecosyst Serv 47:101227. https://doi.org/10.1016/j.ecoser.2020.101227

Weselek A, Ehmann A, Zikeli S, Lewandowski I, Schindele S, Högy P (2019) Agrophotovoltaic systems: applications, challenges, and opportunities. A review. Agron Sustain Dev 39(4):35

Weselek A, Bauerle A, Hartung J, Zikeli S, Lewandowski I, Högy P (2021) Agrivoltaic system impacts on microclimate and yield of different crops within an organic crop rotation in a temperate climate. Agron Sustain Dev 41:59

Yano A, Tsuchiya K, Nishi K, Moriyama T, Ide O (2007) Development of a greenhouse side-ventilation controller driven by photovoltaic energy. Biosyst Eng 96(4):633–641

Yoon C, Choi S, An K-N, Ryu J-H, Jeong H, Cho J (2019) Preliminary experiment of the change of insolation under solar panel mimic shading net. Korean J Agric Forest Meteorol 21:358–365. https://doi.org/10.5532/KJAFM.2019.21.4.358

Yusaf T, Goh S, Borserio J (2011) Potential of renewable energy alternatives in Australia. Renew Sust Energ Rev 15(5):2214–2221

Chapter 17
Integrated Crop Management Practices for Improving Productivity and Environmental Sustainability

Niraj Biswakarma, Badapmain Makdoh, N. C. Gulleibi, Jayanta Layek, Rahul Saikia, Somanath Nayak, Sandesh Rai, R. R. Zhiipao, Namrata Rajbonshi, Koushik Bag, and A. K. Singh

Abstract Intensive production system substantially improves crop productivity and sustains food/nutritional security. However, over the past decades, the climate change and associated problems have intensified the production and resource vulnerability in South Asian agro-ecologies. Consequently, there is an urgent requirement for climate-resilient and cost-effective practices to conserve available natural resources and enhance farm profitability. According to the Food and Agriculture Organization (FAO), integrated crop management (ICM) is a contemporary concept in agriculture aimed at long-term productivity and sustainability. It is a comprehensive approach that combines different farming practices, viz., crop selection, tillage, integrated weed and water management, nutrient management, disease-pest control, and energy management, to maximize crop production without any adverse effect on the environmental quality. Further, the ICM practices help in conserving farm energy by 5–15.6% and thus reduce the need for external inputs and greenhouse gas emissions and enhance crop yield by 10–14% compared to conventional practices. Beside this, ICM practices improve the soil quality through crop residue recycling,

N. Biswakarma · B. Makdoh · N. C. Gulleibi (✉) · J. Layek · R. Saikia · N. Rajbonshi · K. Bag
Agronomy, DSRE, ICAR Research Complex for NEH Region, Umiam, Meghalaya, India

S. Nayak
AICRP on MULLaRP, Central Agricultural University, Imphal, Manipur, India

S. Rai
Department of Agronomy, Uttar Banga Krishi Vishwavidyalaya, Pundibari, West Bengal, India

R. R. Zhiipao
Department of Agronomy, Rajiv Gandhi University, Papum Pare, Arunachal Pradesh, India

A. K. Singh
Department of Soil Science, SAS, Nagaland University, Medziphema, Nagaland, India

S. Babu et al. (eds.), *Agricultural Diversification for Sustainable Food Production*, Sustainability Sciences in Asia and Africa,
https://doi.org/10.1007/978-981-97-7517-0_17

and reduce crop-weed competition, along with insect–pest and disease management. Additionally, it considers postharvest management and horizontal diversification in agriculture for minimizing food loss and maximizing farm income. Studies conducted in India, China, and Nepal have shown a superiority of ICM practices in terms of crop yield, water use efficiency, economics, and soil quality. Thus, the adoption of ICM practices in conventional production system could be viable option for crop diversification, climate resilience, achieving green and circular economy, and agriculture-related United Nations sustainable development goals (SDGs).

Keywords Crop productivity · Climate change · Food security · integrated crop management · Soil health

17.1 Introduction

Nations across the world, predominantly developing countries of South Asian regions, are confronted with issues associated with climate change and food security (Yadav et al. 2023). Further, the exhaustive conventional practices specifically in productive agroecological regions have resulted in a set of production vulnerabilities like deterioration of soil, water ecosystem, declining factor productivity, natural resource base, increasing incidences of diseases and pests, besides chemical contamination of food (Biswakarma et al. 2021). Thus, in the aegis of climate change, the agri-food production system must be diverse climate-smart, along with a supportable agronomic practice, to achieve the United Nations (UN) sustainable development goals (SGDs) of zero hunger, good health and well-being, responsible consumption and production, climate action, and land degradation neutrality. The integrated resource management has become a more crucial to produce greater yields per unit area while conserving natural resources. Further, the growing food demand due to increasing population, rising energy, and production costs requires a paradigm shift and systematic approach to make sustainable use of natural resources. Integrated Crop Management (ICM) concept has been adopted recently in agriculture and is of considerable importance and applicability than the sole approach of soil, water, crops, and pest management (FAO 2012). Literarily, the ICM consists of three words, i.e., (i) integrated—means site-specific and holistic management of the whole production system, (ii) crop—covers all aspects of crop husbandry, and (iii) management—includes planning, setting targets, execution, monitoring, assessment, and achievement of goals. It stands as a significant concept in agriculture, presenting a comprehensive approach to crop production that integrates diverse agronomic practices through crop intensification, minimum reliance on the external chemical inputs such as fertilizers, pesticides, and fossil fuels, maintenance of the landscape, and the enhancement of wildlife habitats (Kumar and Shivay 2008). ICM practices encompass integrated nutrient management (INM), conservation agriculture (CA), integrated weed management (IWM), integrated water management

(IWrM), integrated pest and disease management (IPM & IDM), integrated energy management (IEM), and postharvest management (PHM), among others.

Hence, the Food and Agriculture Organization (FAO) is promoting ICM to ensure the efficient use of natural resources while producing high-quality food on an economic and sustainable foundation. Prior studies have illustrated that the adoption of ICM in crop production improves soil health and yields by 10–14% compared to conventional practices. Further, it also makes the fullest possible use of indigenous technical knowledge (ITKs) with contemporary technologies to deliver sustainable crop production (Bagheri et al. 2019). This not only reduces the burden of purchasing costly external inputs but is also beneficial for small farm holdings due to the emphasis on on-farm resource recycling (Hillocks 2000). Furthermore, the superiority of ICM in terms of crop yields over the farmers' practices had also been reported in Nepal (Regmi and Ladha 2006), China (Wang et al. 2017), and India (Biswakarma et al. 2021).

In regions grappling with the current degradation of the soil-water-plant continuum (Yadav et al. 2021c), there is a pressing need for the implication of ICM practices. Despite this urgency, there remains a huge knowledge gap regarding site-specific ICM tailored to different cropping systems and agro-ecologies. Thus, the imperative lies in developing and disseminating ICM that are socially acceptable, practically feasible, and environmentally resilient. Further, it has been found that the ICM when effectively implemented holds the potential to mitigate production, resource, and climatic vulnerabilities in agriculture (Pooniya et al. 2022). Through integrated management of soil, crops, pests, and energy, ICM can elevate crop productivity, enhance profitability, optimize resource use, conserve energy, reduce greenhouse gas emissions, and promote soil health (Yadav et al. 2019b). Further, the ICM focuses on a holistic and sustainable management system that considers the entire crop production process, from soil preparation and planting to harvest and postharvest activities. Additionally, it seeks to strike a balance between profitable agricultural endeavors and responsible environmental stewardship (Varatharajan et al. 2022). However, currently, only a select few international and national organizations are actively engaged in advancing the ICM having developed specific modules for various cropping systems. Therefore, the present book chapter discusses about the ICM practices with respect to underlying concepts, objectives, aim, rationale, and different components, and their impact on crop productivity, resource use efficiency, and soil health, suggesting an implication of ICM to restore soil and environment quality for the sustainable and green economy (Rathore et al. 2022).

17.2 ICM: An Innovative Approach

The ICM philosophy preserves and enriches nature while providing quality food on a sustainable foundation. It balances food and environmental security by combining indigenous techniques with appropriate modern technology to meet farmers'

Table 17.1 Key elements of integrated crop management

Component	Benefits
Minimum tillage and soil conservation techniques	Low-cost maintenance of soil structure and fertility
Use of biofertilizers, and nitrogen-fixing plants	Improvement in soil fertility
IPM for pests and disease control	Cheap and sustainable plant protection
Crop diversification and crop rotation	Prevent the build-up of pests, diseases, and weeds
Rational use and disposal of plant and animal residues	Resource recycling for better soil, plant, and human health
Maintenance and improvement of ecological diversity	Avoid loss of soil and ecological biodiversity
Minimum use of purchased inputs and nonrenewable fuel resources	Reduce production costs and environmental damage

demands and ecologically improve the productivity of crops with its key elements as suggested in Table 17.1.

17.3 Objectives of ICM

Integrated Crop Management (ICM) is an agricultural method that incorporates several management techniques and practices to grow crops in a sustainable and ecologically friendly way. The major objectives of ICM are

1. Increasing production and productivity of safe and quality agricultural products per unit area.
2. Decreasing the cost of production per unit output.
3. Conserving and enhancing natural resources.
4. Improving soil health and fertility.
5. Enhancing resource-use efficiency.
6. Reducing environmental impact, such as soil contamination with agro-chemicals and greenhouse gas emissions.
7. Managing pests and diseases effectively.
8. Promoting the adoption of climate-resilient practices.
9. Ensuring the economic viability and sustainability of crop production.
10. Providing farmers with socially acceptable and practically feasible crop management practices.

17.4 Rationale of ICMs in the Indian Context

In India, agriculture contributes ~18.2% to the national GDP at current prices and provides a livelihood for a large portion of the population (42.3%). However, over the last three decades, the commercialization of agriculture and a rising emphasis on

Table 17.2 Production and resource threats due to modern agricultural practices

Sustainability issues	References
Receding groundwater table and nitrate (NO_3) pollution in the upper and trans IGP	Mahajan et al. (2012)
Higher requirement of farm inputs and lesser dependence on renewable energy sources	Das et al. (2014a)
Shifting of weeds and increased incidence of new soil-borne pests and diseases	Baghel et al. (2020)
Deterioration of soil physical health due to repeated puddling leading to poor performance of the succeeding upland crops (wheat)	Jat et al. (2009)
Soil organic matter oxidation resulting from intensive tillage and increased temperature under changing climate scenario	Ladha et al. (2003)
Multinutrient deficiencies and nutrient mining of native potassium reserve	Shahane et al. (2020)
Catastrophic effect on human and animal health from in-situ crop residue burning	Abdurrahman et al. (2020)
Greenhouse gas emissions (GHGs), especially methane (CH_4)/nitrous oxide (N_2O)	Pathak et al. (2014)

cereal-based production have resulted in several production, and resource vulnerabilities (Table 17.2). Further, the mechanized farming has resulted in vast quantities of crop residue that are being burned in situ, and India is one of the top burners of crop residues (Bhuvaneshwari et al. 2019). The residue burning has a negative impact on and off farms such as loss of soil organic matter and elementary nutrients, reduction in air quality, and catastrophic effects on human and animal health. Thus, a shift toward resource-efficient ICM practices could address production and resource vulnerabilities, enhance farm profitability, and maintain food security.

17.5 Components of Integrated Crop Management (ICM)

ICM is a whole-farm approach to crop production that is aimed at ensuring efficient and responsible production (Fig. 17.1). The core elements of an ICM are tillage, integrated nutrient management (INM), integrated weed management (IWM), integrated water management (IWRM), integrated pest management (IPM), integrated disease management (IDM), and integrated energy management (IEM) for raising a healthy crop (Varatharajan et al. 2022). ICM practices also allow the recycling of plant and animal residues, the use of biofertilizers, nitrogen-fixing plants, green manures, agroforestry techniques, pest and disease management, and crop diversification (Choudhary et al. 2020). The ICM consists of five interrelated principles, namely food security, environmental sustainability, economic viability, social acceptance, and food safety and quality, which together help in achieving a sustainable development goals in the long run. This multifaceted strategy comprises several key components, each playing a crucial role in fostering a balanced and resilient agricultural system. One fundamental aspect of ICM is crop rotation, where different crops are sequentially cultivated on the same land. This practice helps break pest

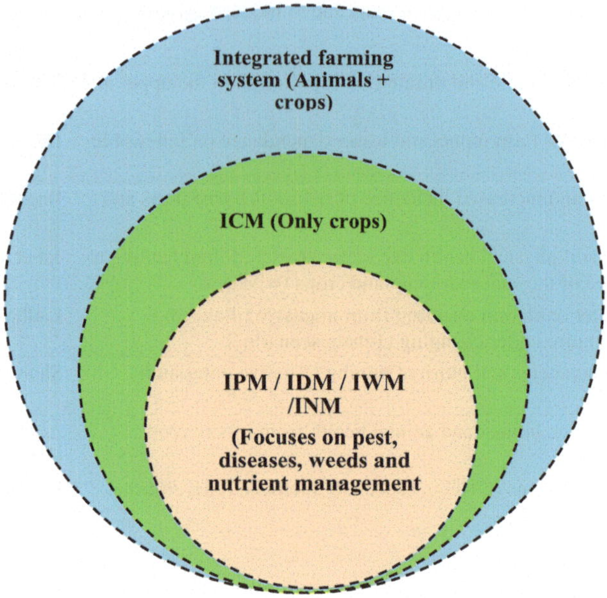

Fig. 17.1 Key components of integrated crop management (ICMs) practices. (Modified from Kumar and Shivay (2008)

and disease cycles, mitigates soil nutrient imbalances, and nurtures a healthier agroecosystem (Altieri 1995). By diversifying crops, farmers can improve soil fertility, reduce the risk of crop-specific pests, and promote sustainable land use.

Soil health management is another important component, involving practices that maintain and enhance soil fertility, structure, and microbial activity. Adopting techniques such as cover cropping, organic matter incorporation, and reduced tillage ensures that the soil remains a thriving medium for plant growth (Bünemann et al. 2018). Healthy soils contribute to improved nutrient availability, water retention, and overall crop resilience. Land configuration and land shaping are another crucial aspect of ICM that involves the manipulation and arrangement of physical features on the agricultural landscape to optimize various factors influencing crop production. Efficient water management is a cornerstone of ICM, focusing on judicious water use through methods like drip irrigation, rainwater harvesting, and soil moisture monitoring. This component aids in conserving water resources, preventing waterlogging or drought stress, and enhancing the efficiency of water and nutrient utilization in agriculture (Kumar et al. 2019). Pest and disease management, under the framework of IPM, integrates various control methods, such as biological, cultural, mechanical, and chemical approaches. By employing a diversified set of strategies, farmers can minimize reliance on chemical pesticides, reduce environmental impact, and foster a balanced predator-prey relationship in the agroecosystem (Gurr et al. 2016). Nutrient management ensures the balanced application of fertilizers based on soil nutrient levels and crop requirements. This component prevents nutrient deficiencies or excesses, improving both crop yield and quality while minimizing environmental pollution (Vanlauwe et al. 2014). Crop residue and weed management involve effective handling of crop residues and implementing measures for weed control. Proper practices in these areas contribute to reduced competition for resources, improved soil structure, and decreased risks of pests and diseases (Mallarino et al. 2019). Conservation tillage, an essential ICM component, minimizes soil disturbance during planting and cultivation. This practice reduces erosion, enhances soil moisture retention, and maintains soil structure, contributing to sustainable land management (Hobbs 2007). Biodiversity conservation emphasizes the cultivation of diverse crops and the preservation of natural habitats on the farm. This supports natural pest control, enhances ecosystem services, and contributes to the overall resilience of the agricultural system (Kremen et al. 2012). Knowledge and information management involve providing farmers with access to the latest information, research findings, and best practices in crop management. This empowers farmers to make informed decisions, adopt innovative technologies, and implement sustainable farming practices (Rejesus et al. 2020).

17.5.1 Tillage Practices

Tillage is an influential aspect of crop production. It involves mechanical manipulation of the soil to create conditions ideal for initial crop establishment and subsequent crop growth (Parihar et al. 2017; Babu et al. 2023a). Manipulating the properties governing soil health affects the soil environment both directly and indirectly (Jat et al. 2020; Das et al. 2022a). Multifaceted benefits can be derived from tilling soil, such as regulation of soil thermal and moisture regimes (Allen et al. 2011; Das et al. 2022b), acceleration of seed germination and root proliferation (Jat et al. 2009), moderation of soil organic matter (SOM) (Six et al. 2001; Singh et al. 2016a; Kumar et al. 2023), reduction of crop-weed interference (Baghel et al. 2020), and augmenting of soil productive capacity through sustainable management of soil and water resources (Prasad et al. 2016). Historically, conventional tillage involved intensive soil disruption through plowing and harrowing to prepare the seedbed. However, the traditional tillage methods have been associated with several drawbacks, including soil erosion, loss of organic matter, and increased fuel consumption (Singh et al. 2021). Hence, its optimization greatly avoids deterioration of soil quality, besides enhancing crop yields (Gathala et al. 2006). As ICM evolved, there has been a paradigm shift toward more sustainable tillage practices. Conservation tillage is an important component of ICM, emphasizing the minimization of soil disturbance through practices such as no-till, reduced tillage, and strip tillage. No-till farming, a prominent example, involves planting crops directly into untilled soil while leaving crop residues on the field surface. This technique has demonstrated its efficacy in retaining soil structure, reducing erosion, and enhancing water conservation (Hobbs 2007). The benefits of conservation tillage within the context of ICM are multifaceted. First and foremost, it contributes to enhance soil health.

Practices such as no-till help preserve soil structure and microbial activity, resulting in improved overall soil health. This not only reduces soil erosion but also facilitates water infiltration and promotes nutrient retention. Further, with respect to water management, conservation tillage plays a crucial role by saving water, maintaining soil moisture levels, reducing water runoff, and mitigating the risk of waterlogging (Lal 1995). Similarly, energy efficiency is another notable advantage associated with conservation tillage, as it reduces fuel consumption and lowers production costs for farmers, besides aligning with environmental sustainability (Hobbs 2007). The weed control is addressed innovatively by conservation tillage through the use of cover crops, crop rotations, and crop residue retention. This strategic approach minimizes the dependence on herbicides, contributing to sustainable farming, and safe food. The farmers transitioning from traditional practices may encounter issues related to weed management, residue decomposition, and the need for adjustments in planting equipment. Additionally, the effectiveness of conservation tillage can vary based on factors such as soil types, climate conditions, and crop choices (Yadav et al. 2019b). These challenges underscore the importance of tailored approaches and continuous adaptation in implementing conservation tillage within ICM practices.

No-till farming, a sustainable agricultural practice, revolutionizes traditional plowing by eliminating soil disturbance and utilizing specialized equipment to plant seeds directly into untilled soil. This approach is of paramount importance in conserving soil structure, reducing erosion, and enhancing water infiltration (Singh et al. 2021). The consequential benefits lead to improved soil health, promoting a more resilient and sustainable agricultural ecosystem (Yadav et al. 2020). The field studies conducted in Cambodia, Laos, and Thailand reported similar yields with wet and dry direct seeding of rice (DSR) relative to the conventional transplanting system (Mitchell et al. 2004; Rickman et al. 2001). However, the adjusted mean of change shows a 12.7–21% lower DSR yield in Pakistan, but 8.6–18.5% higher in Bangladesh and the Philippines than in CT practices (Ladha et al. 2003). Zhu (2008) in China has documented an increase of ~22.5% in grain yield in DSR over transplanted rice. Similarly, in Japan, the average grain yield of dry-DSR was ~11.2 Mg/ha as compared to 9.6 Mg/ha under conventional practices across cultivars and locations (Kato et al. 2009). In India, the performance of DSR in terms of yield varies with location (Gathala et al. 2011). Crop establishment through dry direct seeding resulted in a yield penalty in the northwestern IGP, while this is reversed in the eastern IGP because of variations in precipitation (Jat et al. 2006). The crop and water productivity of different crops under diverse crop establishment techniques is given in Table 17.3. The strip tillage is a precision farming technique that involves tilling narrow strips where seeds are planted, leaving the spaces between the strips untilled. This method is particularly advantageous as it helps retain crop residues between rows, thereby reducing soil erosion and maintaining soil structure. Strip tillage creates a favorable environment for germination while promoting efficient resource use.

Table 17.3 Comparison between different crop establishment techniques (CET) in terms of productivity

Crop/ cropping system	CET	Period (Years)	Percent (%) change over CT		References
			Crop productivity	Water productivity	
Rice	SSNM	2	41.1	–	Regmi and Ladha (2006)
	ICM	2	50.2	–	
Wheat	SSNM	2	55.5	–	
	ICM	2	78.6	–	
Rice–wheat	ZTDSR-ZTW	2	2.1	21.2	Jat et al. (2009)
	CTDSR-CTW	2	−1.3	18.9	
Rice–wheat	RTTR-NTW	2	0.3	−0.85	Saharawat et al. (2010)
	NTTR-NTW	2	0.9	3.44	
	DSR-NTW	2	−2.4	9.74	
Rice	SRI	2	−1.8	51.7	
	Aerobic	2	−21.1	11.3	
Rice–wheat	TPR-ZTW	3	1.3	–	Bhattacharyya et al. (2015)
	MBR + DSR-ZTW-ZTMB	3	4.2	–	
	MBR + DSR-ZTW + RR-ZTMB	3	12.5	–	
Rice	DSR	2	−13.7	21.4	Raj et al. (2017)
	SRI	2	7.6	48.8	
Rice–wheat (ICM)	ZT-DSR + WR + MB–ZTW + RR + MB	5	11.6	14–16	Biswakarma et al. (2021)
	ZT-DSR + WR—ZTW + RR	5	1. 9	–	
	DSR—FIRB	5	−9.0	–	

Note: *ZTDSR-ZTW* zero-till direct drill-seeded rice and wheat after no-tillage, *CTDSR-CTW* conventional-till direct drill-seeded rice, and wheat after conventional tillage, *RTTR-NTW* reduced tilled (unpuddled) transplanted rice–No-till drill sown wheat, *NTTR-NTW* No-till transplanted rice–No-till drill seeded wheat, *DSR-NTW* No-till drill seeded rice–No-till drill seeded wheat, *TPR-ZTW* transplanted rice–zero tilled wheat, *MBR + DSR-ZTW-ZTMB* mungbean residue + dry seeded rice—zero tilled wheat-zero tilled summer mungbean (relay cropping), *MBR + DSR-ZTW + RR-ZTMB* mungbean residue + dry seeded rice– zero tilled wheat + rice residue—ZT summer mungbean (relay cropping), *ZT-DSR + WR + MB– ZTW + RR + MB* zero-till direct-seeded rice with wheat residue—ZT wheat with mungbean and rice residues, *ZT-DSR + WR–ZTW + RR* till direct-seeded rice with wheat residue—zero-till wheat with rice residue, *DSR—FIRB* direct-seeded rice—wheat on furrow irrigated raised bed

17.5.2 Land Configuration

Land configuration includes changing the shape of land surface and seedbeds to create a congenial environment for crop growth and effective utilization of the land to obtain a higher crop yield (Ramesh and Rathika 2009; Yadav et al. 2021c). In deficit and excess rainfall conditions, crop establishment through furrows and ridges conserves and drains excess water, besides regulating soil temperature and aeration (Parihar et al. 2017). A previous study found that raised beds produce higher yields of rice (~6.1%) and wheat (~6.4%) than flatbeds (Sayree and Ramos 1997). In addition, the configuration of the land, using ridges and furrows, contour cultivation, broad beds, and furrows, as well as crop residue retention, reduces surface runoff and leaching losses of nutrients. In a similar study, Chiroma et al. (2006) explored the effects of land configuration coupled with mulching on soil properties such as pore spaces, bulk density (ρb), and penetration resistance and observed significant improvement in the soil's physical health (Babu et al. 2020b).

Land configuration stands as a vital aspect within the framework of ICM. In the pursuit of sustainable farming, efficient land configuration entails meticulous considerations such as topography, water management, and soil erosion control. Several key components underscore the significance of thoughtful land configuration in ICM. Contour farming, a fundamental practice, involves the plowing and planting of crops along the contour lines of the land, perpendicular to the slope. According to Lal (1995), contour farming holds paramount importance in reducing water runoff, minimizing soil erosion, and enhancing water infiltration, particularly on sloping terrains. By following the natural contours of the land, contour farming mitigates the adverse effects of water movement, fostering soil health and sustainable crop production. Terracing, another integral component, entails the construction of steps or platforms on sloping land to create level surfaces for cultivation. Paudel et al. (2015) emphasize the significance of terracing in controlling soil erosion, preventing water runoff, and efficiently utilizing water resources by capturing and distributing rainwater evenly across the terraced fields. Terracing not only enhances water management but also provides a structured and stable environment for crops to thrive.

Ridge and furrow systems involve planting rows of crops on ridges with furrows in between, forming a pattern that aids in soil water conservation and moisture management (Majsztrik et al. 2005; Babu et al. 2020b) (Fig. 17.2). This configuration enhances water infiltration, reduces waterlogging, and promotes root aeration, thereby fostering optimal conditions for crop growth. The structured arrangement of ridges and furrows contributes to improved water-use efficiency and overall crop resilience. Grassed waterways, characterized by natural or constructed channels planted with grass, play a crucial role in preventing soil erosion and managing water runoff. According to the USDA (2006), grassed waterways serve as a protective barrier against gully erosion, stabilize the soil, and filter sediment, contributing to enhanced water quality. These features are instrumental in safeguarding both the land and surrounding water bodies from the detrimental impacts of agricultural

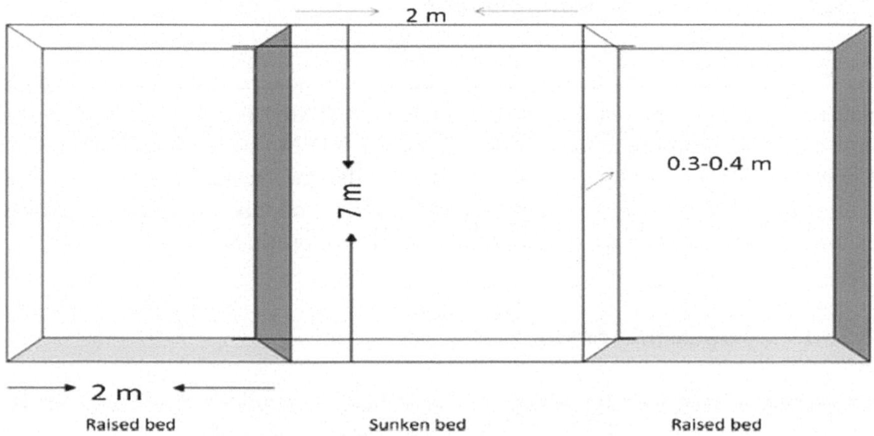

Fig. 17.2 Schematic illustration of alternate raised and sunken beds system (RSB)

activities. Further, the buffer strips mitigate soil erosion, enhance biodiversity, and act as a protective barrier, shielding adjacent water bodies from the potential impacts of agricultural runoff (Osmond et al. 2012). This integration of natural vegetation serves as an environmentally friendly approach within the broader context of ICM. Similarly, the precision agriculture represents a technological advancement in land configuration involving the use of GPS-guided equipment to optimize field-level management in crop farming. The precision agriculture helps farmers to customize input applications, thereby reducing waste and optimizing resource use (Khurana et al. 2018). This precision-driven approach enhances overall resource use efficiency and fosters sustainable agricultural practices aligning with the overarching goal of ICM in ensuring the health and productivity of agricultural landscapes.

A low-elevation area can benefit from the specific land contouring technique known as the Raised and Sunken Bed (RSB), which is created using cutting and filling procedures. In order to reach the target height, this procedure entails removing the topsoil layer from a specific region and redistributing it to a neighboring area (Das et al. 2014b). Adopting RSB land layout offers a way to intensify and diversify crops, which has the potential to improve farmers' livelihoods and socioeconomic well-being. This strategy is mirrored by similar land management techniques like Sorjan in Java and tidal wetland systems in Indonesia, where elevated beds are used for high-value crops and sunken beds are used for rice farming. Farmers in the northeastern state of Meghalaya have used a temporary raised bed agriculture approach, leaving the sunken area (the space between two raised beds) unoccupied (Singh et al. 2011). There have been numerous permutations of permanent or temporary land configurations that have arisen, driven by factors such as soil type, water depth, available resources, and crop selection. Although the efficacy of the raised and sunken bed systems for increasing agricultural yields has previously been studied in select Indian regions, there is a scarcity of study material from the northeastern region.

17.5.3 Soil Fertility Management

The ICM is crucial for addressing food security, soil health, sustainability, and climate resilience. Integrated Nutrient Management (INM) is a key strategy within conventional agriculture to tackle these challenges. INM involves the judicious use of chemical fertilizers alongside organic materials, crop residues, green manures, legumes, and biofertilizers to optimize plant nutrient supply, manage soil fertility, sustain agricultural productivity, and enhance farmer profitability.

17.5.3.1 Components of INM

(a) **Inorganic/synthetic fertilizers:** Balanced use of chemical fertilizers is vital to combat nutrient deficiencies not met by organic sources alone.
(b) **Organic manures:** Derived from animal dung/urine and plant residues, organic manures, with a focus on Farm Yard Manure (FYM), enhance soil properties and build up secondary and micronutrients, contributing significantly to sustainable agriculture.
(c) **Crop residues:** Utilizing crop residues, rich in essential nutrients, can enhance soil quality and productivity. However, challenges like large quantity requirements, use as animal feed, and slow decomposition must be taken into consideration to maximize their benefits.
(d) **Bio-composts:** Advanced technologies, such as biogas and vermicomposting, efficiently address challenges associated with organic manures, providing a rich source of plant nutrients.
(e) **Green manures:** Incorporating undecomposed green plant tissue into the soil improves fertility and is gaining importance due to its positive impact on soil health (Choudhary et al. 2007; Yadav et al. 2021a). Crops like sunnhemp, dhaincha, senji, cowpea, and berseem are commonly used for green manuring. Their deep rooting systems, ability to tolerate diverse agro-climatic conditions, quick growth, and nitrogen-fixing capacity make them ideal choices.
(f) **Legume-based cropping system:** Legumes contribute to sustainable agriculture by fixing atmospheric nitrogen, improving soil structure, and preventing nutrient leaching. Introducing legumes can fulfill 25–50% of nitrogenous fertilizer for subsequent cereals (Subba Rao 1988).
(g) **Biofertilizers:** Biofertilizers play a crucial role in INM, promoting sustainable agriculture by harnessing atmospheric nitrogen fixation, phosphorus solubilization, and the synthesis of growth-promoting biomolecules (Singh et al. 2015). As living microorganisms, biofertilizers, comprising bacteria, algae, and fungi, enhance nutrient availability by converting non-usable forms through biological processes. Nitrogen-fixing biofertilizers (NFBs), like *Rhizobium* sp. and *Azospirillum* sp., convert atmospheric nitrogen for plants, contributing to sustainable nitrogen supply. Phosphate solubilizing biofertilizers (PSBs) such as *Bacillus* spp. and *Pseudomonas* spp. enhance plant accessibility to phosphates

by lowering soil pH. Phosphate mobilizing biofertilizers (PMBs) like mycorrhiza improve phosphorus availability in the soil. Plant growth-promoting biofertilizers (PGPBs) like *Pseudomonas* sp. stimulate root growth and organic matter decomposition, improving nutrient mineralization and crop yield. Potassium solubilizing biofrtilizers (KSBs), including *Bacillus* sp., release potassium by solubilizing silicate minerals. Potassium mobilizing biofertilizers (KMBs) like *Bacillus* sp. mobilize inaccessible potassium forms, while sulfur oxidizing biofertilizers (SOBs), *Thiobacillus* spp. being the predominant, contribute to usable sulfur for plant growth. Biofertilizers, applied to seeds, seedlings, or soil, increase soil fertility and support crop growth and yield. Vesicular Arbuscular Mycorrhizae (VAM), a symbiotic association with plant roots, significantly aids in phosphorus acquisition and solubilization in P-deficient soils. In summary, biofertilizers are valuable tools in sustainable agriculture, enhancing nutrient availability and plant growth while minimizing environmental impact.

(h) **Nonconventional organic sources for plant nutrition:** Materials like sewage sludge, urban wastes, and industrial effluents offer potential plant nutrient sources, but careful management is essential to mitigate potential risks associated with contaminants.

(i) **Site-Specific Nutrient Management (SSNM) approaches under ICM:** SSNM optimizes crop nutrition by dynamically adjusting fertilizer application based on crop needs, location, and season. SSNM enhances Nutrient Use Efficiency (NUE) and emphasizes the optimal use of indigenous nutrient sources. Approaches include plant analysis-based SSNM, soil-cum-plant analysis-based SSNM, decision support systems, and tools like Nutrient Expert®. SSNM efficiently utilizes nutrients, reduces fertilizer wastage, and enhances environmental sustainability, proving essential for modern precision agriculture. Advanced technologies like remote sensing, GIS, and decision support systems play a crucial role in implementing effective SSNM practices.

17.5.3.2 Integrated Crop Management and Soil Quality

Soil quality index, an indicator of soil productivity, can be examined using different soil properties such as physical, chemical, and biological. The soil quality has both intrinsic and dynamic features. According to Doran and Parkin (1994), soil quality is the ability of soil to function normally within its ecosystem boundary, wherein the functions are primarily governed by agronomic management practices and changes in soil condition by land use changes. The soil quality determines the productivity, and any deterioration in the soil quality can provide a precarious indication of ecological degradation and alteration as it functions as a storehouse of essential plant nutrients and water and acts as a sink for pollutant chemicals. The study conducted by Biswakarma et al. (2023) revealed a significant improvement (37.6–69.6%) in soil quality due to the adoption of the conservation agriculture (CA)-based ICMs compared to conventional tillage (CT)-based ICMs (Table 17.4). Furthermore, the

Table 17.4 Effect of integrated crop management practices on the soil quality index (SQI) (±S.D.) at harvest of 6th season rice under the rice–wheat rotation

	Soil quality index (SQI)	
Treatments	0.0–0.15 m	0.15–0.30 m
ICM_1	$0.51^{de} \pm 0.02$	$0.48^{b} \pm 0.03$
ICM_2	$0.43^{ef} \pm 0.09$	$0.31^{b} \pm 0.12$
ICM_3	$0.32^{f} \pm 0.05$	$0.25^{b} \pm 0.10$
ICM_4	$0.29^{f} \pm 0.07$	$0.30^{b} \pm 0.20$
ICM_5	$0.69^{bc} \pm 0.07$	$0.44^{b} \pm 0.06$
ICM_6	$0.64^{cd} \pm 0.02$	$0.47^{b} \pm 0.10$
ICM_7	$0.94^{a} \pm 0.06$	$0.88^{a} \pm 0.11$
ICM_8	$0.83^{ab} \pm 0.03$	$0.82^{a} \pm 0.06$

Adapted from Biswakarma et al. (2023)
Note: $ICM_{1\&2}$ = transplanted rice (TPR) and conventional wheat (CT); $ICM_{3\&4}$ = direct-seeded rice (DSR) and wheat on furrow irrigated raised bed (FIRB); $ICM_{5\&6}$ = zero-till direct-seeded rice with wheat residue (ZTDSR + W*R*) and zero-till wheat with rice residue (ZTW + RR); and $ICM_{7\&8}$ = zero-till DSR with wheat and mungbean residues (ZTDSR + W*R* + MB) and ZT wheat with mungbean and rice residues (ZTW + RR + MB)

total carbon, alkaline phosphatase activity, urease, available nitrogen, and water-stable aggregates from minimum data set in descending order appeared as principal indicators of the soil quality confirmed by the bivariate regression analysis. Overall, the integration of different ICM practices would be a more convincing approach and recommended for restoring the soil quality.

17.5.4 Integrated Weed Management (IWM)

The IWM is a sustainable strategy designed to efficiently control weed populations, mitigating environmental impact and maximizing agricultural productivity (Fig. 17.3). Weeds pose a significant challenge by competing with crops for nutrients, water, and sunlight, resulting in reduced yields and economic losses. IWM strives to maintain weed populations below the economic threshold level. This holistic approach optimizes crop yield while minimizing costs and addressing socio-economic constraints. The integrated weed management (I_{WM}) concept is a framework for the management of weeds in a sustainable, effective manner using appropriate cultural, mechanical, biological, and chemical tactics (Das 2008). Naz et al. (2020) found that the use of pretilachlor (P_E) + 1 hand weeding (H_W) at 30 D_{AS} + brown manuring significantly reduced the weed population during the early growth stage in dry-DSR plots. Soil solarization and glyphosate application at 1.0 kg ai/ha help in reducing the density and dry weight of *C. rotundus* (Kumar et al. 2012). Singh et al. (2007) revealed that the co-culture of rice and sesbania can effectively reduce broadleaf weed density by 76–83%, grasses by 20–33%, and total weed biomass by 37–80% compared to sole rice, besides benefiting rice through

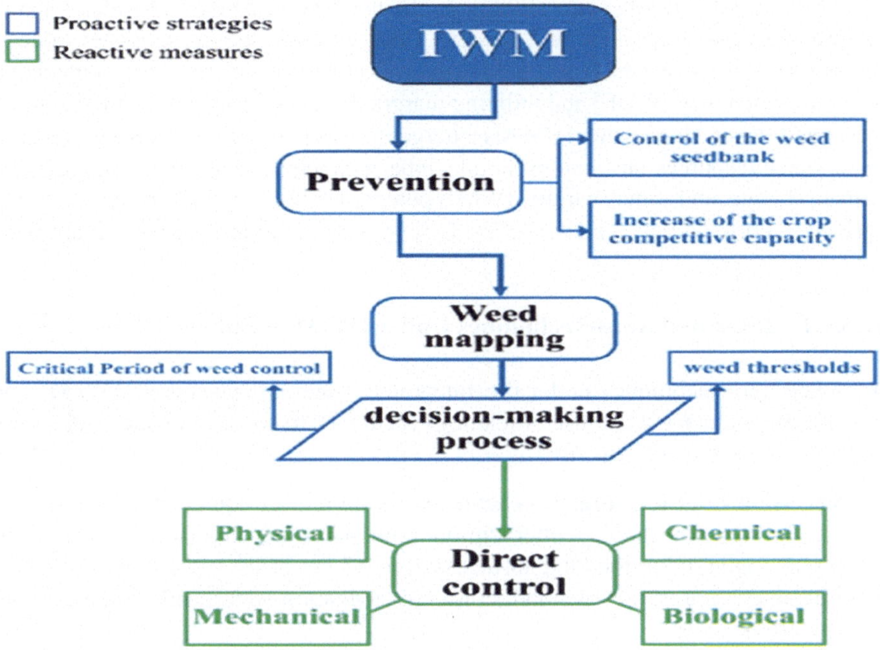

Fig. 17.3 Schematic representations of IWM Approaches

atmospheric N-fixation and reduced soil crust formation (Gopal et al. 2010). Further, the three-year study was conducted by Mishra and Singh (2012) and showed the superiority of ZT technology in managing *A. ludoviciana* and *C. album* infestation, which was due to the quick emergence of wheat seedlings under undisturbed soil conditions (Yaduraju and Mishra 2002). ZT wheat managed the infestation of *Phalaris minor* significantly, but *Rumex dentatus*, *Cynodon dactylon*, and *Cyperus rotundus* were hardly controlled (Chhokar et al. 2009). Nath et al. (2017) also reported the lower weed dry weight and increased WCE and WCI with ZT + residue along with a sequential application of pendimethalin and sulfosulfuron over the conventional plots. In northwestern India, ZT practices assume a greater significance and serve as an effective weed management strategy in rice–wheat rotation. It significantly reduces the *Phalaris minor* intensity ranging from 30% to 40% compared to conventional practices.

17.5.5 Integrated Water Management (IWrM)

In Asia, conventional rice consumes 50% of the total irrigation water (Chakraborty et al. 2017). The availability of water per capita in major rice-growing Asian countries has declined by 34–76% between 1950 and 2000 and is expected to decline by

~18–88% by 2050 (Ladha et al. 2003). In the northwestern states of India, namely Punjab, Haryana, Rajasthan, and western Uttar Pradesh, the groundwater table is decreasing at a mean rate of 4.0 ± 1.0 cm/yr. To address water scarcity, enhancing water use efficiency (WUE) and utilizing marginal waters for irrigation are essential. Integrated Water Management (IWM) emerges as a holistic solution, seeking to optimize water resources sustainably amid global challenges like population growth, climate change, and water scarcity. IWM's potential to boost water efficiency, elevate agricultural productivity, and contribute to long-term environmental sustainability.

17.5.5.1 Improved Agro-Techniques Under IWrM to Enhance WUE

Improved agro-techniques under the Integrated Water Management (**IWrM**) can significantly enhance water use efficiency (WUE). Here are several approaches and technologies that can be useful:

(a) **Precision land leveling:** It is achieved through laser-equipped drag scrapers, exemplifies a mechanical intervention in intensive irrigated farming. This technique, maintaining the land surface within ±2 cm of its average elevation significantly improves crop stand, conserves irrigation water, and enhances input use efficiencies.
(b) **Furrow Irrigated Raised Bed (FIRB) planting:** Crops are sown on 40 cm wide raised beds, with irrigation channeled into 30 cm wide furrows. This technique not only promotes crop diversification but also achieves a 25–35% reduction in irrigation water usage, along with savings in nitrogen and seed rates.
(c) **Alternate furrow irrigation in ridge and furrow planting system:** This system emerges as a viable strategy for water conservation. By irrigating crops in alternate furrows, lateral movement and efficient water distribution across the bed are achieved, significantly reducing surface soil area, evaporation, and deep percolation.

17.5.6 Integrated Pest Management (IPM)

Agriculture is severely constrained by insects/pests, which limit yields and cause enormous losses. The literature indicates that insect pests cause substantial yield losses in rice (25%), wheat (5–10%), and pulses (30%) (Dhaliwal et al. 2015). The ecological approach to insect management, therefore, assumed greater importance due to changes in insect–pest dynamics resulting from changing climate. IPM is a method that focuses on a variety of approaches that combine cultural, mechanical, biological, and chemical options in order to reduce harmful insect–pest populations below economic thresholds. Incorporating IPM principles into crop production does not exclude the use of pesticides completely, but instead uses them along with other tools to manage infestations (Cooper and Dobson 2007). This ensures higher farm profits, as it minimizes pesticide use, provides better quality food, and conserves the environment. Major components of IPM are given in Fig. 17.4.

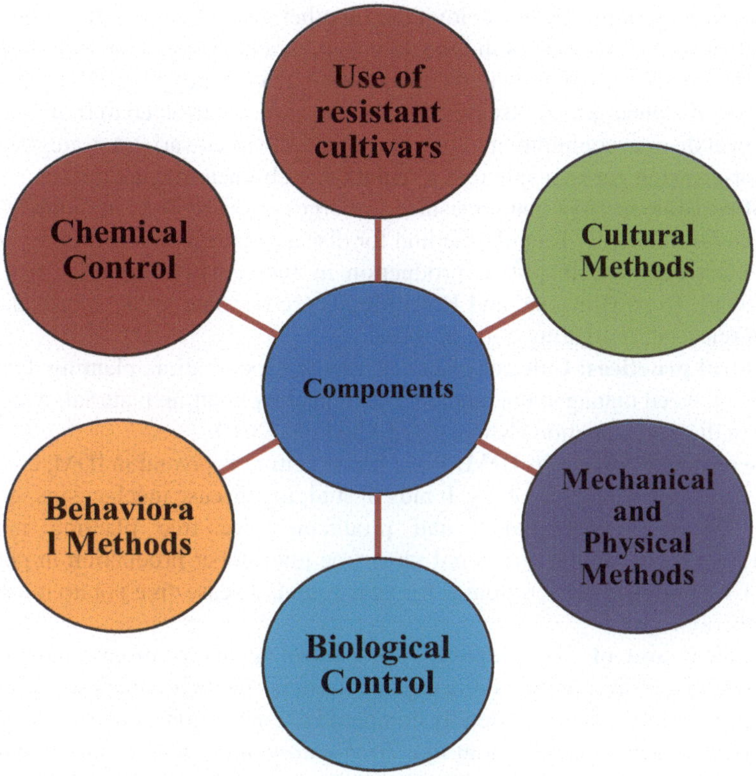

Fig. 17.4 Components of Integrated pest management (IPM)

17.5.7 Integrated Disease Management (IDM)

In tropical and subtropical countries, the favorable environment and inadequate plant protection measures lead to a rising incidence of plant diseases (Heeb et al. 2019). The majority of diseases are caused by fungi, viruses, bacteria, and nematodes. According to Khoury and Makkouk (2010), integrated plant disease management is a decision-based practice that coordinates multiple strategies for controlling pathogens in an ecologically and economically efficient manner.

17.5.7.1 Components for Plant Disease Management

1. **Survey and surveillance for plant disease management:** Disease surveillance, a fundamental component, involves monitoring prevalence, severity, and relevant data for decision-making, including forecasting. Current surveillance mainly aims to optimize chemical application timing, minimizing unnecessary use. Enhancing its efficacy in Integrated Disease Management (IDM) requires comprehensive data on pathogen occurrence, economic threshold levels, meteorological data, and associated risk factors.

2. **Gene deployment:** Gene deployment, another crucial approach, involves frequent cultivar changes at both inoculum sources and distant areas, proving effective in controlling epidemics like brown rust. This strategy, especially pertinent for long-distance spread diseases, has significantly contributed to India's success in wheat disease containment. Over two decades, varied varietal strategies across wheat-growing zones resulted in a remarkable absence of epidemics.
3. **Host resistance:** Host plant resistance emerges as the most economically viable and environmentally friendly method for disease management. Notably, success has been achieved in potato production in India, where varieties like Kufri Giridhari, Kurfi Himsona, and Kufri Megha are resistant to late blight and have been released, mitigating a major threat.
4. **Cultural practices:** Cultural practices, including sanitation, planting date, crop rotation, weed management, and the use of healthy planting material, play a vital role in disease reduction (Khoury and Makkouk 2010).
5. **Role of tissue culture in IDM:** Plant tissue culture is pivotal in IDM, contributing significantly to technology. It aids in studying disease mechanisms, generating disease-resistant plants, and producing virus-free planting material. ICAR-CPRI in Shimla pioneered virus-free microtuber production in potatoes, and CPRI introduced aeroponics for year-round, disease-free potato microtuber production.
6. **Biological control:** The adoption of biocontrol agents for disease management is on the rise, particularly among organic farmers. These products are considered environmentally and user-friendly compared to conventional chemicals. Notable biocontrol agents include fungi like *Trichoderma harzianum*, *Chaetomium globosum*, *Trichoderma viride*, and *Gliocladium virens*, along with the actinomycete *Streptomyces griseoviridis* and the bacterium *Bacillus subtilis*. While biocontrol agents may struggle in specific field conditions, their effectiveness is enhanced when integrated with other management practices.
7. **Chemical control:** While pesticides offer effective plant disease management, their haphazard use poses significant risks. Indiscriminate application may lead to ecological imbalance, pest resurgence, resistance, environmental pollution, and health hazards for humans and animals. Responsible handling is crucial to prevent issues associated with pesticides. For instance, in wheat, a controlled application of propiconazole (Tilt 25EC) at ear head emergence effectively manages Karnal bunt. Similarly, cymoxanil-based fungicides like curzate and dimethmorph+mancozeb are recommended for late blight control in potatoes.

17.5.8 Postharvest Management Under ICM

Integrated Crop Management (ICM) is a holistic approach that encompasses all stages of crop production, from planting to postharvest (Fig. 17.5). While much emphasis is often placed on cultivation practices, postharvest management is equally crucial. Deterioration of fresh agricultural produce after harvest occurs due to

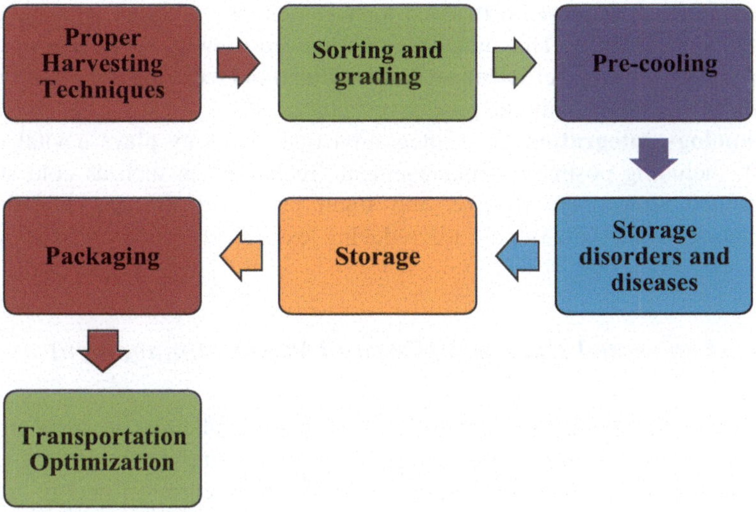

Fig. 17.5 Components of postharvest management under ICM

factors such as moisture loss, physiological disorders, diseases, pest attacks, nutrient loss, root/shoot growth, seed germination, and fiber development. Postharvest management involves the handling, storage, processing, and marketing of agricultural produce after it is harvested. Physiological changes, such as sprouting, rooting, and biochemical changes affecting color, texture, and flavor, contribute to limited storage life. An integrated approach to postharvest management, with appropriate backward linkage at the farm level, ensures quality assurance during subsequent storage and marketing. Despite advances in cultivation techniques, a considerable portion of the world's agricultural produce is lost during the postharvest phase. This not only impacts food security but also results in economic losses for farmers. In the context of Integrated Crop Management, effective postharvest practices are integral to achieving the overarching goals of sustainable agriculture, resource optimization, and minimizing environmental impact.

17.5.8.1 Principles of Postharvest Management Under ICM

1. **Minimizing losses:** One of the primary principles of postharvest management is the reduction of losses. Integrated crop management aims to minimize losses during the transition from the field to the market. This involves adopting techniques that prevent spoilage, reduce physical damage, and address issues related to pests and diseases.
2. **Quality maintenance:** Preserving the quality of harvested crops is crucial for ensuring marketability and consumer satisfaction. ICM emphasizes the implementation of practices that maintain the nutritional content, flavor, and overall quality of produce throughout the postharvest handling process.

3. **Sustainable practices:** Postharvest management should align with the principles of sustainability. This includes adopting eco-friendly packaging materials, implementing energy-efficient storage facilities, and reducing the overall carbon footprint associated with the postharvest phase.
4. **Technology integration:** The integration of technology plays a vital role in ICM, including postharvest management. Technologies such as cold storage, controlled atmosphere storage, and smart packaging systems contribute to extending the shelf life of produce, reducing losses, and ensuring product quality.

17.5.9 Integrated Energy (IEM) and Waste Management

India, despite achieving food self-sufficiency, faces challenges in sustaining and increasing crop yields amid climate variability. With rising demands for food and energy due to population growth, the agriculture sector is currently energy-intensive and relies heavily on nonrenewable sources. Efficient energy use, embracing techniques like minimum tillage and precision machinery, is crucial. Innovations such as zero-till drill, direct-seeded rice, and solar-powered equipment offer water and energy savings. Controlled traffic farming and system of wheat intensification enhance productivity sustainably. To address environmental concerns, precision agriculture integrates GPS and GIS for targeted input applications, conserving resources while maintaining crop yields. ICM incorporates diverse farm machinery and equipment with varying power sources, transitioning from human and animal power to mechanical energy. Enhancing energy efficiency in agriculture involves strategic equipment selection based on field conditions, power source compatibility, operational methods, and fuel-efficient technologies. Farm machinery categories in ICM encompass land preparation and input application, renewable power-driven equipment, and precision farm machinery. Energy dynamics of different CETs are given in Table 17.5.

17.5.9.1 Land Preparation and Input Application in ICM

- **Zero-Till Drill:** In the Indo-Gangetic Plain (IGP), zero-tillage (ZT) has replaced traditional tillage methods, reducing irrigation water usage by 13–33% and offering substantial energy savings. ZT facilitates early sowing, minimizing weed competition and improving soil health, resulting in a 79% energy saving compared to conventional tillage.
- **Happy Seeder:** Addressing challenges in rice–wheat crop rotations, the Happy Seeder combines cutting and lifting rice straw, sowing wheat, and depositing straw as mulch. This innovation reduces establishment costs by 50%, eliminates air pollution from burning, and maintains soil nutrients and organic carbon (Sidhu et al. 2011).

Table 17.5 Comparison between different crop establishment techniques (CET) in terms of energy dynamics

Crop/cropping system	CET	Period (Years)	Percent (%) change over CT			References
			EO	NE	EUE	
Maize–wheat	50% NPK (inorganic fertilizers) + 50% NPK (FYM)	5	–	111.7	15.7	Singh et al. (2016)
	50% NPK (inorganic fertilizers) + 50% NPK (vermincompost)		–	95.1	5.2	
	50% NPK (inorganic fertilizers) + 50% NPK (in-situ green manuring)		–	89.4	0	
Upland rice–mustard	NT-RR	4	−1.52	23.5	62.5	Yadav et al. (2018)
	GM		18.63	26.5	12.2	
	BM		10.85	15.9	8.71	
Wheat	0–20 cm soil water content (SWC)-based water management practice	2	–	−15.5	3.25	Yuan et al. (2018)
	0–40 cm SWC-based water management practice		–	8.12	15.6	
	0–60 cm SWC-based water management practice treatments		–	−0.78	−2.87	
Direct-seeded rice (DSR)	NT	4	1.35	30.2	97.0	Yadav et al. (2019)
	RSM		6.62	7.11	−5.22	
	GLM		23.9	31.2	13.8	
	BMM		12.8	17.4	8.36	
Rainfed rice-based cropping system	Dry-DSR-Toria	3	24.8	−2.08	4	Lal et al. (2020)
	Dry-DSR-Green gram		−19.6	−20.5	−39.4	
	Wet-DSR-Toria		33.4	4.51	5.04	
	Wet-DSR-Green gram		−8.57	−7.50	−5.87	

Note: *EO* energy output, *NE* net energy, *EUE* energy use efficiency, *NT-RR* no-till +100% residue retention, *GM* Gliricidia mulch, *BM* brown manuring mulch, *NT* no-till, *RSM* straw mulch, *GLM* Gliricidia mulch, *BMM* brown manuring mulch

- **Roto-till Drill and Strip-till Drill:** Roto-till drilling, an example of minimum tillage, saves 40% in land preparation costs and increases wheat yield by 5% compared to conventional methods. Strip-till drills further enhance cost savings (61%) and yield (22%), while reducing fuel consumption by up to 57% (Hossain et al. 2012).

- **Bed Planting:** Raised bed planting technology in India allows for intercropping, crop diversification, and significant water savings (30–50%). Broad Bed Furrow (BBF) planting, suitable for rainfed agriculture, not only conserves soil moisture but also reduces energy consumption by 83.32–84.40% (Khambalkar et al. 2014).
- **Laser Land Leveller:** Precision land leveling with lasers creates a consistent slope, saving 25% irrigation water and increasing overall yield compared to non-laser leveled fields (Humphreys et al. 2010).
- **Direct-Seeded Rice (DSR) and System of Rice Intensification (SRI):** DSR reduces labor by 50%, conserves water, and lowers production costs (Choudhary and Suri 2014). SRI, with a square planting pattern, ensures higher yields, resilience against extreme weather, and increased tillering, though it requires careful management of higher labor demands.

17.5.9.2 Solar-Powered Equipment in ICM

- **Solar Dryers:** Solar-powered dryers for grains, fruits, vegetables, and flowers offer an eco-friendly alternative to fossil fuel-based drying methods, reducing costs and environmental impact.
- **Solar-Powered Fencing:** Utilizing solar power for animal control fences ensures efficiency, with voltage requirements ranging from 2 to 11 kV.
- **Solar Pumps:** Solar PV systems power irrigation pumps, providing an eco-friendly alternative to conventional electric or diesel pumps, with economic viability of up to 3 kWp for village water supply and 1 kWp for irrigation.
- **Gasifier and Biomethanation:** Thermochemical conversion of biomass into producer gas through gasifiers presents a sustainable energy solution. Biomethanation offers an eco-friendly approach to organic biomass treatment, producing biogas for energy and nutrient-rich digestate for crop production.

17.5.9.3 Precision Farm Equipment for ICM

- **Yield Monitoring:** Yield monitors, equipped with combined harvesters, provide real-time crop yield mapping, aiding in site-specific nutrient management and treatment strategies.
- **Variable Rate Application System:** Variable rate fertilizer application, guided by fixed yield goals and site-specific soil conditions, enhances fertilizer use efficiency, economic returns, and reduces environmental impact.
- **Optical Crop Sensors:** Optical crop sensors and NDVI algorithms enable precise N application based on plant health and growth, reducing environmental pollution.

17.5.10 Climate Change and ICM

Climate change poses significant challenges to agroecosystems, impacting temperature patterns, precipitation levels, and the frequency of extreme weather events. Changes in temperature and precipitation can alter growing seasons, disrupt crop cycles, and contribute to the increased incidence of pests and diseases. Shifts in climate patterns also affect water availability, leading to droughts or floods in different regions. These changes collectively threaten agricultural productivity, food security, and the overall sustainability of agroecosystems. Climate change is a complex phenomenon, not characterized solely by a gradual warming of the oceans, land, and atmosphere. Instead, it manifests as dynamic alterations in climate patterns, encompassing shifts in precipitation and the frequency and intensity of extreme weather events. Since the late nineteenth century, Earth's average temperature has increased by approximately 1 °C, primarily attributed to heightened CO_2 emissions and other human activities. The emission of CO_2 contributes directly to the greenhouse effect, a natural process crucial in shaping Earth's climate, creating a warm and conducive environment for life to thrive. Major greenhouse gases, including carbon dioxide, methane, water vapor, nitrous oxide, and sulfur hexafluoride, among others, envelop the Earth and intercept solar radiation, elevating atmospheric temperatures.

As depicted in Fig. 17.6, atmospheric CO_2 levels have risen significantly, surpassing 400 ppm from 371 ppm in 2000, significantly impacting Earth's warming. This temperature rise, a consequence of anthropogenic activities such as deforestation, agriculture, land use changes, and urbanization, disrupts natural balance and alters weather patterns. While the greenhouse effect is integral to agriculture, the

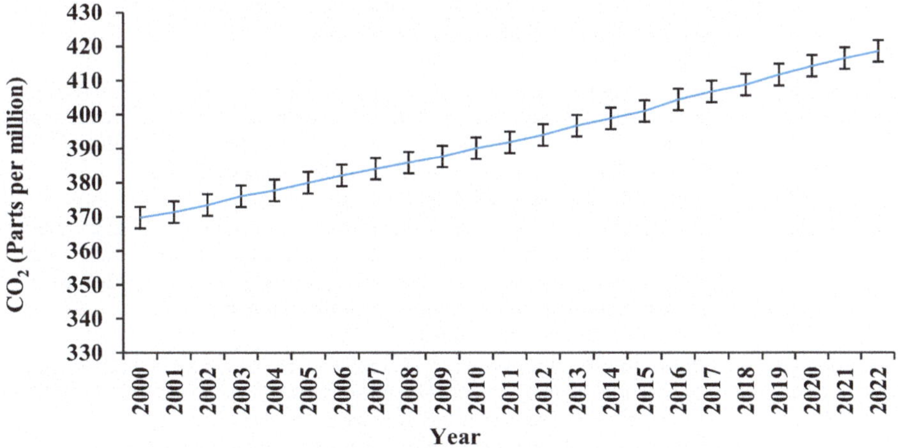

Fig. 17.6 Global annual mean CO_2 emission constructed from daily mean values. (Source: National Aeronautics and Space Administration (NASA))

escalating concentration of atmospheric CO_2 has the potential to modify sunshine, temperature, and rainfall, exerting a direct positive influence on crop growth rates and overall plant and animal productivity (Mahato 2014).

17.5.11 Integrated Crop Management Strategies to Adapt and Mitigate the Impact of Climate Change on Agriculture

Climate-resilient ICM) technologies are crucial for sustaining agricultural productivity in the face of changing climate conditions (Table 17.6). These technologies aim to enhance the adaptability of agricultural systems to factors such as temperature variability, altered precipitation patterns, and the increased frequency of extreme weather events.

Table 17.6 Climate-resilient ICM technologies for sustainable agriculture

Technologies	Importance	References
Crop diversification	Crop diversification involves planting a variety of crops to enhance resilience to climate variability. Diverse crops respond differently to environmental conditions, reducing the risk of total crop failure. This strategy also contributes to improved soil health and pest management.	Lobell et al. (2014) Rathore et al. (2022) Yadav et al. (2021c)
Water management techniques	Efficient water management is crucial in the face of changing precipitation patterns. Techniques such as drip irrigation, rainwater harvesting, and soil moisture monitoring optimize water use. These practices help mitigate the impact of droughts and contribute to sustainable water resource management.	FAO (2012)
Conservation tillage and cover cropping	Conservation tillage and cover cropping minimize soil disturbance, enhance water retention, and promote soil health. By reducing erosion, these practices contribute to climate change mitigation and help maintain soil structure in the face of extreme weather events.	Lal (2015) Yadav et al. (2019)
Agroforestry	Agroforestry integrates trees and shrubs into agricultural landscapes. This practice provides shade, reduces wind and water erosion, and enhances biodiversity. Agroforestry systems contribute to climate resilience by improving microclimates and soil stability.	Jose (2009) Yadav et al. (2021b)
Precision agriculture	Precision agriculture utilizes technology, including sensors and satellite imagery, to optimize resource use. By precisely managing inputs such as water, fertilizers, and pesticides, farmers can enhance efficiency, reduce waste, and adapt to changing climate conditions.	Khurana et al. (2018)

(continued)

Table 17.6 (continued)

Technologies	Importance	References
Integrated Pest management (IPM)	Integrated Pest management combines biological, cultural, and chemical control methods to manage pests sustainably. By promoting natural enemies of pests and reducing reliance on chemical pesticides, IPM contributes to ecosystem resilience in the face of climate change.	Kogan (1998)
Climate-resilient crop varieties	Developing and adopting crop varieties that are adapted to changing climate conditions. Resilient crop varieties can withstand temperature extremes, water stress, and emerging pest and disease patterns, ensuring stable yields in a changing climate	Challinor et al. (2014)
Climate-smart crop rotations	Implementing diversified and climate-smart crop rotations. Crop rotations that consider climate conditions contribute to soil health, reduce pest pressures, and enhance the adaptability of agricultural systems to changing climates	Lal (2015)
Climate information services	Utilizing climate information and advisory services. Importance: Access to timely and accurate climate information allows farmers to make informed decisions regarding planting schedules, irrigation, and other management practices, improving resilience to climate variability	Lobell et al. (2014)
Soil moisture monitoring	Employing soil moisture monitoring systems. Monitoring soil moisture levels helps optimize irrigation schedules, prevent waterlogging or drought stress, and improve overall water use efficiency in response to changing precipitation patterns.	
Knowledge transfer and capacity building	Providing farmers with information and tools to understand and implement climate-resilient practices. Knowledge transfer and capacity building empower farmers to make informed decisions, adapt to changing conditions, and effectively implement sustainable practices.	Schipanski et al. (2016)

17.6 Conclusion

Sustaining a conventional production system by adopting single management practices such as fertilizer, irrigation, and tillage is insufficient to meet the increasing food demand in the South Asian region. This must be reoriented holistically and site-specifically to minimize further resource base degradation and feed the growing population. Therefore, in this book chapter we have discussed the basic and advanced aspects of integrated crop management (ICM) and come to a concrete conclusion that the adoption of ICM practices would be a wise option to obtain higher crop productivity, improve soil's health, and input use efficiencies besides saving energy and water for a long-term benefit. This ICM practice allows the recycling of farm residues and opens up a new possibility to minimize the ongoing issue of crop residue burning, improving air quality, and soil health. Accordingly, the researchers, policymakers, and academicians must focus on more systematic research and multilocations trails by integrating various management practices to make agriculture more productive, conservative, and remunerative in the South Asian agro-ecologies. Further, designing ICM practices for different cropping systems and agro-climatic zones is a need of the hour, as it would help in providing safe, quality, and diversified food, besides improving the livelihood of the poor, and marginal farmers in the South Asian regions.

References

Abdurrahman MI, Chaki S, Saini G (2020) Stubble burning: effects on health & environment, regulations and management practices. Environ Adv 2:100011

Allen DE, Singh BP, Dalal RC (2011) Soil health indicators under climate change: a review of current knowledge. In: Singh BP, Cowie AL, Chan KY (eds) Soil health and climate change. Springer, Berlin/Heidelberg, pp 25–45

Altieri MA (1995) Agroecology: the science of sustainable agriculture. CRC Press, Boca Raton

Babu S, Singh R, Avasthe RK, Yadav GS, Das A, Singh VK, Mohapatra KP, Rathore SS, Chandra P, Kumar A (2020b) Impact of land configuration and organic nutrient management on productivity, quality and soil properties under baby corn in Eastern Himalayas. Sci Rep 10:16129. https://doi.org/10.1038/s41598-020-73072-6

Babu S, Singh R, Avasthe RK, Kumar S, Rathore SS, Singh VK, Ansari MA, Valente D, Petrosillo I (2023a) Soil carbon dynamics under organic farming: impact of tillage and cropping diversity. Ecol Indic 147:109940. https://doi.org/10.1016/j.ecolind.2023.109940

Baghel JK, Das TK, Mukherjee I, Nath CP, Bhattacharyya R, Ghosh S, Raj R (2020) Impacts of conservation agriculture and herbicides on weeds, nematodes, herbicide residue and productivity in direct-seeded rice. Soil Tillage Res 201:104634

Bagheri A, Bondori A, Damalas CA (2019) Modeling cereal farmers' intended and actual adoption of integrated crop management (ICM) practices. J Rural Stud 70:58–65

Bhattacharyya R, Das TK, Sudhishri S, Dudwal B, Sharma AR, Bhatia A, Singh G (2015) Conservation agriculture effects on soil organic carbon accumulation and crop productivity under a rice–wheat cropping system in the western Indo-Gangetic Plains. Eur J Agron 70:11–21

Bhuvaneshwari S, Hettiarachchi H, Meegoda JN (2019) Crop residue burning in India: policy challenges and potential solutions. Int J Environ Res Public Health 16(5):832

Biswakarma N, Pooniya V, Zhiipao RR, Kumar D, Verma AK, Shivay YS, … Swarnalakshmi K (2021) Five years integrated crop management in direct seeded rice–zero till wheat rotation of north-western India: effects on soil carbon dynamics, crop yields, water productivity and economic profitability. Agric Ecosyst Environ 318:107492

Biswakarma N, Pooniya V, Zhiipao RR, Kumar D, Shivay YS, Das TK, … Behara B (2023) Identification of a resource-efficient integrated crop management practice for the rice-wheat rotations in south Asian Indo-Gangetic Plains. Agric Ecosyst Environ 357:108675

Bünemann EK, Bongiorno G, Bai Z, Creamer RE, De Deyn G, De Goede R, … Brussaard L (2018) Soil quality – a critical review. Soil Biol Biochem 120:105–125

Chakraborty D, Ladha JK, Rana DS, Jat ML, Gathala MK, Yadav S, … Raman A (2017) A global analysis of alternative tillage and crop establishment practices for economically and environmentally efficient rice production. Sci Rep 7(1):9342

Challinor AJ, Watson J, Lobell DB, Howden SM, Smith DR, Chhetri N (2014) A meta-analysis of crop yield under climate change and adaptation. Nat Clim Chang 4(4):287–291

Chhokar RS, Singh S, Sharma RK, Singh M (2009) Influence of straw management on Phalaris minor control. Indian J Weed Sci 41:150–156

Chiroma AM, Folorunso OA, Alhassan AB (2006) The effects of land configuration and wood-shavings mulch on the properties of a sandy loam soil in northeast Nigeria. 2. Changes in physical properties. Tropicultura 24(1):33

Choudhary AK, Suri VK (2014) Integrated nutrient-management technology for direct-seeded upland rice (*Oryza sativa*) in Northwestern Himalayas. Commun Soil Sci Plant Anal 45(6):777–784

Choudhary AK, Thakur RC, Kumar N (2007) Effects of organic manures and chemical fertilizers on productivity and profitability in wheat-rice cropping sequence. ORYZA Int J Rice 44(3):239–242

Choudhary AK, Bana RS, Pooniya V, Dass A (2020) Integrated crop management technology for enhance productivity, resource-use efficiency and soil health in legumes. Indian J Agric Sci 90(10):1839–1849

Cooper J, Dobson H (2007) The benefits of pesticides to mankind and the environment. Crop Prot 26(9):1337–1348

Das TK (2008) Weed science: basics and applications. Jain Brothers, New Delhi, pp 1–901

Das TK, Bhattacharyya R, Sudhishri S, Sharma AR, Saharawat YS, Bandyopadhyay KK, Jat ML (2014a) Conservation agriculture in an irrigated cotton–wheat system of the western Indo-Gangetic Plains: crop and water productivity and economic profitability. Field Crop Res 158:24–33

Das A, Patel DP, Ramkrushna GI, Munda GC, Ngachan SV, Buragohain J, Kumar M, Naropongla (2014b) Crop diversification, crop and energy productivity under raised and sunken beds: results from a seven-year study in a high rainfall organic production system. Biol Agric Horticult 30(2):73–87

Das A, Babu S, Datta M, Kumar S, Singh R, Avasthe RK, Rathore SS, Yadav SK, Singh VK (2022a) Restoring soil carbon in marginal land of Indian Himalayas: impact of crop intensification and conservation tillage. J Environ Manag 318:115603. https://doi.org/10.1016/j.jenvman.2022.115603

Das A, Babu S, Singh R, Kumar S, Rathore SS, Singh VK, Datta M, Yadav SK, Wani OA, Yadav D (2022b) Impact of live mulch-based conservation tillage on soil properties and productivity of summer maize in Indian Himalayas. Sustainability 14:12078

Dhaliwal GS, Jindal V, Mohindru B (2015) Crop losses due to insect pests: global and Indian scenario. Indian J Entomol 77(2):165–168

Doran JW, Parkin TB (1994) Defining and assessing soil quality. In: Defining soil quality for a sustainable environment, vol 35. SSSA, Madison, pp 1–21

Food and Agriculture Organization of the United Nations (FAO) (2012) Integrated crop management. Retrieved from http://www.fao.org/3/i2956e/i2956e.pdf

Gathala MK, Ladha JK, Saharawat YS, Pathak H, Gupta RK, Yadav MP (2006) Performance of no-till rice-wheat system in the Indo-Gangetic Plains. In: Proceedings of 2nd international rice congress, New Delhi, 9–13 Oct 2006, p 474

Gathala M, Ladha JK, Saharawat YS, Kumar V, Sharma PK (2011) Effect of tillage and crop establishment methods on physical properties of a medium-textured soil under a seven-year rice-wheat rotation. Soil Sci Soc Am J 75:1851–1862

Gopal R, Jat RK, Malik RK, Kumar V, Alam MM, Jat ML, Mazid MA, Saharawat YS, McDonald A, Gupta R (2010) Direct dry seeded rice production technology and weed management in rice-based systems. Technical Bulletin, International Maize and Wheat Improvement Center, New Delhi, p 28

Gurr GM, Lu Z, Zheng X, Xu H, Zhu P, Chen G, … Heong KL (2016) Multi-country evidence that crop diversification promotes ecological intensification of agriculture. Nat Plants 2(3):1–4

Heeb L, Jenner E, Cock MJ (2019) Climate-smart pest management: building resilience of farms and landscapes to changing pest threats. J Pest Sci 92(3):951–969

Hillocks RJ (2000) Integrated crop management for smallholder farmers in Africa with special reference to coffee in Malawi. Pest Manag Sci 56(11):963–968

Hobbs PR (2007) Conservation agriculture: what is it and why is it important for future sustainable food production? J Agric Sci 145(2):127

Hossain MI, Hossain I, Mamun MAA, Siddiquie NA, Rahman MM, Rahman MS (2012) Two-wheel tractor operated strip tillage seeding equipment for dry land farming. Int J Energy Mach 5(1):35–41

https://www.cabi.org/what-we-do/cabi-centres/integrated-crop-management/

Humphreys E, Kukal SS, Christen EW, Hira GS, Sharma RK (2010) Halting the groundwater decline in north-west India – which crop technologies will be winners? Adv Agron 109:155–217
Jat ML, Chandna P, Gupta R, Sharma SK, Gill MA (2006) Laser land leveling: a precursor technology for resource conservation. Rice-Wheat consortium technical bulletin series 7:48
Jat ML, Gathala MK, Ladha JK, Saharawat YS, Jat AS, Kumar V, … Gupta R (2009) Evaluation of precision land leveling and double zero-till systems in the rice–wheat rotation: water use, productivity, profitability and soil physical properties. Soil Tillage Res 105(1):112–121
Jat ML, Chakraborty D, Ladha JK, Rana DS, Gathala MK, McDonald A, Gerard B (2020) Conservation agriculture for sustainable intensification in South Asia. Nat Sustainability 3(4):336–343
Jose S (2009) Agroforestry for ecosystem services and environmental benefits: an overview. Agrofor Syst 76:1–10
Kato Y, Okami M, Katsura K (2009) Yield potential and water use efficiency of aerobic rice (*Oryza sativa* L.) in Japan. Field Crop Res 113(3):328–334
Khambalkar V, Gajakos A, Karale DS, Kankal U (2014) Performance of broad bed furrow planter in winter season of dryland crops. Int Agric Eng J 23:14–22
Khoury WE, Makkouk K (2010) Integrated plant disease management in developing countries. J Plant Pathol 94:S35–S42
Khurana H, Kumar A, Nathawat MS, Mutum DR (2018) Precision agriculture for sustainable intensification of food production. Sci Total Environ 616-617:1563–1574
Kogan M (1998) Integrated pest management: historical perspectives and contemporary developments. Annu Rev Entomol 43(1):243–270
Kremen C, Iles A, Bacon C, Cane J (2012) Biodiversity and ecosystem services in agroecosystems. In: Lindenmayer DB, Fischer J (eds) Habitat fragmentation and landscape change: an ecological and conservation synthesis. Island Press, Washington, DC, pp 257–273
Kumar D, Shivay YS (2008) Integrated crop management. NISCAIR, New Delhi
Kumar M, Das TK, Yaduraju NT (2012) An integrated approach for management of *Cyperus rotundus* in soybean-wheat cropping system. Crop Prot 33:74–81
Kumar A, Molden D, Horowitz L, Akanda AS (2019) Sustainable agricultural water management across scales: introductory remarks. Agric Water Manag 213:1–4
Kumar A, Behera U, Dhar S, Shukla L, Singh R, Babu S, … Kumar A (2023) Conservation tillage and microbially mediated integrated phosphorus management enhance productivity, profitability, and energy use efficiency of maize. J Agric Sci 1–44. https://doi.org/10.1017/S0021859623000266
Ladha JK, Pathak H, Padre AT, Dawe D, Gupta RK (2003) Productivity trends in intensive rice–wheat cropping systems in Asia. In: Ladha JK et al (eds) Improving the productivity and sustainability of rice–wheat systems: issues and impacts. ASA Spec. Publ. 65. ASA, CSSA, and SSSA, Madison, pp 45–76
Lal R (1995) Tillage systems in the tropics: management options and sustainability implications, vol 71. Food and Agriculture Organization, Rome
Lal R (2015) Restoring soil quality to mitigate soil degradation. Sustainability 7(5):5875–5895
Lal B, Gautam P, Panda BB, Tripathi R, Shahid M, Bihari P, Guru PK, Singh T, Meena RL, Nayak AK (2020) Identification of energy and carbon efficient cropping system for ecological sustainability of rice fallow. Ecol Indic 115:106431
Lobell DB, Roberts MJ, Schlenker W, Braun N, Little BB, Rejesus RM, Hammer GL (2014) Greater sensitivity to drought accompanies maize yield increase in the US Midwest. Science 344(6183):516–519
Mahajan G, Singh K, Gill MS (2012) Scope for enhancing and sustaining rice productivity in Punjab (food bowl of India). Afr J Agric Res 7(42):5611–5620
Mahato A (2014) Climate change and its impact on agriculture. Int J Sci Res Publicat 4(4):1–6
Majsztrik J, Harkess R, Szostkiewicz I (2005) Ridge-tillage improves crop performance in a no-till organic system. Agron J 97(1):31–39
Mallarino AP, Sawyer JE, Castellano MJ, Barker DW, Kovar JL (2019) Crop residue cover effects on corn yield and soil water content in Central Iowa. Agron J 111(4):1839–1846
Mishra JS, Singh VP (2012) Tillage and weed control effects on productivity of a dry-seeded rice-wheat system on a vertisol in Central India. Soil Tillage Res 123:11–20

Mitchell J, Fukai S, Basnayake J (2004) Grain yield of direct seeded and transplanted rice in rain-fed lowlands of South East Asia. In: Proceedings of 4th international crop science congress, 26 Sept–Oct 2004, Brisbane (on CD)
NASA. https://climate.nasa.gov/vital-signs/carbon-dioxide/
Nath CP, Das TK, Rana KS, Bhattacharyya R, Pathak H, Paul S, Meena MC, Singh SB (2017) Weed and nitrogen management effects on weed infestation and crop productivity of wheat-mungbean sequence in conventional and conservation tillage practices. Agric Res 6:33–46
Naz S, Nandan R, Roy DK (2020) Effect of crop establishment methods and weed management practices on productivity, economics and nutrient uptake in direct seeded rice (*Oryza sativa* L.). Int J Curr Microbiol App Sci 9:3002–3009
Osmond DL, Meals D, Hoagland L, Arabi M, Bonta J (2012) Vegetative filter strips: how much soil will they trap? J Soil Water Conserv 67(3):67A–71A
Parihar CM, Jat SL, Singh AK, Majumdar K, Jat ML, Saharawat YS, Pradhan S, Kuri BR (2017) Bio-energy, water-use efficiency and economics of maize-wheat-mungbean system under precision-conservation agriculture in semi-arid agro-ecosystem. Energy 119:245–256
Pathak H, Bhatia A, Jain N (2014) Greenhouse gas emission from Indian agriculture: trends, mitigation and policy needs. Indian Agricultural Research Institute, New Delhi, p 39
Paudel R, Gurung RB, Bajracharya RM (2015) Effects of terracing on soil conservation and agronomic performance of maize in a steep mountainous region of Nepal. Soil Use Manag 31(1):94–102
Pooniya V, Zhiipao RR, Biswakarma N, Kumar D, Shivay YS, Babu S, Das K, Choudhary AK, Swarnalakshmi K, Jat RD, Choudhary RL, Ram H, Khokhar MK, Mukri G, Lakhena KK, Puniya MM, Jat RK, Muralikrishnan L, Singh AK, Lama A (2022) Conservation agriculture based integrated crop management sustains productivity and economic profitability along with soil properties of the maize-wheat rotation. Sci Rep 12(1):1962
Prasad DASHARATH, Rana DS, Babu S, Choudhary AK, Rajpoot S (2016) Influence of tillage practices and crop diversification on productivity and soil health in maize (*Zea mays* L.)/ Soybean (*Glycine max* L.) based cropping systems. Indian J Agric Sci 86(1):96–102
Raj R, Kumar A, Solanki IS, Dhar S, Dass A, Gupta AK, Kumar A, Kumar V, Singh CB, Jat RK, Pandey UC (2017) Influence of crop establishment methods on yield, economics and water productivity of rice cultivars under upland and lowland production ecologies of eastern indo-Gangetic Plains. Paddy Water Environ 15:861–877
Ramesh T, Rathika S (2009) Land configuration techniques for rain fed Alfisols ecosystem-review. Green Farming 2(12):879–881
Rathore SS, Babu S, Shekhawat K, Singh R, Yadav SK, Singh VK, Singh C (2022) Designing energy cum carbon-efficient environmentally clean production system for achieving green economy in agriculture. Sustain Energy Technol Assess. https://doi.org/10.1016/j.seta.2022.102190
Regmi AP, Ladha JK (2006) Enhancing productivity of rice-wheat system through integrated crop management in the Eastern-Gangetic plains of South Asia. J Crop Improv 15(1):147–170
Rejesus RM, Palis FG, Rodriguez DG, Lampayan RM (2020) The role of extension and input supply systems in modernizing developing country agriculture. Agric Syst 179:102749
Rickman JF, Pyseth M, Bunna S, Sinath P (2001) Direct seeding of rice in Cambodia. In: Proceedings of an international workshop, 30 Oct–2 Nov. ACIAR proceedings no. 101, Vientiane
Saharawat YS, Singh B, Malik RK, Ladha JK, Gathala M, Jat ML, Kumar V (2010) Evaluation of alternative tillage and crop establishment methods in a rice–wheat rotation in North Western IGP. Field Crop Res 116(3):260–267
Sayree KD, Ramos OH (1997) Application of raised bed planting to wheat. In: Wheat program, special report number 31, CIMMYT Mexico
Schipanski ME, MacDonald GK, Rosenzweig S, Chappell MJ, Bennett EM, Kerr RB, Blesh J, Crews T, Drinkwater L, Lundgren JG, Schnarr C (2016) Realizing resilient food systems. Bioscience 66(7):600–610
Shahane AA, Shivay YS, Prasanna R, Kumar D (2020) Nutrient removal by rice–wheat cropping system as influenced by crop establishment techniques and fertilization options in conjunction with microbial inoculation. Sci Rep 10(1):21944

Sidhu HS, Singh M, Singh Y, Blackwell J, Singh V, Gupta N (2011) Machinery development for crop residue management under direct drilling. 7-158. In: Proceedings of 5th world congress on conservation agriculture, Brisbane, pp 157–158

Singh S, Ladha JK, Gupta RK, Bhushan L, Rao AN, Sivaprasad B, Singh PP (2007) Evaluation of mulching, intercropping with Sesbania and herbicide use for weed management in dry-seeded rice (*Oryza sativa* L.). Crop Prot 26:518–524

Singh RK, Panda RK, Satapathy KK, Ngachan SV (2011) Simulation of runoff and sediment yield from a hilly watershed in the eastern Himalaya, India using the WEPP model. J Hydrol 405(3-4):261–276

Singh R, Babu S, Avasthe RK, Yadav GS, Chettri TK, Phempunadi CD, Chatterjee T (2015) Bacterial inoculation effect on soil biological properties, growth, grain yield, total phenolic and flavonoids contents of common buckwheat (*Fagopyrum esculentum Moench*) under hilly ecosystems of North- East India. Afr J Microbiol Res 9(15):1110–1117

Singh R, Babu S, Avasthe RK, Yadav GS, Rajkhowa DJ (2016a) Productivity, economic profitability and energy dynamics of rice (*Oryza sativa* L.) under diverse tillage and nutrient management practices in rice-vegetable pea cropping system of Sikkim Himalayas. Indian Journal of Agricultural Sciences 86(3):326–330

Singh RJ, Ghosh BN, Sharma NK, Patra S, Dadhwal KS, Mishra PK (2016b) Energy budgeting and emergy synthesis of rainfed maize–wheat rotation system with different soil amendment applications. Ecol Indic 61:753–765

Singh R, Babu S, Avasthe RK, Meena RS, Yadav GS, Das A, Mohapatra KP, Rathore SS, Kumar A, Singh C (2021) Conservation tillage and organic nutrients management improve soil properties, productivity, and economics of a maize-vegetable pea system in the eastern Himalayas. Land Degrad Dev:1–18. https://doi.org/10.1002/ldr.4066

Six J, Guggenberger G, Paustian K, Haumaier L, Elliott ET, Zech W (2001) Sources and composition of soil organic matter fractions between and within soil aggregates. Eur J Soil Sci 52(4):607–618

Subba Rao NS (1988) Biological nitrogen fixation – recent advancements. Oxford & IBH Pub. Co., New Delhi

USDA (2006) Agricultural waste management field handbook. Natural Resources Conservation Service (NRCS), United States Department of Agriculture, Washington, DC

Vanlauwe B, Wendt J, Giller KE, Corbeels M, Gerard B, Nolte C (2014) A fourth principle is required to define conservation agriculture in sub-Saharan Africa: the appropriate use of fertilizer to enhance crop productivity. Field Crop Res 155:10–13

Varatharajan T, Dass A, Choudhary AK, Sudhishri S, Pooniya V, Das TK, Rajanna GA, Prasad S, Swarnalakshmi K, Harish MN, Dhar S, Singh R, Rishi R, Kumari K, Singh A, Sachin KS, Kumar P (2022) Integrated management enhances crop physiology and final yield in maize intercropped with blackgram in semiarid South Asia. Front Plant Sci 13:975569. https://doi.org/10.3389/fpls.2022.975569

Wang D, Haung J, Nie L, Wang F (2017) Integrated crop management practices for maximizing grain yield of double-season rice crop. Sci Rep 7(1):38982

Yadav GS, Das A, Lal R, Babu S, Meena RS, Saha P, Singh R, Datta M (2018) Energy budget and carbon footprint in a no-till and mulch-based rice–mustard cropping system. J Clean Prod 191:144–157

Yadav GS, Babu S, Das A, Bhowmik SN, Datta M, Singh R (2019a) Soil carbon dynamics and productivity of rice–rice system under conservation tillage in submerged and unsubmerged ecologies of eastern Indian Himalaya. Carbon Manag 10(1):51–62. https://doi.org/10.1080/17583004.2018.1545518

Yadav GS, Das A, Lal R, Babu S, Datta M, Meena RS, Patil SB, Singh R (2019b) Impact of no-till and mulching on soil carbon sequestration under rice (*Oryza sativa* L.)-rapeseed (*Brassica campestris* L. Var. Rapeseed) cropping system in hilly agro-ecosystem of the Eastern Himalayas, India. Agric Ecosyst Environ 275:81–92

Yadav GS, Lal R, Meena RS, Datta M, Babu S, Das A, Layek J, Saha P (2019c) Energy budgeting for designing sustainable and environmentally clean/safer cropping systems for rainfed rice fallow lands in India. J Clean Prod 158:29–37

Yadav GS, Das A, Babu S, Mohapatra KP, Lal R, Rajkhowa DJ (2020) Potential of conservation tillage and altered land configuration to improve soil properties, carbon sequestration and productivity of maize based cropping system in eastern Himalayas, India. Int Soil Water Conserv Res. https://doi.org/10.1016/j.iswcr.2020.12.003

Yadav GS, Babu S, Das A, Datta M, Mohapatra KP, Singh R, Singh VK, Rathore SS, Chakraborty M (2021a) Productivity, soil health, and carbon management index of Indian Himalayan intensified maize-based cropping systems under live mulch-based conservation tillage practices. Field Crop Res. https://doi.org/10.1016/j.fcr.2021.108080

Yadav GS, Kandpal BK, Das A, Babu S, Mohapatra KP, Devi AG, Devi HL, Chandra P, Singh R, Barman KK (2021b) Impact of 28-year-old agroforestry systems on soil carbon dynamics in Eastern Himalayas. J Environ Manag. https://doi.org/10.1016/j.jenvman.2021.111978

Yadav GS, Das A, Kandpal BK, Babu S, Lal R, Datta M, Das B, Singh R, Singh VK, Mohapatra KP, Chakraborty M (2021c) The food-energy-water-carbon nexus in a maize-maize-mustard cropping sequence of the Indian Himalayas: an impact of tillage-cum-live mulching. Renew Sust Energ Rev 151:111602. https://doi.org/10.1016/j.rser.2021.111602

Yadav D, Singh D, Babu S, Madegowda M, Singh DV, Mandal D, Rathore AC, Sharma VK, Singhal V, Kumawat A, Yadav DK, Yadav RK, Kumar S (2023) Intensified cropping reduces soil erosion and improves rainfall partitioning and soil properties in the marginal land of the Indian Himalayas. Int Soil Water Conserv Res. https://doi.org/10.1016/j.iswcr.2023.10.002

Yaduraju NT, Mishra JS (2002) Zero-tillage in rice–wheat cropping system on vertisols in Madhya Pradesh: prospects and problems. In: Malik RK, Balyan RS, Yadav A, Pahwa SK (eds) Herbicide resistance management and zero tillage in rice-wheat cropping system. CCSHAUH, Haryana, pp 117–119

Yuan S, Peng S, Wang D, Man J (2018) Evaluation of the energy budget and energy use efficiency in wheat production under various crop management practices in China. Energy 160:184–191

Zhu L (2008) A report on dry direct seeding cultivation technique of early rice. J Guangxi Agric 23:10–11

Chapter 18
Climate Smart Tools and Approaches in Agriculture for Sustainable Food Production

Sonaka Ghosh, B. Bhargavi, Biswaranjan Behera, Debarati Datta, Sourav Ghosh, Md Yeasin, and Swarna Ronanki

Abstract Food security and agricultural productivity were threatened by population explosion, climate change, and natural resource depletion. Based on population growth and consumption patterns, agricultural production needs to be increased by 70% by 2050 to feed the growing global population. However, the climate change will reduce productivity and greatly impact the production of the agriculture sector (crop, livestock, fisheries, and forestry). The agriculture sector is alone contributing 14% of the global greenhouse gas emissions. To address these issues, agricultural practices must be modified in a more sustainable manner to increase productivity and income. Climate-smart agriculture (CSA) is essential for increasing agricultural productivity and sustainability while lowering GHG emissions. They assist agricultural production systems in withstanding damage and recovering quickly through mitigation and adaptation strategies. Climate-smart tools such as conservation agriculture, precision farming, soil and water conservation measures, residue management, integrated water, nutrient, and pest management strategies, climate information services, and information and communication technologies assist farmers in integrating both traditional and innovative technolo-

S. Ghosh (✉)
ICAR Research Complex for Eastern Region, Patna, Bihar, India

B. Bhargavi
ICAR-Central Research Institute for Dryland Agriculture, Hyderabad, Telangana, India

B. Behera
ICAR-Indian Institute of Water Management, Bhubaneswar, Odisha, India

D. Datta · S. Ghosh
ICAR-Central Research Institute for Jute and Allied Fibres, Barrackpore, West Bengal, India

M. Yeasin
ICAR-Indian Agricultural Statistics Research Institute, New Delhi, India

S. Ronanki
ICAR-Indian Institute of Millets Research, Hyderabad, Telangana, India

S. Babu et al. (eds.), *Agricultural Diversification for Sustainable Food Production*, Sustainability Sciences in Asia and Africa,
https://doi.org/10.1007/978-981-97-7517-0_18

gies and practices and reducing the impact of climate change. Thus, climate-smart tools and practices will play an important role and pave the way forward in increasing agricultural productivity and sustainability while also providing resilience to climate-related risks.

Keywords Climate-smart agriculture · Contingency crop planning · Microirrigation · Rainwater harvesting

18.1 Introduction

Climate-smart agriculture (CSA) is an integrated approach to managing landscapes—cropland, livestock, forests, and fisheries—that addresses interlinked challenges of food security and accelerating climate change (Lipper et al. 2014). CSA includes three outcomes such as increasing agricultural productivity sustainably to support equitable increases in farm incomes, adapting and strengthening the resilience of agricultural systems to climate change at multiple levels, and reducing GHG emissions from agriculture (including crops, livestock, and fisheries) (Lipper and Zilberman 2018; Babu et al. 2022). Climate change and its consequences in the form of extreme weather events are among the most serious threats to food security of a country. Considerable evidence suggests that average global temperatures will increase by 2–4 °C between the present and 2100 under several different GHG emission scenarios (Solomon 2007). High temperatures can shorten crop duration, alter photosynthesis, increase crop respiration rates, and have an impact on pest populations. Climate change hastens mineralization of nutrients, reduces fertilizer use efficiency (FUE), and increases soil evapotranspiration. Climate change has an indirect or direct impact on crops, water, and soil. Based on location, future climate scenarios, and projected year, the estimated implications of both historical and future climate change on cereal crop yields in various regions reveal that yield loss can be up to 35% for rice, 20% for wheat, 50% for sorghum, 13% for barley, and 60% for maize (Khatri-Chhetri et al. 2016). Freshwater availability in Central, South, East, and Southeast Asia is expected to decline by the 2050s, particularly in large river basins (Altieri and Koohafkan 2008). India has become a water-stressed country, with 1544 m^3 of available water, and is on the verge of becoming a water-scarce country (Dhawan 2017). Today, ~38% of the world's land is used for agriculture, with crops accounting for one-third of this total and livestock grazing accounting for the remainder (Pimentel et al. 1995). Agriculture, forestry, and other land use sectors account for roughly one-quarter of net anthropogenic GHG emissions (10–12 Gt CO_2 eq/yr) (IPCC 2014a; Gupta et al. 2016). Agriculture and forestry practices generate 25% of GHG emissions while decreasing carbon sinks and changing hydrological cycles, intensifying the effects of climate change (Altieri and Koohafkan 2008). The growing threat of climate change may significantly impact food production, market price fluctuations, and trade changes (Ghosh et al. 2021). So, it is obvious that sustainable land management is required to protect farmers' food and livelihood security. Food security is becoming increasingly difficult as natural resources are depleted due to overuse and climate change. On a global scale, there are over 475 million farms smaller than two hectares, with small farms controlling 40% to more than 50% of global farmland and producing more than half of the world's food (Samberg et al. 2016). It is estimated that 570 million farmers are responsible for the global food supply, with small farms contributing a major to food production, particularly in low- and middle-income countries, and they are currently threatened by climate change (Herrero et al. 2017). An unforeseen weather event, shorter growing seasons, droughts, temperature changes, and greater exposure to pests and crop diseases pose a considerable challenge to smallholder farmers worldwide, particularly in the tropics, where people rely more on natural resources. In 2017–18, an economic survey warned that "climate change may reduce annual agriculture income by 15–18% and up to 20–25% in unirrigated areas" (Padhee 2018). Humans are vulnerable to health problems as a result of food shortages and nutrient deficiencies caused by insufficient consumption of healthy food. CSA techniques can assist the farmers in adapting to and preparing for consequences, allowing them to maintain and even improve their livelihoods. With the world's population expected to reach more than nine billion by 2050, CSA is critical to ensuring global food security (Taylor 2018). If current trends continue, most of the world's population may face food and water shortages within a few decades. According to several researchers, adopting CSA technology can boost crop yields, increase input efficiency, boost net income, and lower greenhouse gas emissions (Aryal et al. 2020; Ghosh et al. 2019; Yadav et al. 2020; Das et al. 2021a).

To achieve food security and agricultural development goals by 2030, adaptation to climate change and emission reductions will be required. CSA practices can potentially reduce hunger and poverty for future generations due to climate change. It has the potential to transform the current situation and sustain agricultural production from the local to the global levels, especially in a sustainable manner. Management of farms, crops, livestock, aquaculture, and fisheries are some of the aspects that can be integrated into the CSA approach. It aids in resource management and productivity enhancement while increasing resilience; ecosystem and landscape management to conserve ecosystem services and assist farmers and land managers in implementing necessary changes. Various tools of climate-smart agriculture (Fig. 18.1) are summarized below.

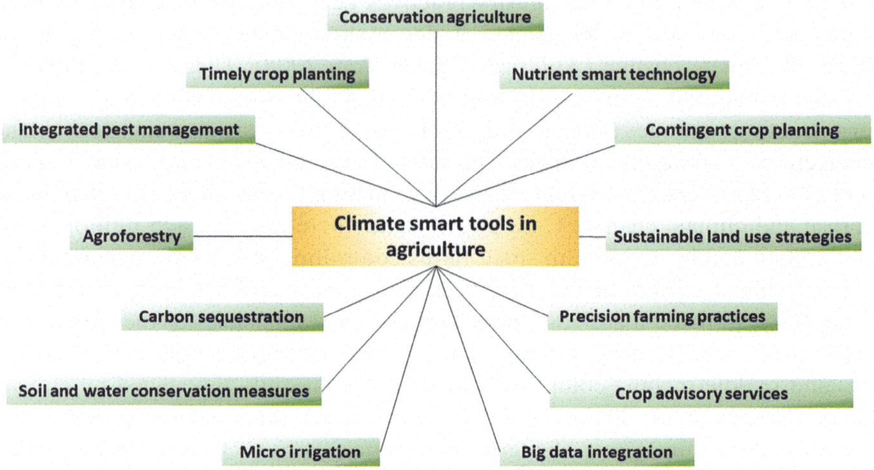

Fig. 18.1 Climate smart tools and approaches in agriculture

18.2 Climate Smart Tools and Approaches

18.2.1 Timely Planting of Crops

Sowing time is an important nonmonetary input that ensures optimum crop productivity and increases resource/input use efficiency. Timely planting and harvesting of crops build the agroecosystem's resilience to climate vagaries and thus provide yield stability (Nyström et al. 2019). The growth phases of a crop can be aligned with favorable climatic conditions by sowing it at the optimum time (Singh et al. 2019). Therefore, timely planting prevents yield losses due to risks associated with suboptimal growth conditions or extreme weather like drought, flood, heat, and cold waves. Effects of climate change in the form of warming temperatures and rainfall variabilities are quite evident, which would negatively impact crop productivity (Kumar et al. 2014). For example, a delay in the sowing of wheat crop in Western IGP pushes the grain-filling stage to the terminal heat stress, which results in a significant yield loss (Lobell et al. 2011). Studies have reported that sowing wheat 1 week earlier than the present sowing time could lead to a yield gain of 5% in Western Indo-Gangetic Plains (Lobell et al. 2013). Besides, it enhances the resource/input use efficiency due to the better response of a healthy crop to the applied inputs. The timely planting of crops provides an opportunity for growing more crops in a year on the same land and paves a path for crop intensification (Acharjee et al. 2019; Avasthe et al. 2020; Brahmanand et al. 2021).

Delayed planting of rice crops during the rainy season does not allow timely sowing of succeeding wheat or other winter season crops and, therefore, becomes a bottleneck for crop intensification. Newport et al. (2020) identified that cropping decisions during the rainy season and irrigation facilities during the winter were the main constraints for the timely sowing of wheat crops in Eastern EIGP. Urfels et al. (2021) classified major constraints of timely sowing crops into two groups: ecosystem factors and social factors. Ecosystem factors include dynamic factors like the onset of monsoon and disease-pest outbreaks and static factors like pre-monsoon water availability and land topography. In contrast, social factors include the timely availability of agro-inputs, labor, and capital. They concluded that the timely availability of irrigation water during premonsoon was the primary factor. In contrast, time to access the agro-inputs was the secondary factor responsible for the delayed planting of rice crops in EIGP. Therefore, suitable policy interventions should be taken for reliance and sustainability of resources required for timely planting of crops to achieve sustainable crop production under climate change scenarios.

18.2.2 Conservation Agriculture (CA)

Ecological footprint and climate change are two major challenges faced by present agricultural practices (Mrunalini et al. 2022). Agriculture has a significant ecological footprint as it shares ~30% of total GHG emissions (IPCC 2014b). Concurrently, it is also affected by climate change, which is a consequence of GHG emissions. Thus, a new agriculture paradigm is needed, which is productive and profitable, enhances resource use efficiency, and provides ecosystem services (FAO 2011). CA is an alternate agriculture system that saves fuel, labor, energy, nutrients, and water and sustainably provides higher productivity and profitability in addition to various ecosystem services (Yadav et al. 2020). Therefore, CA has a reduced ecological footprint and higher resilience to climate change (Das et al. 2021; Babu et al. 2023a). CA is an integration of three practically applicable principles, which are complemented by good agricultural practices for crop production. The three linking principles of CA include no or minimum disturbance of the soil, maintenance of surface residue cover, and crop diversification (Das et al. 2021a; Ghosh et al. 2023). In CA, crop seeds are directly placed in the soil without tilling it; intercultural and harvesting activities are done without soil disturbance and farm traffic to provide a better soil physical environment. At least 30% of the ground surface should be covered with plant residues under CA to enhance the soil's physical, chemical, and biological properties (Yadav et al. 2018). Crop diversification by integrating legume and nonlegume species in CA provides a synergism between component crops and their environment, making the production system more resilient (Singh et al. 2021a; Meena et al. 2022). In an intensified cropping system, the inclusion of cowpea as a live mulch and their integration with no-till (NT) and reduced tillage (RT) changes the different C pools and improves C sequestration potential and overall C storage (Das et al. 2022a) at upper soil surface (0–10 cm soil depth). In India, although ZT is practiced in an area of ~3–3.5 M ha, CA as a concept is practiced in ~1.5–2.5 million ha (Das et al. 2021a). Resource conservation technologies (RCTs) like precision nutrient management (Sapkota et al. 2014), efficient water management (Stagnari et al. 2009), and brown manuring (Behera et al. 2019a) for weed management further make the CA system more productive, profitable, and resource-efficient. For example, the brown manuring practice provides effective weed control in maize and adds carbon and plant nutrients through biomass, which becomes available to the succeeding no-till wheat crop (Behera et al. 2019b). Therefore, CA is a climate-resilient technology that could make a food production system sustainable under changing climate. Integration of minimum tillage along with live mulch of cowpea reduces bulk density and further increases maximum soil water holding capacity at upper surface of the soil compared to the conventional tillage practices in Tripura (Das et al. 2022b). As a consequence, the elimination of tillage and the development of legume-embedded cropping systems resulted in a sustainable management system under organic farming for SOC restoration and soil quality enhancement in the Himalayan environment (Babu et al. 2020, 2023b).

18.2.3 Site-Specific Nutrient Management (SSNM)

Site-specific nutrient management (SSNM) consents farmers to optimize the application of fertilizers, ensuring that crops receive the right amount of nutrients at the right time. This targeted approach improves nutrient uptake by plants, leading to increased crop yields and improved productivity. The majority of Indian soils are deficient in macro and micronutrients, mostly as a result of intensive cropping, which causes nutrient mining or a negative nutrient budget. SSNM approach aims to achieve higher crop productivity and nutrient use efficiency by applying nutrients at optimal rates and times. The excessive use and inappropriate application of fertilizers create environmental issues such as water contamination through eutrophication and greenhouse gas emissions. However, site-specific nutrient management minimizes these negative impacts by reducing fertilizer runoff, leaching, and volatilization. It promotes efficient nutrient utilization, minimizing the risk of nutrient pollution and environmental damage.

SSNM approach advocates sufficient use of fertilizer P and K to avoid soil P and K mining and to allow the best N management. These are determined with a nutrient decision support system (Witt and Dobermann 2004), which maintains scientific principles of the underlying QUEFTS model. Compared to farmer's conventional method of farming practice, SSNM in groundnut increased pod yields by 30% in the Nalgonda district and 27% in the Anantapur district. In Mahbubnagar, Kadapa, and Ranga Reddy districts, SSNM improved castor, cowpea, and maize crop yields by 51%, 41%, and 45%, respectively (Srinivasarao et al. 2010).

18.2.4 Green Manuring

Green manuring is a sustainable agriculture strategy that necessitates planting and integrating certain crops into the soil to increase fertility, structure, and overall health. These crops, also known as green manure or cover crops, are normally produced for a limited time and then plowed back into the soil while still green or shortly after blooming. This process adds organic matter to the soil, enriches it with critical nutrients, increases water retention capacity, and improves overall soil structure. Green manure crops absorb CO_2 from the environment and store it in the soil as organic matter. This process, known as carbon sequestration, serves to reduce climate change by lowering greenhouse gas emissions and increasing the soil's ability to act as a carbon sink. Using a green manure cover crop breaks up heavy compaction to allow rainwater to filter down, thus protecting soil from erosion (Yadav et al. 2021a). Moreover, it allows earthworms and soil microbes to create fertile new topsoil, increasing organic matter content in the soil. The inclusion of cowpea crop as a live mulch between summer and rainy season maize crop has been reported to improve the carbon management index and soil health compared to sole maize crop (Yadav et al. 2021a).

18.2.5 Leaf Color Chart

Among primary nutrients, nitrogen plays an imperative role in photosynthesis and is thus directly related to increase or decrease in total dry matter production. Real-time N management approach can increase the efficiency of nutrients, especially nitrogen, by matching the time of fertilizer application with plant need. A leaf color chart (LCC) is a rapid and reliable monitoring method to measure the relative greenness of a leaf which acts as an indicator of leaf nitrogen status. The nitrogen content of leaf is closely related to photosynthetic rate and biomass production and is a significant indicator indicating crop nitrogen demands (Ahlawat 2008). Need-based nitrogen application using a leaf color chart (LCC) has the potential to reduce N_2O emission and global warming potential by ~11–14% (Bhatia et al. 2012).

18.2.6 Intercropping with Legumes

Intercropping represents an ecological intensification and diversification strategy maximizing land productivity. Sowing crops along with legumes provides stable yields by efficient resource utilization and enhances fertility through atmospheric fixation of nitrogen (Layek et al. 2018; Singh et al. 2021b). High leaf cover provides a cooler microclimate through transpiration, reducing soil evaporation and associated soil temperatures (Miao et al. 2016). Deep-root system of legumes can extract moisture and nutrients from deeper soil layers. Intercropping of low canopy legumes like groundnut, black gram, soybean, and cowpea in wider spaces of maize and sorghum provides ground cover besides reducing the erosional hazards and biological insurance for boosting arable land productivity. Intercropping green gram (Pant Mung 5) with jute has been reported to suppress *Cyperus rotundus* up to 56% due to the smothering effect (Ghorai and Datta 2021).

18.2.7 Integrated Pest and Weed Management

Considering the diversity of weed problems under changing climate scenarios, combination of different weed management strategies should be evaluated to widen the weed control spectrum and efficacy for sustainable crop production. For effective weed management, preventive measures like the use of clean crop seeds and weed-free irrigation canals and bunds should be integrated. Improved agronomic practices, timeliness of operations, fertilizer and water management, and retaining crop residues on the soil surface can help in increasing weed control efficiency (WUE) of applied herbicides and crop competitiveness against weeds. For example, the use of soluble herbicides and minimal disturbance seeders can effectively control ryegrass up to 97% in ZT stubble-retained systems (Crabtree 1999). Similarly, integrating superior genotypes with a high seeding rate and early weed control can lead to yield increment of 40% compared with the combination of weaker genotypes, low-seeding rate, and delayed weed control (Harker et al. 2003). Thus, for sustainable and effective weed management, approaches such as uniform crop establishment, use of cover crops and mulch, crop rotations, stale seedbed practice, and practices for enhanced crop competitiveness with a combination of pre- and postemergence herbicides should be integrated.

Climate-smart pest management (CSPM) aims to reduce GHG emissions intensity per unit of food produced, enhance ecosystem services, and strengthen the resilience of agricultural systems by reducing pest-induced crop losses in the face of climate change. Through implementing CSPM, crop production, extension, research, and policy activities coordinate toward more efficient and resilient food production systems. Foliar disease in winter wheat can be controlled through resistant cultivars and fungicides and is associated with reduced emission with each ton of grain produced (Berry et al. 2008). For example, fungicide application reduced GHG emissions by 11–16% for winter-grown barley and 8–11% for spring barley (Hughes et al. 2011).

18.2.8 Agroforestry

Agroforestry systems are the largest sinks for terrestrial carbon and indirectly affect carbon sequestration by reducing pressure on natural forests. They play a pivotal role in conserving soil health, biodiversity, and enhancing carbon sequestration in soil and plant. The study highlights the effect of agroforestry practice on soil carbon pool, which has indicated a rate of increase of 2–3 Mg C/ha/yr (Garg 1998). The agroforestry systems have been reported to sequestrate carbon about 8.7×10^9 Mg C/yr. in tropical region; thus, they act as an important C reservoir/sink, and in a nominal change in the size of this sink can influence CO_2 emissions and climate change (Yadav et al. 2021b). Estimates of net annual C-sequestration in agroforestry crops indicate the highest values in poplar, eucalyptus, and bamboo. These agri-horticultural agroforestry systems that combine domesticated fruit trees and forest trees have the potential to lift the socio-economic status of farmers. The potential of agroforestry for rural development and employment generation amounts to 5.763 million person-days/year in the Himalayas alone (Dhyani et al. 2005; Rathore et al. 2022).

18.2.9 Green Cover Crops

Cover cropping is a viable, sustainable agricultural practice as it can provide various benefits like preventing soil erosion, weed suppression, providing soil aggregate stability, improving soil hydraulic properties, enhancing C-sequestration, reducing nitrate leaching, and hence reduction of groundwater pollution. Cover cropping may impact soil microbial functionality responsible for important soil ecosystem services. Thus, cover crops can help to improve the resilience of the crop system against climate hazards and minimize soil nutrient losses (Datta et al. 2021).

Research results highlight that soil organic carbon (SOC) at 0–10 cm was greater from cover crops (hairy vetch, rye, mixture of hairy vetch and rye) compared to no cover crops in cotton and sorghum. Predicted data suggests that after using cover crops for 155 years, SOC accumulation would be 16.7 ± 1.5 Mg/ha/year (Poeplau and Don 2015). The average soil C-sequestration rate is 1.3 t CO_2/ha/year, and the net GHG mitigation potential for winter cover crops is 1.9 t CO_2e/ha/year. Cover crops enable warming mitigation through GHG flux by 100–150 g CO_2 e/m^2/year. Nitrate leaching was 40% lower in legume cover crop treatment and 70% lower in nonlegume cover crop treatment than fallow system (Tonitto et al. 2006). Growing perennial ryegrass as a cover crop in barley in Sweden reduced nitrate leaching by less than 5 mg/L compared to 10–18 mg/L without a cover crop (Bergström and Jokela 2001). Oats reduced nitrate concentration by 26%, whereas 48% reduction was found in the rye (Kaspar et al. 2012).

18.2.10 Contingency Crop Planning

Contingency crop planning is fundamentally designed to stabilize crop output under various climatic anomalies, such as the delayed arrival of monsoon, mid-season drought and early withdrawal of rains (terminal drought), tropical storms, floods, heat waves, and cold waves. Abnormalities of weather patterns and extreme climatic events can be managed with enhanced preparedness and real-time implementation of contingency measures, as observed in pilot models in 34 villages across India (Srinivasarao et al. 2015). At Malwa, Madhya Pradesh, during *Kharif* 2013, the ridge/furrow or broad-bed and furrow (BBF) method in deep black soil regions saved soybean crops from excessive waterlogging stress. It resulted in almost normal soybean yields, while crop damage was almost 80–90% under the farmers' practice (flat-bed sowing) (Srinivasarao et al. 2020). In Southern Karnataka, early season drought was managed by modifying the normal variety with alternate finger millet variety (GPU 28) and adopting intercropping with pigeon pea along with conservation furrow, which facilitated better soil moisture storage (Srinivasarao et al. 2020). To manage mid-season droughts (during vegetative crop growth stage), recommended practices include foliar sprays, plant thinning, and lifesaving irrigation with harvested rainwater (Srinivasarao et al. 2013). Bheemappa and Chandranath (2017) reported that, in the northern and northern transition zone of Karnataka, farmers preferred the recommended mono crops over the recommended intercropping systems but were following intercropping of safflower with chickpea and wheat. Delay in crop sowing during *kharif* season reduced productivity levels in several crops like maize (8–29%), groundnut (18–47%), BT cotton (39–60%), and chili (33–55%). Srinivasarao et al. (2020) opined that terminal drought is more critical since the crop duration gets shortened due to forced maturity and interventions, i.e., protective irrigation from harvested rainwater in farm ponds and mulching in different crops gave 39–55% higher yields compared to farmers' practices. In Maharashtra, to combat delayed onset of monsoon during 2014–15, the in situ moisture conservation practices adopted were drip and sprinkler irrigation technologies, increasing planting density with local (desi) varieties of cotton, broad-bed and furrow sowing of soybean, residue mulching in horticultural crops, sowing of hybrid pigeon pea, and use of mungbean instead of soybean (Srinivasarao et al. 2020). Impact of contingency interventions on crop productivity and profitability is given in Table 18.1.

18.2.11 Carbon Sequestration Technologies

Soil organic carbon (SOC) sequestration is the practice of capturing CO_2 from the air into the soil through crop biomass and other organic solids in a form that is not instantly reemitted. Carbon sequestration helps to neutralize emissions from fossil fuel burning and allied carbon-emitting activities, thereby improving the quality of soil and agronomic efficiency over the years. The pace of SOC sequestration is determined by soil texture, soil structure, precipitation, temperature, crop and livestock components, and soil management (West and Post 2002; Lal 2004). In temperate climates, carbon decomposition happens more slowly, with a higher possibility of humus formation, which is a stable part of SOC that leads to carbon sequestration. Crop productivity is positively correlated with SOC content due to enhancement in soil properties (physical, chemical, and biological) and, subsequently, crop productivity (Franzluebbers 2002). The modifications in land management that enhance soil C storage include reduced tillage, no-tillage, modified crop rotations, cover crops, avoiding summer fallow, improved fertilizer and water management, buffer or conservation strips, application of manure and sludge, improved grazing, moisture conservation and rainwater harvesting, agroforestry, and bioenergy plantations (Lal et al. 2015). Lal (1998) projected that practicing reduced

Table 18.1 Impact of contingency interventions on crop productivity and profitability

Intervention	Impact
Climate smart variety	A short-duration variety of finger millet (GPU-28), was superior during early season drought with a higher benefit: cost (B:C) ratio of 2.30 over the long-duration variety MR-1.
Contingent crop	In Punjab, crops like sesame, cluster bean, green gram, black gram, groundnut, and pearl millet gave better insurance during early season drought and higher system productivity. Similarly, in Tamil Nadu, pearlmillet (Bajra) had better performance than other contingent crops like maize, sorghum, and pulses.
Farm pond	In Assam, collecting the harvested rainwater in excavated farm ponds gave 15–20% improved yield with timely transplanting of rice seedlings.
Plant thinning	Under finger millet + pigeon pea intercropping, thinning improved crop yield and benefit: cost ratio by better conservation of soil moisture.
Foliar spray	At Jamnagar, foliar application of Urea and KNO_3 @ 2% increased the yield of chickpea by 26% and 30%, respectively. In Madhya Pradesh, foliar spray of VAM-C 50% SL@ 3.75 L/ha enhanced the yields of soybean, maize, black gram, and horse gram by 15–20% compared to no spray.
Conservation furrow	Adoption of conservation furrow in cotton and soybean conserved soil moisture content and improved yield by 19% and 33%, respectively.
Supplemental irrigation	Lifesaving irrigation enhanced groundnut pod yield and cotton lint yield by 26% and 22%, respectively. In soybean, lifesaving irrigation at pod formation and seed filling stage through stored rainwater provided 55% higher yields than rainfed crops (632 kg/ha).
Flood mitigation	Draining excess water in maize in Jammu and Kashmir increased maize yield by 23%.
Compartmental bunding	In Vijayapura, adoption of compartmental bunding for in-situ moisture conservation increased yield in post-rainy crops like sorghum, sunflower, safflower, and chickpea by 40%, 35%, 38%, and 50% respectively.
Broad-bed and furrow (BBF)/ridge furrow	BBF in black soils of Tamil Nadu increased the productivity of maize by mung bean, barnyard millet, and sorghum by 14–20%.

Source: Das (2008), Srinivasarao et al. (2014, 2020)

tillage on some 400 m ha of cropland by 2020 may lead to total C sequestration of 1500–4900 Mg. West and Post (2002) stated that ZT could rapidly increase soil organic carbon, especially at the soil surface, which could be due to higher soil aggregation (Six et al. 2000). Moreover, conservation tillage offers physical protection to carbon, and the same tillage should be continued to maintain carbon sequestration. Meta-analysis revealed that diversified crop rotations led to improvement in soil organic carbon and microbial biomass as compared to mono-cropping or unscientific diversification, especially upon the inclusion of cover crops (McDaniel et al. 2014). Tiemann et al. (2015) highlighted that increasing rotational diversity positively influences soil carbon accumulation by enhancing the capacity of soil microbes to act on plant residues and protect them in aggregates hastily. Diversified crop rotation and leguminous cover crops, containing carbon compounds that have more resistance to microbe-mediated metabolism, increased the complexity and diversity of soil carbon, rendering increased stability (Wickings et al. 2012). Cover cropping with legumes is one of the significant techniques for SOC sequestration and plays a vital role in mitigating climate change (Lal et al. 2015). Leguminous N-fixing cover crops improve organic C in the soil by degrading their residues. The extensive root system of leguminous cover crops like sun hemp (*Crotalaria juncea*) having extensive root system leads to the larger macro-aggregate formation (Six et al. 2002). Naturally, leguminous crops fix a higher quantity of nitrogen (N) through biological nitrogen fixation (BNF), which, in turn, leads to C sequestration with an average rate greater than the grasses (Martins et al. 2012).

18.2.12 Soil and Water Conservation

Soil conservation has been typically mentioned in relation to keeping soil in place for crop production. Today, the benefits of soil conservation are assessed in terms of how they will help with boosting crop yields, lowering water pollution, and lowering atmospheric GHG concentrations. Overgrazing, deforestation, and poor agricultural practices are each responsible for about 35%, 30%, and 28%, respectively, of soil erosion (FAO 1996). Adopting appropriate soil and water conservation practices is important to cope with progressive climate change and variability. Several such practices are minimum/ZT, laser land leveling, raised bed planting, keeping land consistently covered with crop residues, and agroforestry have been proposed for adaptation to climate change and variability.

18.2.12.1 Rainwater Harvesting

Rainwater harvesting is one of the promising solutions to supplement the scarce surface and underground water resources in areas where the current water supply system is unable to meet demand. It reduces the impact of water availability fluctuations and increases the consistency of agricultural output. Check dams, embankment-style ponds, and dugout ponds are examples of rainwater harvesting structures. Engineering and agronomic practices are advantageous for rainwater harvesting. Agronomic techniques such as contour cultivation, mulching, trench plantation, furrow irrigation, deep tillage, contour farming, raised bed techniques of cultivation, and ridges may be used for arable land at the farm level. Contrarily, engineering methods include broad-bed and furrow systems, contour trenches, graded bunds, bench terraces, and contour trenches.

Rainwater harvesting from roofs of houses and rainwater harvesting from sloped roofs of low-cost and hi-tech polyhouses are becoming more crucial, particularly in India's northeastern hill states. In countries like China, Brazil, Australia, and India, rainwater harvesting is encouraged and promoted. IPCC (2014a) advocated that rainwater harvesting structures are extremely important for mitigating the impact on agriculture and increasing agricultural productivity.

18.2.12.2 Minimum Tillage/Zero-Tillage (ZT)

ZT was found to be beneficial for enhancing agricultural output, soil health, water use, and farmers' income (Gupta and Seth 2007; Yadav et al. 2021c; Ghosh et al. 2022). Zero-tillage wheat, in the Indian IGP, has been the most successful resource-conserving technology in rice–wheat systems to date. When compared to traditional tillage, ZT wheat uses residual water more efficiently, which minimizes the need for irrigation in wheat (Erenstein 2007). According to reports, ZT can reduce input costs by \$40–\$50/ha, water use by 13–33%, fuel consumption by 75%, and tractor operation time by 81% (Malik et al. 2002). Compared to conventional tilling, zero tillage with residue retention reduces evaporative moisture loss and preserves soil moisture (Jat et al. 2009; Yadav et al. 2019b).

18.2.12.3 Laser Land Leveling

Field unevenness results in inadequate use of provided water and nutrients, which results in uneven crop stand, poor crop performance, and lower farmer profitability (Rickman 2002). Farmers have traditionally leveled their fields with scrapers or leveling boards pulled by tractors or draught animals, but these methods are cumbersome, slow, and less precise. Laser-guided land leveling is the cutting-edge technique used to level or slope the field. Flood irrigation is a common practice in IGP's rice–wheat system; however, poor management and uneven fields cause a loss of irrigation water (10–25%) during application on the farm, resulting in lower crop yields, higher irrigation costs, and inefficient resource use (Kahlown et al. 2002). Improved crop stand and crop production (up to 30%), enhanced nutrient use efficiency, and a reduction in labor requirements for weeding rice fields from 21 to 5 d/ha are additional advantages of laser land-leveling (Jat et al. 2006).

18.2.12.4 Raised Bed Planting

Flood irrigation results in inefficient use of irrigation water and nitrogen, especially when the fields are uneven in topography. Farmers in many parts of the world have been using raised bed planting techniques for ages (Govaerts et al. 2007), and in India, it is majorly followed in wheat-based cropping systems (Ram et al. 2005). Briefly, the raised planting technique involves seeding in 2–3 rows on raised beds that are 70–90 cm wide. Typically beds are of 15–30 cm height and furrows of 30 cm wide. Irrigation water is applied to the furrows between the beds. Raised bed planting has the potential to save irrigation water (up to 30–50% in wheat) coupled with enhanced water use efficiency and nitrogen use efficiency (Hobbs and Gupta

2001; Kukal et al. 2005; Jat et al. 2005). Other advantages include decreased crop lodging, mechanical weeding, reduced seed rate, and rainwater conservation (Gupta et al. 2000; Fahong et al. 2004). Based on research findings from India and other countries, the raised bed planting system is being assessed and recommended for many crops to deal with the weather aberrations.

18.2.13 Microirrigation

In response to the rapidly declining irrigation water potential and rising demand for water from other sectors, various strategies have been established to conserve water and improve the current water usage efficiency in global agriculture. Microirrigation (drip and sprinkler methods of irrigation) is one such technique that was only majorly adopted in Indian agriculture. Microirrigation boosts crop output and water efficiency while lowering cultivation costs. According to studies conducted by numerous institutes, the microirrigation technique increases yield for some crops by up to 100% while using 40–70% less water than surface irrigation. The water-intensive crops like sugarcane, bananas, and vegetable crops are proven to save more water than other crops. The other benefits of microirrigation are lower weed growth and reduced groundwater exploitation (Viswanathan and Bahinipati 2015). Initiatives can be taken to increase water productivity for agriculture, and the water saved can be used to irrigate additional land or to meet environmental and social needs. At the field level, several factors including crop type and spacing between them, kind of microirrigation technology, soil type, climate, and geo-hydrology need to be taken into account for actual water savings through the use of microirrigation systems. Saline water can also be used in this approach, as the salt only accumulates at the borders of the wetting zone and has no effect on crop development (Sivanappan 2014). Only 7.7 million hectares of land is microirrigated, of which 3.37 million hectares use drip irrigation and 4.36 million hectares with sprinkler irrigation. India, however, is thought to have a total potential of 69.5 million hectares for microirrigation (FICCI 2016).

18.3 Climate Information Services

Climate change and its associated risks, i.e., drought, floods, heat, and cold waves, result in reduced agricultural productivity and farm profitability. Thus, timely information on weather parameters, market intelligence, and provision of agro-advisories will enable minimizing adverse effects of climate change. In this context, climate information services (CIS) play a major role in adaptation to climate change. They are defined as services that assist in the decision-making of individuals/organizations by providing climate information in a systematic way (Lugen 2017). They provide timely information on climate and weather-related data and thus assist in decision-making and policy planning (Vaughan et al. 2018). Quality services of CIS will empower the decision makers to manage activities in a better way in key sectors like agriculture, energy sector, public health, and disaster risk management, thus paving a way for the development of the nation. Climate information services provide scientific knowledge to individuals in order to manage the risks associated with climate variability and thus help the public fraternity to cope with increased climate change effects and become more resilient. Major goal of CIS is to provide individual and location-specific climate information to the end users and assist in decision support tools (Krauss and von Storch 2012). CIS will forecast the upcoming seasons in order to enable timely decisions, project long-term trends to guide policy making and strategic planning, and assist in tracking the climate-related uncertainties for timely risk management (Vaughan et al. 2018).

18.3.1 Weather-Based Crop Advisory Services

Localized CIS in agriculture provides information regarding climate, its impacts on crops, livestock, fisheries, and management practices to be followed to prevent, reduce, and/or manage risks. This kind of weather-based crop advisory information will assist farmers in making better management decisions to reduce climate-related risks. They enable the farmers to select profitable cropping patterns, decide on optimum sowing dates, method of planting, application of plant protection chemicals, and time and stage of harvesting. These optimized and better crop management practices will cutoff the production costs and ultimately result in increased crop productivity and reduced crop losses and thus increase the resilience to climate change. In India, medium-range weather forecasts in various agro-climatic zones are used to provide weather-based crop advisories to farmers. The economic value of the advisories prepared from agromet information is obvious in terms of its impact on cutting off the cost of external farm inputs, increasing crop yields, and enhancing profitability of farming communities (Balaji and Craufurd 2011). India Meteorological Department (IMD) along with the Indian Council of Agricultural Research, state universities, and Federal and State Ministries of Agriculture provide agromet advisory services (AAS) bulletins weekly twice to the end users. IMD provides a link between national agrometeorological institutions and those involved in the transfer of agromet information and advisories.

18.3.2 ICT-Based Agro-advisories

Information and communication technologies (ICTs), i.e., mobile phones, audio-visual aids, kisan call centers, and internet services, are a promising field for the dissemination of CIS and CSA technologies and practices to farmers. ICT-based climate information and agro-advisory services will help the farmers to choose appropriate technology, better crop management practices, and marketing approaches (Ospina and Heeks 2012). For example, timely information about monsoons will enable the farmers to optimize their sowing dates in order to avoid seasonal droughts during crop growth period, particularly in rain-fed areas (Raghuvanshi et al. 2018). ICTs strengthen the physical preparedness of livelihood systems for climate change-related events through applications such as geographic information and positioning systems and modeling applications. ICT plays a remarkable role in improving the productivity of agricultural crops. Some successful ICT applications in India are Warana, Agisnet, Dristee, Agmarknet, E-Chaupal, Kisan call centers, Lokmitra, E-Post, Gramdoot, Dyandoot, Tarahaat, Dhan, Akshaya, Tata Kisan Kendra, etc. (Singh et al. 2015).

18.4 Big Data Integration

Big data is an important tool for assessing farm-level decisions, policy making, and marketing measures in order to improve agricultural production in a sustainable way. Big data refers to vast, diversified, complex, distributed datasets collected from a variety of sources. Traditional data analysis might also have large number of observations, but just a few variables are used to describe single phenomena. Big data transforms this paradigm by emphasizing that when data and sources are huge and diverse, that may be used to describe complicated events more accurately (Halevy et al. 2009). Large volume, variety, and velocity (3V's) are three fundamentals of big data. Agricultural data are enormous in size, generated from various cropping patterns as well as numerous agricultural operations from planting to marketing. This data possesses high value at many phases of cultivation, such as soil pH, soil nutrient content, moisture content, humidity, irrigation requirements, disease and insect infestation, and other associated management practices. Agricultural data possesses all elementary properties of big data. Basic concept of big data analysis is the employment of innovative technologies for data integration, storage, processing, visualization, and analysis (Mayer-Schönberger and Cukier 2013). The process of extracting information from big data consists of: (i) data collection and storage, (ii) data preprocessing, (iii) data integration, (iv) statistical analysis, and (v) inference and conclusion. A process of big data analysis is depicted in Fig. 18.2.

Climate change perpetuates challenges of food security and sustainability by increasing uncertainty and danger of natural calamities in agriculture. Modern technology provides an opportunity to transform traditional agriculture into CSA. The transition from traditional agriculture to climate-smart agriculture requires a large quantity of data from all stages of crop development and its relation to environment. Major drawbacks of implementation of CSA technologies are the lack of spatial and temporal specificity of climate change data. Big data has the capacity to produce the spatial and temporal specificity as well as the timeliness and scalability of agricultural data that is required for CSA. Big data analytics helps to improve strategic and tactical decision-making with spatial and temporal specificity, which is mandatory

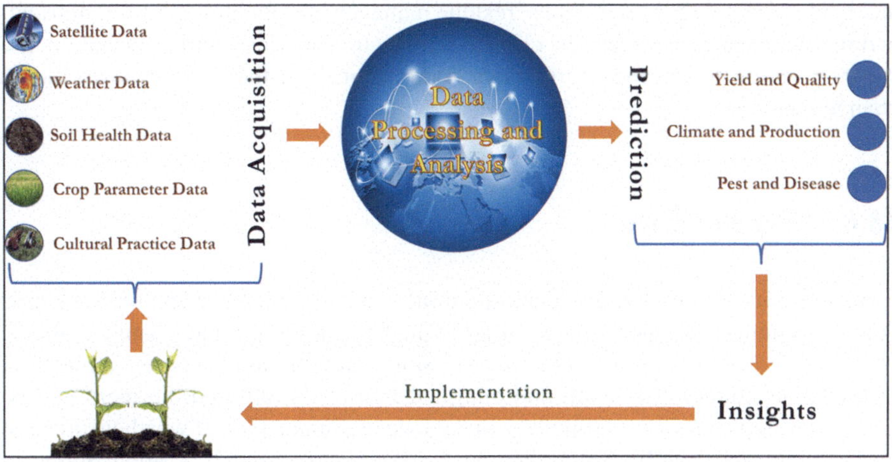

Fig. 18.2 Big data analysis in agriculture

part of climate-smart agriculture. Application of big data can be summarized in broad areas, such as

- developing a predictive model for climate change with resilience farm planning and practice
- hastening and improving the precision of plant breeding for climate resilience, and
- delivering specialized knowledge support at the farm level for climate change adaptation

Despite the huge benefits of big data applications in agriculture, there is still a controversy over their relevance to Indian agriculture. The use of agricultural big data is still at a premature stage, which needs more investment in infrastructure, training related personnel, improving farmers' technological knowledge, raising awareness about the benefits of big data technologies, and changing farmers' attitudes towards technology adoption.

18.5 Conclusion

The purpose of this chapter was to highlight (i) the necessity of climate-smart agriculture and (ii) some significant climate-smart techniques to address difficulties to be faced with food security, mitigation, and adaptation of climate change, which are closely associated with agriculture. Because climate change affects agricultural production and food systems, a strategy for transforming agricultural systems is urgently needed to help with global food security and poverty reduction. Farmers can use climate-smart tools such as conservation agriculture, precision farming, soil and water conservation techniques, residue management, integrated water, nutrient, and pest management strategies, climate information services, and information and communication technologies to ensure food security in the face of new climate change realities.

18.6 Way Forward

Climate-smart agricultural practices and tools can help transform and reorient agricultural systems to support food security in the face of new climate change realities. However, more research into precise irrigation, nutrient, and pesticide application technologies suitable for small and marginal farmers is required. Farmers will be able to develop contingent crop plans in the face of climate risks if weather forecasts are accurate and timely. Strengthening agro-advisories and ICT applications to quickly disseminate data to farmers is critical. To develop abiotic stress-tolerant crop varieties that can adapt to future climate change, novel breeding and biotechnological approaches must be strengthened. Many agricultural technologies and practices, including those eligible for CSA, are not reaching their full potential impact due to low levels of adoption among farmers in developing countries. There is a need to demonstrate real impact beyond the plot or site level to more people across larger areas, as well as on institutions and policies, which drives the interest in scaling up.

References

Acharjee TK, Van Halsema G, Ludwig F, Hellegers P, Supit I (2019) Shifting planting date of Boro rice as a climate change adaptation strategy to reduce water use. Agric Syst 168:131–143. https://doi.org/10.1016/j.agsy.2018.11.006

Ahlawat RPS (2008) Reorientation of agronomic research. In: Sharma et al (eds) Souvenir, national symposium on new paradigms in agronomic research, pp 19–21)

Altieri MA, Koohafkan P (2008) Enduring farms: climate change, smallholders and traditional farming communities, vol 6. Third World Network (TWN), Penang

Aryal JP, Sapkota TB, Rahut DB, Jat ML (2020) Agricultural sustainability under emerging climatic variability: the role of climate-smart agriculture and relevant policies in India. Int J Innov Sustain Dev 14(2):219–245. https://doi.org/10.1504/IJISD.2020.106243

Avasthe RK, Babu S, Singh R, Yadav GS, Kumar A (2020) Productivity and profitability assessment of organically grown vegetables embedded in rice-based cropping sequences in Sikkim Himalayas, North East India. J Environ Biol 41:111–117

Babu S, Singh R, Avasthe RK, Yadav GS, Das A, Singh VK, Mohapatra KP, Rathore SS, Chandra P, Kumar A (2020) Impact of land configuration and organic nutrient management on productivity, quality and soil properties under baby corn in Eastern Himalayas. Sci Rep 10:16129. https://doi.org/10.1038/s41598-020-73072-6

Babu S, Das A, Singh R, Mohapatra KP, Kumar S, Rathore SS, Yadav SK, Yadav P, Ansari MA, Panwar AS, Wani OA, Singh M, Ravishankar N, Layek J, Chandra P, Singh VK (2022) Designing an energy efficient, economically feasible, and environmentally robust integrated farming system model for sustainable food production in the Indian Himalayas. Sustainable Food Technol. https://doi.org/10.1039/d2fb00016d

Babu S, Singh R, Avasthe RK, Rathore SS, Kumar S, Das A, Layek J, Sharma V, Wani OA, Singh VK (2023a) Conservation tillage and diversified cropping enhance system productivity and eco-efficiency and reduce greenhouse gas intensity in organic farming. Front Sustain Food Syst 7:1114617. https://doi.org/10.3389/fsufs.2023.1114617

Babu S, Singh R, Avasthe RK, Kumar S, Rathore SS, Singh VK, Ansari MA, Valente D, Petrosillo I (2023b) Soil carbon dynamics under organic farming: impact of tillage and cropping diversity. Ecol Indic 147:109940. https://doi.org/10.1016/j.ecolind.2023.109940

Balaji V, Craufurd P (2011) Using information and communication technologies to disseminate and exchange agriculture-related climate information in the Indo-Gangetic Plains, for the CCAFS Project Report, 2011

Behera B, Das TK, Rathi N (2019a) Carry-over effect of brown manuring supplemented with nitrogen on productivity and profitability in succeeding wheat. Indian J Agric Sci 89(4):152–157

Behera B, Das TK, Ghosh S, Parsad R, Rathi N (2019b) Effects of brown manure species, seed rate and time of application of 2, 4-D on weed control efficiency, productivity and profitability in maize. Indian J Weed Sci 51(4):393–397. https://doi.org/10.5958/0974-8164.2019.00082.0

Berry PM, Kindred DR, Paveley ND (2008) Quantifying the effects of fungicides and disease resistance on greenhouse gas emissions associated with wheat production. Plant Pathol 57(6):1000–1008. https://doi.org/10.1111/j.1365-3059.2008.01899.x

Bergström LF, Jokela WE (2001) Ryegrass cover crop effects on nitrate leaching in spring barley fertilized with 15NH415NO3. J Environ Qual 30(5):1659-1667

Bhatia A, Pathak H, Jain N, Singh PK, Tomer R (2012) Greenhouse gas mitigation in rice–wheat system with leaf color chart-based urea application. Environ Monit Assess 184(5):3095–3107. https://doi.org/10.1007/s10661-011-2174-8

Bheemappa A, Chandranath HT (2017) Adoption of contingency crop planning by the farmers of North Karnataka. J Farm Sci 30(1):87–94

Brahmanand PS, Behera B, Srivastava SK, Singandhupe RB, Mishra A (2021) Cultivated land utilization index vis-a-vis cropping intensity for crop diversification and water resource management in Odisha, India. Curr Sci 120(7):1217–12241217. https://doi.org/10.18520/cs/v120/i7/1217-1224

Crabtree W (1999) Stubble, harrows, ryegrass, soil residual herbicides and no-till sowing. 1999 Western Australian GRDC Crop Updates-Weed Update, pp 35–37

Das TK (2008) Weed science: basics and applications. Jain Brothers, New Delhi. 901p

Das TK, Ghosh S, Das A, Sen S, Datta D, Ghosh S, Raj R, Behera B, Roy A, Vyas AK, Rana DS (2021a) Conservation agriculture impacts on productivity, resource-use efficiency and environmental sustainability: a holistic review. Indian J Agron 66:S111–S127

Das TK, Ghosh S, Datta D, Sen S, Roy A, Raj R, Behera B, Sharma AR (2021) Conservation agriculture for efficient resource use, sustainable crop production and better environment. Indian Farming 71(11):07–11

Das A, Babu S, Datta M, Kumar S, Singh R, Avasthe RK, Rathore SS, Yadav SK, Singh VK (2022a) Restoring soil carbon in marginal land of Indian Himalayas: impact of crop intensification and conservation tillage. J Environ Manag 318:115603. https://doi.org/10.1016/j.jenvman.2022.115603

Das A, Babu S, Singh R, Kumar S, Rathore SS, Singh VK, Datta M, Yadav SK, Wani OA, Yadav D (2022b) Impact of live mulch-based conservation tillage on soil properties and productivity of summer maize in Indian Himalayas. Sustain For 14:12078

Datta D, Ghosh S, Saha R, Nath CP (2021) Cover crops: potential and prospects in conservation agriculture. In: Saha D, Barman MS (eds) Conservation agriculture and climate change: impact and adaptation. New India Publishing Agency, New Delhi, p 496. 110 034. ISBN 978-93-90591-08-4

Dhawan V (2017) Water and agriculture in India. In: Background paper for the South Asia expert panel during the Global Forum for Food and Agriculture (GFFA), p 15

Dhyani SK, Sharda VN, Sharma AR (2005) Agroforestry for sustainable management ofsoil, water and environmental quality: looking back to think ahead. Range Manage Agrofor 26(1):71–83

Erenstein O (2007) Adoption and impacts of zero tillage as a resource conserving technology in the irrigated plains of South Asia, vol 19. CIMMYT, International Water Management Institute, Colombo

Fahong W, Xuqing W, Sayre K (2004) Comparison of conventional, flood irrigated, flat planting with furrow irrigated, raised bed planting for winter wheat in China. Field Crop Res 87(1):35–42

Federation of Indian Chambers of Commerce and Industry (2016) Accelerating growth of Indian agriculture: micro irrigation an efficient solution. Strategy paper. Prospects of micro irrigation in India. Federation of Indian Chambers of Commerce and Industry. http://www.grantthornton.in/globalassets/1.-member.../india/.../micro-irrigation-report.pd

Food and Agriculture Organization (1996) Our land our future. Food and Agriculture Organization and United Nations Environment Programme, Rome

Food and Agriculture Organization (2011) Save and grow, a policymaker's guide to sustainable intensification of smallholder crop production. Food and Agriculture Organization, Rome

Franzluebbers AJ (2002) Water infiltration and soil structure related to organic matter and its stratification with depth. Soil Tillage Res 66(2):197–205. https://doi.org/10.1016/S0167-1987(02)00027-2

Garg VK (1998) Interaction of tree crops with a sodic soil environment: potential for rehabilitation of degraded environments. Land Degrad Dev 9(1):81–93. https://doi.org/10.1002/(SICI)1099-145X(199801/02)9:1<81::AID-LDR267>3.0.CO;2-R

Ghorai AK, Datta D (2021) Crop diversification in CA/RCTs under climate change perspective: special reference to jute and allied fibres. In: Saha D, Barman MS (eds) Conservation agriculture and climate change: impact and adaptation. New India Publishing Agency, New Delhi, p 119. 110 034. ISBN 978-93-90591-08-4

Ghosh S, Das TK, Sharma DK, Gupta K (2019) Potential of conservation agriculture for ecosystem services: a review. Indian J Agric Sci 89(10):1572–1579

Ghosh S, Prasad JVNS, Ravisankar N, Naik MR, Srinivasarao C (2021) Farming system research for resource conservation and improved livelihoods in India. In: Srinivasarao C et al (eds) Agricultural research, technology and policy: innovations and advances. ICAR-National Academy of agricultural research management (NAARM), Hyderabad, pp 163–190

Ghosh S, Das TK, Shivay YS, Bandyopadhyay KK, Sudhishri S, Bhatia A, Biswas DR, Yeasin M, Ghosh S (2022) Weeds response and control efficiency, greengram productivity and resource-use efficiency under a conservation agriculture-based maize-wheat-greengram system. Indian J Weed Sci 54(2):157–164

Ghosh S, Das TK, Nath CP, Bhatia A, Biswas DR, Bandyopadhyay KK, Yeasin M, Raj R (2023) Weed seedbank, above-ground weed community and crop yields under conventional and conservation agriculture practices in maize–wheat–mungbean rotation. Weed Res 63(4):270–281

Govaerts B, Sayre KD, Lichter K, Dendooven L, Deckers J (2007) Influence of permanent raised bed planting and residue management on physical and chemical soil quality in rainfed maize/wheat systems. Plant Soil 291(1–2):39–54. https://doi.org/10.1007/s11104-006-9172-6

Gupta R, Seth A (2007) A review of resource conserving technologies for sustainable management of the rice–wheat cropping systems of the indo-Gangetic plains (IGP). Crop Prot 26(3):436–447. https://doi.org/10.1016/j.cropro.2006.04.030

Gupta RK, Hobbs PR, Salim M, Malik RK, Varma MR, Pokharel TP, ... Tripathi J (2000) Research and extension issues for farm-level impact on the productivity of the rice-wheat systems in the Indo-Gangetic Plains of India and Pakistan. No. REP-8551. CIMMYT

Gupta DK, Bhatia A, Kumar A, Das TK, Jain N, Tomer R, Malyan SK, Fagodiya RK, Dubey R, Pathak H (2016) Mitigation of greenhouse gas emission from rice–wheat system of the Indo-Gangetic plains: through tillage, irrigation and fertilizer management. Agric Ecosyst Environ 230:1–9. https://doi.org/10.1016/j.agee.2016.05.023

Halevy AY, Norvig P, Pereira FC (2009) The unreasonable effectiveness of data. IEEE Intell Syst 24:8–12

Harker KN, Clayton GW, Blackshaw RE, O'Donovan JT, Stevenson FC (2003) Seeding rate, herbicide timing and competitive hybrids contribute to integrated weed management in canola (*Brassica napus*). Can J Plant Sci 83(2):433–440. https://doi.org/10.4141/P02-073

Herrero M, Thornton PK, Power B, Bogard JR, Remans R, Fritz S, Gerber JS, Nelson G, See L, Waha K, Watson RA, West PC, Samberg LH, van de Steeg J, Stephenson E, van Wijk M, Havlík P (2017) Farming and the geography of nutrient production for human use: a transdisciplinary analysis. Lancet Planet Health 1(1):e33–e42. https://doi.org/10.1016/S2542-5196(17)30007-4

Hobbs PR, Gupta RK (2001) Rice–wheat cropping systems in the Indo-Gangetic plains: issues of water productivity in resource conserving technologies. In: Proceedings of the water productivity workshop, International Water Management Institute, Nov 2001, pp 12–14

Hughes DJ, West JS, Atkins SD, Gladders P, Jeger MJ, Fitt BD (2011) Effects of disease control by fungicides on greenhouse gas emissions by UK arable crop production. Pest Manag Sci 67(9):1082–1092. https://doi.org/10.1002/ps.2151

IPCC (2014a) 2014: Synthesis report. Contribution of working groups I, II and III to the fifth assessment report of the Intergovernmental Panel on Climate Change Core Writing Team. In: Pachauri RK, Mayer LA (eds) Climate change. IPCC, Geneva, p 151

IPCC (2014b) Fifth assessment synthesis report (longer report) of Intergovernmental Panel on Climate Change. In: Climate change. Cambridge University Press, Cambridege

Jat ML, Singh S, Rai HK, Chhokar RS, Sharma SK, Gupta RK (2005) Furrow irrigated raised bed planting technique for diversification of rice-wheat system of Indo-Gangetic Plains. J Japan Assoc Int Coop Agric For 28(1):25–42

Jat ML, Chandna P, Gupta R, Sharma SK, Gill MA (2006) Laser land leveling: a precursor technology for resource conservation. Rice-Wheat Consortium Technical Bulletin Series 7:48

Jat ML, Gathala MK, Ladha JK, Saharawat YS, Jat AS, Kumar V, Sharma SK, Kumar V, Gupta R (2009) Evaluation of precision land leveling and double zero-till systems in the rice-wheat rotation: water use, productivity, profitability and soil physical properties. Soil Tillage Res 105(1):112–121. https://doi.org/10.1016/j.still.2009.06.003

Kahlown MA, Gill MA, Ashraf M (2002) Evaluation of resource conservation technologies in rice, wheat system of Pakistan. Pakistan Council of Research in Water Resources, Islamabad

Kaspar TC, Jaynes DB, Parkin TB, Moorman TB, Singer JW (2012) Effectiveness of oat and rye cover crops in reducing nitrate losses in drainage water. Agric Water Manag 110:25–33. https://doi.org/10.1016/j.agwat.2012.03.010

Khatri-Chhetri A, Aryal JP, Sapkota TB, Khurana R (2016) Economic benefits of climate-smart agricultural practices to smallholder farmers in the Indo-Gangetic Plains of India. Curr Sci 110(7):1251–1256

Krauss W, von Storch H (2012) Post-normal practices between regional climate services and local knowledge. Nat Cult 7(2):213–230. https://doi.org/10.3167/nc.2012.070206

Kukal SS, Humphreys E, Yadvinder-Singh TJ, Thaman S (2005) Performance of raised beds in rice–wheat systems of northwestern India. Evaluation and performance of permanent raised bed cropping systems in Asia. Aciar proceedings, p 121

Kumar R, Gautam HR (2014) Climate change and its impact on agricultural productivity in India. J Climatol Weath Forecast 2(1):1-3

Lal R (1998) The potential of U.S. cropland to sequester carbon and mitigate the greenhouse effect. Ann Arbor Press, Chelsea

Lal R (2004) Soil carbon sequestration impacts on global climate change and food security. Science 304(5677):1623–1627. https://doi.org/10.1126/science.1097396

Lal R, Negassa W, Lorenz K (2015) Carbon sequestration in soil. Curr Opin Environ Sustain 15:79–86. https://doi.org/10.1016/j.cosust.2015.09.002

Layek J, Das A, Mitran T, Nath C, Meena RS, Yadav GS, Shivakumar BG, Kumar S, Lal R (2018) Cereal+ legume intercropping: an option for improving productivity and sustaining soil health. In: Legumes for soil health and sustainable management. Springer, Singapore, pp 347–386

Lipper L, Zilberman D (2018) A short history of the evolution of the climate smart agriculture approach and its links to climate change and sustainable agriculture debates.

In: Natural resource management and policy. Springer, Cham, pp 13–30. https://doi.org/10.1007/978-3-319-61194-5_2

Lipper L, Thornton P, Campbell BM, Baedeker T, Braimoh A, Bwalya M, Caron P, Cattaneo A, Garrity D, Henry K, Hottle R, Jackson L, Jarvis A, Kossam F, Mann W, McCarthy N, Meybeck A, Neufeldt H, Remington T, Sen PT, Sessa R, Shula R, Tibu A, Torquebiau EF (2014) Climate-smart agriculture for food security. Nat Clim Chang 4(12):1068–1072. https://doi.org/10.1038/nclimate2437

Lobell DB, Schlenker W, Costa-Roberts J (2011) Climate trends and global crop production since 1980. Science 333(6042):616–620. https://doi.org/10.1126/SCIENCE.1204531

Lobell DB, Ortiz-Monasterio JI, Sibley AM, Sohu VS (2013) Satellite detection of earlier wheat sowing in India and implications for yield trends. Agric Syst 115:137–143. https://doi.org/10.1016/j.agsy.2012.09.003

Lugen M (2017) Climate services for development in Burkina Faso: institutions, use and needs for national planning. KLIMOS working paper no. 13. Klimosacropolis, Brussels. https://ees.kuleuven.be/klimos/papers/wp13_climate_services_in_burkina_faso.pdf

Malik RK, Yadav A, Singh S, Malik RS, Balyan RS, Banga RS, Bellinder R (2002) Herbicide resistance management and evolution of zero tillage: a success story, Research bulletin, pp 1–43

Martins MR, Angers DA, Corá JE (2012) Carbohydrate composition and water-stable aggregation of an oxisol as affected by crop sequence under no-till. Soil Sci Soc Am J 76(2):475–484. https://doi.org/10.2136/sssaj2011.0110

Mayer-Schönberger V, Cukier K (2013) Big data: a revolution that will transform how we live, work, and think. Houghton Mifflin Harcourt, Boston

McDaniel MD, Tiemann LK, Grandy AS (2014) Does agricultural crop diversity enhance soil microbial biomass and organic matter dynamics? A meta-analysis. Ecol Appl 24(3):560–570. https://doi.org/10.1890/13-0616.1

Meena OP, Sammauria R, Gupta AK, Gupta KC, Behera B, Saxena R, Yadav MR, Singh P, Meena RK, Raza MB, Anil AS, Lal MK (2022) Energy-carbon footprint vis-à-vis system productivity and profitability of diversified crop rotations in semi-arid plains of North-West India. J Soil Sci Plant Nutr 22(2):2026–2041. https://doi.org/10.1007/s42729-022-00791-2

Miao Q, Rosa RD, Shi H, Paredes P, Zhu L, Dai J, Gonçalves JM, Pereira LS (2016) Modeling water use, transpiration and soil evaporation of spring wheat–maize and spring wheat–sunflower relay intercropping using the dual crop coefficient approach. Agric Water Manag 165:211–229. https://doi.org/10.1016/j.agwat.2015.10.024

Mrunalini K, Behera B, Chandana P, Patnaik GP, Modi RU, Saraswat A, … Kumar N (2022) Legumes to reduce ecological footprints for climate-smart cropping systems. In: Advances in legumes for sustainable intensification. Academic, London, pp 403–420

Newport D, Lobell DB, Balwinder-Singh, Srivastava AK, Rao P, Umashaanker M, Malik RK, McDonald A, Jain M (2020) Factors constraining timely sowing of wheat as an adaptation to climate change in Eastern India. Weather Clim Soc 12(3):515–528. https://doi.org/10.1175/WCAS-D-19-0122.1

Nyström M, Jouffray JB, Norström AV, Crona B, Søgaard Jørgensen P, Carpenter SR, Bodin Ö, Galaz V, Folke C (2019) Anatomy and resilience of the global production ecosystem. Nature 575(7781):98–108. https://doi.org/10.1038/s41586-019-1712-3

Ospina AV, Heeks R (2012) ICT-enabled responses to climate change in rural agricultural communities. Climate Change, Innovation and ICTs Project-Strategy Brief 2:5

Padhee AK (2018) Climate-smart policies for Indian agriculture. Indian Farming 68(10):61–63

Pimentel D, Harvey C, Resosudarmo P, Sinclair K, Kurz D, McNair M, Crist S, Shpritz L, Fitton L, Saffouri R, Blair R (1995) Environmental and economic costs of soil erosion and conservation benefits. Science 267(5201):1117–1123. https://doi.org/10.1126/science.267.5201.1117

Poeplau C, Don A (2015) Carbon sequestration in agricultural soils via cultivation of cover crops – a meta-analysis. Agric Ecosyst Environ 200:33–41. https://doi.org/10.1016/j.agee.2014.10.024

Raghuvanshi R, Raj S, Bhattacharjee S (2018) Climate smart agriculture and advisory services: approaches and implications for future. In: Discussion paper 1. Center for Agricultural Extension Innovations, Reforms and Agripreneurship (CAEIRA), Rajendranagar

Ram H, Yadvinder-Singh JT, Humphreys E, Dhillon SS, Kumar K, Kler DS (2005) Performance of upland crops on raised beds in northwestern India. In: Evaluation and performance of permanent raised bed cropping systems in Asia. Australian Centre for International Agricultural Research, Canberra, pp 41–58

Rathore SS, Babu S, El-Sappah AH, Shekhawat K, Singh VK, Singh RK, Upadhyay PK, Singh R (2022) Integrated agroforestry systems improve soil carbon storage, water productivity, and economic returns in the marginal land of the semi-arid region. Saudi J Biol Sci 29:103427

Rickman JF (2002) Manual for laser land leveling, Rice-wheat consortium technical bulletin series 5, vol 110, New Delhi, pp 12–24

Samberg LH, Gerber JS, Ramankutty N, Herrero M, West PC (2016) Subnational distribution of average farm size and smallholder contributions to global food production. Environ Res Lett 11(12):124010. https://doi.org/10.1088/1748-9326/11/12/124010

Sapkota TB, Majumdar K, Jat ML, Kumar A, Bishnoi DK, McDonald AJ, Pampolino M (2014) Precision nutrient management in conservation agriculture based wheat production of Northwest India: profitability, nutrient use efficiency and environmental footprint. Field Crop Res 155:233–244. https://doi.org/10.1016/j.fcr.2013.09.001

Singh KM, Kumar A, Singh RKP (2015) Role of information and communication technologies in Indian agriculture: an overview. https://doi.org/10.2139/ssrn.2570710. Available at SSRN 2570710

Singh A, Dhillon BS, Sidhu AS (2019) Productivity of chickpea (Cicer arietinum) under different sowing date and seed rate in South West Punjab. Int J Curr Microbiol Appl Sci 8(05):1419-1425

Singh R, Babu S, Avasthe RK, Meena RS, Yadav GS, Das A, Mohapatra KP, Rathore SS, Kumar A, Singh C (2021a) Conservation tillage and organic nutrients management improve soil properties, productivity, and economics of a maize-vegetable pea system in the Eastern Himalayas. Land Degrad Dev:1–18. https://doi.org/10.1002/ldr.4066

Singh R, Babu S, Avasthe RK, Yadav GS, Das A, Mohapatra KP, Kumar A, Singh VK, Chandra P (2021b) Crop productivity, soil health, and energy dynamics of Indian Himalayan intensified organic maize-based systems. Int Soil Water Conserv Res 9(2):260–270. https://doi.org/10.1016/j.iswcr.2020.11.003

Sivanappan RK (2014) Micro irrigation potential in India. In: Goyal M (ed) Sustainable microirrigation: principles and practices. Apple Academic Press (Taylor & Francis Group), Boca Raton, pp 87–120

Six J, Elliott ET, Paustian K (2000) Soil macroaggregate turnover and microaggregate formation: a mechanism for C sequestration under no-tillage agriculture. Soil Biol Biochem 32(14):2099–2103. https://doi.org/10.1016/S0038-0717(00)00179-6

Six J, Callewaert P, Lenders S, De Gryze S, Morris SJ, Gregorich EG, Paul EA, Paustian K (2002) Measuring and understanding carbon storage in afforested soils by physical fractionation. Soil Sci Soc Am J 66(6):1981–1987. https://doi.org/10.2136/sssaj2002.1981

Solomon S (2007) The physical science basis: contribution of working group I to the fourth assessment report of the Intergovernmental Panel on Climate Change. In: Climate Change, 2007. Intergovernmental Panel on Climate Change (IPCC), Geneva, p 996

Srinivasa Rao C, Venkateswarlu B, Wani SP, Sahrawat KL, Dixit S, Kundu S, Devi KG, Rajesh C, Pardhasaradhi G (2010) Productivity enhancement and improved livelihoods through participatory soil fertility management in tribal districts of Andhra Pradesh. Indian J Dryland Agric Res Dev 25(2):23–32

Srinivasarao C, Ravindra-Chary G, Mishra PK, Nagarjuna Kumar R, Maruthi Sankar GR, Venkateswarlu B, Sikka AK (2013) Real time contingency planning: initial experiences from Aicrpda. All India Coordinated Research Project for Dryland Agriculture, Central Research Institute for Dryland Agriculture, Indian Council of Agricultural Research, Hyderabad

Srinivasarao C, Ravindra-Chary G, Mishra PK, Subba Reddy G, Sankar GRM, Venkateswarlu B, Sikka AK (2014) Rainfed farming – a compendium of doable technologies. All India Coordinated Research Project for Dryland Agriculture, Central Research Institute for Dryland Agriculture, Indian Council of Agricultural Research, Hyderabad

Srinivasarao C, Venkateswarlu B, Sikka AK, Prasad YG, Chary GR, Rao KV, Gopinath KA, Osman M, Ramana DBV, Maheswari M, Rao VUM (2015) District agriculture contingency plans to address weather aberrations and for sustainable food security in India, CRIDA-NICRA bulletin 2015/3. ICAR-Central Research Institute for Dryland Agriculture, Natural Resource Management Division

Srinivasarao C, Rao KV, Gopinath KA, Prasad YG, Arunachalam A, Ramana DBV, Chary GR, Gangaiah B, Venkateswarlu B, Mohapatra T (2020) Agriculture contingency plans for managing weather aberrations and extreme climatic events: development, implementation and impacts in India. In: Advances in agronomy, vol 159. Academic, Amsterdam. https://doi.org/10.1016/bs.agron.2019.08.002

Stagnari F, Ramazzotti S, Pisante M (2009) Conservation agriculture: a different approach for crop production through sustainable soil and water management: a review. In: Organic farming, pest control and remediation of soil pollutants. Springer, Dordrecht, pp 55–83

Taylor M (2018) Climate-smart agriculture: what is it good for? J Peasant Stud 45(1):89–107. https://doi.org/10.1080/03066150.2017.1312355

Tiemann LK, Grandy AS, Atkinson EE, Marin-Spiotta E, McDaniel MD (2015) Crop rotational diversity enhances belowground communities and functions in an agroecosystem. Ecol Lett 18(8):761–771. https://doi.org/10.1111/ele.12453

Tonitto C, David MB, Drinkwater LE (2006) Replacing bare fallows with cover crops in fertilizer-intensive cropping systems: a meta-analysis of crop yield and N dynamics. Agric Ecosyst Environ 112(1):58–72. https://doi.org/10.1016/j.agee.2005.07.003

Urfels A, McDonald AJ, van Halsema G, Struik PC, Kumar P, Malik RK, Poonia SP, Balwinder-Singh, Singh DK, Singh M, Krupnik TJ (2021) Social-ecological analysis of timely rice planting in Eastern India. Agron Sustain Dev 41(2):14. https://doi.org/10.1007/s13593-021-00668-1

Vaughan C, Dessai S, Hewitt C (2018) Surveying climate services: what can we learn from a bird's eye view? Weather Clim Soc 10(2):373–395. https://doi.org/10.1175/WCAS-D-17-0030.1

Viswanathan PK, Bahinipati C (2015) Exploring the socio-economic impacts of microirrigation system (MIS): a case study of public tube wells in Gujarat, Western India. SAWAS 5(1):1–25

West TO, Post WM (2002) Soil organic carbon sequestration rates by tillage and crop rotation: a global data analysis. Soil Sci Soc Am J 66(6):1930–1946. https://doi.org/10.2136/sssaj2002.1930

Wickings K, Grandy AS, Reed SC, Cleveland CC (2012) The origin of litter chemical complexity during decomposition. Ecol Lett 15(10):1180–1188. https://doi.org/10.1111/j.1461-0248.2012.01837.x. Johnson N (ed)

Witt C, Dobermann A (2004) Toward a decision support system for site-specific nutrient management. In: Increasing productivity of intensive rice systems through site rice systems through site-specific nutrient management. IRRI, Science Publishers, Enfield, pp 358–395

Yadav GS, Das A, Lal R, Babu S, Meena RS, Saha P, Singh R, Datta M (2018) Energy budget and carbon footprint in a no-till and mulch-based rice–mustard cropping system. J Clean Prod 191:144–157

Yadav GS, Das A, Lal R, Babu S, Datta M, Meena RS, Patil SB, Singh R (2019b) Impact of no-till and mulching on soil carbon sequestration under rice (*Oryza sativa* L.)-rapeseed (*Brassica campestris* L. Var. Rapeseed) cropping system in hilly agro-ecosystem of the Eastern Himalayas, India. Agric Ecosyst Environ 275:81–92

Yadav GS, Babu S, Das A, Mohapatra KP, Singh R, Avasthe RK, Roy S (2020) No-till and mulching enhance energy use efficiency and reduce carbon footprint of a direct-seeded upland rice production system. J Clean Prod 271:122700. https://doi.org/10.1016/j.jclepro.2020.122700

Yadav GS, Babu S, Das A, Datta M, Mohapatra KP, Singh R, Singh VK, Rathore SS, Chakraborty M (2021a) Productivity, soil health, and carbon management index of Indian Himalayan intensified maize-based cropping systems under live mulch-based conservation tillage practices. Field Crop Res 264:108080. https://doi.org/10.1016/j.fcr.2021.108080

Yadav GS, Kandpal BK, Das A, Babu S, Mohapatra KP, Devi G, Chandra P, Singh R, Barman KK (2021b) Impact of 28-year-old agroforestry systems on soil carbon dynamics in Eastern Himalayas. J Environ Manag 283:111978. https://doi.org/10.1016/j.jenvman.2021.111978

Yadav GS, Das A, Kandpal BK, Babu S, Lal R, Datta M, Das B, Singh R, Singh VK, Mohapatra KP, Chakraborty M (2021c) The food-energy-water-carbon nexus in a maize-maize-mustard cropping sequence of the Indian Himalayas: an impact of tillage-cum-live mulching. Renew Sust Energ Rev 151:111602. https://doi.org/10.1016/j.rser.2021.111602

The manufacturer's authorised representative in the EU is Springer Nature Customer Service Centre GmbH, Europaplatz 3, 69115 Heidelberg, Germany. If you have any concerns regarding our products, please contact ProductSafety@springernature.com

Printed and bound by CPI Group (UK) Ltd, Croydon, CR0 4YY
01/10/2026
02225914-0015